Lecture Notes in Electrical Engineering

Volume 260

For further volumes:
http://www.springer.com/series/7818

Yueh-Min Huang · Han-Chieh Chao
Der-Jiunn Deng · James J. (Jong Hyuk) Park
Editors

Advanced Technologies, Embedded and Multimedia for Human-centric Computing

HumanCom and EMC 2013

Volume I

Editors
Yueh-Min Huang
ES Department
National Cheng Kung University
Tainan
Taiwan, R.O.C.

Han-Chieh Chao
Institute of Computer Science
and Information Engineering
National Ilan University
Ilan City
Taiwan, R.O.C.

Der-Jiunn Deng
Department of Computer Science and
Information Engineering
National Changhua University of Education
Changhua City
Taiwan, R.O.C.

James J. (Jong Hyuk) Park
Department of Computer Science and
Engineering
Seoul University of Science and
Technology (SeoulTech)
Seoul
Republic of Korea (South Korea)

ISSN 1876-1100 ISSN 1876-1119 (electronic)
ISBN 978-94-007-7261-8 ISBN 978-94-007-7262-5 (eBook)
DOI 10.1007/978-94-007-7262-5
Springer Dordrecht Heidelberg New York London

Library of Congress Control Number: 2013943942

Printed on acid-free paper

Springer is part of Springer Science+Business Media (www.springer.com)

Contents

Part VI Embedded Computing

Part XIII All-IP Platforms, Services and Internet of Things in Future

Part XIV Networking and Applications

Part I
HumanCom

An Interface for Reducing Errors in Intravenous Drug Administration

Frode Eika Sandnes and Yo-Ping Huang

Abstract Input errors occur with drug infusion pumps when nurses or technicians incorrectly input the prescribed therapy through the control panel. Such number copying tasks are cognitively and visually demanding, errors are easily introduced and the process is often perceived as laborious. Stressful working conditions and poorly designed control panels will further add to the chance of error. An alternative scheme is proposed herein, termed intravenous prescription phrase, based on dictionary coding. Instead of copying number sequences the user copies sequences of familiar words such that the cognitive load on the user is reduced by a factor of five. The strategy is capable of detecting errors and is easy to implement.

Introduction

The operation of medical devices, in particular the input of numeric values has received much attention [1]. Intravenous drug administration involves the input of rate, dose, time and volume of drugs prescribed by doctors and input errors can be lethal. Studies have shown that error rates in intravenous drug administration can be as high as 50–80 % [2, 3]. The technical complexity of menu structures constitute one problem with such medical devices is [4]. Another key issue is the copying of digits such as when placing phone calls. Often this involves a number copying task where the number is read from some source such as a phone directory

F. E. Sandnes (✉)
Institute of Information Technology, Oslo and Akershus University College of Applied Sciences, N-0130 Oslo, Norway
e-mail: eika.sandnes@hioa.no

Y.-P. Huang
Department of Electrical Engineering, National Taipei University of Technology, 10608 Taipei, Taiwan, Republic of China
e-mail: yphuang@ntut.edu.tw

Y.-M. Huang et al. (eds.), *Advanced Technologies, Embedded and Multimedia for Human-centric Computing*, Lecture Notes in Electrical Engineering 260,
DOI: 10.1007/978-94-007-7262-5_1,

and input using the numeric keypad [5]. Airlines often use booking references which users input on self-service check-in terminals in the airport to obtain their boarding cards. These higher base numbers, such as base 36, often comprise combinations of letters and digits.

Shannon [6] identified that a decimal digit is approximately equal to 3 1/3 bits. Similarly, base 36 numbers are equal to 5.2 bits. One may argue that the increased information capacity gained through large base-digits does not justify the increased complexity of the copying tasks. One reason is that certain letters and digits look similar such as 0—O, 1—l, 2—Z, 5—S, 6—G and 8—B. Differentiating such symbols is even more difficult when the user has reduced visual acuity and given certain fonts. High base numbers appear as random strings. The lack of internal structure means that users are unable to resolve ambiguous looking characters. This problem is well known in the optical character recognition literature [7].

It has been found that number input errors are caused by either motor slips where the fingers do not perform as expected, recall slips where numbers are remembered incorrectly, or perception slips where the source number is read incorrectly [8]. These errors can result in digit substitutions, insertions or omissions. Most digit and text input studies operates with typical error rates of 5–10 %, which means that up one in every 10 digits are likely to be incorrect. A study of the frequency of digits in drug infusion pumps showed that 0 is the most frequent digit, followed by 1 and 5, while 3, 4, 6, 7 and 8 are comparatively rare [9]. Numbers ranged in 1–5 digits in length with 3 digits being the most common. One class of errors termed "out by 10 errors" is caused by incorrectly entering a decimal point or a zero [10, 11]. Out of 10 errors greatly affect the magnitude of numbers and will have serious consequences.

Several studies have investigated various number input interfaces including touch based [12] numeric keypads, displays with individual incremental up-down buttons, incremental left–right/up-down buttons or 5-button interfaces [13] that are often attached to mobile equipment in ambulances, hospital wards, etc., and results show that the slower incremental up-down interfaces leads to fewer errors than numeric keypads [14]. Another approach is to avoid the input of digits altogether where the prescription is printed on bar-codes and input using bar-code readers [15]. However, bar codes require printing and they are less flexible, as the information cannot be communicated orally. This study focuses on the manual input of prescriptions.

Number Copying Challenges

Number copying tasks are problematic for several reasons. Miller's limit on humans' short term memory of 7 ± 2 pieces of information [16] is often cited and subsequent studies have narrowed this limit down to a maximum of 5–7 pieces [17]. Studies involving memorizing phone numbers have shown that recall performance rapidly degrades beyond 6 digits [5]. A digit copying task therefore has to be split up into several read-input cycles, each cycle requiring alternating the visual attention between the source and the target. This is difficult if the source digits are not grouped such as 16 digit credit cards numbers presented groups of four. Without grouping additional effort is required to locate where to resume when shifting attention between the source and the target. Another issue is a lack of standard interface layout. If there is a mismatch in both information sequence and information location on the printed prescription and the target it will harder for users to pair source and target.

Second, digit copying tasks are error-prone and parity checks are unable to capture all errors.

Third, digit copying is perceived as time-consuming and laborious. Many portable laptop computers do not have numeric keypads and it has been demonstrated that the input of digits on QWERTY keyboards without numeric keypads are significantly slower than using full keyboards with numeric keypads [18]. It is occasionally necessary and practical to communicate number information from one individual to another orally via phone, such as if receiving a prescription from a medical doctor via phone. Each digit will have to be read out one by one and often repeated several times to confirm the correctness of the data. This further reduces efficiency and increases the chances of error and frustration.

Memory Aids

One popular memory aid is to remember number sequences associated with word input on mobile keypad, for example 2,326 m as ADAM. Instead of recalling the individual digits where each digit counts as one piece of information, the word is remembered counting as one piece of information. Unfortunately, not all number sequences have corresponding words as 0 and 1 are not assigned letters.

This paper proposes to use a fixed dictionary for encoding and decoding. Digit sequences are split into groups, where each group is converted to a linguistic word. The user is presented with a word sequence instead of a number sequence and will thus be able to copy more information per copy-input cycle. Group sizes of 5 digits are chosen as this relies on wordlists with 100,000 entries.

For example, 01234 56789 is first split into two groups of 5 digits, namely 01234 and 56789. Each number is then looked up in the wordlist (see Fig. 1). The digits 01234 could correspond to the English word "stir" and 56789 correspond to "galumphs" and the word sequence "stir galumphs" is presented to the user, who has the memory capacity to simultaneously hold both words in working memory while copying the words to the target. Next, the words are identified in the wordlist on the receiving end and the indices of the two words identified.

The wordlist can be organized such that the magnitude of the number correlates with the word length such that smaller numbers with many prefix zeros are

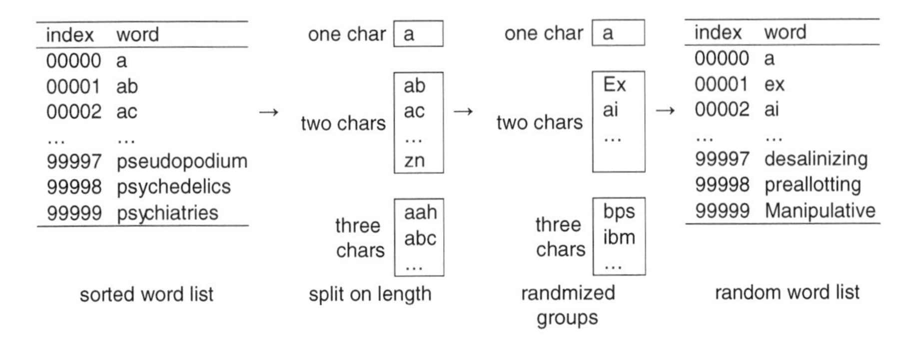

Fig. 1 Composing the error tolerant wordlist

assigned shorter words and larger number are assigned longer words. Another strategy would be to assign the most frequently used numbers shorter words.

A key advantage of linguistic words is that readers read words as whole units, while digit sequences are read individually. The reader recognizes the height signature of lower case letters. This greatly adds to the reading speed. Moreover, users are able to spot simple errors due to the internal word structure. Visual perception errors are reduced since words are read as one unit rather than individual unrelated units. Recall errors are also reduced for the same reason.

Error Detection

The literature on spell checking classifies input errors as deletions, insertions or replacements [19]. This strategy proposes two levels of error detection. The first level of error detection catches misspelled words not found in the wordlist.

Next, the wordlist is organized such that similar words are associated with dissimilar digit sequences. The Levenshtein distance is often used to measure the distance between words [20], but the Hamming distance can be used if the words have equal lengths. The following scheme is employed. The list of 100,000 words is first sorted into groups of equal length and words in each group shuffled into random order. Finally, the groups are recombined into one list organized according to increasing word lengths while exhibiting a random internal structure.

For example, the word "buys" can easily be mistyped as "buts" by substituting the character y with t. The Hamming distance between these words are 1. The corresponding numbers for the two words are 1225 and 3022. These numbers have a Hamming distance of 3 and a numeric distance of 1797. With running numbers the two numbers would be 1352 and 1354, respectively, yielding a Hamming distance of 1 and a numeric distance of 2. Figure 1 illustrates the wordlist construction process.

A second level of error detection is achieved by introducing a parity check [21] by summing each 5-digit chunk multiplied by unique factors and computing the desired modulus 100,000 of the total. By multiplying each number with a factor the parity check will detect if elements are swapped. A modulus 10 scheme misses 10 % of the errors, while modulus 100,000 only misses 0.001 % of the errors.

The linguistic digit representation can also assist detecting errors when comparing numbers. Imagine a customer comparing the dates 23-06-2012 and 23-08-2012, coded as 23062 and 23082, respectively. Each date comprises three information parts, day, month and year. If the user focuses on the day he or she may overlook the difference in month, that is, June versus August since the shape of 6 is similar to 8. This will not happen when comparing their linguistic representations "thrives" and "marxist".

Information Coding Schemes

This section illustrates how to code values with the proposed scheme. Each unit should be fitted to a chunk to avoid overlap across chunks.

Large numbers with more than 5 digits are split into several chunks. Some values with more than 5-digits can be reduced to 5-digit chunks without information loss. Dates are often described with eight digits, namely day, month and year. One simplification is to use a modulo-10 single digit for the year. This will work as most applications operate within a limited time scope. This date format allows dates to be specified using a single linguistic word which is useful for making comparisons. Comparisons across years are simplified further if the year is dropped altogether.

Numbers comprising 5 digits or less can simply be represented using heading zeros such as small quantities, drug doses, etc. Analysis of commercially available equipment showed that numbers frequently represent volumes in ml or rates in ml/hr with rarely more than three digits and one digit after the decimal point. Such numbers can therefore be represented using the least significant digit of the 5-digit sequence to represent the digit after the decimal point and the four most significant digits to represent the digits on the left of the decimal point. For example 0.1 would be 00001, 10 or 10.0 would be 00010, 450 would be 04500, etc.

However, most drug infusion pumps do not actually have the precision indicated by the operating panels. Some equipment specifies an error in dosage of about 12 % from the set value. It is thus pointless to distinguish between 450.1 and 450.0 ml.

Pairs of related parameters with similar magnitudes can be combined into 5 digit numbers as a type of double floating point numbers. A parameter is represented using its two most significant digits, that is A and B, for example AB0, AB or A.B, and 3 bits are used to indicate the position of the decimal point from 10-1 to 104 represented by the values from 0 to 6, respectively. For example to code the parameter pair 2.0 and 1.5 using this scheme one would get 02015. The first digit 0 indicates the decimal point in the most leftmost position. To code the number pair 45 and 35 the decimal is moved one position to the right and the first digit is therefore 1 yielding the value 14535. Similarly, the pair 450 and 350 can be coded as 24535.

Materials and Methods

An English wordlist published by the SIL International Linguistics Department containing 109,582 entries was used. The entries were sorted according to increasing word length and the 100,000 shortest words were kept. The average word length is 8 and the maximum words length 12 characters. The size of the final wordlist was less than 1 Mb making it applicable to devices with limited memory. A proof of concept coder and decoder were implemented in Microsoft Excel. Excel was chosen in order to demonstrate that it is simple to implement. Several drug infusion pump user manuals were studied to acquire information about common intravenous drug administration practices and prescription parameters, namely Curlin Medical's 4,000 Plus and 4,000 CMS Ambulatory Infusion Systems, Abbott Laboratories' PLUM A+, Eureka's IP and LF infusion pumps and BodyGuard's 323 Multi-Therapy ambulatory infusion pump. Common parameters are listed in Table 1. Of the less obvious parameters titration limit specifies the maximum infusion rate, keep vein open rate specifies the rate of fluid transmitted when the devices is in an open state. This measure is specified as the overall rate is achieved by injecting drugs in smaller doses, and this specifies the rate for each dose.

Table 1 Common drug infusion pump parameters

Parameter	Resolution	Max	Unit	Type	Flag no.
Bag volume	1	9,999	ml	N/A	
Concentration	0.1	999	mg/ml	Required	
Start time	00:01	23:59	Hours: minutes	Optional	1
Titration limit	0.1	400	ml/hour	Optional	2
Amount to infuse	0.1	9,999	ml	Required	
Rate of injection	0.1	400	ml/hr	Required	
Vein open rate	0.1	10	ml/hour	Optional	2
Loading dose	0	50	ml	PCA	3
Bolus dose	0	50	ml	PCA	4
Delta rate	0	50	ml	PCA	5
Delta time	0	60	Minutes	PCA	5
Max bolus dose	0	100	ml	PCA	4
No. bolus dose/hr	0	15	Doses	PCA	4
Min bolus interval	0	60	Minutes	PCA	4
Up-ramp	00:01	99:59	Hours: minutes	Optional	6
Down-ramp	00:01	99:59	Hours: minutes	Optional	6

The following parameters apply to patient controlled analgesia therapies: Loading dose indicates an initial dose of drugs infused at the beginning of a therapy. Bolus dose indicates the amount of drugs in lm. Delta rate and delta time specifies the gradual increase of a dose per time unit. Maximum bolus dose indicates the maximum dose that can be demanded by the patient, minimum bolus interval specifies the minimum time interval between consecutive bolus doses and number of boluses allowed per hour specifies an upper limit to how many bolus doses the patient are allowed per hour. Up-ramp and down-ramp parameters can be used to gradually increase and decrease doses.

This parameter list comprise six optional parameters, therefore a check word can be introduced to indicate both which optional parameters that are included in the message and a parity symbol for error detection purposes. The two most significant digits are used to indicate the parameters included and the three least significant digits constitute a parity symbol.

Results and Discussion

The following example illustrates how the proposed strategy is applied to the following simple fictitious prescription:

Concentration	1.5 mg/ml
Amount to infuse	75.0 ml
Rate of injection	15.0 ml/hr

The corresponding number sequence for this prescription is:
00695 00015 00750 00150
Here 00695 is the control value where the first two heading zeros indicate that no optional parameters are present. The parity value 695 is computed as the modulus 1,000 of the sum of the three parameters computed with weights 3, 2 and 1. The corresponding word prescription phrase sequence is
but md lip pub
Setting three digits may not seem hard, but it would probably require three read-input cycles. However, under stress one may misread one of the digits or swap the numbers that are relatively similar in structure and magnitude.

The corresponding four word phrase "but md lip pub" is easier to remember, and can be input in one read-input cycle. Moreover, any mistakes would be detected, for instance if the word "but" is incorrectly input as "byt" since y is next to u on the QWERTY-keyboard. The letter sequence "byt" would be detected as an invalid word.

Next, imagine the word "bit" input instead of "but" as the character i is next to u on the keyboard. The word bit gives parity 456, which indicates that it is a mistake somewhere since it does not match the parity value 695.

If two of the words are accidentally swapped, for instance md and lip, giving the incorrect phrase "but lip md pub", the parity becomes 430 which does not match 695.

If we include an extra word accidentally, for example md twice, giving "but md md lip pub" the error is detected as there are no flags indicating optional parameters. Also, the parity values do not match. Next, consider the following prescription:

Concentration	1.5 mg/ml
Titration limit	30.0 ml/hr
Amount to infuse	75.0 ml
Rate of injection	20.0 ml/hr
Vein open rate	10 ml/hr
Loading dose	5.0 ml
Bolus dose	1.0 ml
Delta rate	1.0 ml
Delta time	5 min
Max bolus dose	10 ml
No. bolus dose/hr	2
Min bolus interval	30 min
Up-ramp	03:00
Down-ramp	06:00

This prescription involves 14 data items with 27 digits and 5 separators that possibly would have to be input through an intricate procedure involving different menu screen in 14 read-input cycles. This prescription is represented using the following string:
37983 00015 00300 00750 00200 0010 00050 00010 00010 00005 00010 00002 00030 00300 00600
All but the first parameters are included meaning that all bit flags apart from the first bit are set, giving 37. The parity value of the parameters is 983 using weights from 13 to 1 from left to right. The corresponding drug prescription phrase is thus
flipper md fen lip lap us od us et us ai on fen doz

The resulting 15 word phrase can be copied in three read-input cycles, as "flipper md fen lip lap", "us od us et us" and "ai on fen doz".
The two examples illustrate a very general and flexible approach. However, some values occur repeatedly such as 10 resulting in monotonous phrases with several similar ("us"). Since most drug infusion pumps do not actually have the precision indicated by the operating panels the final example illustrates the use combined parameter pairs for obtaining shorter intravenous prescription phrases. The following related parameters from the previous example can be paired, titration limit (30 ml/hr) and keep vein open rate (10 ml/hr) coded as 13010, loading dose (5 ml) and bolus dose (1 ml) coded as 05010, delta dose (1 ml/hr) and delta time (5 min) coded as 10105, max bolus dose (10 ml), bolus doses per hour (2) and minimum bolus interval (30 min) coded as 10230, and finally up-ramp (03:00) and down

ramp (6:60) coded as 11836. The resulting string is thus37271 00015 00750 00200 13010 05010 10105 10230 11836
which gives the following intravenous prescription phrase
cappers md lip lap vegans divvy nitty babes milton
This phrase is easily remembered in two steps, namely "cappers md lip lap" and "vegans divvy nitty babes milton". This optimized phrase does not compromise accuracy.

Conclusions

A strategy for reducing the errors in intravenous drug administration is proposed. It comprises a memory aid that simplifies manual number copying tasks. The strategy converts the prescription data comprising digits sequences to sequence of linguistic words. Ordinary drug prescriptions can therefore be memorized for short durations. In addition the strategy allows users to easily verify the correctness of information as it is easier to compare linguistic words than digit sequences.

References

1. Oladimeji P (2012) Towards safer number entry in interactive medical systems. In: Proceedings of the 4th ACM SIGCHI symposium on engineering interactive computing systems (EICS '12), ACM, New York, pp 329–332
2. Barber N, Taxis K (2004) Incidence and severity of intravenous drug errors in a German hospital. Eur J Clin Pharmacol 59:815–817
3. Taxis K, Barber N (2003) Ethnographic study of incidence and severity of intravenous drug errors. BMJ 326:684
4. Nunnally M, Nemeth CP, Brunetti V, Cook RI (2004) Lost in menuspace: user interactions with complex medical devices. IEEE Trans Syst Man Cybern Part A 34:736–742
5. Raanaas RK, Nordby K, Magnussen S (2002) The expanding telephone number part 1: Keying briefly presented multiple-digit numbers. Behav Inf Technol 21:27–38
6. Shannon CE (2001) A mathematical theory of communication. SIGMOBILE Mob Comput Commun Rev 5:3–55
7. Kahan S, Pavlidis T, Baird HS (1987) On the recognition of printed characters of any font and size. IEEE Trans Pattern Anal Mach Intell PAMI-9 2:274–288
8. Wiseman S, Cairns P, Cox A (2011) A taxonomy of number entry error. In: Proceedings of the 25th BCS conference on human-computer interaction (BCS-HCI '11) British Computer Society, UK, pp 187–196
9. Wiseman S (2011) Digit distributions, What digits are really being used in hospitals? In: Proceedings of the fourth York doctoral symposium on computer science, The University of York, pp 61–68
10. Thimbleby H, Cairns P (2010) Reducing number entry errors: solving a widespread, serious problem. J R Soc Interface 7:1429–1439
11. Thyen AB, McAllister RK, Councilman LM (2010) Epidural pump programming error leading to inadvertent 10-fold dosing error during epidural labor analgesia with Ropivacaine. J Patient Saf 6:244–246

12. Isokoski P, Koki M (2002) Comparison of two touchpad-based methods for numeric entry. In: Proceedings of the SIGCHI conference on human factors in computing systems (CHI '02), ACM, New York, pp 25–32
13. Cauchi A (2012) Differential formal analysis: evaluating safer 5-key number entry user interface designs, EICS'12. In: Proceedings of the 4th ACM SIGCHI symposium on engineering interactive computing systems, ACM, New York, pp 317–320
14. Oladimeji P, Thimbleby H, Cox A (2011) Number entry interfaces and their effects on error detection, LNCS, vol 6949. Springer, Heidelberg, pp 178–185
15. Meyer GE, Brandell R, Smith JE, Milewski FJ, Brucker P, Coniglio M (1991) Use of bar codes in inpatient drug distribution. Am J Health Syst Pharm 48:953–966
16. Miller GA (1956) The magical number seven, plus or minus two: some limits on our capacity for processing information. Psychol Rev 63:81–97
17. Simon HA (1974) How big is a chunk? Science 183:482–488
18. Sandnes FE (2010) Effects of common keyboard layouts on physical effort: implications for kiosks and Internet banking. In: Proceedings of Unitech 2010, Tapir Academic Publishers, Trondheim, pp 91–100
19. Kukich K (1992) Techniques for automatically correcting words in text. ACM Comput Surv 24:377–437
20. Navarro G (2001) A guided tour to approximate string matching. ACM Comput Surv 33:31–88
21. Wagner NR, Putter PS (1989) Error detection decimal digits. Commun ACM 32:106–110

A Vocabulary Learning Game Using a Serious-Game Approach

Kanako Nakajima and Tatsuo Nakajima

Abstract It is always hard to keep motivated while doing something we must do but we do not want to. However, gamers put so much time into their favorite games, just because its fun. Games have many tricks to keep attracting people, and nowadays these gimmicks are included into education-games. However, not many of the education-games in the markets are fun enough to keep users motivated for playing. In this paper we address this conflict, propose a better education-game created based on a popular smartphone game, and evaluate the improvement of the motivation through playing the game we offer. As a conclusion, we discovered that the examinees are motivated through the experiment using the education-game we created; however, these motivations are passive as they are not actively willing to do, but rather not mind doing it. Supplementations to shift these passive motivations to active motivations are considered in our future work.

Keywords Serious game · Education

Introduction

It is always hard to keep motivated while doing something we must do but we do not want to, such as studying and working. On the other hand, there is something we enjoy doing it although we do not have to: games. Games are loved by many; in the research held in 2009 in Japan, 34 % of all Japanese citizens, and 64 % of Japanese elementally students play games regularly [1], and those gamers are willing to play games for there lives [2]. Good games have power to attract people, thus applying game mechanics to education has already been investigated for

K. Nakajima (✉) · T. Nakajima
Department of Computer Science and Engineering, Waseda University, 3-4-1 Okubo Shinjuku, Tokyo 169-8555, Japan
e-mail: kanako.n@dcl.cs.waseda.ac.jp

Y.-M. Huang et al. (eds.), *Advanced Technologies, Embedded and Multimedia for Human-centric Computing*, Lecture Notes in Electrical Engineering 260,
DOI: 10.1007/978-94-007-7262-5_2,

several times. However, it is doubtful that the education games produced in the market today work well enough. According to *the Annual Video Game Industry Report*, the peak of the education game market was in 2006–2007 in Japan, with the release of *Brain Age: Train Your Brain in Minutes a Day! and English Training: Have Fun Improving Your Skills!*. *Brain Age: Train Your Brain in Minutes a Day!*, the one sold very well, is full of game mechanics, and even though game was not the main part of *English Training: Have Fun Improving Your Skills!*, it included some of game parts and sold well. Many education games are released after the hit of *Brain Age*, but most of them were not successful. The reason of the failure of the education game industry is based on misusing the concept of "game". According to the *Annual Video Game Industry Report In 2008*, "many of the education game titles in the market today are just reprints of the existing books, thus they can reduce the cost of software development" [1]. These software are sold as they are games, but most of the times there is almost no game mechanics are included. Thus, it could not continue the boom of education games.

In this paper, we produced a English vocabulary learning game, Vocab Draw, by adding some educational elements to the game instead of adding game mechanics to education; with this approach, we believe that the gaming attributes remain while learning materials are provided.

Preliminary Survey

Preliminary Survey has been done for learning examinees' attitude towards games, especially the educational ones. The samples for this survey are 10 males and 7 females, all around 20-year-old.

Eighty eight percent, 15 of 17 samples are active game players. For those game players, such a question is asked: *when they quit playing the game they are playing if it is not fun*. None of the examinees answered they play until the end whether it is good or bad, and 1/3 of them answered they quit playing immediately when they find out the game is boring. 2/3 of them play for a while even tough they cannot find fun in it, waiting for the game to gets better by any change. With this result, it can be said that amusingness is important for games to make players continue playing it.

However, as its discussed in the previous section, recent education games are not attracting people well. 9 of 17 examinees have had experience of playing education games. The education games they played were able to be divided into three groups, brain training, English learning, and other. Taking close look at the brain training and English learning games, the willing to play the game again towards those two types were different; those who played brain training games are willing to play the game again, because it is fun. On the other hand, those who played English learning game are also willing to play the game again, but only to learn english, even though the game was boring. However, it is doubtful that boring game can attract people for long time even if it is efficient for learning.

This assumption can be verified by the popularity of education games today. By the *Annual Video Game Industry Report*, the sales of the education games in 2011 were 300 thousands, 0.56 % of all genres of games, the lowest genre out of 13 others. Also, according to our survey, the popularity of education game was the lowest as well. Even though the players say they will play the education games for their goods, while other interesting games are out there, it is hard for them to pick a boring educational game. And if they cannot continue playing the game, the efficiency of learning would become waste. Thus, we put focus on making the game interesting, rather than improve the efficiency.

Proposal of a Vocabulary Learning Game

Draw Something

As we developed an interesting English vocabulary learning game, we used an existing popular game as a reference. The game is called *Draw Something*, published by OMGPOP. This game is distributed for iPhone and Android, and 20 million people have played free and non-free version in total on Android. The game is originally provided in English, but since it is for native English speakers there is no learning elements of English.

The basic rule is simple; two players draw and exchange their drawings, and make guess what other players have drawn. First of all, a player chooses an opponent and begins the game. Then a word selection scene is shown, and he chooses a word he might be able to draw. A level is set to each word, and the reward to each level is different. The player must choose an appropriate word from the words he can draw, while considering the rewards he can get. After the player has chosen the word, it is time for him to draw. The drawing is sent to the opponent he chose in the beginning, and then the opponent can guess what he has drawn. If the opponent corrects the answer, then their continuity is counted up as combo and some coins, which depends on the word level, are rewarded to the players. And then the drawing is now sent from the opponent to the player, and they keep exchanging the drawings to add up the combos. This series of action can be done in parallel, with many opponents at a time.

Vocab Draw

The rules we applied to Vocab Draw, our novel English vocabulary learning game, is exactly same as Draw Something. However, we added some learning elements to Draw Something in order to make it as an attractive education game. Also, we

Fig. 1 Draw-part1

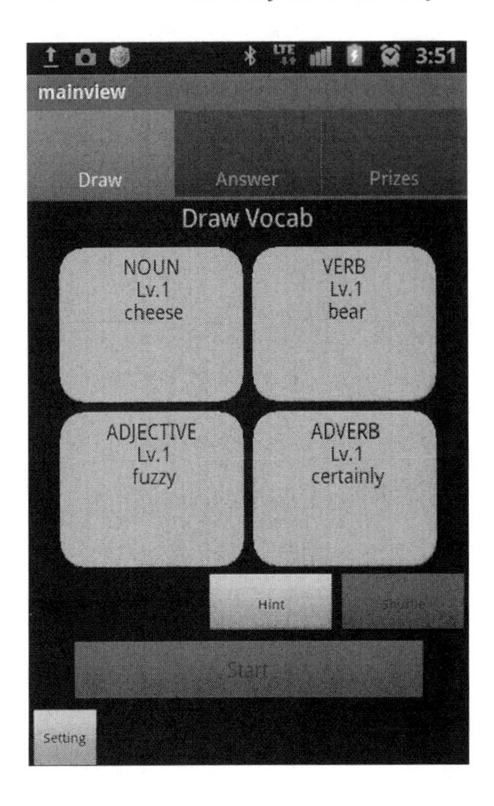

added a beneficial change to Draw Something; so players can keep their motivation. In this section, the difference between Draw Something and Vocab Draw is presented.

Interface Change As several changes are made from Draw Something to Vocab Draw, the interface was upgraded. The game part is decided into two main parts, one is Draw-part and another is Answer-Part as shown in Figs. 1 and 2. Each part starts with a word selection part, followed by actual Drawing- and Answering-parts, shown in Figs. 3 and 4.

Improvement from Draw Something At first, we changed a way of drawing exchanging from one-to-one to one-to-many. In Draw Something, a player was be able to send to only one opponent per drawing, and when his opponent answered the game continued. In Vocab Draw this rule is discontinued because with this method the player must wait until the opponent answers, in order to keep playing. This mechanism is shown in the Fig. 5, when exchanging the drawing one by one, if an opponent's answer is slow or given up because he decreased the motivation to play the game, the player cannot get the feedback to his drawing. In such case, it is worried that the player's motivation is also declined. Instead in Vocab Draw, the relationship between players are reconsidered as one-to-many from one-to-one in Draw Something. As shown in Fig. 6, if the drawing is distributed to many rather than one, the chance of getting feedback is increased.

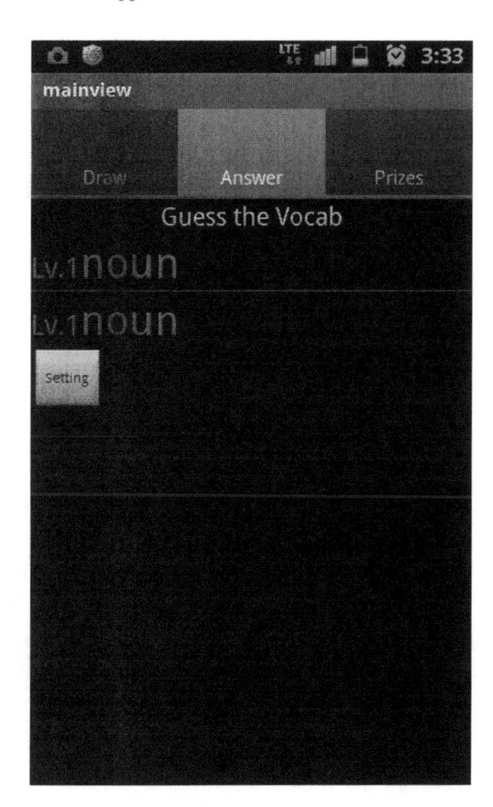

Fig. 2 Answer-part1

Learning Elements In order to include educational functions into Vocab Draw, materials listed below are considered:

1. Player Level
2. Coverage of fundamental part of speech
3. Japanese translation.

Firstly, the concept of level is given to each player. A player sets a level and the system shows the vocabularies according to that level. With this level, a player can learn vocabularies effectively as he tries the problems he can actually understand. In Figs. 1 and 2, the level is set as one.

Secondly, vocabularies are chosen by four fundamental parts of speech: noun, verb, adjective, and adverb. These parts of speech are considered fundamental, based on the English Note, which is published by the Ministry of Education in Japan as the base of English teaching material for Japanese elementary students [3]. In English Note, vocabularies provided for elementary students are nouns, verbs, adjectives, adverbs, and prepositions. Although preposition is as important as other five parts of speech, since it is used to show the relationship between other parts of speech, and we considered that teaching its meaning alone is pointless and confusing for the learners/players; thus, we included the first five parts of speech only.

Fig. 3 Draw-part2

Thirdly, Japanese translations are shown to improve the understanding of the vocabularies. These Japanese translations are, not shown in the first time, and they are shown by pressing Hint Button as shown in Fig. 1. If the player cannot draw any of the words provided, they can shuffle the words to get another set; an important thing is that they cannot shuffle the words until they check the Japanese translation, so there would be more chance to see English and Japanese together to learn the vocabularies. Once the player starts to draw with a hint Japanese meaning is provided as shown in Fig. 3, if he/she starts without with a hint, then there would be no translation shown.

Evaluation

After the preliminary survey, examinees were asked to play both Draw-part and Answer-part of Vocab Draw once for each. First of all, 82 % of the examinees responded with positive feedback for the question whether he/she wants to play Vocab Draw again. The specific numbers are as follows: 6 % are strongly willing to play again, 35 % are willing to play again, 41 % can play again, 18 % are not

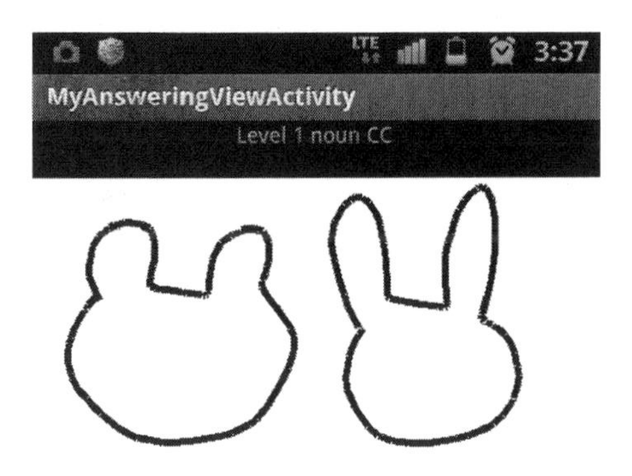

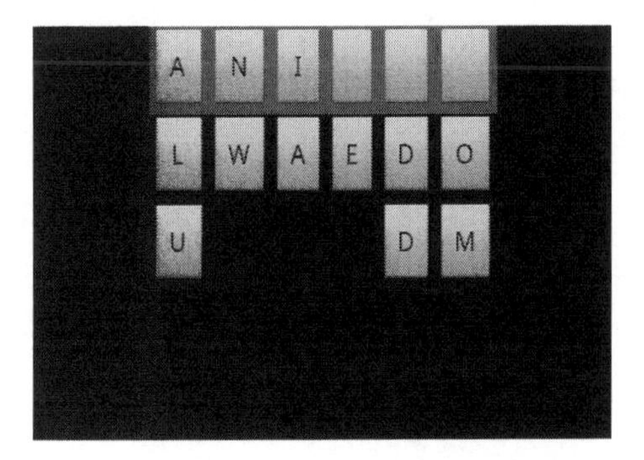

Fig. 4 Answer-part2

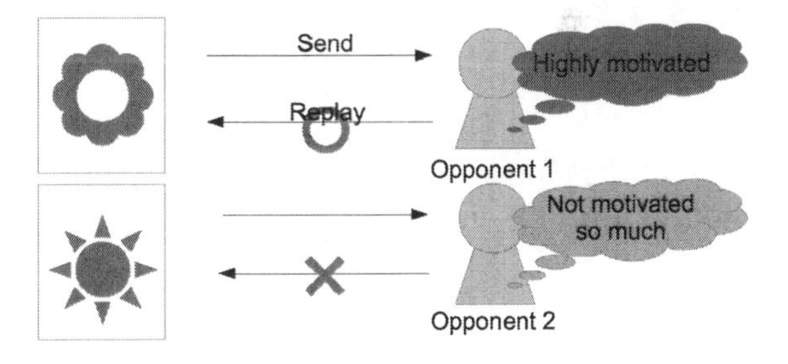

Fig. 5 One-to-one (draw something)

willing to play again. In addition, paying a close attention to the ones who gave negative posture for learning English in the preliminary survey, it is clear that their attitude changed via the usage of Vocab Draw. 72.7 % of those who are not willing to put effort into learning english, 78.6 % of those who hate learning vocabularies, and also 72.7 % of those who are not doing any extra work for learning vocabularies have improved their motivation towards English learning if it is with Vocab Draw. Asking why they want to play it again, most of the answers were because of the amusingness of the game, shown in the Fig. 7. By this result, it is proven that the amusingness would improve the attitude of learners, even if they hate learning

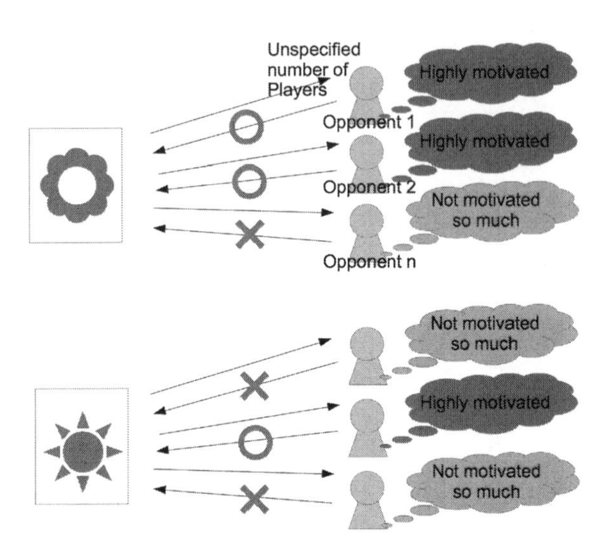

Fig. 6 One-to-many (vocab draw)

it. However, half of the positive feedback are as weak as that they CAN play the Vocab Draw again, and only one player answered that he strongly wants to play it again. From this it can be said that motivations can be improved with the amusingness as much as they do not leave the game, but it is not enough to lead them to strong attitude or willingness to play. Our future direction is to change the examinees' attitude from that they CAN play it again to they WANT to play it again.

We also asked the examinees to feedback freely with pros and cons of the application. For educational elements, some opinions in Table 1 is provided by the examinees. Most of the positive opinions were about the usage of drawing. Four examinees answered that it is easy to see the vocabulary as an image, thus it is helpful to learn it better. As another opinion, one examinee considered that Vocab Draw is more like an usual game rather than a boring education game. This opinion is important because it was our final goal.

On the other hand, some issues became distinct. Our biggest issue is that some of the words are difficult to draw in a picture. Especially the words which are not noun, it is difficult for the players to associate the word with an image. If many

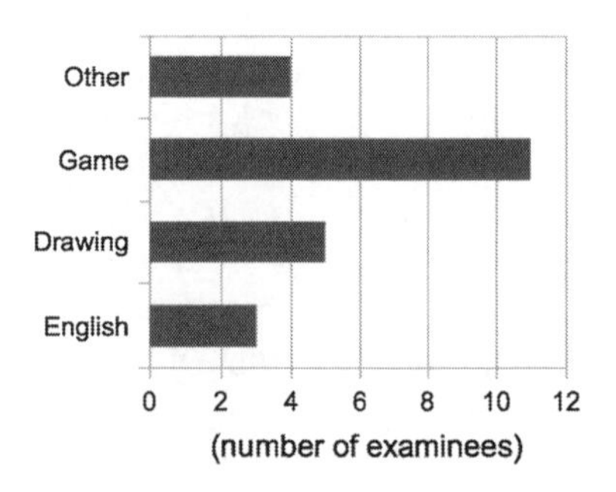

Fig. 7 Features of vocab draw motivate the examinees to play again (multiple answers)

Table 1 Feedback of vocab draw

Number of people	Feedback
Positive feedbacks	
4	Easy to see the vocabulary as an image
3	Helpful to learning vocabularies
1	Does not give the feeling of learning
1	More chance to play with English
1	Might be memorable since playing while guessing
Negative feedbacks	
4	Many words are difficult to draw
1	Words should be used in text, rather than just translated alone
1	Learning efficiency is low, since long time is taken to draw
1	It is more like a testing rather than learning

people cannot draw the word except noun, the number of those words would be reduced in the answering part as well, resulting less opportunity to experience the word. Thus, in our experiment, almost everyone chose a noun to draw. If we keep using these four parts of speech, we must re-consider the easier way to draw, or the sets of the vocabularies.

Conclusion and Future Work

As the conclusion, we have succeeded to encourage the examinees to learn English vocabularies with our novel game, Vocab Draw. Their motivation to play it again was as high as 82 %, however, not all of them were strongly passionate to do so.

In order to change the players attitude from passive to active motivations, introduction of the Bartle's personality classification [4] should be considered. Bartle's personality classification is probably the most famous classification of game player personality, originally applied to MMORPG players. The game players are divided into four personalities of achievers, explorers, socializers, and killers. Each personality has different aim to feel joy in a game, and he plays the game to accomplish his aim. A list of their basic aims are provided in the Table 2. One goal might be a perfect fit for one type, but occasionally it wouldn't be an amenity to another type at all. Great games tend to support all of these goals shown in Table 2, so it could attract any game players. It is proven by many top selling

Table 2 Bartle's personality classification

Player type	Aim to motivate the player
Achiever	Achieving
Explorer	Curiosity
Socializer	Co-operating
Killer	Competition

Table 3 Possible new functions for vocab draw

Player types	Functions to implement	Aiming goals
Achiever	Gallery	Collect all the words
Explorer	Stages	Open up every stages
Socializer	Team play	Co-operate with others
Killer	Ranking	Becoming a top player in the ranking

games that this classification is useful, but it is not proven that it would work to educational games as well yet. Therefore, in our future work, we would like to validate its usefulness in an educational game by adding some new features listed in Table 3.

References

1. The Annual Video Game Industry Report: Media Create (2007–2012)
2. McGonigal J (2011) Reality is broken: why games make us better and how they can change the world. Penguin Books, City of Westminster
3. Kamiya N, Hasegawa N, Hasebe N, Machida N (2011) Part of speech ratio and kinds of verbs in "english note": scientific approaches to language 1(9):233–258
4. Bartle RA (1996) Hearts, clubs, diamonds, spades: players who suit muds, http://www.mud.co.uk/richard/hcds.htm

An Improved Method for Measurement of Gross National Happiness Using Social Network Services

Dongsheng Wang, Abdelilah Khiati, Jongsoo Sohn, Bok-Gyu Joo and In-Jeong Chung

Abstract Studies on the measurement of happiness have been utilized in a variety of areas; in particular, it has played an important role in the measurement of society stability. As the number of users of Social Network Services (SNSs) increase, efforts are being made to measure human well-being by analyzing user messages in SNSs. Most previous works mainly counted positive and negative words; they did not consider the grammar and emotion. In this paper, we reorganize the mechanism to harness the advantages of (a) Part-Of-Speech (POS) tagging for grammatical analysis, and (b) the SentiWordNet lexicon for the assignment of sentiment scores for emotion degree. We suggest a modified formula for calculating the Gross National Happiness (GNH). To verify the method, we gather a real-world dataset from 405,700 Twitter users, measure the GNH, and compare it with the Gallup well-being release. We demonstrate that the method has more precise computation ability for GNH.

Keywords Social network service (SNS) · Happiness · Well-being · Gross national happiness (GNH)

D. Wang (✉) · A. Khiati · I.-J. Chung
Department of Computer and Information Science, Korea University, Seoul, Korea
e-mail: dswang2011@korea.ac.kr

A. Khiati
e-mail: le_zakkaz@korea.ac.kr

I.-J. Chung
e-mail: chung@korea.ac.kr

J. Sohn
Service Strategy Team, Visual Display, Samsung Electronics, Seoul, Korea
e-mail: jongsoo.sohn@samsung.com

B.-G. Joo
Department of Computer and Information Communications, Hong-Ik University, Seoul, Korea
e-mail: bkjoo@hongik.ac.kr

Y.-M. Huang et al. (eds.), *Advanced Technologies, Embedded and Multimedia for Human-centric Computing*, Lecture Notes in Electrical Engineering 260,
DOI: 10.1007/978-94-007-7262-5_3,

Introduction

Gross National Happiness (GNH) was initially proposed in 1972 by the former King of Bhutan, who wanted to promote the development of people's mental health and happiness according to Bhutan's Buddhist culture [1]. Recently, GNH is utilized in a variety of areas as an evaluative indicator of the measure of happiness in a nation and its regions. Moreover, it plays a predominant role in measuring personal happiness, according to a diversity of studies [2]. Gallup, a representative organization to measure GNH, annually measures happiness for each nation by conducting survey tracking [3].

Evaluating the GNH based on Social Network Services (SNSs) was attempted initially by Kramer [4]. He assumed and proved that the more positive words people used in their updates, the happier they were, and vice versa. Thus, he used LIWC2007 corpus [5] to classify positive and negative words and count them respectively. Additionally, he proposed a formula to normalize the happiness scores, and applied this to his research. Subsequently, various studies were conducted using Kramer's method. However, Kramer's method has two problems.

- Lack of a Part-Of-Speech (POS) analysis: It is hard to distinguish the sentiment polarity. Considering the statements "I like my family" and "She looks like her sister," the term 'like' in the former statement conveys a positive sentiment, whereas its use in the latter one conveys an objective description.
- Lack of sentiment degree analysis: Words that convey emotions may express varying degrees of a particular sentiment. For example, although both 'bad' and 'terrible' are negative words, they convey different degrees of the quality, i.e., 'terrible' conveys a greater degree of negativity than 'bad'.

In order to overcome these barriers, we make use of a POS [6] tagger to grammatically analyze a user's message. This enables us to clarify the meaning of each word and then classify positive and negative words clearly. For example, 'like' is commonly regarded as a positive word; however, in the sentence "she looks like somebody," it is an objective description without any sentimental meaning, which can be identified if the POS of words is provided. Further, POS can be applied to analyze negative phrases such as 'not good' or 'not bad.' Secondly, we employ the *SentiWordNet* [7] lexicon to assign a sentiment scores (a real number from -1 to 1) to each emotion word. These sentiment scores can be used not only in the classification of positive and negative words but also in the utilization of the degree of sentiment, thus overcoming the drawbacks of Kramer's approach. Finally, we suggest the improved formula to measure GNH based on Kramer's method and the sentiment score.

In order to verify the proposed method, we establish Gallup's survey tracking as ground truth, and then compare the method to the results of the *Word Count* method using the same datasets for eight cities in the United States. We show that the suggested method is comparable to that of Gallup and better than the previous *Word Count* methods.

Related Works

The progress of health care in Bhutan was reported in [8], which provided a quantitative meaning to the measurement of GNH for the government. The authors of [9] emphasize the conception GNH and propose a GNH framework to assess the health impact where the health, environmental, and economic impact can be estimated collaboratively.

Survey tracking has been a predominant method to measure happiness and the current survey method employs a self-report methodology [3], which has been found to be reasonably accurate in measuring the well-being of an individual [2]. Gallup [3], has studied human nature and behavior for more than 75 years. He released a well-being index based on a daily tracking survey of 1,000 American adults.

Winton performed a comprehensive investigation from different views of measurement and concluded that no measurement is definitely perfect and the best approach is to use a range of measures [1]. SNS-based measurement is evolving to be an innovative and significant approach for measuring GNH. Some researchers employ the *Word Count* strategy to measure GNH, which extracts words from user's messages in SNSs and then classifies positive and negative words by using the LIWC2007 corpus (as depicted in Fig. 1). For instance, Facebook is used by Kramer to track the well-being of the population in the U.S. that uses Facebook [4]. He assumed that the more positive words people use in their status updates, the happier they are, and vice versa.

For example, the user's message "It's a happy day." gets a positive score of 1/4, and a negative score of 0/4. Not only Facebook, the biggest social media, but also Twitter, the second biggest one, was used to measure happiness. Daniele [10] attempted to measure the Gross Community Happiness from Tweets using a method similar to Kramer's, and validated the feasibility of using Twitter even though Twitter has many of its own unique characteristics. However, the *Word Count* method is more likely to result in inaccurate and incorrect analyses since it is difficult to represent the degree of emotion conveyed using each word and infeasible to analyze the POS of each word.

Various studies were conducted to derive insights using national happiness indices. In [11], an evaluation was made based on 336,802 unique users and 12,781,243 Tweets to explore and visualize topics in Twitter. The drawback of the method is that it analyzes only the frequency of words without considering the POS and the relationship between words.

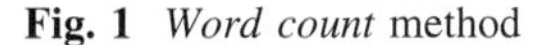

Fig. 1 *Word count* method

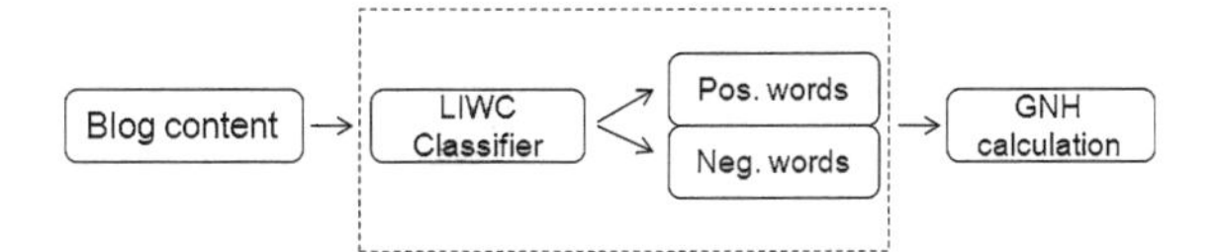

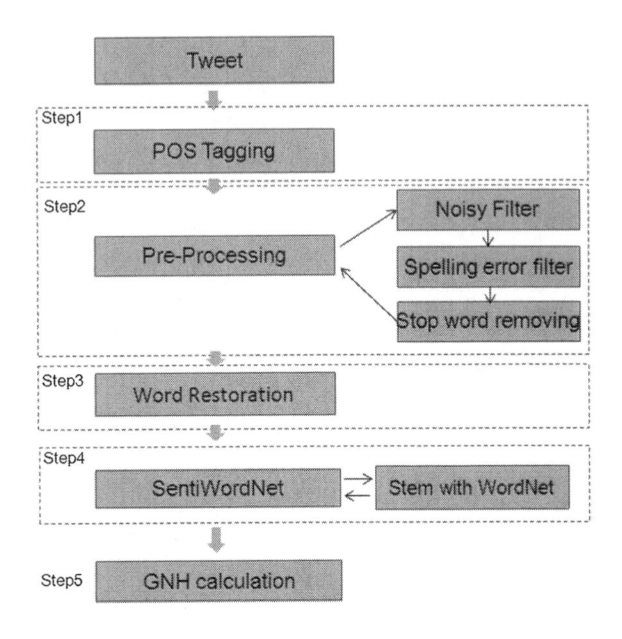

Fig. 2 Overview of the suggested method for GNH measurement

Improved Method for GNH Measurement

In this paper, we proposed a modified version of the GNH measurement method based on the frequency of words in SNS messages. We enhance the measurement method of the GNH by adding POS tagging and utilizing the *SentiWordNet* lexicon. The suggested method evaluates the sentiment scores, as shown in Fig. 2.

The proposed method consists of five steps: (1) POS tagging, (2) preprocessing, (3) restoration of the original form of each word, (4) assignment of a sentiment score for each word, and (5) computation of GNH based on the suggested formula.

Step 1: POS Tagging. POS tagging divides each sentence into words and restores the original form of each word using *WordNet*. It recognizes the POS, such as noun, adjective, and verb, and tags words respectively. Thus, it reduces the error rate of the classification of positive and negative words and increased the ease in processing synonyms.

For the case depicted in Fig. 3, the result of the tagging is:

$\{(actually{:}\,R), (not{:}\,R), (that{:}\,D), (stressed{:}\,A), (\ldots{:},), (life{:}\,N), (is{:}\,V), (goood{:}\,A)\}$

R denotes an adverb; *D*, determiner; *A*, adjective; ‘,’, punctuation; *N*, common noun; and *V*, verb [6].

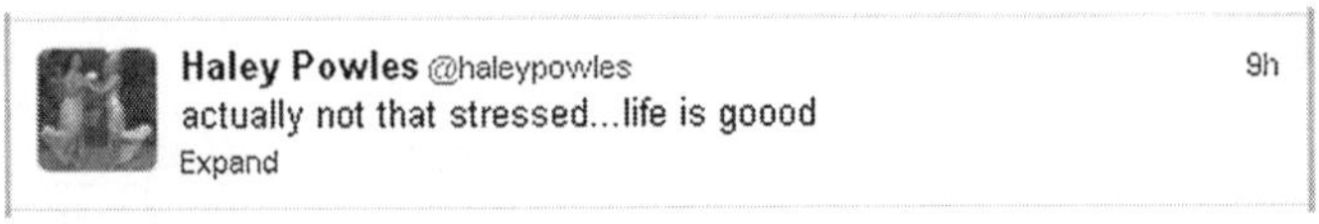

Haley Powles @haleypowles 9h
actually not that stressed...life is goood
Expand

Fig. 3 Example of a tweet

Step 2: Preprocessing. The preprocessing step encompasses (1) noise filtering, (2) spelling error filtering, and (3) stop word removal. Noise filtering removes symbols such as URLs, @, RTs, emoticons, and punctuations. Words that are often skipped and filtered out by search machines, such as 'a,' 'is,' and 'the,' are removed directly.

For the case depicted in Fig. 3, '…' is filtered according to sub-step (1). 'that' and 'is' are also filtered during sub-step (3).

Step 3: Word Restoration. This step restores the original, condensed form of each word associated with subjectivity and sentiment, which may have been intentionally lengthened, such as 'Goooood' and 'Coooooool.'

For the case depicted in Fig. 3, 'goood' is condensed back to 'good' so that it can retrieve a sentiment score from *SentiWordNet*.

Step 4: Sentiment Score Assignment. *SentiWordNet*, a lexical dictionary developed based on *WordNet* [12], describes terms and their semantic relationships. It consists of approximately 117,660 synsets (207,000 word-sense pairs). Each term in its synset is described using a triple of positive, negative, and neutral (objective) sentiment scores. For example, {pos. = 0.35, neg. = 0.65, neutral = 0}, which sums up to 1 (i.e., pos. + neg. + neutral = 1).

WordNet and *SentiWordNet* are applied to represent the sentiment score for each word as a real number between −1 and 1. In addition, the sentiment score of an emotion word that belongs to negative sentence (appears after 'not,' 'never,' 'none,' 'no,' 'hardly,' 'seldom,' etc.) is reversed.

For the case depicted in Fig. 3, after the three subsequent steps are completed, 'actually,' 'not,' 'stressed,' 'life,' and 'good' remain. The adverb 'not' expresses a negation, which reverses the sentiment score of the emotion word after it. The sentiment score of 'stressed,' which is −0.34, is reversed as +0.34. The adjective 'good' gets a sentiment score of 0.48 and the noun 'life' gets a 0.0.

Step 5: Formula Calculation. Formulae (1) and (2) are defined to calculate the sentiment score of a message. p_i is the positive score of the message, and n_i is the negative score of the message. N indicates the number of positive words in formula (1), and the number of negative words in formula (2); $score_w$ is the *SentiWordNet* score of each word.

$$p_i = \sum_{p_w=1}^{N} \frac{score_w}{\text{Number of total words}} \tag{1}$$

$$n_i = \sum_{n_w=1}^{N} \frac{score_w}{\text{Number of total words}} \tag{2}$$

For example, consider the Tweet "the weather is good, but I am sad." The preprocessing result is {0, 0, 0, 0.478, 0, 0, 0, −0.416} as the scores of 'good' and 'sad' are 0.478 and −0.416 respectively. By applying formulae (1) and (2), the positive score p_i is 0.478/8, and the negative score n_i is −0.416/8 respectively. For the case depicted in Fig. 3, $p_i = \frac{0.87}{7} = 0.11$, $n_i = \frac{0.0}{7} = 0$.

$$H_i = \frac{p_i - \mu_p}{\sigma_p} - \frac{n_i - \mu_n}{\sigma_n} \qquad (3)$$

Further, we put this score into the formula (3) [4] in order to normalize the happiness score (in practice, for the control of scale, we aggregate all Tweets of a user daily as 'one Tweet'). In the formula (3), H_i is the happiness score for user i at a specific time, where p_i and n_i represent the positive and negative sentiment scores; μ_p and μ_n represent the average positive and negative sentiment scores across all users during a specific time; and $\sigma_p(\sigma_n)$ are the corresponding standard deviations. As described above, we compute the happiness score for each user in a particular SNS using the suggested steps and formulae.

Performance Study

For the validation and evaluation of the suggested method, we implemented a system and conducted an experiment. In the experiment, with the aid of Twitter Application Programming Interface (API), we crawled 2,072,329 user profiles across 8 cities in the United States. In order to guarantee that the users are active, we refined the dataset with some strategies, and obtained 405,700 user profiles as the experimental dataset. We collected up to 300 Tweets created by each user in 2011. As a result, we acquired approximately 60 million Tweets from 405,700 users.

We applied the *Word Count* method and the suggested method to the same dataset for the measurement of the GNH results and compared the two results. Given that the Gallup is ground truth, Fig. 4 illustrates the accuracy rates of the *Word Count* method and our suggested measurement with respect to Gallup. The x-coordinate represents the 12 months in 2011, and the y-coordinate is the degree

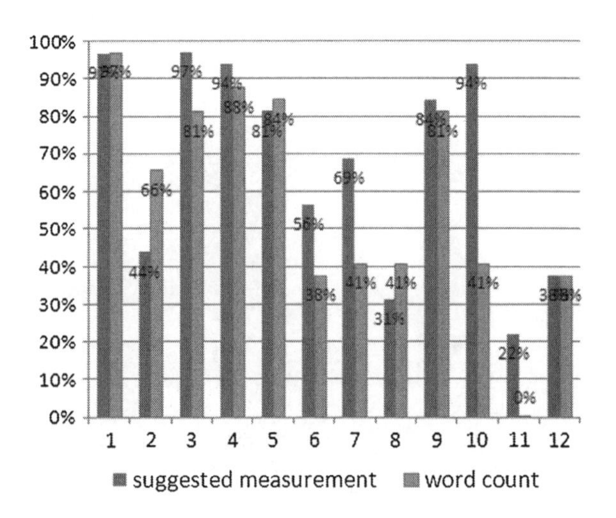

Fig. 4 Accuracy distribution in each month

of proximity to Gallup. Except for February, May, and August, we find a stronger correlation with Gallup during all other months, and the overall improvement is by approximately 10 %.

Compared to the survey tracking method, SNS-based GNH measurement is able to deal with a huge amount of data while consuming a smaller amount of time. Additionally, the suggested measurement is more accurate and credible than the existing *Word Count* method.

Conclusion

GNH measurement based on data in SNSs has been studied recently. However, previous methods consider only the polarity during classification leading to ambiguous and unreliable results. Accordingly, we consider the emotional degree in the computation of GNH and reorganize the mechanism to be more accurate. First, we employ (a) POS tagging and (b) emotion degree assignment and detection of dual affirmation and negation (score reversing) to obtain a more precise sentiment score for each word. We then validate the proposed method in the experiment by providing a comparison with the *Word Count* method. The experiment illustrated that the suggested method shows better results over existing methods and demonstrated the feasibility of considering the degree of emotion words.

In comparison with the questionnaire-based survey tracking, the suggested SNS-based method for the measurement of GNH was able to reduce the time and cost drastically. Moreover, the method facilitates the establishment of policies and strategies for customer marketing in enterprises and societies based on the computed happiness scores.

In future, based on the experimental results, the proposed method can be applied to analyze the relationship between the happiness score and other mass media such as online news and articles with the aid of semantic web technology.

References

1. Bates W (2009) Gross national happiness. Asian-Pac Econ Lit 23:1–16
2. Diener E, Diener M, Diener C (1995) Factors predicting the subjective well-being of nations. J Pers Soc Psychol 69:851–864
3. Walker SS, Schimmack U (2008) Validity of a happiness implicit association test as a measure of subjective well-being. J Res Pers 42:490–497
4. Kramer ADI (2010) An unobtrusive behavioral model of "gross national happiness." In: 28th international conference on human factors in computing systems, ACM, Atlanta, Georgia, USA, pp 287–290
5. James W, Pennebaker CKC, Ireland M, Gonzales A, Booth RJ (2007) The development and psychometric properties of LIWC2007. LIWC.Net, Austin, TX

6. Gimpel K, Schneider N, O'Connor B, Das D, Mills D, Eisenstein J, Heilman M, Yogatama D, Flanigan J, Smith NA (2011) Part-of-speech tagging for Twitter: annotation, features, and experiments. In: 49th annual meeting of the association for computational linguistics: human language technologies: short papers. vol 2. Association for Computational Linguistics, Portland, Oregon, pp 42–47
7. Sebastiani AEAF (2006) SentiWordNet: a publicly available lexical resource for opinion mining. In: Language resources and evaluation (LREC), pp 417–422
8. Tobgay T, Dorji T, Pelzom D, Gibbons RV (2011) Progress and delivery of health care in Bhutan, the land of the thunder dragon and gross National happiness. Trop Med Int Health 16:731–736
9. Pennock M, Ura K (2011) Gross National happiness as a framework for health impact assessment. Environ Impact Asses 31:61–65
10. Quercia D, Ellis J, Capra L, Crowcroft J (2012) Tracking "gross community happiness" from Tweets. In: ACM 2012 conference on computer supported cooperative work. ACM, Seattle, Washington, USA, pp 965–968
11. Brew A, Greene D, Archambault D, Cunningham P (2011) Deriving insights from National happiness indices. In: 2011 IEEE 11th international conference on data mining workshops. IEEE Computer Society, pp 53–60
12. Miller GA (1995) WordNet: a lexical database for English. Commun ACM 38:39–41

Advanced Comb Filtering for Robust Speech Recognition

Jeong-Sik Park

Abstract This paper proposes a speech enhancement scheme that leads to significant improvements in recognition performance when used in the Automatic Speech Recognition (ASR) front-end. The proposed approach is based upon adaptive comb filtering. While adaptive comb filtering reduces noise components remarkably, it is rarely effective in reducing non-stationary noises due to its uniformly distributed frequency response. This paper proposes an advanced comb filtering technique that adjusts its spectral magnitude to the original speech, based on the gain modification function, an Minimum Mean Squared Error (MMSE) estimator.

Keywords Advanced comb filtering · Gain modification function · Minimum mean squared error · Robust speech recognition

Introduction

Considerable efforts have been directed toward reducing various kinds of noises including additive and channel noise, with the goal of improving speech quality and intelligibility. Although various algorithms for speech enhancement have been developed over the last several decades, ASR systems in speech-driven interfaces such as mobile devices require further effective reduction of non-stationary noises.

The adaptive comb filtering firstly presented by Frazier enhances noisy speech with the pitch period of a voiced sound [1]. Although adaptive comb filtering has

J.-S. Park (✉)
Department of Intelligent Robot Engineering, Mokwon University, Daejeon, South Korea
e-mail: parkjs@mokwon.ac.kr

Y.-M. Huang et al. (eds.), *Advanced Technologies, Embedded and Multimedia for Human-centric Computing*, Lecture Notes in Electrical Engineering 260,
DOI: 10.1007/978-94-007-7262-5_4,

been applied to noise reduction with low complexity and high capacity [2, 3], it is vulnerable to non-stationary noises due to uniformly distributed frequency response of comb-filter as well as the difficulty in estimating the accurate pitch period. This paper introduces some problems of conventional adaptive comb filtering and proposes an improved comb filtering process employing an Minimum Mean Squared Error (MMSE) estimator.

Section Advanced comb filtering explains the principles of adaptive comb filtering and the modified comb filtering. And, sections Estimation of speech absence probability and Improvement of comb-filter by the gain modification function present experimental results and conclusions, respectively.

Advanced Comb Filtering

The comb filtering is based on the observation that wave-forms of voiced speech are periodic, corresponding to the fundamental frequency [1]. As noise components generally exist between the harmonics of speech spectrum, they are reduced by a comb-filter, which passes only the harmonics. The basic operation of a comb-filter can be explained by considering its impulse response as:

$$h(n) = \sum_{k=-L}^{L} \alpha_k \times \delta(n - T) \tag{1}$$

where $\delta(n)$ is an unit impulse function, T corresponds to the pitch period, and the length of the filter is $2L + 1$. α_k is the filter coefficient that is associated with the intelligibility of the filtered speech, satisfying $\sum_{k=-L}^{L} \alpha_k = 1$.

A comb-filter can significantly reduce noise components existing between the harmonics while preserving the speech sounds. Nevertheless, comb filtering occasionally results in poor performance in certain cases. First, the filter may degrade the quality of clean speech, as regularly repeated frequency response can distort the speech signal where the fundamental frequency is changed continuously. Second, severe noise may also cause distortion due to inaccurate pitch estimation.

This study proposes an approach to compensate for the distortions, using the Gain Modification Function (GMF), an MMSE estimator. The GMF denotes the probability of speech-presence in the frequency bin and consists of the Speech Absence Probability (SAP), priori SNR, and posteriori SNR [4]. A frequency bin with low SAP and high SNR represents the speech region rather than the non-speech region, indicating a high value of GMF. The key issue in this work is adjusting the frequency response of the comb-filter to match the local characteristics of the clean speech spectrum on the basis of GMF estimated in each frequency bin.

Estimation of Speech Absence Probability

First, we estimate the SAP given by [5]:

$$q_l(\omega) = \alpha \cdot q_{l-1}(\omega) + (1 - \alpha) \cdot I_l(\omega) \tag{2}$$

which denotes the SAP of l-th frame in the frequency bin ω. $I_l(\omega)$ is a hard-decision parameter that determines whether speech is present in the frequency bin. The decision is made from the posteriori SNR (γ_ω^l) and a threshold (γ_{TH}) as:

$$I_l(\omega) = \begin{cases} 0 \ (\gamma_\omega^l \geq \gamma_{TH}) \\ 1 \ (\gamma_\omega^l < \gamma_{TH}) \end{cases} \tag{3}$$

The frequency bin, where $I_l(\omega)$ equals 0, refers to the speech-present region. In the speech-absent region, $q_l(\omega)$ is close to 1, thus giving high SAP. In [5], the parameters γ_{TH} and α are set as constants (0.8 and 0.95, respectively) based on informal listening tests. As such, the fixed value of γ_{TH} may mislead (3), thus resulting in an incorrect decision of a speech-present/absent region, particularly for heavily damaged speech. We update the parameter continually, considering the posteriori SNR changed per frame and frequency bin, as shown in:

$$\gamma_{TH}^l(\omega) = \sum_{k=l-\beta}^{l-1} \frac{\gamma_\omega^k}{\beta} \tag{4}$$

The average of γ_ω estimated in the previous β frames is expected to advance the hard-decision and thus improve the estimate of SAP applied to GMF. Our experiments determined that the smallest number of β is 5 while preserving the performance. And we confirmed that α doesn't influence the performance a lot and used a fixed value of 0.95.

Improvement of Comb-Filter by the Gain Modification Function

The proposed approach is based on a proposition that the GMF can adjust the frequency response of the comb-filter to the spectrum of the original speech, depending on the speech absence probability for each frequency bin. The adaptive comb filtering modified by the MMSE estimator is summarized as:

$$\widehat{A}_l(\omega) = G_l(\omega) \times A_l(\omega) \tag{5}$$

$A_l(\omega)$ is the spectral energy of the adaptive comb-filter in the frequency bin ω for l-th frame and $G_l(\omega)$ is the gain modification function estimated by [5]:

$$G_l(\omega) = \frac{\Lambda_l(\omega)}{1 + \Lambda_l(\omega)} \approx 1 - q_l(\omega) \tag{6}$$

where $\Lambda_l(\omega)$ is a likelihood ratio calculated from the SAP and $G_l(\omega)$ is close to 1 in the speech present region. According to (5), $G_l(\omega)$ degrades the spectral energy of the comb-filter in the frequency bins corresponding to the speech-absent regions, where $G_l(\omega)$ goes to 0. This modification gives further correct estimate for the frequency response of comb-filter.

It should be noted that the direct use of $G_l(\omega)$ ranging from 0 to 1 may degrade the spectral energy over all of the frequency bins including the speech region. To prevent signal distortion induced by this property, we modify $G_l(\omega)$ based on the SAP estimator as shown in (7).

$$\widehat{G}_l(\omega) = \{1 - (q_l(\omega) - \varepsilon)\} \times G_l(\omega) \tag{7}$$

where $q_l(\omega)$ gives the GMF a limited range from 0 to $(1 + \varepsilon)$.

By applying $\widehat{G}_l(\omega)$ instead of $G_l(\omega)$ in (5), the modified GMF degrades the spectral energy of the comb-filter over the region where $q_l(\omega)$ is higher than ε. On the other hand, it retains or emphasizes the energy over the region where $q_l(\omega)$ is equal to or lower than ε. By experiments, we concluded that $\varepsilon = 0.35$ provides the best performance. Based on the modified GMF, the comb-filter is expected to further remove the spectral noise components, eliminating signal distortions.

Experiments and Results

Experimental Environments

We performed the ASR experiments on the Aurora 2 database, following the procedures for training Hidden Markov Model (HMM) under the clean condition and recognizing data by using the HTK software tools [6, 7]. The training and testing set consist of 8,440 and 4,004 digit strings, respectively. In addition, the test data is divided into three sets according to noise types and channel condition. We obtained 12 Mel Frequency Cepstral Coefficients (MFCCs) and a log energy with their first and second derivatives from all training and testing data. Each word was modeled by a simple left-to-right 16-state three-mixture whole word HMM. A three-state six-mixture silence model and a one-state six-mixture pause model were also used.

Recognition Results

We performed two kinds of experiments to compare the performance of our proposed technique to several comb filtering techniques introduced previously.

Table 1 Comparison of WERs (%) for the speech processed by several adaptive comb filtering

	Data set			Avg.	WER reduction
	Set A	Set B	Set C		
Original	44.76	50.35	39.22	44.78	–
BACF	42.14	48.56	37.10	42.60	4.86
FACF	41.42	46.76	37.26	41.85	6.54
MACF	38.82	43.38	35.41	39.21	12.45

Table 2 Comparison of WERs (%) over clean and high SNRs

	SNR(dB)			Avg.	WER reduction
	clean	20 dB	15 dB		
Original	1.68	4.83	17.07	7.86	–
BACF	4.05	6.06	13.94	8.02	−1.99
FACF	3.54	5.98	13.70	7.74	1.53
MACF	2.21	5.25	11.00	6.15	21.71

Table 1 shows results for the first experiment. We performed recognition experiments on four kinds of speech data, one of which is the raw data itself ("original") and the others were processed by the basic adaptive comb-filter ("BACF"), the fully adaptive comb-filter ("FACF") and the modified comb-filter ("MACF"), respectively. FACF was known as the representative method for enhancing the basic comb-filter, by improving the fixed filter coefficients [3]. We report the performance as the Word Error Rate (WER) averaged over Signal-to-Noise Ratios (SNRs) from 0 to 20 dB. In all test sets, the comb-filtered data outperforms the raw speech data due to noise reduction. Compared with the original speech, three kinds of com-filters lead to WER reduction of 4.86, 6.54, and 12.45 %, respectively. MACF shows a significant improvement, 8 and 6.3 % WER reduction, over BACF and FACF, respectively.

Although the above results verify that the proposed approach improves BACF and FACF successfully, more reliable results are given in Table 2, where BACF has even higher WER than Original or MACF over clean and high SNRs (15–20 dB). This poor accuracy results from the distortions of clean speech due to comb filtering. By the way, MACF leads to remarkably 45.4 and 37.6 % WER reduction compared with BACF and FACF, respectively, in clean environment. This result demonstrates that MMSE estimator-based comb filtering guards against signal distortions significantly while previous comb filtering techniques don't prevent from the distortions.

Conclusion

We proposed a speech enhancement scheme that combines MMSE estimator and conventional adaptive comb filtering to protect against signal distortions caused by comb filtering. The proposed method adjusts the spectral magnitude of the comb-filter to the spectrum of the clean speech, based on gain modification function. This approach degrades the spectral energy of comb-filter in the frequency region where no speech components exist while retaining or emphasizing the energy of comb-filter in the speech-present region. We applied our speech enhancement scheme to the ASR front-end based on Aurora 2 database. Recognition results showed that the modified comb filtering yields superior performance compared to no-processing (baseline) and other comb filtering techniques, presenting WER reduction of 12.5 % to baseline. The results also confirmed that our approach notably improves the distortions of clean speech, which is the most serious weakness of comb filtering. Further work will be aimed at verifying our approach with experiments on another data sets and applying to ASR system in speech-driven interfaces.

Acknowledgments This work was supported by the NAP (National Agenda Project) of the Korea Research Council of Fundamental Science and Technology.

References

1. Frazier RH, Samsam S (1976) Enhancement of speech by adaptive filtering. In: Proceedings of IEEE International Conference on ASSP, Philadelphia. PA, pp 251–253
2. Nehorai A, Porat B (1986) Adaptive comb filtering for harmonic signal enhancement. IEEE Trans Acoust Speech Signal Process 34(5):1124–1138
3. Veeneman DE, Mazor B (1989) A fully adaptive comb filter for enhancing block-coded speech. IEEE Trans Acoust Speech Signal Process 37(6):955–957
4. Ephraim Y, Malah D (1985) Speech Enhancement using a minimum mean-square error log spectral amplitude estimator. IEEE Trans Acoust Speech Sig Process 33(2):443–445
5. Malah D, Cox RV, Accardi AJ (1999) Tracking speech-presence uncertainty to improve speech enhancement in nonstationary noise environments. In: Proceedings of ICASSP, pp 201–204
6. Hirsch HG, Pearce D (2000) The AURORA experimental framework for the performance evaluations of speech recognition systems under noisy conditions. Proc ICSLP 4:29–32
7. Young S, Evermmana G, Gales M, Hain T et al (2006) The HTK Book for HTK Version 3.4., Cambridge University Engineering Department,Cambridge

Fall Detection by a SVM-Based Cloud System with Motion Sensors

Chien-Hui (Christina) Liao, Kuan-Wei Lee, Ting-Hua Chen, Che-Chen Chang and Charles H.-P. Wen

Abstract Recently, fall detection has become a popular research topic to take care of the increasing aging population. Many previous works used cameras, accelerometers and gyroscopes as sensor devices to collect motion data of human beings and then to distinguish falls from other normal behaviors of human beings. However, these techniques encountered some challenges such as privacy, accuracy, convenience and data-processing time. In this paper, a motion sensor which can compress motion data into skeleton points effectively meanwhile providing privacy and convenience are chosen as the sensor devices for detecting falls. Furthermore, to achieve high accuracy of fall detection, support vector machine (SVM) is employed in the proposed cloud system. Experimental results show that, under the best setting, the accuracy of our fall-detection SVM model can be greater than 99.90 %. In addition, the detection time of falls only takes less than 10^{-3} s. Therefore, the proposed SVM-based cloud system with motion sensors successfully enables fall detection at real time with high accuracy.

Keywords Fall detection · Kinect · Support vector machine (SVM) · Cloud computing

C.-H. (Christina) Liao (✉) · K.-W. Lee · T.-H. Chen · C.-C. Chang · C. H.-P. Wen
Department of Electrical and Computer Engineering, National Chiao Tung University, 1001 University Road, Hsinchu, Taiwan, Republic of China
e-mail: liangel.cm97g@g2.nctu.edu.tw

K.-W. Lee
e-mail: behigheveryday.eed99@nctu.edu.tw

T.-H. Chen
e-mail: tinghua000@gmail.com

C.-C. Chang
e-mail: boy76229@yahoo.com.tw

C. H.-P. Wen
e-mail: opwen@g2.nctu.edu.tw

Y.-M. Huang et al. (eds.), *Advanced Technologies, Embedded and Multimedia for Human-centric Computing*, Lecture Notes in Electrical Engineering 260,
DOI: 10.1007/978-94-007-7262-5_5,

Introduction

Nowadays, various health-care systems have been rapidly developed to take care of the increasing aging population. Particularly, since falls are dangerous and even fatal to the elder, fall detection [1–4] has became an important topic for elder healthcare. Most previous works used accelerometers, gyroscopes and cameras to distinguish falls from normal behaviors. However, these techniques encountered some challenges from privacy, accuracy, convenience and runtime to its practicality.

Many of previous works [5–9] used tri-axial accelerometers or gyroscopes to detect fall accidents. However, these wearable accelerometers and gyroscopes are inconvenient to users and also hard to accurately differentiate falls from normal behaviors such as sitting down, lying and hunkering. On the other hand, using cameras [10–12] to detect falls can achieve higher accuracy than using accelerometers and gyroscopes. However, cameras will collect massive data, either useful or useless. Thus, heavy image-processing is required, which makes it hard to be applied for many real-time applications.

Recently, cloud applications for human-centered computing (HCC) arise rapidly. Compared to traditional approaches [1–4], a cloud system intends to handle only useful data for efficiency. Figure 1 shows an examples for illustration. However, a cloud system for HCC still comes with three problems: *big data*, *massive communication* and *heavy computation*. To overcome these three problems to enable a real-time system for highly-accurate and highly-private fall detection, this work uses *motion sensors* (i.e. Microsoft Kinects [13]) to collect data from human beings and applies *support vector machine* (SVM) models [14] in a cloud system for fall detection. Motion sensor devices like Microsoft Kinect have many advantages and are outlined as below:

1. An image of a human being can be transformed into skeleton points, providing better privacy. So our fall-detection systems can even be installed in the bathroom where accidents often occur for the elder.
2. Motion data from multiple users can be captured simultaneously.
3. Data size of skeleton points is much smaller than that of the image.

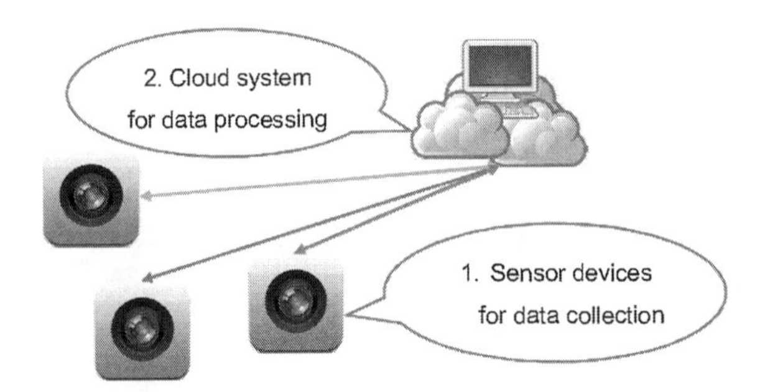

Fig. 1 Cloud systems with sensor devices

Developing a SVM model to determine human behaviors on the cloud system with motion sensors can be widely used at home, hospital, kindergartens, and etc. Our experimental results show that using Kinect as a sensor device, our SVM-based cloud system can reach greater than 99.90 % accuracy under the best setting for fall detection. In addition, using the linear kernel or polynomial kernel for the SVM learning, the response time (for prediction) takes less than 10^{-3} s on average, making such system suitable for real-time applications.

The rest of this paper is organized as follows. Section SVM-based Cloud Systems with Motion Sensors for Fall Detection describes the architecture of our SVM-based cloud system with motion sensors. In this section, Microsoft Kinect is introduced and the proposed SVM model for fall detection is also elaborated. In section Experimental Results, experimental results show the training time and the compression ratio of the proposed SVM model. Furthermore, performance including accuracy and response time (for prediction) for the application of fall detection is also demonstrated. Finally, section Conclusion concludes this paper.

SVM-Based Cloud Systems with Motion Sensors for Fall Detection

Along with the rapidly development of cloud computing and growing popularity of sensor devices, a cloud system with sensor devices have been prevailed in many applications. However, a cloud system combined with sensor devices will encounter three major issues including (1) *big data* (2) *massive communication* and (3) *heavy computation* during applications. To optimize the performance of such system for real-time fall detection, these three challenges need to be taken care of properly.

The idea of data compression is employed to deal with the first two problems (i.e. big data and massive communication). Motion sensors like Microsoft Kinect applies data compression to convert an image to skeleton points and thus are used as the sensor devices in our fall-detection cloud system. Moreover, to alleviate heavy computation, support vector machine (SVM), a machine-learning algorithm to extract only important data points, is also incorporated in this work. Therefore, Fig. 2 shows the overall architecture of our cloud system where motion sensors and the SVM model are the two key components.

Skeleton-Based Motion Sensor

In the proposed cloud system, motion sensors like Microsoft Kinect are responsible for extracting motion data from human beings. Compared to camera sensors which need to process huge amount of pixels, to wait for long uploading time and to perform complex pattern recognition, motion sensors leave only three-dimensional

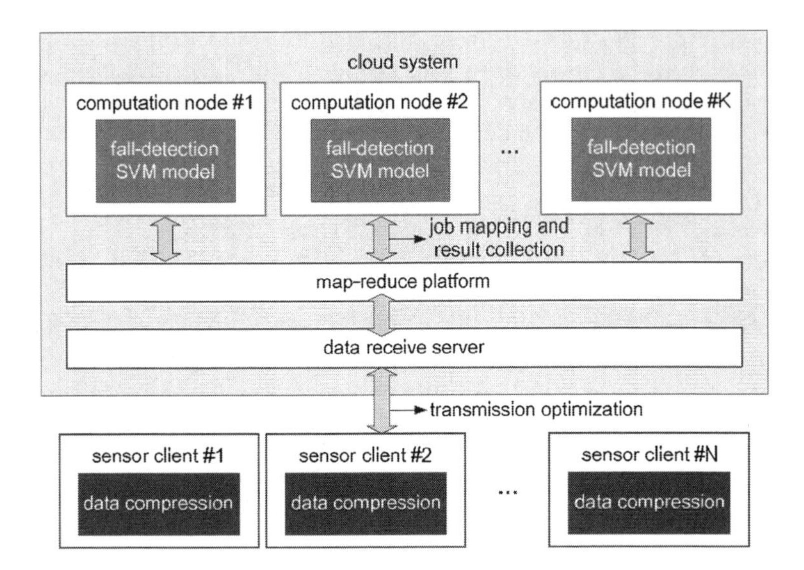

Fig. 2 Architecture of the SVM-based cloud system with motion sensors

Table 1 Comparison between cameras, accelerometers/gyroscopes and motion sensors

Sensor device	Cameras	Accelerometers/gyroscopes	Motion sensors
Data transmission loading	Large	Small	Small
Accuracy (%)	93.3	90.0	97.0
Application flexibility	High	Low	High
Convenience	High	Low	High

skeleton points to represent a human being, and thus compress data, increase data-uploading efficiency and lower computation greatly. Table 1 shows the comparison among cameras [12], accelerometers/gyroscopes [9] and motion sensors from different perspectives. As a result, the motion sensor is the best candidate for the sensor device in our fall-detection cloud systems.

Microsoft Kinect is one new type of motion sensors and can represent the body structure of a human being as twenty three-dimensional skeleton points shown in Fig. 3. Compared to the other types of motion sensors, such as ASUS Xtion [15], Kinect can capture skeleton points more accurately with a wider angle. Furthermore, Kinect supports up to thirty time-frames per second where each time-frame contains twenty points (x, y, z)'s (each point contains three coordinate information, x, y and z). Therefore, Kinect motion sensors can leave only $30 \times 20 \times 3 \times 4 = 7{,}200$ bytes data per second, several-order smaller than the typical image data size.

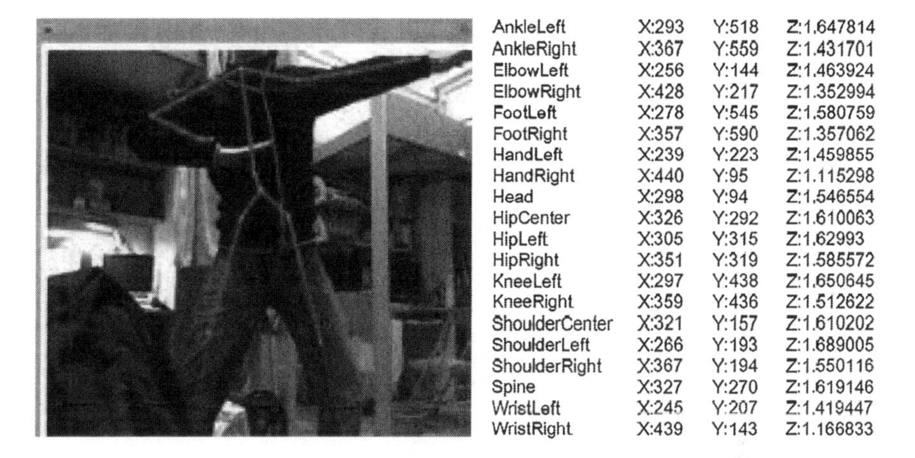

AnkleLeft	X:293	Y:518	Z:1.647814
AnkleRight	X:367	Y:559	Z:1.431701
ElbowLeft	X:256	Y:144	Z:1.463924
ElbowRight	X:428	Y:217	Z:1.352994
FootLeft	X:278	Y:545	Z:1.580759
FootRight	X:357	Y:590	Z:1.357062
HandLeft	X:239	Y:223	Z:1.459855
HandRight	X:440	Y:95	Z:1.115298
Head	X:298	Y:94	Z:1.546554
HipCenter	X:326	Y:292	Z:1.610063
HipLeft	X:305	Y:315	Z:1.62993
HipRight	X:351	Y:319	Z:1.585572
KneeLeft	X:297	Y:438	Z:1.650645
KneeRight	X:359	Y:436	Z:1.512622
ShoulderCenter	X:321	Y:157	Z:1.610202
ShoulderLeft	X:266	Y:193	Z:1.689005
ShoulderRight	X:367	Y:194	Z:1.550116
Spine	X:327	Y:270	Z:1.619146
WristLeft	X:245	Y:207	Z:1.419447
WristRight	X:439	Y:143	Z:1.166833

Fig. 3 Twenty three-dimensional skeleton points captured by Microsoft Kinect

Support Vector Machine

Support vector machine (SVM) is an advanced algorithm, which is widely used for machine learning problems [14] and has three major characteristics:

- SVM can find a global optimal solution for a convex problem easily.
- SVM can compute high-dimensional data effectively.
- SVM can use only a small subset to derive a decision boundary for a collection of data where each sample in the subset called a support vector.

SVM is often used to derive a function as the decision boundary with minimal errors and a maximal margin to separate data in a multi-dimensional space. An illustration of separating the data set with many possible decision boundaries is

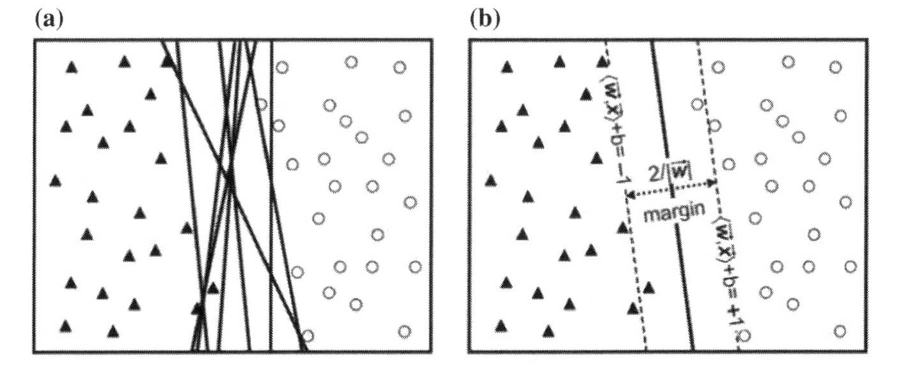

Fig. 4 Linear decision boundaries for a two-class data set [16]

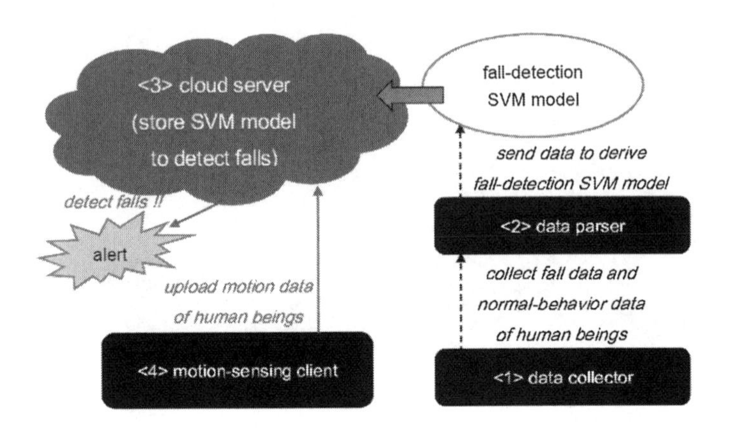

Fig. 5 Overall training framework

shown in Fig. 4a [16] where Fig. 4b indicates a boundaries with the minimal errors and the maximal margin.

SVM can be applied to classification problems and derives a smooth function that minimizes slacks through three steps: (1) primal-form optimization, (2) dual-form expansion, and (3) kernel-function substitution. The nature of the classification problem is first presented by the primal form. Then, through kernel functions, Lagrange method transforms from the primal form into the dual form. In this paper, fall data and normal-behavior data are both used to train SVM models with four different kernel functions (linear, polynomial, radial and sigmoid). After deriving the SVM models, it can be used to distinguish fall accidents from normal behaviors of human beings.

Data Collection and Fall-Detection Model

To build a real-time SVM-based cloud system with motion sensors, this work proposes a training framework consisting of a data collector, a data parser, cloud severs and motion-sensing clients as shown in Fig. 5.

1. Data collector: a data collector in Fig. 6a is a software implemented to collect motion data (either falls or normal behaviors) from Kincet devices. Then, these data will be sent to the data parser. The time-length of a datum can be defined at user's discretion. In our experiment, each datum contains motion information of three seconds long.
2. Data parser: this tool reads the skeleton data including fall and normal-behavior ones into the cloud server for the SVM training to derive models.
3. Cloud server: it applies the pre-built SVM models to first distinguish falls from normal behaviors immediately and then return the result to the clients. If a fall

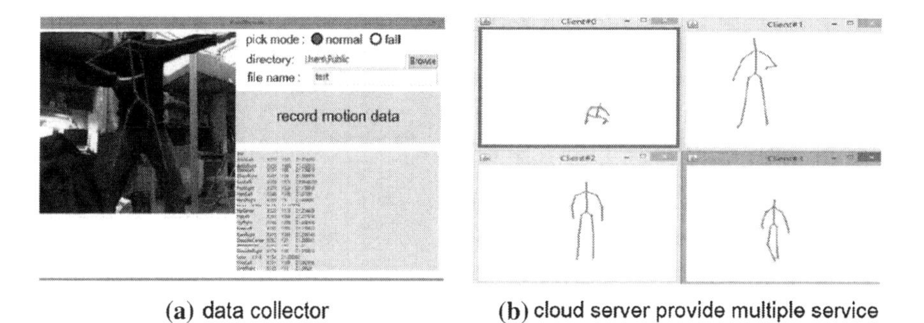

(a) data collector (b) cloud server provide multiple service

Fig. 6 Data collector and cloud server in our training framework

accident is detected, the cloud server will return an alert signal to the client as shown in Fig. 6b.

4. Motion-sensing client: every 0.5/1/1.5 s, it sends out a set of motion data of three seconds long to the cloud server for fall detection.

After using the data collector to collect a set of fall samples and normal-behavior samples, a half of the data is used to train fall-detection SVM models. Then, the other half is used to evaluate the accuracy of the SVM model.

Experimental Results

Our framework was implemented in C# for the client and in Java for the server. It ran on a Linux machine with a Intel Core i5 (3.6 GHz) processor and 2 GB memory. 320 fall data and 320 normal-behavior data were collected through our data collector. Each datum contains three-second long skeleton-points as the motion information. A half of the fall data and normal-behavior data was used to train the SVM model while the other half was used to validate the SVM model. In the experiments, the training time, compression ratio, accuracy and fall-detection time of the SVM models under different kernel functions (linear, polynomial, radial basis and sigmoid) and different update periods (0.5, 1.0 and 1.5 s) of input data are presented. The default parameters provided by the LIBSVM library [17] of the SVM-model kernel functions were used in this paper.

Table 2 compares the SVM-model training time and SVM-model compression ratio under different kernel functions and different update periods of input data. Training SVM models with the linear and polynomial kernel functions are preferable over the other two due to better compression. Particularly, training by the radial basis kernel function results in no compression. In addition, using a longer update period of input data (i.e. 1.5 s) can derive the model faster than using a shorter period (0.5 s) on the same data. However, the compression ratio of a

Table 2 Training-time and compression-ratio comparison on different settings

Kernel function	Training time (s)			Compression ratio (%)		
	Update period of input data			Update period of input data		
	0.5 s	1.0 s	1.5 s	0.5 s	1.0 s	1.5 s
Linear	17.6	6.8	4.7	0.70	1.19	1.79
Polynomial	17.0	6.9	4.8	0.82	1.39	1.85
Radial basis	2620.0	226.0	104.0	100.00	100.00	100.00
Sigmoid	39.0	15.6	10.9	2.88	5.68	8.39

Table 3 Fall-detection accuracy and time comparison on different settings

Kernel function	Fall-detection accuracy (%)			Fall-detection time (10^{-3}s)		
	Update period of input data			Update period of input data		
	0.5 s	1.0 s	1.5 s	0.5 s	1.0 s	1.5 s
Linear	99.89	99.82	99.79	1.06	0.94	0.86
Polynomial	99.90	99.84	99.81	1.00	0.91	0.87
Radial basis	99.27	98.55	97.84	89.52	47.03	32.09
Sigmoid	99.27	98.55	97.84	3.02	2.96	3.06

shorter update period is better than that of a longer update period for most of the cases in Table 2.

Table 3 compares the accuracy and fall-detection time of SVM models under different kernel functions and different update periods. All models can reach more than 97 % accuracy. Particularly, using SVM models with the linear and polynomial kernel functions can achieve higher accuracy (≧99.79 %) for all cases. In addition, the SVM models derived from the linear and polynomial kernel functions only take less than 10^{-3} s to detect falls. Therefore, based on the experimental results, the linear and polynomial kernel functions are recommended to be used for deriving the fall-detection SVM model. To sum up, our experiments demonstrate that the proposed SVM-based cloud system with motion sensors can run at real time and detect fall behaviors accurately.

Conclusion

For enabling a real-time highly-accurate fall-detection cloud system, skeleton-based motion sensors (i.e. Microsoft Kinect) and support-vector-machine (SVM) learning are employed. Using skeleton trace of motions can provide high privacy, high convenience, high application flexibility and effective data compression. Moreover, the SVM models derived from the linear and polynomial kernel functions can result in high accuracy (≧99.79 %) and take less than 10^{-3} s for fall detection. Through the experimental results, the proposed SVM-based cloud system with motion sensors has been successfully demonstrated to be able to detect falls at real time with high accuracy.

References

1. Tomii S, Ohtsuki T (2012) Falling detection using multiple doppler sensors. In: IEEE 14th international conference on e-health networking, applications and services (Healthcom), pp 196–201
2. Popescu M, Hotrabhavananda B, Moore M, Skubic M (2012) VAMPIR- an automatic fall detection system using a vertical PIR sensor array. In: 6th international conference on pervasive computing technologies for healthcare (Pervasive Health), pp 163–166
3. Lustrek M, Kaluza B (2009) Fall detection and activity recognition with machine learning. informatica (Slovenia), pp 197–204
4. Yu M, Rhuma A, Naqvi SM, Wang L, Chambers J (2012) A posture recognition-based fall detection system for monitoring an elderly person in a smart home environment. IEEE Trans Inf Technol Biomed 16(6):1274–1286
5. Cheng J, Chen X, Shen M (2013) A framework for daily activity monitoring and fall detection based on surface electromyography and accelerometer signals. IEEE J Biomed Health Inf 17(1):38–45
6. Tong L, Song Q, Ge Y, Liu M (2013) HMM-based human fall detection and prediction method using tri-axial accelerometer. IEEE Sens J 13(5):1849–1856
7. Lan C-C, Hsueh Y-H, Hu R-Y (2012) Real-time fall detecting system using a tri-axial accelerometer for home care. In: international conference on biomedical engineering and biotechnology (iCBEB), pp 1077–1080
8. Li Q, Stankovic JA, Hanson MA, Barth AT, Lach J, Zhou G (2009) Accurate, fast fall detection using gyroscopes and accelerometer-derived posture information. In: Sixth international workshop on wearable and implantable body sensor networks, pp 138–143
9. Liu C-H, Hsieh S-L (2011) A fall detection system using accelerometer and gyroscope. Master thesis, Tatung University
10. Ozcan K, Mahabalagiri AK, Casares M, Velipasalar S (2013) Automatic fall detection and activity classification by a wearable embedded smart camera. IEEE J Emerg Selected Topics Circuits Syst, PP(99): 1–12
11. Yu M, Naqvi SM, Rhuma A, Chambers J (2012) One class boundary method classifiers for application in a video-based fall detection system. IET Comput Vision 6(2):90–100
12. Lei C-W, Wan T-P (2011) Using distributed video coding and decoding-friendly encoder design for video in the cloud. Master thesis, National Kaohsiung First University of Science and Technology, Taiwan
13. Microsoft kinect, http://www.microsoft.com/en-us/kinectforwindows/
14. Vapnik VN (1995) The nature of statistical learning theory. Springer, New York
15. ASUS xtion, http://www.asus.com/Multimedia/Xtion_PRO/
16. Peng H-K, Huang H-M, Kuo Y-S, Wen H-P (2012) Wen: statistical soft error rate (SSER) analysis for scaled CMOS designs. ACM transactions on design automation of electronic systems (TODAES), 17(1): 9:1–9:24
17. Chang C-C, Lin C-J, LIBSVM- A library for support vector machines, http://www.csie.ntu.edu.tw/~cjlin/libsvm/

A Study on Persuasive Effect of Preference of Virtual Agents

Akihito Yoshii and Tatsuo Nakajima

Abstract A virtual agent can have a graphical appearance and give users non-verbal information such as gestures and facial expressions. A computer system can construct intimate relationship with and credibility from users using virtual agents. Although related work have been discussing how to make users feel better about computers, a room for discussing effectiveness of letting users choose their favorite characters still exists. If a user's favorite agent is more believable than not favorite one for him/her, the computer system can construct better relationship with the user through an agent. In this paper, we have examined an effect of making user's favorite agent selectable on his/her behavior by conducting an experiment. We divided participants into four groups according to these conditions and ask them to have a conversation with an agent. As a result, we found a possibility of increasing credibility of an agent from users by letting them choose their favorite one.

Introduction

Persuasion and Agents

Persuasion is an attempt to encourage an individual to change his/her behaviors or attitudes [1]. Researches on persuasive computer systems have accelerated as the prevalence of computers and the Internet.

A. Yoshii (✉) · T. Nakajima
Department of Computer Science and Engineering, Waseda University, Tokyo, Japan
e-mail: a_yoshii@dcl.cs.waseda.ac.jp

T. Nakajima
e-mail: tatsuo@dcl.cs.waseda.ac.jp

Y.-M. Huang et al. (eds.), *Advanced Technologies, Embedded and Multimedia for Human-centric Computing*, Lecture Notes in Electrical Engineering 260,
DOI: 10.1007/978-94-007-7262-5_6,

Using a virtual agent as a persuader is one of methods of constructing closer relationship between a computer and a user. In this paper, we use a word "virtual agent" as the meaning of visual entity represented graphically by a computer; specifically a character which can have a conversation with its user. Related work on using conversational agents as a user interface has also been exists. For example, Bickmore et al. have discussed a model of dialogue building trust from users mentioning a *relational agent* which uses verbal and nonverbal conversational strategies same as human uses [2].

Credibility of Computer Systems

Whether a user credits a computer or not also influences on outcomes of persuasion from the computer; for example, Fogg has described credibility in persuasion [1].

Credibility can be used as a clue of whether one can believe computers or other individuals. For example, credibility which is called *surface credibility* is one of four types of credibility described in [1]. This kind of credibility comes from first impression of surface traits such as appearance [1].

A room for discussion of making users' impression of computers better in an aspect of credibility still exists. Attempts of giving users favorable impression have existed and the discussed agents have varied from text-based to graphical ones (for example, [2] has summarized related work). In addition, a computer system which user can choose and interact with his/her favorite agent also exists [3]. However, a comparison between the case an agent can be chosen by a user and an agent is fixed regardless of a user's preference can give a profound aspect. That is, if the agent just meets a user's preference or a character from a user's favorite anime or games, s/he is expected to interact with the agent with eagerly by enhanced persuasive effect.

Our Purpose

In our research, we examine a degree of persuasiveness of conversation with an agent considering two cases. One case is that a user can choose an agent according to his/her preference and the other is that a user can not choose his/her favorite agent.

We constructed an agent system using existing software and conducted an experiment using our system. In the experiment, we asked male participants who belong to one of four groups combining two different conditions to talk with a female agent; after the conversation, we interviewed them about the talk with the agent.

Based on the results of the experiment, we will also discuss an effectiveness of reflecting preferences of a user to selection of an agent; in an aspect of designing credible persuasive computer systems which incorporate with agents.

Related Work

In this section, we introduce related work on an agent and discuss the relationship with our case.

An Appearance of an Agent

Zanbaka et al. have examined effect of gender and appearance of a virtual agent as a persuader [4]. In this research, an experiment has been conducted where three kinds of agents persuade users about the same topic. One agent is a picture of a real person (human), another is a CG based human-like agent (virtual human) and the other is a CG based not-human-like agent (virtual character).

As a result, a user has been more persuaded by an agent with different gender than with same although significant effects of appearance and gender on persuasion did not exist. In addition, although users had positive impression toward the virtual agents, a virtual character can be perceived as bolder than other types of an agent.

Affiliation Need

Katagiri et al. have mentioned a need to establish and maintain affinitive relationships with others, called as affiliation need. They have examined construction of relationship between an agent and a user based on affiliation need using an exhibition guidance system incorporates with an agent [3]. With this system, a user can receive explanations and recommendations of exhibition transferring an agent from his/her portable device to an information terminal besides the exhibition.

A user can choose his/her favorite agent out of nine kinds of agents. An agent tells the user that it will wait for him/her at the exhibition which has been recommended by the agent after the fourth visit to exhibitions. If the user goes to the recommended exhibition, the agent appreciates him/her; on the other hand, if the user does not, the agent complains about it. The study has confirmed whether these reactions of the agent induce affiliation need from the user and affect later behavior of the user.

Consequently, the result has showed that those who had visited more than four exhibitions had changed their behavior after the fourth visit. In this experiment, participants have been divided into two groups and asked to walk around the exhibitions using the guidance system. Each member of one group finished after four visits of exhibitions while a member who belongs to the other group was able to visit more than four exhibitions. Although they have said that they need more participants in order to obtain statistically clearer results, they concluded that the interaction from the agent had had an effect on behavior of a user.

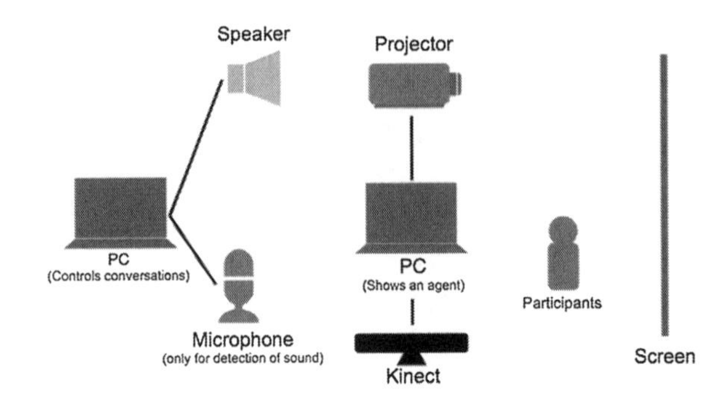

Fig. 1 An overview of the conversation system

System Design and Materials

We conducted an experiment in order to examine how will a user responds to an agent when we let them choose their favorite agent. Before the study, we constructed a conversation system which incorporates with an agent and interacts with users. We will describe details of this system in this section.

Overview

Figure 1 shows an overview of our conversation system. A user can have an oral conversation with a virtual agent which can speak in synthesized voice. An appearance of an agent and speech were presented in a wizard-of-oz style. More precisely, an agent was controlled by gestures of the researcher using a Kinect and the each sentences of speech were stepped by key input on a shell.

An agent and the speech components were deployed on different computers separately because of the difference of operating systems. An agent was controlled via a PC on which Microsoft Windows 7 was installed and the speech was generated on a virtualized Debian Linux (Squeeze) environment of another computer.

Agent

An agent was a 3D character which has been generated by software which is called as MikuMikuDance (MMD)[1]. The MMD has been originally developed a software tool to enable authors to generate a music video using 3D virtual character with synthesized vocal and music notes. We chose a Kinect to control behaviors of an agent by gestures using a plugin[2] instead of programming numerically.

---

[1] http://www.geocities.jp/higuchuu4/index_e.htm

[2] http://www.xbox.com/ja-JP/kinect

We used 3D character models bundled with books which were published by Shinyusha featuring the MMD. Such models have been provided by many people sometimes based on anime characters or games and they can be loaded to the MMD.

Speech of an Agent

The speech of an agent was synthesized by OpenHRI[3]. This software is a collection of components for Human Robots Interaction including speech synthesis and recognition. We used the speech synthesis feature based on Open JTalk[4], which is supported in the OpenHRI.

We prepared a component which receives text input from a shell and command speech related components in order to generate synthesized speech. The OpenHRI provides each features such as speech recognition or speech synthesis as components. These components can be connected to each other graphically via an input port and an output port using RT System Editor which was installed on Eclipse.

We chose a female voice because all of the character models we had prepared were female and we used the same voice for all characters. The conversations are constructed partially based on social dialogue [5]. For example, we used "empathetic statements" (line 3 in Fig. 2) and "prompting for self-disclosure by the participant" (line 5 in Fig. 2).

A: An agent / **B**: A participant
***** or -- : variable parts
Sentence no. from 13 to 16 are conditional branch

1. A: Hello, my name is --. Nice to meet you.
2. B:*****
3. A: Thank you for coming all the way here. I'm glad to see you.
4. B:*****
5. A: What is ** san's hobby?
6. B:*****
7. A: That's great! Please tell me more next time!
8. B:*****
9. A: I like music. I often listen to especially classical music.
10. A: Well, let me move to the main topic. Today I'd like to talk about exercise, all right?
11. B:*****
12. A: By the way, do ** san exercise regularly? Say, walking, cycling...
13. *B: 1. Yes, 2. Sometimes, 3. Not so much*
14. *A1: Really? Then you must be healthy!*
15. *A2: Well, do you walk or ride a bike? I heard that your brain become active if you change your route to different one. You may come up with new idea.*
16. *A3: Well, do you know this story? Your brain becomes active if you have some exercise. You may be able to refresh your mind and spend comfortable time if you exercise weekend.*
17. B:*****
18. A: Come to think of it, exercising reduces your stress and improves depressed feelings.
19. A: Exercise is surprisingly good for mental side. Have you refreshed when you get some exercise?
20. B:*****
21. A: I also started to walk as refreshment these days.
22. A: And then, I'm becoming fond of walking where I haven't been to and I found a nice cafe the other day.
23. B:*****
24. A: Oh, sorry, I have to go out. Although I can talk with you for short time, I will appreciate if you remember what I said.
25. B:*****
26. A: See you again!
27. B:*****

Fig. 2 The script of conversation

---

[3] http://openhri.net/?lang=en

[4] http://open-jtalk.sourceforge.net

Table 1 Groups of the participants (the unit is [person(s)])

	(A) person(s)	(B) person(s)
1	3	1
2	1	1

The microphone was used only for indicator of sound levels because the researcher attempted not to hear the speech of a participant in order to let them talk without pressure.

Experiment

An experiment consists of three parts. These parts include a pre-questionnaire, a conversation session with an agent and an interview.

Participants and Tasks

We recruited 6 participants (5 Japanese, 1 Chinese) and divided into four groups according to two conditions (Table 1). One condition is whether a participant can choose his/her favorite agent from the book (a) or not (b); the other is whether s/he has a specific favorite character, game or anime regularly (1) or not (2). As the Table 1 shows, we named each groups as combination of an alphabet and a number; for example, the group whose members do not have specific preference (2) and they can select their favorite agent (a) is group A-2. One participant could not fully understand the content of the conversation and we inquired mainly the impression of the agent and imaginary-based opinion.

Gender of the participants was unified to be male so as to exclude an effect of difference of participants' gender. All of the agents were female and according to Zanbaka et al. female speakers are more persuasive toward male participants than male speakers [4].

In the main study, each participants had conversations related to exercise with an agent. The conversations took place in Japanese and look like Fig. 2 as an English translated form.

During the conversations, the researcher heard music via earphones in order to hide the details of the participants' talk letting them talk more naturally.

Interview

Participants were asked to give an open-style interview with the researcher about the conversation and the agent after the conversation. In the interview, we firstly asked participants to tell us about the entire conversation freely and then we interviewed them according to following topics.

- How did you feel about the conversations with the agent?
- Did you really like the agent? Why?
- Did you have an interest with exercise? Why?
- Do you have a favorite character?

Results and Discussion

In this section, we will show the result of the experiment and discuss them. We will use fragments derived from comments of participants and these are edited for explanation.

Conversation with an Agent

From the results of the experiment, five participants felt the speech of an agent was unnatural. Such perception relates to all of or a part of the timing, the voice quality and the manner of speaking. For example, longer time lags or overlaps between speech of a participant and an agent have occurred. This was mainly because of manual conversation control without hearing the participant's talk. Besides, according to four participants, intonations of the speech was "machine-like".

Synthesized voice can have a negative effect on credibility of an agent with less reduction of persuasive effect. However, if we can increase credibility of agents using characteristics such as a visual appearance or a personality, negative effects of synthesized voice can be reduced. As for perception of synthesized speech, Stern et al. have compared synthesized speech and human speech. Their results have shown that synthesized speech was rated less knowledgeable and truthful. However, in terms of the persuasion, a significant difference between human speech and synthesized speech did not exist [6].

Favorableness of an Agent

In an aspect of the favorableness of an agent, a significant difference between the groups was not found. One of members of the group A-1 said that "Although I am not fun of a specific character, the agent which I have selected was favorable". On the other hand, a member of group A-2 said that "I did not feel especially about the agent". These results suggest that users' regular preference does not significantly affect the immediate favorableness of an agent. We still have to examine other characteristics such as a personality and a voice of an agent in order to reduce unnatural impression from a user.

Other comments from four participants suggested the possibility of positive impression by a selectable character. We asked the participants how they thought that they can choose their favorite agents; as for those who are in group B, we asked them to imagine the situation. In addition, we also asked them what they think if an agent of our conversation system were replaced by their "regular" favorite characters of games, anime or any other media. Comments we received were "I am not feel like talking with the agent if she were not my favorite", "Choosing from many agents was difficult. But if I could not choose an agent, she may make me less impressed" and "I did not have special feelings toward the agent this time, but I may listen to my favorite character more". On the other hand, one of members of group B told us that he will have same feelings even if he could choose his favorite agent.

Persuasiveness of an Agent

Two participants were affected by the agent's persuasion. Both of them had chosen "I do not exercise regularly" on the pre-questionnaire about exercise. One participant said that "I have got to know new facts about exercise and I may remember them and exercise someday." On the other hand, four participants were not persuaded. The comment of such participant was "I will not change my behavior unless I start a conversation with an agent on my own will".

This result suggests relationship between a topic of a conversation and participants' current exercise behavior have to be considered while the experiment with significant number of participants is expected. Specifically, we did not assign the participants to a group according to exercise behavior of them; in addition, the conversation which we have prepared contained same persuasion for all participants except for a conditional branch on lines from 13 to 16 in Fig. 2.

Each option of the questionnaire about exercise can refer to a behavior model. According to Prochaska et al. behavior can be divided into five stages from *precontemplation* level to *maintenance* [7]. Among these stages, *precontemplation* is a stage for those who do not intend to change their behavior and *contemplation* is for those who are seriously considering changing their behavior. In addition, different processes of change are needed in order to move from one stage to another smoothly. For example, giving information about a target behavior to a *precontemplation* or *contemplation* individual is *consciousness-raising*. Informing participants of positive aspect of exercising can be consciousness-raising for those who answered "I do not exercise regularly" to our questionnaire; however, not to those who exercising regularly.

Conclusion

We discussed an effect of letting a user choose his/her favorite agent on behavior of the user. In addition, we also conducted an experiment constructing an agent system with which a user can have a conversation in order to examine the effect of a favorite agent. In the study, we divided participants into four groups according to two conditions. One condition was whether a participant can choose his favorite agent and the other condition was whether a participant has a "regular" favorite characters, game or anime in his/her daily life.

As a result, making an agent selectable by a user according to his/her preference has a possibility of increasing credibility of an agent system. However, we still have to adopt personalities, voices and the contents of conversations to the appearance of an agent in order to reduce unnatural feelings from participants.

Acknowledgments The authors would like to thank those who participated our experiment and authors of components and materials.

References

1. Fogg BJ (2003) Persuasive technology. Morgan Kaufmann Publishers, Burlington
2. Bickmore T, Cassell J (2001) Relational agents: a model and implementation of building user trust. In: Proceedings of the SIGCHI conference on human factors in computing systems. CHI '01, New York, USA, ACM 396–403
3. Katagiri Y, Takahashi T, Takeuchi Y (2001) Social persuasion in human-agent interaction. In: Second IJCAI (ed) Workshop on knowledge and reasoning in practical dialogue systems, IJCAI-2001. Morgan Kaufman Publishers, Menlo Park, pp 64–69
4. Zanbaka C, Goolkasian P, Hodges L (2006) Can a virtual cat persuade you? the role of gender and realism in speaker persuasiveness. In: proceedings of the SIGCHI conference on human factors in computing systems. CHI '06, New York, USA, ACM, pp 1153–1162
5. Schulman D, Bickmore T (2009) Persuading users through counseling dialogue with a conversational agent. In: persuasive '09: Proceedings of the 4th international conference on persuasive technology, New York, USA, ACM, pp 1–8
6. Stern SE, Mullennix JW, Dyson Cl, Wilson SJ (1999) The persuasiveness of synthetic speech versus human speech. Human factors: the journal of the human factors and ergonomics society 41(4), pp 588–595
7. Prochaska JO, Norcross JC, DiClemente CC (1994) Changing for good. William Morrow, an imprint of Harper Collins Publishers

Human Computing for Business Modelling

Yih-Lang Chen and Yih-Chang Chen

Abstract The importance of linking information technology to business goals and objectives is addressed by the concept of BPR. However, many methodologies of BPR do not address the implementation issues of the information systems development and the combination of this with the implementation of BPR in detail. What we need is an effective method to analyse the dynamic behaviour of the organization and to evaluate the alternative choices of solutions to problems. In this paper, we introduce the concept of participative process modelling where we view system development and BPR as cooperative activities which involve different groups of people with different competencies, viewpoints and requirements. We will also describe a framework, based on the principles of Empirical Modelling, aims to give a comprehensive view of the real-world situations so as to allow all participants to experience what it is, or would be like, to work within such situations and therefore draw on their tacit and non-explicit knowledge and experience. We propose the application of EM to provide a practical way of implementing participative BPR.

Keywords Business Modelling · Empirical Modelling · Experience-based · BPR

Y.-L. Chen
Department of Food and Beverage Management, National Kaohsiung University of Hospitality and Tourism, Xiaogang, Kaohsiung 81271, Taiwan, Republic of China
e-mail: chenyl@mail.nkuht.edu.tw

Y.-C. Chen (✉)
Department of Information Management, Chang Jung Christian University, Guei-Ren, Tainan 71101, Taiwan, Republic of China
e-mail: cheny@mail.cjcu.edu.tw

Y.-M. Huang et al. (eds.), *Advanced Technologies, Embedded and Multimedia for Human-centric Computing*, Lecture Notes in Electrical Engineering 260,
DOI: 10.1007/978-94-007-7262-5_7,

Introduction

The role of information technology as the enabler for organizational rethinking has been emphasized in much business modelling and business process reengineering (BPR) literature. However, when coming to the actual implementation of the BPR project, the implementation issues of IT are usually ignored or addressed by just picking an off-the-shelf application and changing the business processes to fit it. We have pointed out in [1] that businesses and business processes are such complex systems that the developers and users require appropriate models to understand the behaviour of such systems whether in order to design new systems or to improve the existing ones. In defining the requirements for BPR and its support systems, it is essential to have a broad understanding of the organizational environment in order to make appropriate decisions about what changes to make or which parts to retain. In this paper, we will investigate the potential of computer models as a suitable technique for business modelling. We start by introducing of the concepts of participative process modelling. After the presentation of our SPORE framework, we will assess its potential as a medium for participation in business modelling.

Participative Process Modelling

Participative process modelling can be interpreted in two different ways: *modelling of participative processes* and *participative modelling of processes*. When we think about how processes are conceived in the business area, there are two kinds of actions involved: the manual actions by human being and the automated actions by systems. These can be described in terms of two kinds of agency. The first is associated with the *internal* agents which enact the process and which are responsible for the state change. Their interaction can be referred to as a 'participative process'. The interactions of the internal agents may be governed by strict rules and reflect different degrees of autonomy. The second is associated with the *external* agents whose observation and comprehension of the interactions within a process rely on the integration of many agent viewpoints. Their interaction can be referred to 'participative modelling'. The external agents are typically responsible for designing and managing the process, but may also act simply as observers.

The distinction between internal and external agents can help to refine our thinking about the issue of BPR. The basic tenet of BPR is to gain competitive advantages by rethinking business processes and the use of IT for the redesigned processes. Such rethinking can be made from two directions. External rethinking is made from the *external* observers of the business system and its products or services. The aim is to make the business more responsive and effective. *Internal* rethinking is made from the viewpoints of component agents of the process.

The aim is to find the barriers in that process or any difficulty these members may meet, and thus look for new ways to make the process more efficient. The work of BPR thus should combine both kinds of rethinking: to provide the personalized empowered environment for individuals and to develop the supporting technology for the automation of business processes and the collaborative environment.

But in most BPR contexts, the creation or modification of computer systems to support the newly designed business processes is not suitable or even has a negative impact on the business. This is because the complex issues of human factors are not taken sufficiently into consideration during the BPR work. Such rethinking is mainly based on the external viewpoints and the redesigned processes are framed in terms of preconceived interactions to serve particular purposes. So the models are designed for describing 'what' the business process looks like, but cannot express 'why' the process has a certain form, and the motivation and intents behind the activities. Without considering or understanding these factors, it will be difficult to integrate existing systems or to design new systems to meet the changing needs. Although the design of the software system does radically impact the design of the business, two important points are still ignored by conventional BPR approaches:

- The software system is still a software system with all the old problems of versioning, integration, complexity, etc.
- The rules governing good architectural practices in software are not the same as the rules governing good business structures.

The SPORE Framework

SPORE (Situated Process of Requirements Engineering) is a human-cantered framework which was proposed by the second author [2]. It is problem-oriented because the requirements in this framework are viewed as the solutions to the problems identified in the application domain. The requirements are developed in an open-ended and situated manner. It is open-ended because such requirements cannot be completely specified in advance; and it is situated because the context presented in SPORE is closely connected to the referent in the real-world domain. The SPORE framework is depicted in Fig. 1. Three kinds of inputs of SPORE are: *Key problems* of the domain which are identified by the participants in seeking to address the requirements of the proposed system. *Relevant contexts*, act as motives and constraints for the participants in creating the outputs. *Available resources* are used by participants to facilitate the creation of SPORE outputs.

There are also four kinds of outputs from SPORE. The main one is *provisional solutions* to the identified problems which are developed by participants on the basis of the available resources and the current contexts. The other three outputs, which include *new contexts*, *new resources* and *new problems*, combine with their earlier versions and in turn form the new inputs to the model for creating the next

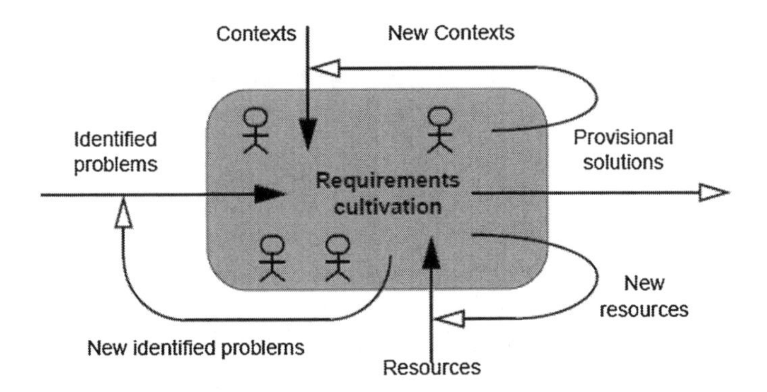

Fig. 1 The SPORE Framework

outputs. That is, all these contexts, resources and problems are modifiable and extensible in SPORE. Thus participants can develop requirements in a situated manner to respond to the rapid change in the contexts, resources as well as the problems themselves.

The models that feature in a requirements specification in conventional software development are usually too abstract for it to be possible to get a detailed understanding of the model by simply reading it. This situation may be even worse if the application domain is a complex system. Therefore having a way for the participants to experience and visualize the behaviours defined by the model is essential for system development. The SPORE framework involves constructing computer-based artefacts to be used to explore and integrate the insights of individual participants in an interactive manner. The artefacts created are ISMs and are based on the principles and tools of Empirical Modelling [3]. In EM the knowledge of participants is constructed in an experimental and not a declarative manner. That is, the insight of the participant is expressed by the coherence between what he expects in his mind and his experiments with the ISMs and the external referents. His insight can also be extended through 'what-if' experiments. Any introduction of new definitions, or redefinition of existing definitions, will evoke changes of state in the models and these in turn will affect the state of his mind through the visual interface.

Through our distributed EM tools the interaction of an individual participant can be propagated to other artefacts and thus affect the views and insights of other participants. In this way the participants can collaboratively interact with each other through their artefacts. This distributed EM tool supplies the framework for the collaborative environment in which the shared understanding of the key problems and their solutions (i.e. the requirements) can be established.

One of the most important benefits of interacting with the ISM is that it makes the individual insights and the shared understanding visible and communicable. This overcomes the disadvantage of invisibility and incommunicability of shared understanding based on conventionally test-based models. This experimental interaction can also keep the requirements synchronized with the shared understanding among participants, which evolves faster than textual specifications. That is, the way in which the evolution of computer models and the individual insights are synchronized allows the participants to 'see' other participants' viewpoints and to 'communicate' with them by interacting with their own artefact.

EM for Business Modelling

In today's business environment, information systems are becoming more interconnected to each other and are increasingly involved in complex business processes. That is, the systems used in an organization need to cooperate with human beings to achieve the organizational goals. So the information system development occurs in the context of legacy systems and business processes, which involves the issues of systems comprehension and BPR. In determining the requirements of information systems, it is necessary to understand the organizational environment so that the proposed systems can work well together with human beings. What we need is an open-ended and flexible approach to modelling the organizational environment and the behaviour of the actors and the support system. In EM, the agents, and how they relate to each other, are characterized in terms of observations. Through the construction of the computer model, the understanding and analysis of agency can be made concrete and amenable.

As mentioned earlier, a model should have both the *indicative* properties referring to properties of the existing environment and the *optative* properties referring to properties of the future environment in which the proposed system is to operate. In the developing procedure of EM (cf. Figure 2), the elicitation activity (the arrow of new input) identifies the indicative properties of the existing system (the ISM) and its environment from observation of the real-world situations. It provides the support for producing the conceptual model and generating goals from the interactions and experience of the observed system and model. Through this procedure, the interrelations between the abstract requirements (and goals) and the artefact of real-world referent can be established, and these can be used for the elicitation of system requirements.

Since the conceptual models of participants are too complicated and difficult to represent in notations which all participants can understand, we propose the cooperative validation of the conceptual models within the SPORE framework. This is carried out by participants with different roles who can elaborate different behaviours through ISMs in an interactive experiment. During the experiment, different patterns of behaviour can be explored.

In EM, the construction of ISMs is closely linked to comprehension. Thus the interactions in the real-world domain and the development and validation of ISMs are interdependent. During the validation process (the arrow of test and experiments), the indicative factors identified during the elicitation activity will be extended and adapted to fulfil the additional optative properties. This validation is

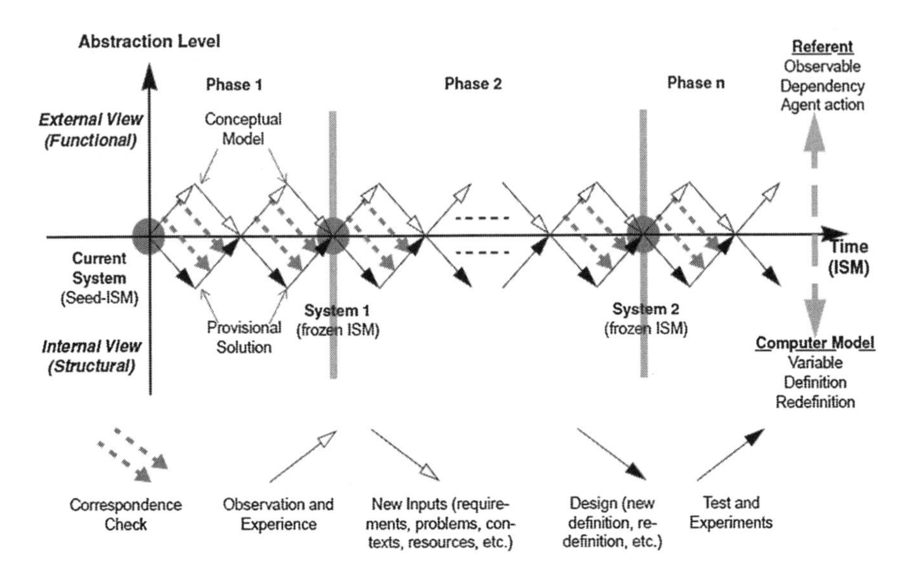

Fig. 2 The unified development procedure in empirical modelling

mainly achieved by correspondence checks. Once a reliable correspondence between the states of ISMs and those observed in the referent is established, the interaction with ISMs can serve as a representation of the understanding of real-world domain. Thus, through the correspondence checks, the modelers will gain insights into both the existing reality (the indicative properties) and the new requirements or new goals (the optative properties) which in turn form the new inputs for the next phase of system development. The open-ended characteristic of ISMs provides a way of integrating the elicitation and validation. Prior to defining and establishing the reliable patterns of state change, the interactions with ISMs, similar to activities in our everyday life, have an experimental character and are the primary means to improve our understanding of the domain.

The Characteristics of EM in Business Modelling

From our previous discussion, we infer that EM has great potential for system development and BPR. The following summarizes the characteristics of experimental interactions with ISMs, as these apply to business modelling:

- EM is an unified activity rather than the traditional development lifecycle. It can potentially address BPR from the perspectives of both the internal and external agents, and empower and allow participants to consider the meaning of the task at hand rather than becoming embroiled in peculiarities of its implementation.

- Interacting with ISMs within the SPORE framework allows participants to experiment with any entity of the business system. This means that the behaviour of both computer systems and human components can be identified and thus incorporated in a natural way. Thus the various enablers of BPR, such as information systems or human resource management strategies, can be investigated in essentially the same experimental manner.
- Under the 'open development' paradigm, participants can set up problematic scenarios, realize them visually and correct them easily and inexpensively using a 'what-if' strategy. Furthermore, the computer-based character of ISMs renders the changes and updates to the existing system apparent and makes the model maintainable and reusable.
- EM helps the participants to communicate their ideas and assess the impact of proposed changes/alternatives immediately. Participants can be guided towards a shared understanding and consensus for decision-making through the continuous evaluation, communication and checks for consistency in the distributed ISMs.
- The visibility and communicability of the interaction within the SPORE framework increases the participants' understanding of roles and relationships amongst others, as well as the effects of individual activities. This lets the participants gain feedback from the results of experiments which is used not only as a basis for comparing the alternative solutions but also as a means of evaluating their validity.
- EM allows the participants to obtain a 'global' view of the effects of 'local' changes made by individual artefacts. This assists the identification of implicit dependencies between parts of the business system.

Conclusions

Many processes in science and engineering are precisely prescribed by theories and equations, but for business processes it is difficult to consider the process and its associated real-world factors (including human factors) in an abstract and unified way. Our best line of attack is to develop the rich and flexible concepts and models to directly involve the potential users and incorporate their activities into the analysis and design of the new system or business. In this context, subject to the modeler having an adequate construal and sufficient understanding of the situation, EM has the advantage of allowing the states of the real world to be modelled in an open-ended fashion so that any new factors considered relevant can be taken into account. This kind of experimental interaction with the computer model seems to be the appropriate technique for the purpose of business process modelling. It can also be used for experimentation purposes and to help decision making during the modelling procedure. By the EM approach singular conditions can be explicitly modelled and human intervention, which is essential when modelling the scenarios in business, is possible throughout the modelling process. To sum up, EM provides flexible and human-cantered support for the design and construction of a wide range of interactive multi-user systems.

References

1. Chen YC (2012) Innovative intelligent computing for software development: an empirical modelling approach. In: international symposium on computer, consumer and control, pp 898–901
2. Chen YC (2001) Empirical modelling for participative business process reengineering. Ph.D. thesis, department of computer science, The University of Warwick, UK
3. Empirical modelling (EM) research group of the University of Warwick website, http://www.dcs.warwick.ac.uk/modelling
4. Jacobson I, Ng PW, McMahon PE, Spence I, Lidman S (2013) The essence of software engineering. Addison-Wesley, Boston
5. Warboys B, Snowdon B, Greenwood RM, Seet W, Robertson I, Morrison R, Balasubramaniam D, Kirby GNC, Mickan K (2005) An active-architecture approach to COTS integration. IEEE Softw 22:20–27

Teaching and Learning Foreign Languages in a Digital Multimedia Environment

Vladimir Kryukov, Alexey Gorin and Dmitry Mordvintsev

Abstract The paper describes the structure and functions of a digital multimedia environment as a means of teaching and studying foreign languages. The authors suggest an LMS-centered environment consisting of language labs, interactive whiteboard rooms, multimedia rooms, a videoconferencing hall, a webinar platform, and multimedia repositories, with all the components being linked together by an e-learning platform. The article is intended for experts in information technology and second language teaching methodology, as well as for all those interested in the problems of computer-assisted language learning.

Keywords Computer-assisted language learning · e-Learning · Blended learning · Multimedia environment · Language lab · Videoconferencing · Webinar

Introduction

The concept of the development and application of information technology at the Vladivostok State University of Economics and Service is based on the model of e-campus. The e-campus is aimed at increasing the efficiency of the business processes at the University by incorporating its information infrastructure and corporate information systems.

V. Kryukov (✉) · A. Gorin · D. Mordvintsev
Vladivostok State University of Economics and Service, Vladivostok, Russia
e-mail: vladimir.kryukov@vvsu.ru

A. Gorin
e-mail: aleksey.gorin@vvsu.ru

D. Mordvintsev
e-mail: dmitriy.mordvintsev@vvsu.ru

Y.-M. Huang et al. (eds.), *Advanced Technologies, Embedded and Multimedia for Human-centric Computing*, Lecture Notes in Electrical Engineering 260,
DOI: 10.1007/978-94-007-7262-5_8,

In the era of economic, political and cultural globalisation, the University pays special attention to teaching foreign languages as a means of improving the competitiveness of its graduates in today's global marketplace.

This brought about the idea of elaborating an environment for teaching and learning foreign languages on the basis of digital multimedia technology.

A Digital Multimedia Environment for Teaching and Learning Foreign Languages

The digital multimedia environment at the Vladivostok State University of Economics and Service comprises the following key elements: language labs, interactive whiteboard rooms, multimedia rooms, a videoconferencing hall, a webinar platform, and multimedia repositories. All these elements are linked together by the University's e-learning platform.

The multimedia environment is used to deliver foreign language courses in a variety of ways [1], both in class (language labs, multimedia and interactive whiteboard rooms) and via asynchronous (e-mail, forums) and synchronous (chat rooms, webinars, and video conferences) communication channels. Such a combination of different modes of delivery, known as blended learning, represents a real opportunity to create engaging learning experiences for each and every student [2].

E-Learning Platform

Description. The e-learning platform is based on modular object-oriented dynamic learning environment (MOODLE) software which is a free open-source e-learning solution [3]. Its features are typical of an e-learning platform. Instructors can upload study materials to be viewed and/or downloaded by students, give them assignments for further submission, grade their work, publish news and announcements, organise chats and forums, administer tests and quizzes online, etc.

The main reason for choosing Moodle was its high interoperability, modular structure and open source code, which allowed for easy integration into the University's e-campus of which it is now an integral part. Instructors use their domain accounts to log in; Moodle's database is synchronised with the databases storing information about students, student groups, study subjects, schedule, etc. A special module made it possible to integrate Moodle with the University's webinar platform.

Usage. The e-learning platform is the core of the University's digital multimedia environment. Instructors create multimedia-rich electronic courses which contain training materials in the form of hyperlinks to resources in the University's digital repository and embedded videos from the video portal. Being available on the Internet and on the intranet, the e-learning platform is extensively used both in the language labs, the multimedia rooms, and the interactive whiteboard rooms and as a resource for students' independent work.

Experience and problems. The main inconvenience connected with the usage of the e-learning platform is that, if an instructor wants to include in their course a study material from the repository, they have to copy a link to the material in the repository and then paste it into a Moodle course. Similarly, including a video from the video portal requires copying and pasting the embed code.

It should be mentioned that improving the usability of the e-learning platform is of paramount importance, as one of the University's most significant goals is to further develop computer-assisted learning technologies by encouraging instructors to design interactive multimedia-rich e-learning materials that arouse students' interest and increase their academic performance and motivation.

Future plans. The University's IT personnel are now working on a Moodle module to fully integrate the digital repository and the video portal into Moodle.

Language Labs

Description. The University has three software-based language labs with 15, 25, and 30 workstations. The software Dialog Nibelung 2.0 (produced by the Russian manufacturer Lain Ltd.) installed in the labs uses the local network to link computers in the class [4]. This local network can be used for transmission of audio and video materials, various text documents and other kinds of files, and for full control of students' PCs from the teacher's workplace.

The language instructor can arrange student workplaces by groups or pairs for discussion, assign different tasks to different groups or pairs, listen in or talk to a selected student, demonstrate his/her screen to students, show a student's screen as an example to other students, record audio materials, automatically monitor students' PCs, view the selected/all students' screens, remotely control students' PCs, transmit/receive various documents to/from students, organise chat sessions, etc.

Obviously, the functions of the software are standard for the software of this kind. However, the main advantage of the solution is that it is quite inexpensive (less than $200 per workplace, excluding the cost of PCs), although it has all features necessary for teaching and learning foreign languages through discussion and extensive use of multimedia. At the same time, it allows lessons on many other subjects to be efficiently conducted, and the students' knowledge to be evaluated with the help of the integrated test system.

Usage. The language labs are used for teaching both general English and English for specific purposes (ESP), as well as for training interpreters. The largest lab (30 workstations) is also a venue for students' independent work. Not less than six hours daily are allocated for this purpose.

Experience and problems. Being purely software tools, the language labs cannot always provide the sound quality necessary for teaching and learning foreign languages [5]. Sound is sometimes lost, echoed or distorted, which complicates communication. Besides, the computers are not equipped with webcams, although this would make the language labs suitable for conducting webinars.

Future plans. To solve the problem above, the University is planning to equip the labs with gigabit switches and more perfect headsets. Webcams will also be purchased. It is also planned to replace PCs with zero clients to reduce maintenance costs and improve the efficiency of the IT infrastructure.

Multimedia Repositories

Multimedia repositories are facilities to store multimedia content used in the language labs and multimedia and interactive whiteboard rooms as well as those uploaded to the e-learning platform.

Multimedia content is stored in the University's digital repository which is intended for storing, searching and granting access to digital study materials for teaching and learning foreign languages.

Video materials are stored on the University's video portal which also makes it possible to store and search data. Videos can be viewed on the portal directly or be embedded into Moodle courses.

Multimedia and Interactive Whiteboard Rooms

The university has more than 200 multimedia rooms equipped with projectors and projection screens. Instructors use VGA outlets to connect their laptops to the projector. The multimedia rooms make it possible to enrich foreign language classes with multimedia content, with the only problem being the lack of stationary sound systems. This makes instructors carry portable speakers, which is inconvenient.

More than 20 rooms are equipped with interactive whiteboards, three of which are used for teaching and learning foreign languages. The University has a wide variety of digital multimedia resources produced by the world's leading publishers.

Apart from this, all students are given netbooks and can connect to the e-campus during class, which enables them to access the digital multimedia environment from virtually anywhere on campus.

Videoconference Hall

Description. The videoconferencing hall is intended for conducting local and international videoconferences. It holds 80 delegates and makes it possible to conduct conferences and seminars in 3 different languages with simultaneous interpretation.

Delegates' workplaces are equipped with delegates' consoles having microphones and earphones, if necessary. When a delegate's microphone is activated, one of the six video cameras installed along the perimeter of the room focuses on the delegate and transmits the image to the opposite side.

All videos are projected to the central screen or to the smaller side screens, with the latter option being used if the presenter and their power point are to be demonstrated simultaneously. All videos are stored on the server for further processing and elaboration of video lectures and tuition courses.

Usage. As practice shows, in approximately 80 % of cases the hall is used to deliver remote lectures for the University's branches and representative offices in other cities. Lectures are often delivered to both remote students and local students sitting in the hall.

The remaining 20 % of the time is occupied by various conferences and seminars, including lectures by foreign professors.

Experience and problems. Undoubtedly, the University needs a hall of such capacity. However, the large number of workplaces leads to some difficulties in its usage. The main problem is that the hall is overloaded owing to a wide variety of the functions it performs. Large-scale events are often preferred to lessons in small groups. At the same time, the usage of such powerful and costly equipment is not always justified while working with small groups.

Another problem is the lack of special soundproof booths for simultaneous interpreters whose workplaces are now located in the next room, together with the videoconference communication operator's workplace. This creates additional difficulties for two interpreters and an operator working in a small room.

One of the disadvantages of the hardware is that it is impossible to connect remote participants to different simultaneous interpretation channels. In other words, all remote participants can hear translation to only one language.

Besides, the videoconference hall does not fully meet the requirements of the training of simultaneous interpreters, as it is not intended for practical classes. All operation modes are designed for actual lectures, seminars and conferences only.

Future plans. By the end of next year, it is planned to significantly expand the presence of IT and multimedia in teaching and learning foreign languages. To meet this goal, the University developed projects of a specialist facility for teaching simultaneous interpretation and an additional facility for videoconferencing and webinars. The experts did their best to consider both teachers' and students' requirements and previously gained experience of working with similar technologies.

Webinar Platform

Description. The webinar platform is based on the BigBlueButton web conferencing system [6], and it is fully integrated into the University's IT infrastructure; instructors and students use their domain accounts to log on and access webinars. Instructors can also conduct webinars within the e-learning platform, including record and playback of sessions.

Instructors can arrange the date and time of a webinar, control the number of its participants, transmit audio and video, demonstrate documents of various formats (.ppt(x),.doc(x),.xls(x),.pdf,.jpg,.png,.pdf,.djvu,.gs, etc.), use whiteboard features, and broadcast their desktop for all students to see. It is also possible to communicate with students via the chat.

Usage. The webinar platform is used to give classes to students of the University's remote branches and makes it possible for instructors and experts from other cities and countries to deliver lectures to the University's students.

Experience and problems. The use of webinars for teaching and learning foreign languages is limited by the high cost and relatively low quality of internet connectivity for end-users.

Future plans. To solve the problem, the University is planning to arrange additional webinar facilities and to equip the language labs with webcams, so that students could be able to participate in webinars irrespective of the quality of their home internet service.

Challenges for Developing the Digital Multimedia Environment

The initial development of the LMS-centered multimedia environment faced some integration problems with the University's existing information infrastructure (particularly, with Nginx Web Server). There was also a number of third-party Moodle plugins having interoperability issues with Microsoft SQL Server.

However, the biggest challenge was connected with designing online materials. Due to the lack of financing, instructors used to confine themselves to designing the simple multiple-choice quizzes and fill-in-the-blank exercises instead of multimedia-rich study materials with carefully chosen video, sound, graphics, photographs and animations [1], which led the University to develop a set of financial incentives to address the challenge. Research [7] has shown that the described situation is typical of many educational institutions.

Future Installations

Language Lab for Training Simultaneous Interpreters

A language lab for training simultaneous interpreters was designed on the basis of the hardware and software Sanako Lab 100 STS. The lab includes 21 workplaces (16 delegates, 4 interpreters and the teacher), a multimedia projector, a personal computer, an interactive graphics tablet, a document camera, and equipment for recording digital audio.

The lab can be used both as a regular language lab and a simultaneous interpretation lab. Interpreters' workplaces are separated from the lab by a soundproof glass partition. A similar glass partition forms two booths, each for two interpreters. This makes it possible to simulate real working conditions.

The document camera is connected to the multimedia projector and makes it possible to demonstrate images of printed materials, such as A4 sheets, book leaves, or printed texts. The interactive graphics tablet makes it possible to process any image and save all notes and corrections made into it. All this can be done at the teacher's workplace.

Videoconference and Webinar Room

The room has 34 workplaces (11 videoconference delegates, 1 teacher, and 22 participants), an information display system (an interactive whiteboard, a projector, and 4 plasma TV sets), and an audio system.

The delegates' and the teacher's workplaces are equipped with microphone consoles. When a microphone is activated, of the two video cameras automatically focuses on the speaker and transmits the image to the opposite side and to the internal displays, if necessary. Signal from any device in the room can be transmitted to any display. Displays are located in such a way that each delegate or participant has visual access to all the materials displayed.

Participants' workplaces are equipped with cloud-based terminals, webcams and headsets. This enables the participants to participate in different webinars simultaneously. Besides, the teacher who delivers the webinar can demonstrate it to all participants present in the room. In the future, it is planned to move one of the existing language labs to this room, which will significantly increase its functionality.

The room is planned to be used for conducting classes in small groups, delivering lectures to remote students and recording educational videos. This will make the usage of the existing conference hall more efficient and productive, and the two facilities will complement each other.

Conclusion

In conclusion, it should be said that the digital multimedia environment at the Vladivostok State University of Economics and Service incorporates engagement and effectiveness, allowing for learning through a mixture of technological features possible in online and offline course delivery.

Foreign language teaching and learning in a blended environment promotes the high quality and efficiency of second language acquisition and creates a real opportunity for each and every individual to study a foreign language in the right time and at the right place.

References

1. Clarke A (2001) Designing computer-based learning materials. Gower Publishing, Aldershot
2. Thorne K (2003) Blended learning. Kogan Page, London
3. Moodle.org: Open-source community-based tools for learning, https://moodle.org
4. Dialog nibelung language labs, http://www.dialog.su/en/
5. Davies G, Bangs P, Frisby R, Walton E (2012) Guidance for MFL heads of department and ICT managers on setting up and using digital labs and multimedia ICT suites, http://www.camsoftpartners.co.uk/docs/CILT_Digital_Labs.htm
6. BigBlueButton web conferencing system, http://bigbluebutton.com
7. Beatty Ken (2010) Teaching and researching computer-assisted language learning, Pearson Education, London

An Interactive Web-Based Navigation System for Learning Human Anatomy

Haichao Zhu, Weiming Wang, Jingxian Sun, Qiang Meng, Jinze Yu, Jing Qin and Pheng-Ann Heng

Abstract This paper presents an interactive web-based anatomy navigation system based on the high-resolution Chinese Visible Human (CVH) dataset. Compared with previous anatomy learning software, there are three new features in our navigation system. First, we directly exploit the capabilities of graphics hardware to achieve real-time computation of large medical dataset on the web. In addition, various visualization effects are supplied to enhance the visual perception of human model. Second, to facilitate user interaction, we design a set of user-friendly interface by incorporating the Microsoft Kinect into the system, and the users can navigate the Visible Human with their hand gestures. Third, in order to eliminate the unreliable bottleneck: network transmission, we employ a progressive strategy

H. Zhu (✉) · W. Wang · J. Sun · Q. Meng · J. Yu · J. Qin · P.-A. Heng
Department of Computer Science and Engineering, The Chinese University of Hong Kong, Hong Kong, China
e-mail: hczhu@cse.cuhk.edu.hk

W. Wang
e-mail: wangwm@cse.cuhk.edu.hk

J. Sun
e-mail: jxsun@cse.cuhk.edu.hk

Q. Meng
e-mail: qmeng@cse.cuhk.edu.hk

J. Yu
e-mail: jzyu@cse.cuhk.edu.hk

J. Qin
e-mail: jqin@cse.cuhk.edu.hk

P.-A. Heng
e-mail: pheng@cse.cuhk.edu.hk

J. Qin · P.-A. Heng
Shenzhen Institutes of Advanced Technology, Chinese Academy of Sciences, Beijing, China

Y.-M. Huang et al. (eds.), *Advanced Technologies, Embedded and Multimedia for Human-centric Computing*, Lecture Notes in Electrical Engineering 260,
DOI: 10.1007/978-94-007-7262-5_9,

to transmit the data between the server and the client. Experimental results validate the advantages of the proposed navigation system for learning human anatomy, indicating its great potential in clinical applications.

Keywords Anatomy navigation system · Chinese Visible Human · Graphics hardware · User interaction · Network transmission

Introduction

Due to fast development and new enhancement of network services, such as fast net-work transmission and high security mechanisms, web-based applications have attracted more and more attentions. Medical training and diagnosis systems based on web also emerged rapidly in recent years, where the doctors or experts from different locations can collaborate on specific tasks through the Internet. Major challenges in designing a successful web-based application include how to achieve real-time computation for large dataset on the web, how to distribute the workload between the server and the client so that the overall performance can be maximized, and how to design user-friendly interface to facilitate the user interaction.

Various web-based software and tools for medical applications have been proposed, and their impact on medical education is still growing [1]. Poliakov et al. [2] presented a server-client approach that enables the user to visualize, manipulate, and analyze the brain imaging dataset through the Internet. However, this approach may be degraded under bad network condition or if there are massive client requests at the same time. In [3], Mahmoudi et al. developed several interactive web-based tools for 2D and 3D medical image processing and visualization. A wide range of methods, such as registration and segmentation, are incorporated, and the client can directly use these functionalities without any plug-in installation. As hardware acceleration is not exploited, better performance is still expected for real-time applications. A web-based navigation system for Visible Human is also developed in [4], but important anatomical details are lost due to relatively low resolution of the rendered sections.

This paper extends our previous work [5] to provide an interactive web-based navigation system for users to learn the human anatomy through the Internet. Specifically, we fully exploit the functionalities of graphics hardware to achieve real-time computation on the web by utilizing the WebGL [6] API, which supports direct access to Graphics Processing Units (GPU) from the browser. Various visualization effects, e.g., surface rendering and translucent rendering, are also supported in the navigation sys-tem. Moreover, a set of user-friendly interface is designed to facilitate the user inter-action, where the Microsoft Kinect is employed to track and recognize the user' hand gestures. Lastly, to eliminate the unreliable bottleneck: network transmission, we adopt a progressive strategy to transmit the data between the server and the client. The server first sends a coarse image to the client without any delay, and finer images are later transmitted progressively if the user is interested on certain anatomical structures.

Methods

The high-resolution CVH dataset [7], with a total size about 72 GB, is employed to test the proposed anatomy navigation system. There are 3,640 slices in the dataset (0.5 mm interval) and each slice has a resolution of 3,872 $\times$ 2,048. Compared with other imaging modalities, such as CT, MRI and US, major advantage of the CVH dataset lies on its high-resolution cross-sectional images. Hence various anatomical details are preserved, which is of great value for learning human anatomy.

The overview of our anatomy navigation system is shown in Fig. 1 and there are two major parts: the server and the client. We distribute the workload between the server and the client so as to maximize the overall performance. The client visualizes the human model and tracks the user' hand gestures to determine the cross-sectional plane. The server computes the corresponding cross-sectional anatomical image and sends the results to the client progressively. In the following subsections, we detail the processing of major components.

WebGL-Based Rendering

After receiving the human model from the server, the client renders the human body with various visualization effects. However, it is not an easy task to achieve

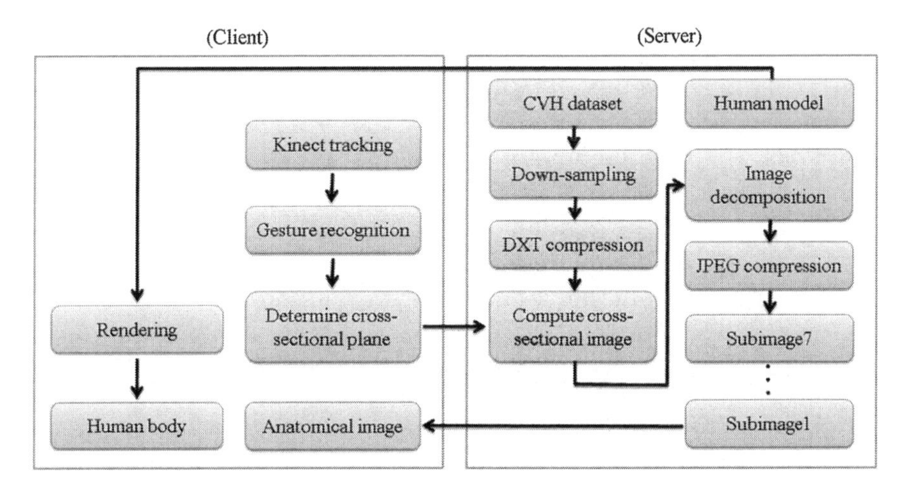

Fig. 1 Overview of the proposed anatomy navigation system

real-time rendering for complex translucent objects on the web. Traditionally, to achieve correct translucent effect with alpha blending, it is required to first sort the geometry primitives according to their depth distances and then render the primitives from front to back(or from back to front) sequentially. However, the sorting operation will be very time-consuming if the models are very complex. In [8], Everitt presented a GPU-accelerated depth peeling algorithm to solve this problem without any sorting operation. In this algorithm, each unique depth in the scene is extracted into one layer, which will be later blended in depth-sorted order to produce the final image. In practice, the algorithm is implemented with a depth test to extract the currently frontmost layer (including both color and depth information) in every rendering pass.

Unfortunately, the original depth peeling algorithm cannot be directly applied on the web due to limited supports of current WebGL standards. Adapting to the features of WebGL, we modify the algorithm to provide translucent visualization effect on the web by utilizing multiple framebuffer and texture objects to store the color and depth information of each layer separately. The major difference between the modified and original depth peeling algorithm is that we need to render the scene twice now in order to extract each layer of the scene, one for color information and the other one for depth information. After all the layers are peeled and blended correctly, we obtain the final translucent image.

User Interaction

Traditionally, the keyboard or mouse is usually employed as input device for human–computer interaction. However, the user may need to learn lots of commands in a complex training system, which requires long time practice. To facilitate the interaction process, we design a set of user-friendly interface based on the Microsoft Kinect, and allow the user to interact with the computer using their body gestures. The Kinect is a real-time video motion detecting device and is employed here to track and recognize the user' hand gestures. The hand gestures are then used to determine the position and direction of the cross-sectional plane, which will be sent to the server to compute the corresponding cross-sectional image. Currently, four hand gestures are defined in our system as follows:

- **Translation**: users can translate the position of the cross-sectional plane with the movement of one hand (Fig. 2a).
- **Rotation**: users can rotate the direction of the cross-sectional plane with a circular motion of two hands (Fig. 2b).
- **Scaling**: users can zoom in or out the cross-sectional anatomical image by adjusting the distance of two hands (Fig. 2c).
- **Screenshot**: users can screen shot the cross-sectional anatomical image by holding the fist (Fig. 2d).

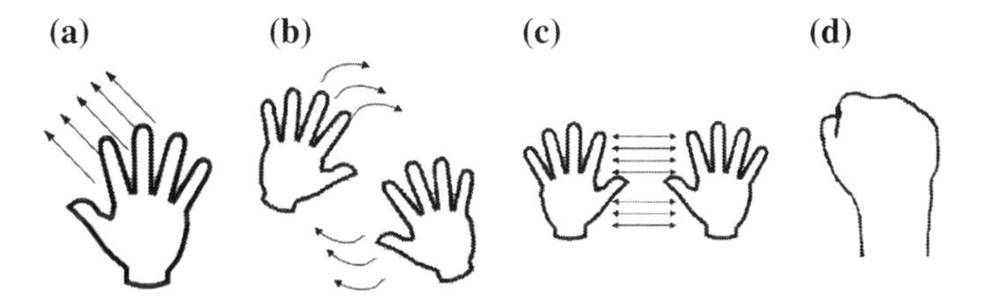

Fig. 2 **a** Translation. **b** Rotation. **c** Scaling. **d** Screenshot

Cross-Sectional Computation

Given the information about the cross-sectional plane, the server can compute the corresponding anatomical image. In order to accelerate the computation, several preprocessing steps are performed to compress the CVH dataset. In details, the dataset is first down-sampled without any visual quality degradation. Then we detect image blocks that contain meaningful contents and discard the other regions. Finally, the image blocks are compressed with DXT algorithm [9] to further reduce the data size. The advantages of DXT algorithm lie on its low quality loss and fast decompression speed. Please refer to [10] for more details.

After the above processing, the dataset is reduced to about 1 GB that is small enough to be loaded into computer memory. Because the calculation for each pixel within the cross-sectional plane is independent, the performance can be accelerated by parallelizing the computation. Moreover, to achieve smooth resultant images, 3D interpolation is employed to sample the pixels that do not lie in the dataset.

Progressive Data Transmission

To reduce the burden of data transmission, we adopt a progressive strategy to transmit the data between the server and the client. The idea is motivated by the Adam7 algorithm [11], where an image is decomposed into several small subimages with different levels of fidelity. Specifically, as shown in Fig. 3, we decompose the original cross-sectional image with resolution 3,872 × 2,048 into seven subimages as: 484 × 256, 484 × 256, 968 × 256, 986 × 512, 1,972 × 512, 1,972 × 1,024 and 3,972 × 1,024. During the transmission, the coarsest subimage is first sent to the client without any delay. If the user is interested on that anatomical image, the server will progressively send the finer subimages to the client to recover the original image.

To decrease the transmission time, we further compress the seven subimages to reduce the image size. There are lots of techniques for image compression, such as Discrete Cosine Transform (DCT) [12], Huffman entropy encoding [13], and

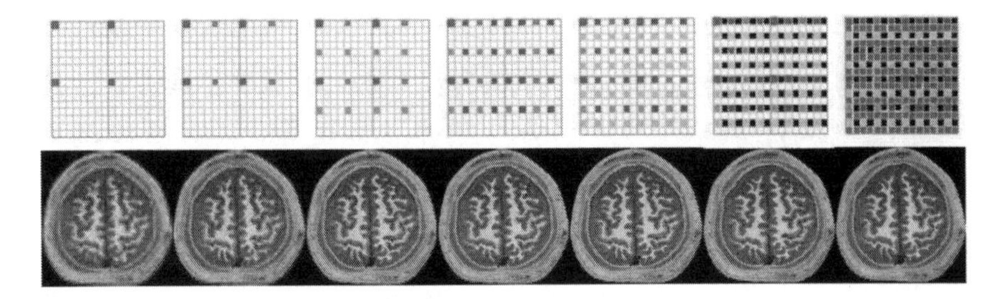

Fig. 3 Representation of image decomposition

wavelet [14, 15]. As JavaScript [16] is still not fast enough for online decompression, here we compress the images with JPEG format that is supported by most web browsers. The performance gain attribute to the JPEG compression is analyzed in the next section.

Experiments

We use the CVH dataset to demonstrate the proposed anatomy navigation system. The output for the user includes two parts. The first part is for the visualization of the human model (left of Fig. 4). The human bone is rendered with high opacity while the skin is visualized with translucent effect. The second part is for the display of the anatomical image (right of Fig. 4). The high-resolution cross-sectional image clearly shows the inner structures of the human body, and it will be very useful for anatomy teaching and learning. Lastly, a photo of our anatomy navigation system is presented in Fig. 5, where the user stands in front of the computer and uses her hand gestures to navigate the Visible Human.

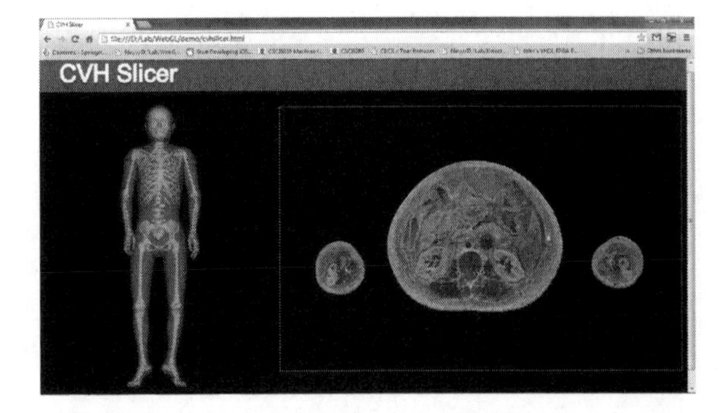

Fig. 4 Screenshot of our anatomy navigation system

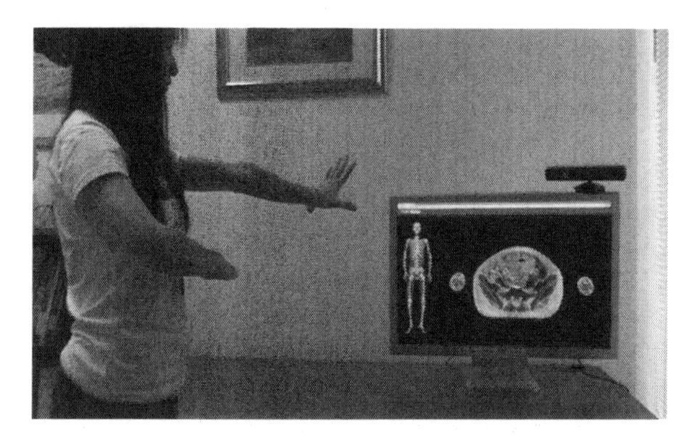

Fig. 5 A photo of our anatomy navigation system

Analysis of Data Transmission

In this section, we analyze the performance for transmitting data between the server and client to demonstrate the advantages of our progressive transmission strategy. The resolution of the cross-sectional image is 3,872 × 2,048, with 32 bits per pixel; hence the image size is about 30.25 MB. Assuming the network bandwidth of the server is unlimited and the impact of Round Trip Time (RTT) can be neglected, it takes about 121 s to transmit the image from the server to the client for bandwidth of 2 Mbit/s.

Even for bandwidth of 100 Mbit/s, it still needs 2.42 s, which is not fast enough for interactive applications.

With our progressive transmission, the resolution of the coarsest image is 484 × 256 and the image size is about 484 KB. After the JPEG compression, the data size is further reduced to about 25 KB and it only needs 0.1 s to transmit the coarsest image for bandwidth of 2 Mbits/s. The transmitting time is even shortened to 0.02 and 0.002 s for bandwidth of 10 Mbits/s and 100 Mbits/s, respectively.

In Table 1, we compare the data transmitting time among three different networks bandwidth: 2 Mbit/s, 10 Mbit/s and 100 Mbit/s. Both the average and maximum time for transmitting the coarsest subimage, the total seven subimages and the original uncompressed image are measured in the table. It can be found that the transmitting time of our progressive transmission is much lower than that for transmitting the original uncompressed image. It is also observed that the actual transmitting time is a bit larger than the above theoretical analysis. This is because that the data transmission is not only influenced by the bandwidth of the server but also the routers in the Internet. However, we still achieve real-time performance even with a moderate network bandwidth.

Table 1 Data transmitting time of three network bandwidth

Bandwidth	Coarsest subimage		Seven subimages		Original image	
	Average (s)	Maximum (s)	Average (s)	Maximum (s)	Average (s)	Maximum (s)
2 Mbits/s	0.11537	0.15816	6.8	10.7	136	156
10 Mbits/s	0.02114	0.02649	1.43	5.3	29	70
100 Mbits/s	0.00232	0.00292	0.124	0.157	4	10

User Evaluation

To validate the advantages of the Kinect-based user interface, we invite twenty students to experience and evaluate the navigation system. These students all do not have much knowledge about human anatomy, with age from 18 to 25. The experiment is conducted in the following way. First, the students are given ten anatomical images showing some common organs and structures, and are taught how to identify these organs and structures. Then these students are asked to located these organs and structures using the proposed anatomy navigation system with different interacting devices. Specifically, the twenty students are divided into two groups, ten for each group. The first group use the keyboard and mouse to navigate the Visible Human while the second group use the Kinect for interaction. We count the total time for each person to locate all the organs and structures. Lastly, we average the time for each group since to provide a statistical measure in accomplishing the task. The average time is 60.8 and 28.4 s for the first and second groups, respectively. The experiment results indicate that the Kinect-based navigation system improves the learning and training process as compared with traditional interacting devices. The reason is that the user can freely control their hand gestures using the Kinect, which facilitates the human–computer interaction.

Conclusion

An interactive web-based navigation system for learning human anatomy is presented in this paper. To achieve real-time performance, we exploit the WebGL API to directly invoke the functionalities of GPU from the browser. To provide user-friendly interface, we allow the user to interact with the computer using their hand gestures. To improve the data transmission, we employ a progressive strategy to transmit the data between the server and the client. However, the functionalities of graphics hardware have not been fully exploited yet due to limited supports of current WebGL standards. In future work, we will continue to design new web-based algorithms for online medical applications, and incorporate them into the proposed anatomy navigation system.

References

1. John NW (2007) The impact of Web3D technologies on medical education and training. Comput Educ 49:19–31
2. Poliakov AV, Albright E, Hinshaw KP, Corina DP, Ojemann G, Martin RF, Brink-ley JF (2005) Server-based approach to web visualization of integrated three-dimensional brain imaging data. J Am Med Inform Assoc 12:140–151
3. Mahmoudi SE, Asl AA, Rahmani R, Faghih-Roohi S, Taimouri V, Sabouri A, Soltanian-Zadeh H (2010) Web-based interactive 2D/3D medical image processing and visualization software. Comput Methods Programs Biomed 98:172–182
4. Hersch RD, Gennart B, Figueiredo O, Mazzariol M, Tarraga J, Vetsch S, Messerli V, Welz R, Bidaut L (2000) The visible human slice web server: a first assessment. In: Proceedings IS&T/SPIE conference on internet imaging, vol 3964, pp 253–258
5. Wang W, Meng Q, Qin J, Wei M, Chui YP, Heng PA (2013) An interactive web-based anatomy navigation system via WebGL and Kinect NextMed/MMVR20, poster
6. Marrin C (2011) WebGL Specification, Khronos WebGL Working Group
7. Zhang SX, Heng PA (2004) The Chinese visible human (CVH) datasets incorporate technical and imaging advances on earlier digital humans. J Anat 204:165–173
8. Everitt C (2001) Interactive order-independent transparency. Technical report, NVIDIA Corporation
9. Renambot L, Jeong B, Leigh J (2007) Real-time compression for high-resolution content In: Proceedings of the access grid retreat
10. Meng Q, Chui YP, Qin J, Kwok WH, Karmakar M, Heng PA (2011) CvhSlicer: an interactive cross-sectional anatomy navigation system based on high-resolution chinese visible human data. Stud Health Technol Inform 163:354–358
11. Greg R (2011) PNG: the definitive guide, O'Reilly and Associates, Inc
12. Rao KR (1976) Orthogonal transforms for digital signal processing. In: IEEE international conference on ICASSP, pp 136–140
13. Van LJ (1976) On the construction of Huffman trees. In: Proceedings of the 3rd international colloquium on automata, languages and programming, pp 382–410
14. Cohen A, Daubechies I, Feauveau JC (1992) Biorthogonal bases of compactly supported wavelets. Information Technology
15. Davis GM, Nosratinia A (1998) Wavelet-based image coding: an overview. Appl Comput Control Sig Circ 1:205–269
16. Di BM, Ponchio F, Ganovelli F, Scopigno R (2010) Spidergl: a Javascript 3d graphics library for next-generation WWW. In Web3D, pp 165–174

Community Topical "Fingerprint" Analysis Based on Social Semantic Networks

Dongsheng Wang, Kyunglag Kwon, Jongsoo Sohn, Bok-Gyu Joo and In-Jeong Chung

Abstract Community analysis of social networks is a widely used technique in many fields. There have been many studies on community detection where the detected communities are attached to a single topic. However, an overall topical analysis for a community is required since community members are often concerned with multiple topics. In this paper, we propose a semantic method to analyze the topical community "fingerprint" in a social network. We represent the social network data as an ontology, and integrate with two other ontologies, creating a Social Semantic Network (SSN) context. Then, we take advantage of previous topological algorithms to detect the communities and retrieve the topical "fingerprint" using SPARQL. We extract about 210,000 Twitter profiles, detect the communities, and demonstrate the topical "fingerprint". It shows human-friendly as well as machine-readable results, which can benefit us when retrieving and analyzing communities according to their interest degrees in various domains.

Keywords Community · Topical fingerprint · Community detection · Social semantic network (SSN) · Semantic network (SN)

D. Wang (✉) · K. Kwon · I.-J. Chung
Department of Computer and Information Science, Korea University, Seoul, Korea
e-mail: dswang2011@korea.ac.kr

K. Kwon
e-mail: helpnara@korea.ac.kr

I.-J. Chung
e-mail: chung@korea.ac.kr

J. Sohn
Service Strategy Team, Visual Display, Samsung Electronics, Suwon, Korea
e-mail: jongsoo.sohn@samsung.com

B.-G. Joo
Department of Computer and Information Communications, Hong-Ik University, Seoul, Korea
e-mail: bkjoo@hongik.ac.kr

Y.-M. Huang et al. (eds.), *Advanced Technologies, Embedded and Multimedia for Human-centric Computing*, Lecture Notes in Electrical Engineering 260,
DOI: 10.1007/978-94-007-7262-5_10,

Introduction

Community analysis is widely used in many areas such as personal services, politics, commercial advertising and marketing. Many methods are investigated to detect communities through topological information such as density based algorithms [1] and modulatory algorithms [2]. Also, some topic-oriented methods utilize the profiles of actors as well as the topological information to attach a topic for the detected communities. However, it is unreasonable for a community to be explained by a single topic because the community members are generally concerned with many distinguishable interests or topics in various domains.

In this paper, we analyze the topical "fingerprint" of social network communities. First of all, we crawl a part of the Twitter, and represent the user profiles and their relationships in an ontology. For the crawled graph, we map the influential nodes, specifically celebrities, to the *WordNet Domain* [3] by the Spreading Activation (SA) mechanism. Subsequently, we detect communities through the "*Louvain* method" [2]. Finally, for the detected communities, we can retrieve the celebrities associated with each community, aggregate their domains and average them into degrees of interest. In this way, a topical "fingerprint" for each community can be retrieved and analyzed.

We extract about 210,000 Twitter users' profiles, transform them into an ontology, and then merge them with *YAGO* [4] and the *WordNet Domain* ontology. The experiment shows a topical "fingerprint" that is human-understandable. Moreover, since the domains are mapped to the *WordNet Domain* ontology, the topical "fingerprint" is machine-readable and supports semantic searching of communities. We can retrieve and analyze communities according to their degrees of interest for various domains.

Related Works

Community Detection and Analysis

Recently there has been a lot of research into community detection. A density-based clustering approach in [1] employs two distance functions to validate the advantage and limitation of them, respectively. Newman propose a significant algorithm to partition social network graphs of links and nodes into sub graphs, and an associated concept, modularity, which has also attracted a large amount of attention for development in the study of community detection [5]. Since the main drawback is that this algorithm is time-consuming, Vincent suggests the modified version of the algorithm to make it faster, giving rise to what is known as the "*Louvain* method" [2]. The method iteratively optimizes the modularity in a local way, and aggregates nodes of the same community. The modularity gain

ΔQ obtained by moving an isolated node i into a community C can be computed by the following Eq. (1) [2].

$$\Delta Q = \left[\frac{\sum_{\text{in}} + 2k_{i,\text{in}}}{2m} - \left(\frac{\sum_{\text{tot}} + k_i}{2m}\right)^2\right] - \left[\frac{\sum_{\text{in}}}{2m} - \left(\frac{\sum_{\text{tot}}}{2m}\right)^2 - \left(\frac{k_i}{2m}\right)^2\right] \quad (1)$$

In Eq. (1), $\sum_{\text{in}}$ is the sum of the weights of the links inside C, $\sum_{\text{tot}}$ is the sum of the weights of the links incident to nodes in C, k_i is the sum of the weights of the links incident to node i, $k_{i,\text{in}}$ is the sum of the weights of the links from i to nodes in C, and m is the sum of the weights of all the links in the network.

The topic-oriented method in [6] combines both social object clustering and link analysis. The entropy based method in [7] combines topological and semantic information to detect communities more accurately. The authors of [8] build community detection on a Semantic Network (SN), and prove that this method is faster, more effective, and has robust benefits. Another semantic web technology based method, called *SemTagP* [9], labels each community as tags used by people as well as relationships inferred between tags.

"Fingerprints" in Social Networks

There are some studies about a personal "fingerprint" in social networks. An interesting commercial one is called as *PeerIndex,*[1] which divides all topics into eight different areas. Figure 1 shows the topical "fingerprint" in these areas. Gunther argues that a user ID can be forged but his or her habits cannot be detected. Therefore he attempts to depict the users' behavior "fingerprints" through tracking users' internet behavior and express the detected features through a semantic pattern [10].

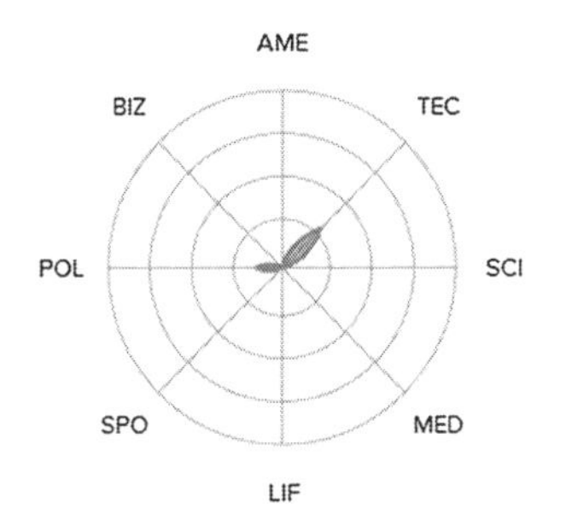

AME - arts, media, entertainment
TEC - technology, internet
SCI - science, environment
MED - health, medical
LIF - leisure, lifestyle
SPO - sports
POL - news, politics, society
BIZ - finance, business, economics

Fig. 1 *PeerIndex* personal fingerprint

[1] http://www.peerindex.com

Community Topical "Fingerprints" Based on SSN

The Social Semantic Network (SSN) we work on includes: *YAGO*, the *WordNet Domain* ontology, and the Twitter graph generated by us (i.e. nodes and relations between users). The *WordNet Domain* ontology is manually generated from the *WordNet Domain* Hierarchy[2] which is a lexical resource. In section Domain Mapping for Celebrities (or Hubs), we discuss how celebrities are classified and mapped into the *WordNet Domain* ontology. In section Community Topical "Fingerprint", we show how semantic search works. We can formalize the progress into three steps:

- Step 1: First of all, we store social network data in an ontology, and integrate this ontology with *YAGO* and the *WordNet Domain* ontologies. We then map celebrities into the *WordNet Domain* ontology by performing SA on the knowledge base, creating a huge SSN context.
- Step 2: Secondly, we detect the communities using the existing "*Louvain* method".
- Step 3: Finally, we analyze the community's topical "fingerprint" by retrieving celebrities followed by the community members, aggregating their domains and averaging by community member amount.

User profiles are presented as instances of Friend-Of-A-Friend (FOAF) concepts. The knowledge base is integrated with the social network. The social network within the Semantic Network (SN) is called the Social Semantic Network (SSN) [11]. We demonstrate the overall graph as a three layer architecture, as shown in Fig. 2. At the bottom are common actors who are following celebrities in the second layer (influential nodes). The celebrities are mapped to the *WordNet Domain* ontology in the third layer.

Definition (Social Semantic Network, *SSN*): A set *SSN* means the overall three layer Social Semantic Network, and we express it as $SSN = (SN, A, H, D, R_k, R_m)$ where:

- Set *SN* indicates the first two layers (*L*1 and *L*2); the set *L*1 indicates common actors and *L*2 expresses celebrities or influential nodes ($L1 \subseteq SN$ and $L2 \subseteq SN$). The set *L*3 expresses the domains in the third layer.
- Common actors in the first layer can be expressed as $L1 = (A, R_k)$ where $A = \{a_1, a_2, \ldots, a_i, \ldots, a_n\}$ where $|A| = n$ is called the node set, and R_k is called the edge set with 'the *knows*' relationship on *A*. For example, if an actor a_i has 'the *knows*' relationship with a_{i+1}, then $(a_i, a_{i+1}) \in R_k$.
- Celebrities in the second layer can be expressed as the set *H*, and are denoted by $H = \{h_1, h_2, \ldots, h_j, \ldots, h_m\}$ with $|H| = m$.

[2] http://wndomains.fbk.eu

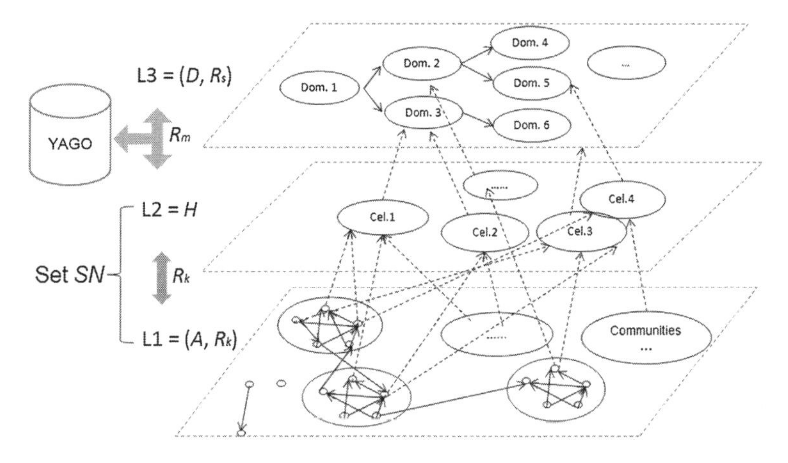

Fig. 2 Three layer architecture of SSN

- Domain hierarchy in layer three can be expressed as $L3 = (D, R_s)$, where $D = \{d_1, d_2, \ldots, d_k, \ldots, d_o\}$ indicates the set of domains with $|D| = o$, and R_s expresses the set of *subdomain* relations on *D*. For example, if the d_k has a *subdomain* relation of d_{k+1}, then $(d_k, d_{k+1}) \in R_s$.
- It is noted that the mapping edge set from *L*1 to *L*2 also employs the set R_k, namely, the relationship of *knows*. For example, if an actor $a_i (a_i \in A)$ follows a celebrity $h_j (h_j \in H)$, then $(a_i, h_j) \in R_k$.
- The mapping edge set from *L*2 to *L*3 employs the set R_m. If a celebrity h_j is mapped to a domain d_k, then $(h_j, d_k) \in R_m$.

Definition (A set of communities, *C*): A set *C* of communities indicates the partitions or subsets of the set *L*1 where $C = \{c_1, c_2, \ldots, c_p, \ldots, c_q\}$ with $|C| = q$ and $C_p = (A', R_k')$. For a specified community, the number of members within this community is expressed as $|C_p|$, which is also called the community size.

Domain Mapping for Celebrities (or Hubs)

This part of our work focuses on the mapping from *L*2 to *L*3, as shown in Fig. 2. The knowledge base is a SN where the knowledge is represented in patterns of interconnected nodes and arcs [13]. Celebrities are indicated by those nodes that are very influential and often have a large amount of followers. They are also significant since their trending or words affect their numerous followers and the whole social network. We regard the actors who have more than ten thousand followers as celebrities [14]. It is meaningful to map them into their corresponding domains so that we can know what kind of celebrities normal users are interested

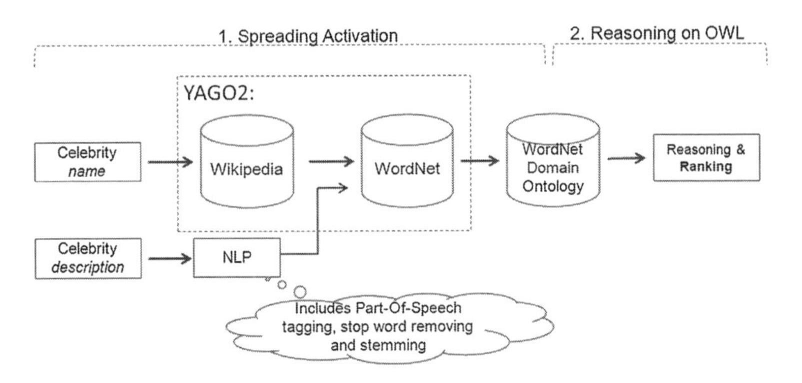

Fig. 3 Spreading activation of celebrity's profiles [12]

in and are following. The automatic mapping method is based on SA on the SN. SA is a model that simulates the memory mechanism of human brain based on a knowledge graph [15]. We apply SA technology on our knowledge base consisting of *YAGO* and the *WordNet Domain* ontology.

As shown in Fig. 3, for a Twitter celebrity, we take the celebrity's *name* and *description* as the input and perform a series of SA from nodes to their neighboring nodes along a constrained direction on the SN. The celebrity's domain can be mapped to the *WordNet Domain* ontology by reasoning the SA results based on the ontology.

Community Topical "Fingerprint"

We are left with a huge interconnected SSN after the mapping process discussed in section Domain Mapping for Celebrities (or Hubs). We set a series of Boolean values based on the three layer architecture, which is shown in Fig. 2.

For the mapping from $L1$ to $L2$, We set a Boolean value of $b_{i,j}$, if an actor $a_i(a_i \in A)$ follows a celebrity $h_j(h_j \in H)$, that is, $(a_i, h_j) \in R_k$, then $b_{i,j} = 1$, otherwise, $b_{i,j} = 0$.

For the mapping from $L2$ to $L3$, we set a Boolean value of $b_{j,k}$, if $(h_j, d_k) \in R_m$, then $b_{j,k} = 1$, otherwise, $b_{j,k} = 0$.

We set a Boolean value of $b_{i,p}$ to judge whether an actor belongs to a community. If $a_i \in C_p$, then $b_{i,p} = 1$, otherwise, $b_{i,p} = 0$.

Definition (The Topical Fingerprint, l_k): For a community, we calculate the interest degrees of each basic domain by aggregating the mapping amount from $L1$ to $L3$ and averaging by the community members, called as community size. For a domain d_k, we calculate the interest degree of a community C_p as follows:

```
PREFIX FOAF:<http://xmlns.com/foaf/0.1/>
PREFIX iis:<http://iis.korea.ac.kr/20130107/wordnet2.owl#>

SELECT ?dom {
        ?user FOAF:belong2Comm FOAF:community.
        ?user FOAF:knows ?hubs.
        ?hubs iis:belong2wordnetDomain ?dom
}
```

Fig. 4 Semantic search

$$l_k = \frac{\text{aggregate amount of } d_k}{\text{size of } C_p} = \frac{\sum_{i=1}^{n} \sum_{j=1}^{m} b_{i,j} \times b_{j,k} \times b_{i,p}}{|C_p|} \quad (2)$$

In Eq. (2), l_k is the interest level of domain d_k for the specified community C_p. The parameters such as i, j, k and p are explained in former definitions. In this way, we calculate the interest degrees of each basic domain for a specified community.

In this equation, l_k represents the aggregate mapping number associated with a specified community for each domain from $L1$ to $L3$. The equation can be easily computed by a single SPARQL as:

As shown in Fig. 4, the parameter *?dom* indicates total domain aggregation. *?user FOAF:belong2Comm FOAF:community* expresses the retrieval of users who belong to a specified community. *?user FOAF:knows ?hubs* retrieves the celebrities (or hubs) the user is following. *?hubs iis:belong2wordnetDomain ?dom* searches the various domains the celebrities are mapped to. In this way, we get the aggregate of the various domains. Then we aggregate each domain amount, average them by community size $|C_p|$, and get the "fingerprint" l_k.

Experiment

For the evaluation of the suggested method, we conduct an experiment with the aid of the Twitter Application Programming Interface (API). The SN (a large ontological data set) is merged and manipulated with the aid of the Jena framework.[3] We randomly fetch 210,000 profiles from Twitter and transform them into a RDF triple. From this data, we filter 15,879 celebrities and map them to the corresponding *WordNet Domain* ontology.

We adopt the "*Louvain* method", and the modularity we calculated for this social graph is 0.57, as we know that Twitter does not have a high degree of "Socialization" [16]. The distribution of community sizes is listed in Table 1 (1,719 communities in total).

[3] http://jena.apache.org/

Table 1 Distribution of the number of community members

Members	0–100	100–200	200–300	300–400	400–500	500–600	600–700	700–800	>800
Number of community	1,649	16	9	6	4	7	2	1	25

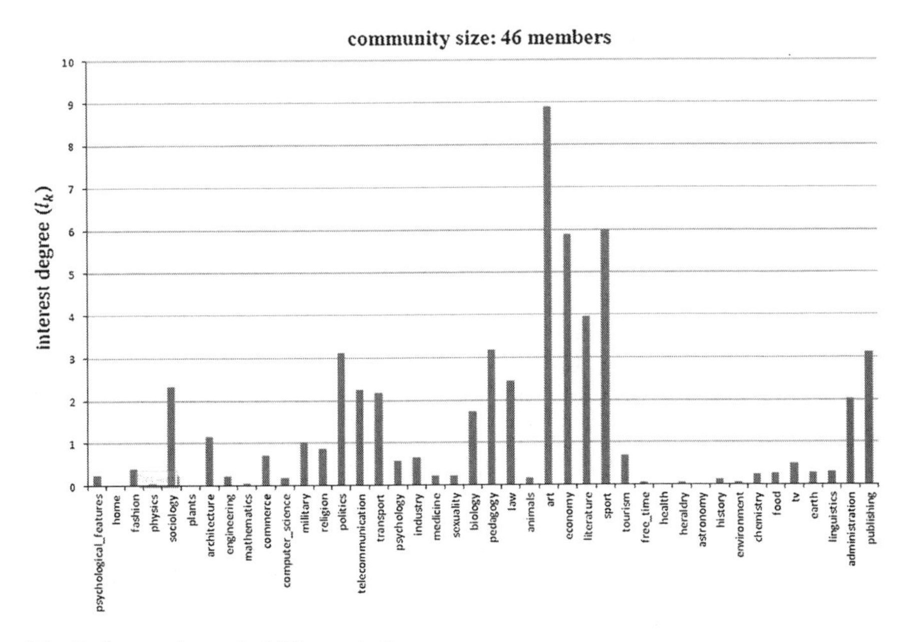

Fig. 5 Community topical "fingerprint"

We implement our method and demonstrate one of the running results. One of the communities has 46 members, and its topical "fingerprints" are illustrated in Fig. 5. As shown in Fig. 5, the average following of the *art* domain celebrities is the highest, followed by the *sport* and *economy* ones.

Conclusion

We presented a topical "fingerprint" analysis method for social network communities that takes into account the fact that community members are generally concerned with various topics in different domains. The method performs semantic search on the SSN to analyze the "fingerprint" of the community. First of all, we create a SSN context where we map the celebrities or hubs to the *WordNet Domain* ontology. Secondly, we take advantage of the previous "*Louvain* method" to detect communities. Finally, we retrieve "celebrities" or "hubs" the community is concerned with, aggregate their domains, and average by the number of community members. We extracted about 210,000 Twitter profiles, detected the communities, and analyzed their "fingerprints". It demonstrates human-friendly as well as machine-readable results. In this way, we can retrieve and analyze communities according to their degree of interest in various domains.

Acknowledgment This research was partially supported by Korea University.

References

1. Subramani K, Velkov A, Ntoutsi I, Kroger P, Kriegel HP (2011) Density-based community detection in social networks. In: 5th IEEE international conference on internet multimedia systems architecture and application, pp 1–8
2. Blondel V, Guillaume JL, Lambiotte R, Lefebvre E (2008) Fast unfolding of communities in large networks. J Stat Mech Theory Exp 2008:P10008
3. Bentivogli L, Forner P, Magnini B, Pianta E (2004) Revising the wordnet domains hierarchy: semantics, coverage and balancing. In: Proceedings of the workshop on multilingual linguistic resources
4. Suchanek FM, Kasneci G, Weikum G (2008) YAGO: a large ontology from Wikipedia and WordNet. J Web Semant 6:203–217
5. Newman MEJ (2004) Analysis of weighted networks. Phys Rev E 70:056131
6. Zhao Z, Feng S, Wang Q, Huang JZ, Williams GJ, Fan J (2012) Topic oriented community detection through social objects and link analysis in social networks. Knowl Based Syst 26:164–173
7. Cruz JD, Bothorel C, Poulet F (2011) Entropy based community detection in augmented social networks. In: International conference on computational aspects of social networks, pp 163–168
8. Xia Z, Bu Z (2012) Community detection based on a semantic network. Know Based Syst 26:30–39
9. Ereteo G, Gandon F, Buffa M (2011) SemTagP: semantic community detection in Folksonomies. In: IEEE/WIC/ACM international conference on web intelligence and intelligent agent technology, pp 324–331
10. Lackner G, Teufl P, Weinberger R (2010) User tracking based on behavioral fingerprints. In: Heng SH, Wright R, Goi BM (eds) Cryptology and network security, vol 6467. Springer, Heidelberg, pp 76–95
11. Mika P (2004) Social networks and the semantic web. In: Proceedings of IEEE/WIC/ACM international conference on web intelligence, pp 285–291
12. Wang D, Kwon K, Chung I (2013) Domain classification for celebrities using spreading activation and reasoning on semantic network. In: 5th international conference on ubiquitous and future networks
13. Sowa JF (2006) Semantic networks. Encyclopedia of Cognitive Science. Wiley, New Jersey
14. Lim KH, Datta A (2012) Following the follower: detecting communities with common interests on twitter. In: Proceedings of the 23rd ACM conference on hypertext and social media
15. Anderson JR (1983) A spreading activation theory of memory. J Verbal Learn Verbal Behav 22:261–295
16. Kwak H, Lee C, Park H, Moon S (2010) What is twitter, a social network or a news media? In: Proceedings of the 19th international conference on World Wide Web

Content Recommendation Method Using FOAF and SNA

Daehyun Kang, Kyunglag Kwon, Jongsoo Sohn, Bok-Gyu Joo and In-Jeong Chung

Abstract With the rapid growth of user-created contents and wide use of community-based websites, content recommendation systems have attracted the attention of users. However, most recommendation systems have limitations in properly reflecting each user's characteristics, and difficulty in recommending appropriate contents to users. Therefore, we propose a content recommendation method using Friend-Of-A-Friend (FOAF) and Social Network Analysis (SNA). First, we extract user tags and characteristics using FOAF, and generate graphs with the collected data, with the method. Next, we extract common characteristics from the contents, and hot tags using SNA, and recommend the appropriate contents for users. For verification of the method, we analyzed an experimental social network with the method. From the experiments, we verified that the more users that are added into the social network, the higher the quality of recommendation increases, with comparison to an item-based method. Additionally, we can provide users with more relevant recommendation of contents.

Keywords Social network analysis (SNA) · Content recommendation · Friend-of-a-friend (FOAF)

D. Kang (✉) · K. Kwon · I.-J. Chung
Department of Computer and Information Science, Korea University, Seoul, Korea
e-mail: internetkbs@korea.ac.kr

K. Kwon
e-mail: helpnara@korea.ac.kr

I.-J. Chung
e-mail: chung@korea.ac.kr

J. Sohn
Service Strategy Team, Visual Display, Samsung Electronics, Suwon, Korea
e-mail: jongsoo.sohn@samsung.com

B.-G. Joo
Department of Computer and Information Communications, Hong-Ik University, Seoul, Korea
e-mail: bkjoo@hongik.ac.kr

Y.-M. Huang et al. (eds.), *Advanced Technologies, Embedded and Multimedia for Human-centric Computing*, Lecture Notes in Electrical Engineering 260,
DOI: 10.1007/978-94-007-7262-5_11,

Introduction

With the proliferation of users through Web 2.0, it is important for users to participate in a community, and create and share their contents with each other. Community-based websites such as Facebook,[1] Twitter,[2] Del.icio.us,[3] Pinterest,[4] etc. are propagating dramatically, and users in the websites share their favorites, blogs, photos, music, videos, and other content with each other. A report in 2013 described that about 83 % of people of age 18–29, and 77 % of people of age 30–49, use online social networks [1]. In spite of the great success of many community-based websites, few of them consider the characteristics of each user for content recommendation, thus making it difficult to provide users with more relevant contents.

Accordingly, we propose a content recommendation method using Friend-Of-A-Friend (FOAF)[5] ontology and Social Network Analysis (SNA) [2], to reflect the content and characteristics of the user. The method is based on Web Ontology Language (OWL) for the representation of each user profile, and employs SNA for content recommendation. The method consists of three main phases: FOAF data collection, data integration and graph generation, and SNA and content recommendation.

For the validation of the method, we collected the experimental dataset over 2,676 nodes from three different websites, and then evaluated the performance of the method, using the hit and recall ratio. From the result, we showed the method has a higher recall ratio of 0.103 than that of the item-based one, 0.02, and the quality of contents rises, as the number of users increases.

Related Works

Content recommendation in fields such as e-commerce, and community-based websites plays a significant role in providing relevant items for user purchase, and helping users to make a proper decision on the selection of good contents such as music and news [3]. Nowadays, the amount of Internet content has increased drastically, as users of online networks are able to directly create and easily share their contents. Accordingly, user selection of pertinent contents on the websites is of great importance, and research on recommendation methods has been continuously conducted. Representative recommendation methods of contents are

---

1 http://www.facebook.com

2 http://www.twitter.com

3 http://del.icio.us

4 http://pinterest.com/

5 http://www.foaf-project.org/

categorized into three types, namely collaborative methods, content-based methods, and hybrid methods.

Collaborative recommendation methods are widely used in Social Network Services (SNSs) [4]. The methods have advantages, such as low computation complexity on the web server, and easier implementation, since a large number of users directly create and share their contents. Therefore, much research on the utilization of collaborative methods has been conducted in a variety of areas [5, 6]. In [5], authors used the iterative polling method for recommendation. Authors in [6] took advantage of linked-data, and proposed the open recommendation method. Although these methods are suitable for the recent trend of Web 2.0, they have difficulties in considering a diversity of relations between users, or a user and contents for recommendation, to gather each user's characteristics dynamically, and to manage their profiles efficiently. Furthermore, the quality of content for recommendation decreases significantly, when the number of users or participants to evaluate contents is not large enough [7].

Content-based recommendation methods are divided into two categories: item-based [8, 9], and user-based [10–13] recommendation methods. The item-based method [8, 9] takes advantage of contents that users like or liked in the past, and then recommends items to users. In [9], authors suggested a content recommendation method, by measuring each similarity between items. Since the method did not consider user's interests and their relations, it was not able to provide relevant recommendations according to each user's concerns. In addition, the authors in [8] extracted tags of user interests; however, they did not consider user relations in social networks. As most item-based recommendation methods mainly focus on contents to be recommended for users, they have drawbacks, in that they make it difficult to consider each user's interests, and the contents themselves.

In contrast, user-based recommendation [10, 11, 13] methods mainly concentrate on finding users, their relations, and socially common interests. In [12], authors researched the method to score items, and predict another item's score. However, most user-based methods, such as [10–13], have a common disadvantage, in that they do not make full use of characteristics and details for the recommendation of contents. In addition, the quality of recommended contents could be potentially lower, if each user profile does not have enough information [7].

Recently, research on hybrid recommendation methods, which combine both content-based and collaborative methods, has been conducted, to complement each individual method's shortcomings. The authors in [14] suggested a recommendation system for online research papers, based on research topic ontologies, Quickstep and Foxtrot. The author in [15] demonstrated a hybrid news recommendation method, using an aggregated user profile. Moreover, in [16], the authors proposed a system for the recommendation of relevant restaurants, by using domain knowledge that is related to restaurants. Their proposed methods [14, 16] are not easy to comprehensively apply to other fields, since their research has been conducted to recommend contents in a domain-specific area.

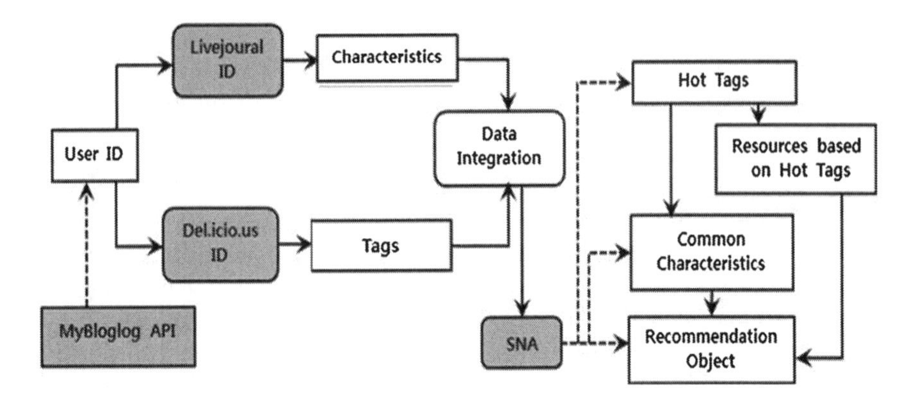

Fig. 1 Overall architecture of the suggested recommendation method

Content Recommendation Using FOAF and SNA

The proposed method can be divided into three main steps: (1) data collection, (2) data integration and graph generation, and (3) SNA and content recommendation. Figure 1 demonstrates the overall architecture of the suggested recommendation method.

Data Collection

Most community-based websites provide open Application Programming Interfaces (APIs). A user FOAF profile can be obtained using the API, and we can obtain abundant information of each user. A dataset is divided into two types: characteristic data, such as user interest or taste, and tag data, as keywords for user content. We firstly provide three definitions for the proposed method, as follows.

Definition 1 (*A set A of actors*) An actor is a user who uses a web or social network service. A set of actors, A is defined as users who use the website. A set A is denoted by $A = \{a_1, a_2, \ldots, a_i, \ldots, a_n\}$, where $|A| = n$, and a_i is the i-th element in the set A.

Definition 2 (*A set C of characteristics*) A characteristic of an actor includes nationality, age, and interest, etc. from FOAF ontology. A set C is represented as $C = \{c_1, c_2, \ldots, c_j, \ldots, c_p\}$, where $|C| = p$, and c_j is the j-th element in the set C.

Definition 3 (*A set T of tags*) Each tag in web documents or contents stands for keywords for them. The tags are crawled by the websites, and are defined as a set $T = \{t_1, t_2, \ldots, t_k, \ldots, t_q\}$, where $|T| = q$, and t_k is the k-th tag in the set T.

The suggested method takes advantage of the MyBlogLog API to crawl each user's ID (A). Afterwards, with the collected FOAF data of users, we extract a set of tags (T) from Del.icio.us website, and a set of characteristics (C) from LiveJournal.

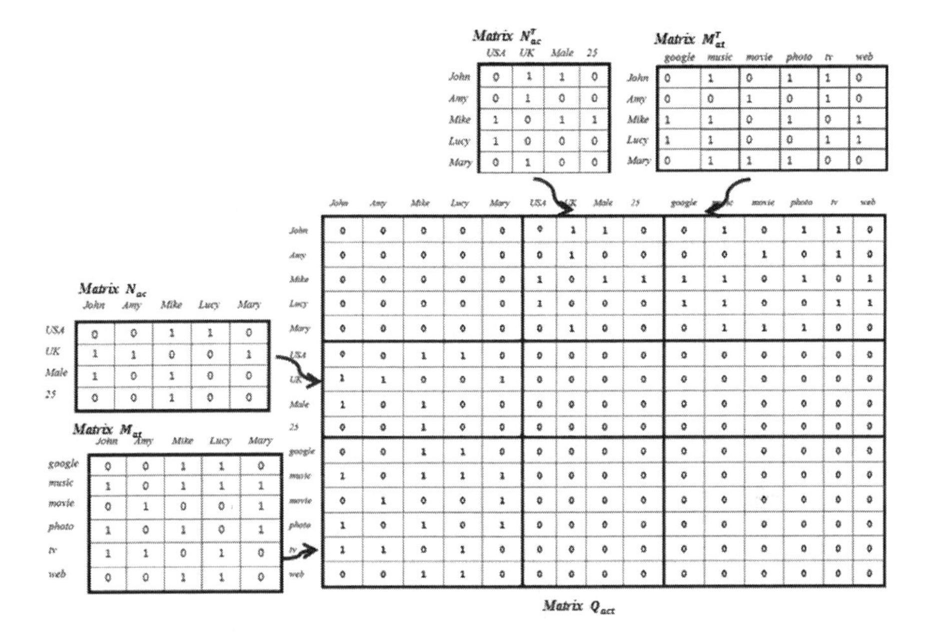

Fig. 2 Two-dimensional matrices, N_{ac}, M_{at}, Q_{act}

Data Integration and Graph Generation

From three collected datasets, A, T and C, the proposed method creates two-dimensional matrices, N_{ac}, M_{at}, Q_{act}, as shown in Fig. 2.

Definition 4 A matrix N_{ac} is a two-dimensional matrix, in which each dimension represents a user (a) and a characteristic (c), respectively. Each element of a matrix N_{ac} is represented by n_{xy}, with a relation of the x-th user and the y-th characteristic. If there is a relation between the x-th user and the y-th characteristic, n_{xy} is 1, and otherwise 0.

Definition 5 A matrix M_{at} is a matrix whose dimensions describe a user (a) and a tag (t). Each element in M_{at} is denoted by m_{ij} with the i-th user and the j-th tag. m_{ij} is 1 when there is a relation between the user and tag, and otherwise 0.

Definition 6 A matrix Q_{act} is a matrix which combines two matrices, N_{ac} and M_{at}, and is defined as follows (see the middle figure in Fig. 2).

$$\text{A Matrix } Q_{act} = \left\{ \begin{matrix} 0 & N_{ac}^T & M_{at}^T \\ N_{ac} & 0 & 0 \\ M_{at} & 0 & 0 \end{matrix} \right\}$$

A matrix Q_{act} is mainly used for the generation of a social network graph. According to Definition 6, the matrix Q_{act} consists of $N_{ac}, M_{at}, N_{ac}^T, M_{at}^T, 0$. Since the matrix is symmetric, the method computes only two matrices, N_{ac}, M_{at}, thus preventing exponential increase of the computation time. M_{at}, N_{ac} and Q_{act} are converted into graphs, since they represent relations between two dimensions, such as relations between a user and a tag, and a user and a characteristic. Accordingly, three matrices are considered as graphs, which represent the relations of users, characteristics, and tags. Two constructed graphs from M_{at} and N_{ac} are represented by two sub-graphs of the generated graph from a matrix Q_{act}, since a matrix Q_{act} is the combined matrix of M_{at} and N_{ac}. We formalize the three graphs as follows.

Definition 7 (A graph AT) A graph AT is composed of a set V_{AT} of nodes and a set E_{AT} of edges, such that each edge $e \in E_{AT}$ is associated with an unordered pair of nodes. V_{AT} consists of a set A of users, and a set T of tags. A graph AT is denoted by $AT = (V_{AT}, E_{AT})$, where $V_{AT} = A \cup T$, and $E_{AT} = \{(a,t) \in A \times T | (a \in A, t \in T)\}$.

Definition 8 (*A graph AC*) A graph AC consists of a set V_{AC} of nodes and a set E_{AC} of edges, such that each edge $e \in E_{AC}$ is connected with an unordered pair of nodes. V_{AC} consists of a set A of users, and a set C of characteristics. A graph AC is denoted by $AC = (V_{AC}, E_{AC})$, where $V_{AC} = A \cup C$, and $E_{AC} = \{(a,c) \in A \times C | (a \in A, c \in C)\}$.

Definition 9 (*A graph ACT*) A graph ACT consists of a set V_{ACT} of nodes and a set E_{ACT} of edges, such that each edge $e_{ij} \in E_{ACT}$ is connected with an unordered pair of nodes, e_i and e_j. V_{ACT} is composed of a set A of users, a set C of characteristics, and a set T of tags. A graph ACT is the combination of two graphs AC and AT, and is denoted by $ACT = (V_{ACT}, E_{ACT})$, where $V_{ACT} = V_{AT} \cup V_{AC}$, and $E_{ACT} = E_{AT} \cup E_{AC}$.

Figure 3 demonstrates an example of a social network graph, ACT which is the combination of two graphs AT and AC. In the Fig. 3, the left graph is graph AC, the right graph is graph AT, and the middle graph is the combined graph ACT. The graph ACT has nodes that represent all users, tags, and characteristics, and edges that represent their relations. We denote users as ○, tags as △, and characteristics as □, respectively.

Social Network Analysis and Content Recommendation

Step 1 (**Measurement of degree of centrality**). The degree of centrality [17] is a SNA method used for computation of each centrality of a tag in the suggested method. The larger the value of degree centrality of a node is, the more influential the node is in the social network graph. A node that has a high value of degree centrality usually plays an important role as a hub for other nodes in the graph [2].

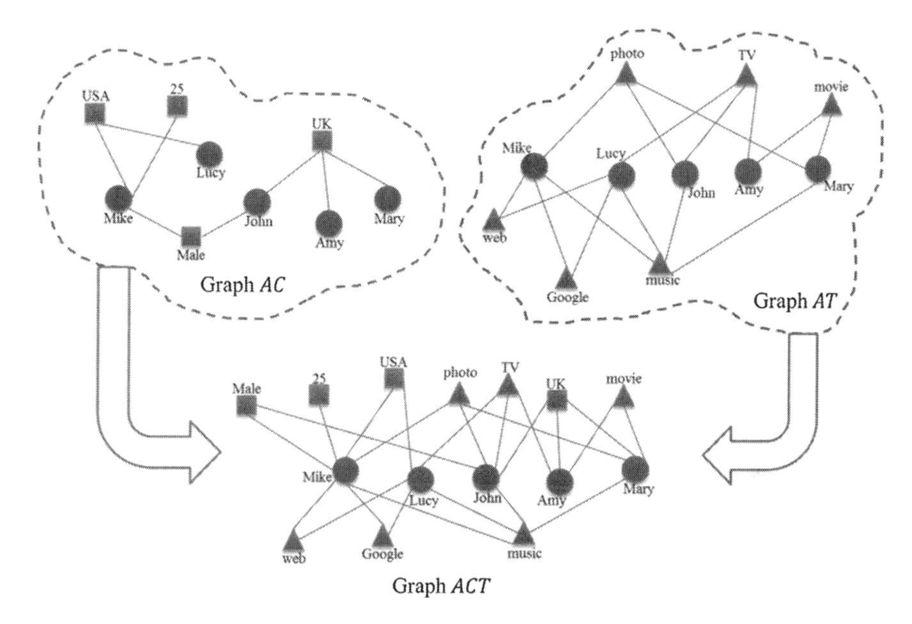

Fig. 3 An example of a social network graph, *ACT*

The degree of centrality C_D for a node is computed with Eq. (1). First, we get the number of nodes that are connected with node i in a set V_{ACT} of a graph ACT. We then calculate each degree of centrality for each node in the graph ACT, with Eq. (1).

$$C_D(i) = \frac{\sum_{j \in V_{ACT}} e_{ij}}{|V_{ACT}| - 1} \quad (0 \leq C_D(i) \leq 1) \tag{1}$$

Step 2 (**Selection of hot tags**). A tag that a lot of users use is defined as a hot tag. As each hot tag is dependent on the size of the social network, the reflection rate β_t of a tag t is used for a more reliable experiment. As β_t approaches 0, more hot tags are generated in the graph, and they have a lower degree of centrality, and vice versa. After computing the β_t of the tag in the graph, the threshold θ_t is calculated as Eq. (2). In the method, a tag that is higher than the threshold θ_t is selected as a hot tag.

$$\theta_t = \frac{|T| + |A|}{|V_{ACT}|} \times \beta_t \quad (0 \leq \beta_t \leq 1) \tag{2}$$

Step 3 (**Extraction of common characteristics**). A common characteristic is defined as a feature that the majority of users use properties of, such as *foaf:interest* of FOAF ontology. The common characteristic depends on the size of the social network, so that a reflection rate β_c for a characteristic c is defined. A reflection rate β_c for a characteristic c has the same properties as a reflection rate β_t. After computation of a reflection rate β_c for each characteristic c, a threshold θ_c for c is computed as Eq. (3).

$$\theta_c = \frac{|C| + |A|}{|V_{ACT}|} \times \beta_c \quad (0 \leq \beta_c \leq 1) \tag{3}$$

Step 4 (Recommendation of selected content for users). Finally, we find relevant users and contents to recommend. A user to recommend refers to one who has the same characteristics, but who does not have hot tags. A content to be recommended indicates one that will be recommended to users. After a user who has the same common characteristics chooses hot tags, contents connected with selected hot tags are recommended for the user using the proposed method.

Performance Study

Experimental Dataset

For the evaluation of the proposed method, we selected 18 users who use three SNS at the same time, and then collected user IDs from MyBlogLog, their tag data from Del.icio.us, and their characteristics from FOAF ontology in LiveJournal. We arbitrarily selected 18 users, and extracted 98 characteristics from LiveJournal, and 2,560 tags from Del.icio.us. We then conducted experiments using the tools, *Pajek* and *UCINET6*. On the collected dataset, we measured the degree of centrality for each tag, and obtained 30 hot tags (Table 1) with a reflection rate $\beta_t = 0.2$, and 36 common characteristics (Table 2) for each user with a reflection rate $\beta_t = 0.2$.

Table 1 List of selected hot tags

Business	Technology	Blog	Photo	Work	Politics
Download	Tool	Online	News	Twitter	Cards
Music	Health	E-book	Job	Learning	Free
Google	Marketing	Education	Money	Web	Networking
Software	Video	Art	Books	Advertising	Funny

Table 2 List of extracted common characteristics

Writing	80's	Female	UK	Apple	US
Reading	70's	Jazz	Discovery	Novel	50's
Art	Travel	Nature	Google	Literature	Love
Food	Advertising	Photography	Movies	Music	Prayers
Religion	CA	Painting	Singing	Cooking	Education
Charities	Internet	Male	Business	Religion	Education

Evaluation

For the verification of the proposed method, comparison with an item-based recommendation method [18] is conducted. The item-based recommendation method is an approach in which contents are assessed by users' recommendations and votes. Those that have a high score are then recommended to users [19]. For a more precise comparison of the two methods, a hit-ratio and a recall ratio in [18, 20] are adopted, as follows.

$$\text{(Proposed method)} \quad \text{Hit Ratio}(a_i) = |T_{a_i} \cap ht_{a_i}|/|T_{a_i}| \tag{4}$$

$$\text{(Item-based method)} \quad \text{Hit Ratio}(a_i) = |I_{a_i} \cap i_{a_i}|/|I_{a_i}| \tag{5}$$

Equation (4) shows the way to calculate the hit-ratio of the recommended hot tag ht_u of a set T_u of tags in a social network for an i-th actor a_i, where Eq. (5) computes the hit-ratio of the item i_u of a set I_u of items in the social network for an i-th actor a_i. Both equations range from 0 to 1 as a real number, and indicate the number of tags or items that are mapped on all tags or items in a social network.

$$\text{Recall Ratio} = \frac{\sum_{i=1}^{n} \text{Hit Ratio}(a_i)}{n} \times 100\,\% \tag{6}$$

Equation (6) demonstrates a recall ratio for n users, and represents the percentage of the summation of hit-ratios of recommended hot tags and items in a set of documents. For a comparison of the two methods, we calculated each recall ratio of tags or items, using two methods.

In Table 3, the item-based recommendation method does not grasp relevant tags, since the method suggests contents based only on tags in each user document. In contrast, the proposed method shows a higher recall ratio than the existing one, since the method not only extracts tags in each user's document, but also takes advantage of selected hot tags, using SNA and FOAF ontology.

Figure 4 demonstrates the comparison result of two methods as the number of users increases, based on Table 3. The recall ratio is an indicator that shows the number of hit results on a query for information retrieval, and points out that the higher the recall ratio, the more precise the ratio of relevant results. Thus the suggested recommendation method has a better performance than the conventional item-based one. Furthermore, as the number of users increases, the recall ratio also increases proportionally, as shown in Fig. 4. As a result, we are able to expect a much higher recall ratio, if the proposed content recommendation method is applied to a real social network.

Table 3 Comparison of the two recall ratios for the two recommendation methods

No	User	Tags	Item-based method		Proposed method	
			Item	Recall	Hot tags	Recall
1	Brajeshwar	864	4	0.005	26	0.030
2	Tuluum	246	9	0.037	24	0.098
3	Riverred	118	1	0.009	21	0.178
4	aries_hu	270	4	0.015	16	0.059
5	jamieolender	118	4	0.034	20	0.169
6	askjimcobb	182	3	0.016	12	0.065
7	valeriovillari	70	1	0.014	18	0.257
8	smoky8	280	1	0.004	19	0.068
9	NOLArising	26	0	0.000	2	0.077
10	Tucats	83	3	0.036	7	0.084
11	Fakonig	268	2	0.007	25	0.093
12	Winehiker	677	10	0.015	25	0.037
13	Jtfmulder	73	2	0.027	8	0.110
14	Blogindonesia	78	2	0.026	12	0.154
15	Leonbasin	485	5	0.010	29	0.060
16	Xian	563	4	0.007	29	0.051
17	Donssite	119	1	0.008	8	0.067
18	Clixpert	10	1	0.100	2	0.200
Average		251	3	**0.020**	16	**0.103**

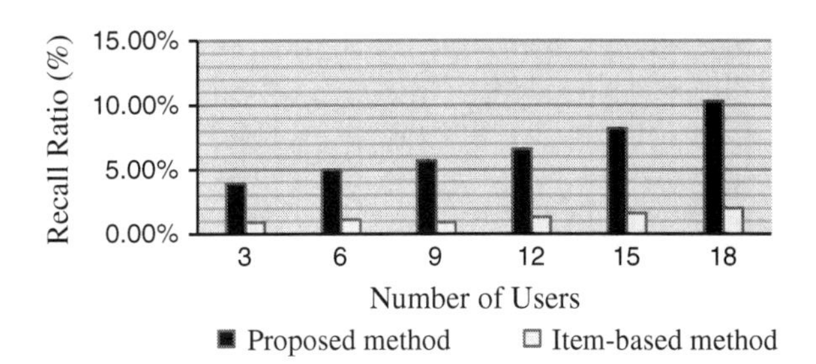

Fig. 4 Comparison of the two methods according to the number of users

Conclusions

Recently, community-based websites have dramatically developed. However, there are still several challenges in recommending pertinent contents to users in SNS, because existing methods mainly focus on the utilization of user interests that are in user account information, and not user characteristics. Therefore, we proposed an enhanced content recommendation method in online social network sites, using FOAF and SNA. The method extracts personal profiles and tags from FOAF in community-based websites, analyzes them, and then provides users with recommendations of contents. Furthermore, the method makes it easier to share contents with much enriched personal information, such as user characteristics, and tags in a social network.

For the verification of the proposed method, we collected the data over 2,676 nodes from three different websites, and then evaluated the performance of the method, using the hit ratio and recall ratio. From the experimental result, we showed that the suggested method has a better performance in recall ratio, than the item-based recommendation method. Moreover, we determined that the quality of contents rises proportionally, as the number of users increase. Therefore, the proposed method increases the reliability of recommended services, and the service providers are able to suggest more relevant contents for users. Additionally, the method is directly applicable to real social network services, blogs, etc., and feasible for sharing contents based on personalized recommendation.

For future research directions, we will conduct more experiments with a much larger size of social network, extract more characteristics of each user, and utilize them. We believe that these efforts can contribute to social network service providers, as well as a variety of other fields, such as e-commerce, e-business, and online marketing.

References

1. Duggan M, Brenner J (2013) The demographics of social media users: 2012, PewResearchCenter
2. Scott J (1991) Social network analysis. Sociology 22(1):109–127
3. Musiał K, Kazienko P, Kajdanowicz T (2008) Social recommendations within the multimedia sharing systems. WSKS'08, LNAI 5288:364–372
4. Zhou J, Luo T (2010) A novel approach to solve the sparsity problem in collaborative filtering. In: International conference on networking, sensing and control, pp 165–170
5. Jeong B, Lee J, Cho H (2009) An iterative semi-explicit rating method for building collaborative recommender systems. Expert Syst Appl 36(3):6181–6186
6. Heitmann B, Hayes C (2010) Using linked data to build open, collaborative recommender systems. Association for the advancement of artificial intelligence
7. Hwang S, Wei C, Huang Y, Tang Y (2010) Combining co-authorship network and content for literature recommendation. In: Pacific Asia conference on information systems
8. Li X, Guo L, Zhao Y (2008) Tag-based social interest discovery. In: International World Wide Web conference committee
9. Deshpande M, Karypis G (2004) Item-based top-n recommendation algorithms. ACM Trans Inf Syst 22:143–177
10. Ali-Hasan N, Adamic LA (2007) Expressing social relationships on the blog through links and comments. In: Proceedings of international conference on Weblogs and Social Media
11. Schwartz MF, Wood DCM (1993) Discovering shared interests using graph analysis. Commun ACM 36(8):78–89
12. Wang P, Ye H (2009) A personalized recommendation algorithm combining slope one scheme and user based collaborative filtering. In: International conference on industrial and information systems
13. Hassan A, Radev D, Cho J, Joshi A (2009) Content based recommendation and summarization in the blogosphere. In: 3rd international ICWSM conference

14. Campos LMD, Fernández-Luna JM, Huete JF, Rueda-Morales MA (2010) Combining content-based and collaborative recommendations: a hybrid approach based on Bayesian networks. J Approximate Reasoning 51(7):785–799
15. Mannens E, Coppens S, Pessemier T, Dacquin H, Deursen D, Sutter R, Walle R (2013) Automatic news recommendations via aggregated profiling. Multimedia Tools Appl 63:407–425
16. Bogers, T, Bosch AVD (2009) Collaborative and content-based filtering for item recommendation on social bookmarking websites. In: ACM workshop on recommender systems and the social web
17. Casciaro T (1998) Seeing things clearly: social structure, personality, and accuracy in social network perception. Soc Networks 20:331–351
18. Karypis G (2001) Evaluation of item-based top-N recommendation algorithms. In: 10th international conference on information and knowledge management
19. Lenhart A, Purcell K, Smith A, Zickuhr K (2010) Social media and mobile internet use among teens and young adults. Pew Internet and American Life Project
20. Ji A, Yeon C, Kim H, Jo G (2007) Collaborative tagging in recommender systems. AI 2007, LNAI, vol 4830, pp 377–386

The Design of an Information Monitoring Platform Based on a WSN for a Metro Station

Dong Chen, Zhen-Jiang Zhang and Yun Liu

Abstract Due to the safety and comfort problems of subway stations, we propose a system that can monitor environmental information by constructing a fundamental information monitoring platform based on a WSN. The basic idea is to deploy a large number of sensor nodes with the ability of sensing, computing and communicating in the station to form a wireless network and the information collected by different nodes will be sent in real time to the backend database. Then, we can acquire all kinds of monitoring information from the platform after a series of processing steps conducted by the platform. If any abnormal data ware found, the corrective measures will be taken immediately. The main features of our proposed platform are the design of the system's functional structure and the design of the software and hardware modules. Also, we propose to incorporate a dormancy mechanism to save energy.

Keywords Monitoring platform · Sensor nodes · WSN

D. Chen · Z.-J. Zhang (✉) · Y. Liu
School of Electronic and Information Engineering, Beijing Jiaotong University, Beijing 100044, China
e-mail: zhjzhang1@bjtu.edu.cn

D. Chen
e-mail: 12120202@bjtu.edu.cn

Y. Liu
e-mail: liuyun@bjtu.edu.cn

D. Chen · Z.-J. Zhang · Y. Liu
Key Laboratory of Communication and Information Systems Beijing Municipal Commission of Education, Beijing Jiaotong University, Beijing 100044, China

Y.-M. Huang et al. (eds.), *Advanced Technologies, Embedded and Multimedia for Human-centric Computing*, Lecture Notes in Electrical Engineering 260,
DOI: 10.1007/978-94-007-7262-5_12,

Introduction

The urbanization process has resulted in the rapid development and construction of urban rail transit systems. However, due to the low level of the information and integration, traditional management approaches of construction and operation have not adapted to the requirements of such systems. With more and more rail operation rely on the network, informational and intelligent collection methods are urgently needed to acquire environmental information. Railway stations have encountered many difficulties in the process of constructing the infrastructure of railway stations. The problems associated with the aging of the structure have become increasingly prominent as the service life has been extended. The underground structure of urban subway tunnels is subject to ground water erosion and vibration loading imposed by the trains. So, safety monitoring work, which is difficult to accomplish, has become very important. Because the rail networks are located underground, most stations may have many limits when they are compared with other types of space, Such as insufficient daylight, the lack of fresh air, air pollution, and high temperature [1]. All of these factors have a significant effect on the comfort of the people who use the stations. Thus, work designed to monitor the comfort levels is also important [2].

In order to improve the safety and comfort index of subway stations, we have established an omni-directional, real-time, intelligent digital monitoring platform, based on a wireless sensor network, Java programming, and database technologies, to monitor basic information related to the safety and comfort of rail stations and networks [3]. The establishment of such a platform is important because it will help provide the required information to monitor the rapidly-expanding development of rail systems, enhance the efficiency of the systems, provide a scientific approach that will enhance the management of the systems, and improve the ability to deal with emergences. In addition, the proposed platform also will increase operational safety, improve comfort levels, guarantee the safety of people's lives and property, and promote economic development [4].

Function Design

The structure of the proposed system's function menu is shown in Fig. 1:

Collection of Related Information

The information collected by different monitoring nodes will be sent in real time to the database, and management personnel can log on the screen to check the collected information. With the ability to scroll through updated fundamental

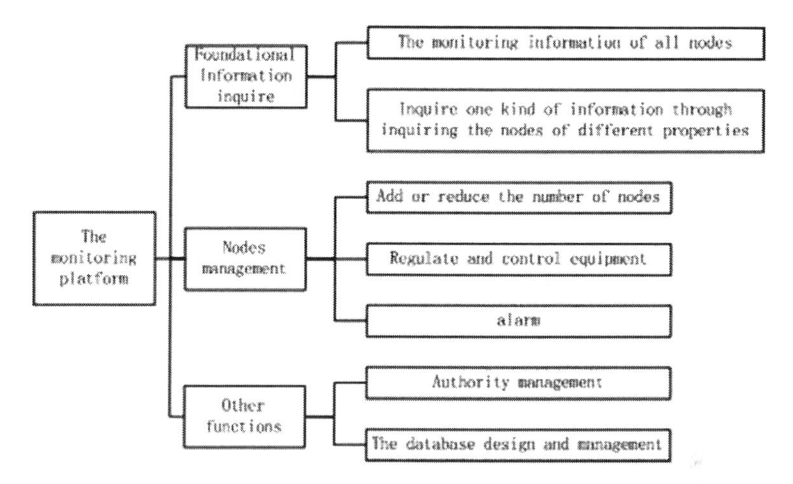

Fig. 1 The structure of the proposed system's function menu

information, management personnel can check and monitor information at any time [5]. There are two ways to make inquiries of the information, i.e., make an overall inquiry, which does not consider the properties of the nodes according to the time sequence of the data that are transferred and make an inquiry for only one kind of information, which focuses on one specific kind of information from the nodes that are being monitored, according to the different properties of the nodes [6]. For example, if users wish to focus on information concerning the temperatures in the system, they can check only the node's temperature information.

Node Management

The requirement of data accuracy is not always constant for the monitoring platform. For example, highly accurate data will be required when the metro is operating in one of its peak period. Conversely, the accuracy of the data can be somewhat less when there are fewer passengers using the metro. Based on the above analyses, the number of working nodes can be adjusted flexibly according to the different accuracy requirements of the data. This approach can reduce costs and save the node's energy. In addition, it can prolong the service life of the node. Control nodes can regulate equipment when they receive orders from the local control center. If monitoring nodes collect abnormal data or the administrator does not react to such data in a reasonable amount of time, the alarm function will be triggered when the values of the abnormal data value exceed the threshold that has been set.

Other Functions

Authority management

In this system, in order to prevent illegal users from getting into the system and to ensure that only lawful users can access system, we propose the authority management function to ensure that the system can be operated safely. The core mission of authority management is to give users relevant privileges according to different levels of access. User will be authorized for different levels of access, ranging from just visiting to revising data to deleting data. Also, security verification and authority distribution are the major divisions of safety management that can achieve operation that distributes the appropriate levels of authority to users who has entered the management system.

System Architecture

The system consists mainly of some sensor nodes and control nodes, a gateway node, a local monitoring center and a remote control center [7].

Sensor nodes are responsible for collecting the temperature and humidity of the air, the temperature of the operating equipment, light intensity, and other important environmental information. All of these sensor nodes must be deployed appropriately. Some of them will be deployed in a key position in the train, and some will be deployed in the tunnel. Due to limited lifespan of the batteries, the lifespan of the nodes and the communication distances are limited to some extent.

Control nodes are used to receive the control information from the local monitoring center or remote control center. They can control the ventilation and heating systems thereby regulating and controlling the environmental parameters of operational equipment. Control nodes use DC power as the power supply.

Gateway nodes serve as important information transfer stations. They are responsible for transferring the information collected by the sensors to the local monitoring center and for transferring commands to the control nodes. The data acquired by the sensor nodes will be transmitted by multi-hop routing through the other adjacent nodes. In the transmission, the monitoring data may be processed by many nodes, and the information will reach the gateway nodes through multi-hop routing. The gateway nodes will transmit the received data directly to the local monitoring center. Therefore, the computers in the monitoring center will make a series of operations including processing, storage, and analysis. Also, they will execute various kinds of control algorithms and send out corresponding control instructions to enhance the monitoring and regulation of the equipment. The local monitoring center can interact with the remote control center through the Internet, which will facilitate remote monitoring. The diagram of system architecture is shown in Fig. 2:

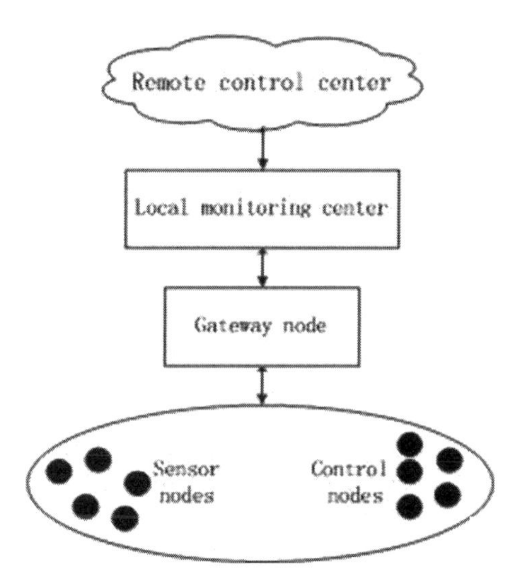

Fig. 2 The system architecture

Software Design

Design of Software for the Sensor Node

The software of the sensor node is used to collect the environmental information and process it simply. Then, Radio Frequency (RF) module transmits the processed data to the gateway nodes through a wireless system. In order to improve the energy efficiency of the nodes, the sensor node's working model is divided into three working states, i.e., dormant, awakened, and normal. In the dormant state, the processor stops working, and the RF module is in a low current-receiving state. After information has been received from the gateway node or neighboring nodes, the sensor node will judge whether it is the destination node or not. If it is, the sensor node will convert into the working state; if it is not, the sensor node will forward the information and return to the dormant state again. The flow diagram for the software that is used to control the sensor nodes is shown in Fig. 3.

Design of the Software for the Gateway Node

Gateway nodes have many functions, Such as creating a wireless network, receiving the data collected by the sensor nodes, setting the properties of the network nodes, transmitting data to the local monitoring center, receiving commands from the monitoring center and forwarding them the commands to control nodes. Figure 4 shows the design of the software for the gateway nodes.

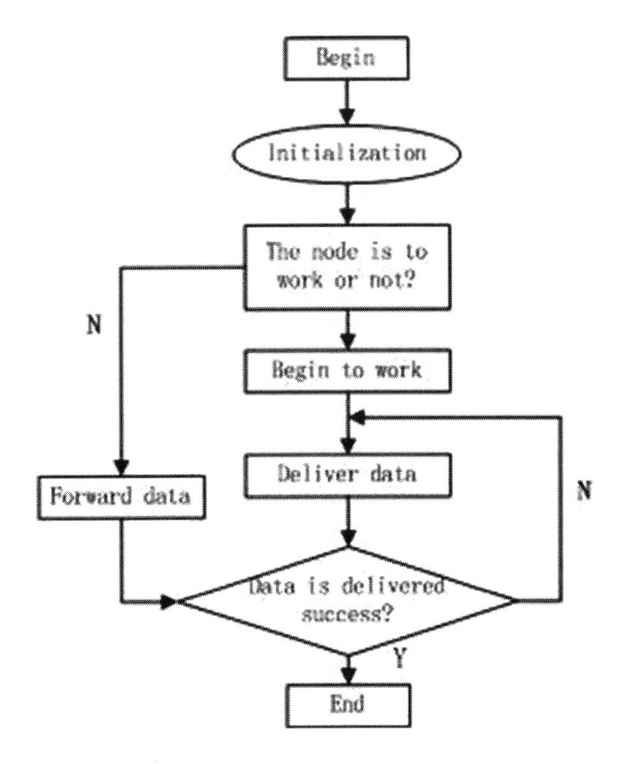

Fig. 3 Flow diagram for the software

Hardware Module

Selection and Management of the Sensor Nodes

The hardware part of a sensor node includes the sensor module, a microprocessor, and a wireless communication module, the power module, and storage and display equipment.

For the microprocessor and the wireless communication module, the TI's system-on-a-chip: CC2430 WAS selected, because it can improve performance and meet the requirements of low cost and low power consumption when using the 2.4 GHz ISM band based on ZigBee. CC2430 has a structure in which ZigBee RF, memory and the microcontroller are integrated into a signal chip. It uses a MCU(8051) of eight bits, 32/64/128-kb programmable flash memory, 8-kb RAM, and it also contains an ADC, several Timers, AES128 collaborative processer, a 32-kHz timer of crystal oscillator of the sleep mode, Power on reset, Brown out detection and 21 programmable I/O pins. But the sensor nodes are made of various kinds of sensor chips. Sensor modules can sense temperature, humidity, and the concentration of carbon dioxide. Temperature and humidity sensors use the I2C bus digital sensor, SHT75. It has two major advantages, i.e., small size and low energy consumption, and its temperature range and accuracy are $-40\mathrm{T} \sim 123\mathrm{T}$

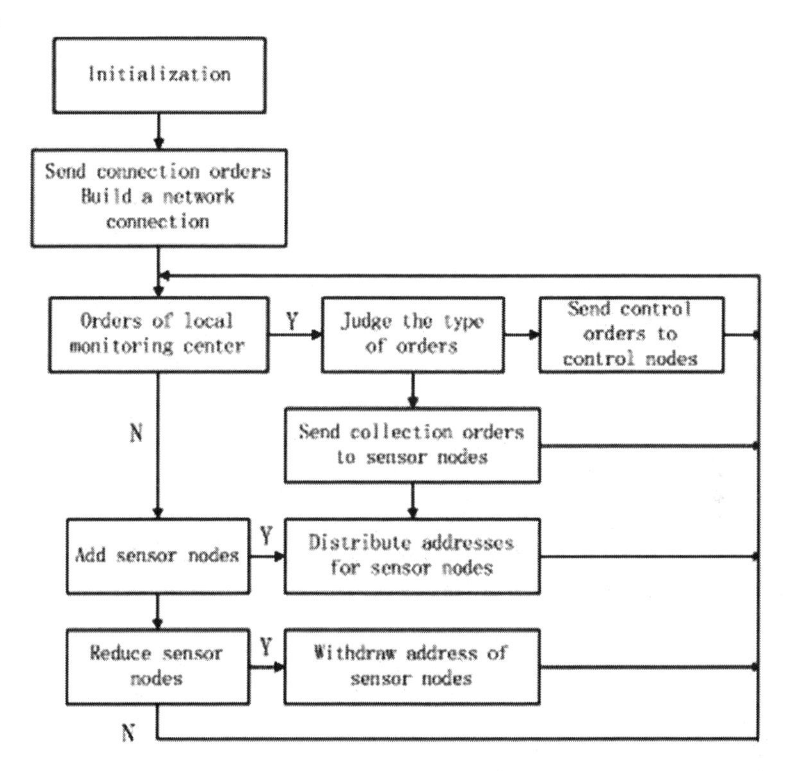

Fig. 4 Software flow diagram of the gateway nodes

centigrade and ±0.3T centigrade, respectively, whereas the relative humidity range and accuracy are 0–100 % and ±1.8 TRH respectively.

The carbon dioxide sensor uses a metal oxide semiconductor CO_2 sensor, i.e., SB-AQ6A (The FIS's product). The analytical range of the SB-AQ6A for CO_2 is 400–3,000 ppm and it is used to control concentrations in the air conditioning and ventilation systems. Furthermore, it has the advantages of low cost, no maintenance, and a longer service life than optical analyzers.

Selection and Management of the Control Node

The hardware in the control node consists mainly of a microprocessor, a wireless communication module, the power module, and the driving module. The control node has the same type of micro-processer and wireless communication module as the sensor node. The power module uses DC power. The driver module consists mainly of a relay and a photo-electric coupler, and it controls the dehumidifier, air heater, and air cooler through the relay module when it receives orders from the monitoring center.

Selection and Management of the Gateway Node

The gateway node includes mainly a microprocessor, a wireless communication module, a power module, a storage module, RJ45 ethernet interface module, an RS232 string-line interface module, and a USB interface module.

Management of the Energy of the Nodes

In the research described in this paper, the sensor nodes used 3.3-V batteries to supply power. The service life of the sensor and the transmission distance are restricted by the capacity of the batteries, so power consumption is an important consideration [8]. Thus, a dormancy mechanism can be used to reduce the power requirements; this mechanism closes the wireless communication module and the data acquisition module to save energy when the sensor nodes have no information collection tasks and are not forwarding data for other nodes. In this dormancy mechanism, only a neighboring area of the sensor nodes is active when a sensor task occurs, and the active area will move along with the data transmitted to the gateway node, as this occurs, the active nodes in the last active area can return to the dormancy mode to save energy.

Application Prospects

Wireless sensor networks, combined with a variety of advanced technologies—provides a new approach for obtaining process information. Also, they can provide data that researchers need, automatically and in real time. In addition, they have no adverse impacts on the normal operation of the train and the convenience of this mode of travel. Furthermore, the monitoring data are relatively accurate.

So, wireless sensor networks are a feasible way to increase the use of digital processes and equipment to assess the conditions on a subway train, thereby improving the operational safety of the train as well as the comfort level of the passengers.

In this paper, we reported the development of an information monitoring platform based on a wireless sensor network to significantly improve the conditions associated with rail traffic. Thus, the proposed system could provide reference for the use of WSNs to enhance the comfort and safety of rail traffic in China.

Acknowledgments This research is supported by National Natural Science Foundation of China under Grant 61071076, Beijing Science and Technology Program under Grant Z121100007612003, the Beijing Municipal Natural Science Foundation under Grant 4132057.

References

1. Mo L, Yunhao L (2007) Underground structure monitoring with wireless sensor networks. In: Proceedings of 6th international symposium on information processing in sensor networks, 2007. IPSN 2007
2. Akyildiz IF et al (2002) A survey on sensor networks. Commun Mag IEEE 40(8):102–114
3. Ulema M (2004) Wireless sensor networks: architectures, protocols, and management. In: Network operations and management symposium, 2004. NOMS 2004. IEEE/IFIP
4. Chen CW, Wang Y (2008) Chain-type wireless sensor network for monitoring long range infrastructures: Architecture and protocols. Int J Distrib Sens Netw 4(4):287–314
5. Pereira V et al (2011) A taxonomy of wireless sensor networks with QoS. In: New technologies, mobility and security (NTMS), 2011 4th IFIP international conference
6. Sharma P, Bhadana P (2010) An effective approach for providing anonymity in wireless sensor network: detecting attacks and security measures. Int J Comput Sci Eng 1(5):1830–1835
7. Sikka P, Corke P, Overs L (2004) Wireless sensor devices for animal tracking and control. In: Local computer networks, 2004. 29th annual IEEE international conference
8. Lee DS, Liu YH, Lin CR (2012) A wireless sensor enabled by wireless power. Sensors 12(12):16116–16143

An Improved Social Network Analysis Method for Social Networks

Jongsoo Sohn, Daehyun Kang, Hansaem Park, Bok-Gyu Joo and In-Jeong Chung

Abstract Recently, Social Network Service (SNS) users are rapidly increasing, and Social Network Analysis (SNA) methods are used to analyze the structure of user relationship or messages in many fields. However, the SNA methods based on the shortest distance among nodes is time-consuming in measuring computation time. In order to solve this problem, we present a heuristic method for the shortest path search using SNS user graphs. Our proposed method consists of three steps. First, it sets a start node and a goal node in the Social Network (SN), which is represented by trees. Second, the goal node sets a temporary node starting from a skewed tree, if there is a goal node on a leaf node of the skewed tree. Finally, the betweenness and closeness centralities are computed with the heuristic shortest path search. For verification of the proposed method, we demonstrate an experimental analysis of betweenness centrality and closeness centrality, with 164,910 real data in an SNS. In the experimental results, the method shows that the computation time of betweenness centrality and closeness centrality is faster than the traditional method. This heuristic method can be used to analyze social phenomena and trends in many fields.

J. Sohn (✉)
Service Strategy Team, Visual Display, Samsung Electronics, Suwon, South Korea
e-mail: jongsoo.sohn@samsung.com

D. Kang · H. Park · I.-J. Chung
Department of Computer and Information Science, Korea University, Seoul, South Korea
e-mail: internetkbs@korea.ac.kr

H. Park
e-mail: park11232000@korea.ac.kr

I.-J. Chung
e-mail: chung@korea.ac.kr

B.-G. Joo
Department of Computer and Information Communications, Hongik University, Seoul, South Korea
e-mail: bkjoo@hongik.ac.kr

Y.-M. Huang et al. (eds.), *Advanced Technologies, Embedded and Multimedia for Human-centric Computing*, Lecture Notes in Electrical Engineering 260,
DOI: 10.1007/978-94-007-7262-5_13, © Springer Science+Business Media Dordrecht 2014

Keywords Social network (SN) · Social network analysis (SNA) · Betweenness centrality · Closeness centrality · Heuristic approach

Introduction

Recently, online social network services are becoming popular with users, along with the expansion of Web 2.0-based services and the widespread use of smart devices. Online SNSs are online community services which enables users to communicate with each other, share information, and expand their human relationships [1]. In an SNS, each relation between users is represented by a simple graph, which consists of nodes and edges. As online SNS users are increasing rapidly, SNSs are actively utilized in enterprise marketing, analysis of social phenomena, trends, and so forth [2, 3].

Meanwhile, SNA is a way of analyzing social relationships among users in an SN. Through the SNA, it is possible to measure relationships between members, degree of intimacy, and intensity of connection, and to detect communities. The following are conventional SNA methods: degree centrality, betweenness centrality, and closeness centrality [4]. In the degree centrality analysis, the shortest path is not considered; however, it is used as a crucial factor in betweenness centrality, closeness centrality, and other SNA methods [4]. In previous works, the computation time was not time-consuming, due to the small size of the SN [5]. But, finding the path needs significant time to process data, since the number of nodes consists of online SNSs. For instance, if the number of nodes in an online SNS is n, the maximum number of its link is $n(n-1)/2$. This indicates that it is too expensive to analyze an SN; for example, if the number of nodes is 10,000, the number of links is 49,995,000.

Therefore, we propose a heuristic method for searching the shortest path among users in an SN graph. Moreover, we devise an enhanced method with addition of the best-first search, to reduce the computation time, and search the path rapidly, in an online SNS of huge size.

To verify the proposed method, we crawled 160,000 user IDs from online SNSs and constructed a graph out of them. Then, we compared this with previous methods, which are the best-first search and breadth-first search, in the time taken to search nodes and analyze SNSs. The suggested method took 240 s (7.4 times faster) to search nodes, where the breadth-first search took 1,781 s. Moreover, the method for SNA is 6.8 times and 1.8 times faster than the betweenness centrality analysis and closeness centrality analysis, respectively.

The suggested method shows the possibility of analyzing a large sized SN with better performance in time. Consequently, the method improves the efficiency of SNA and is used to determine social trends or phenomena.

Related Works

Social Network Analysis

Due to the popularity of Web 2.0 and the wide propagation of smart devices, there has been a diversity of research studies using SN [6, 7]. SNA represents the relationships between users in a graph. The most popular methods of SNA are degree centrality, closeness centrality, and betweenness centrality. Degree centrality is the index to figure out the importance in the whole SN, by measuring a node directly connected to other nodes [8]. Degree centrality computes the related degree of other users to a user v, when there is the user v of n users. It is represented by Eq. (1).

$$C_D(v) = \frac{deg(v)}{n-1} \quad (1)$$

Closeness centrality represents how closely related one user is to another user. If there are two users P_i and P_k of n users, this formula is represented by $C_C(P_k)$, as shown in Eq. (2).

$$C_C(P_k) = \left[\sum_{i=1}^{n} d(P_i, P_k)\right]^{-1} \quad (2)$$

In (2), the subscript C is an abbreviation of the word Closeness, $d(P_i,\ P_k)$ is the number of the shortest path from node i to node $k(i \neq k)$, and n is the number of users. Betweenness centrality is an index to measure how well a node performs as a mediator in the SN. It is computed by formula (3). In (3), the subscript B of C_B is Betweenness, and δ_{st} is the number of the shortest path between two nodes s and t. $\delta_{(v)}$ is the number of passing v in the path. If the betweenness centrality is large, it has an effect on the information flow in the SN.

$$C_B(v) = \sum_{s \neq v \neq t \in V} \frac{\delta_{(v)}}{\delta_{st}} \quad (3)$$

The Shortest Path Search Method

Breadth-First Search (BFS) [8] and Depth-First Search (DFS) [9] are ways to transform a graph to state space, in order to explore the shortest path from the starting node to the goal node. BFS is a strategy to search nodes from nodes located at a close level, to nodes located at a distant level, in sequence. DFS is a recursive search method, which explores from the start node, to the node of distant level. DFS has less efficiency on a huge graph in that it needs to know how to search each node level.

The Best-First search is based on a heuristic method, to measure the shortest path between two nodes in graph. The Best-First search is an algorithm using a heuristic method, which is formalized by human experience. In [5], the authors proposed a method to estimate the shortest path in a large graph, using Monte Carlo Simulation. The proposed method in [5] measures the distance of two nodes, by comparing complete enumeration and Monte Carlo sampling in a large graph, which consists of 10,000 nodes. However, the method of [5] has a limitation on being applied to an SN, which graph changes frequently, because it requires the process of extracting a model. Another method in [10] suggested a self-organizing strategy, to detect the shortest path on complex networks. Likewise, there are the same shortcomings, as described in [5, 11].

Improved Social Network Analysis Method

Closeness centrality and betweenness centrality, excepting degree centrality, in the typical SNA method are based on computation of the shortest path [12]. Thus, much time to compute closeness centrality and betweenness centrality is required in a large SN. Due to this problem, we came up with a method to add preprocessing steps with the Best-First search method, in order to compute the shortest path in an online SN. In an online SN, only a few users have a great number of connections; however, most other users have few connections. Hence, the users who have many connections perform a hub role in the SN [13]. Thus, finding users who have many connections in advance is desirable for the enhancement of search probability, since those are more influential on others in the SN. For this reason, the degree of node v_n is used as a heuristic evaluation function in a graph $G = (V, E)$, consisting of a set V of node v and set E of connections, when there are an existing n users, such as formula (4).

$$f(n) = \text{The number of degree}(v_n) \tag{4}$$

If a heuristic evaluation function is applied to the best-first search, such as formula (4), the shortest path is computed with comparison to a BFS. The heuristic evaluation function as shown in formula (1), however, does not rapidly increase path search performance, because its worst case occurs when nodes are skewed in state space. Thus, in this paper, the best-first search is utilized, after setting the root node of a skewed tree as a sub goal, if the search path contains a skewed tree.

Figures 1 and 2 show phrases of the proposed shortest path method. In Fig. 1, 'Start' and 'Goal' indicate a start node and a goal node, respectively. The values in parenthesis of each node are the evaluation value gained through formula (4) for each node. Before finding the path between 'Start' and 'Goal', a pertinent node is assigned in a search of the last part of a skewed tree, if the goal node is located in a

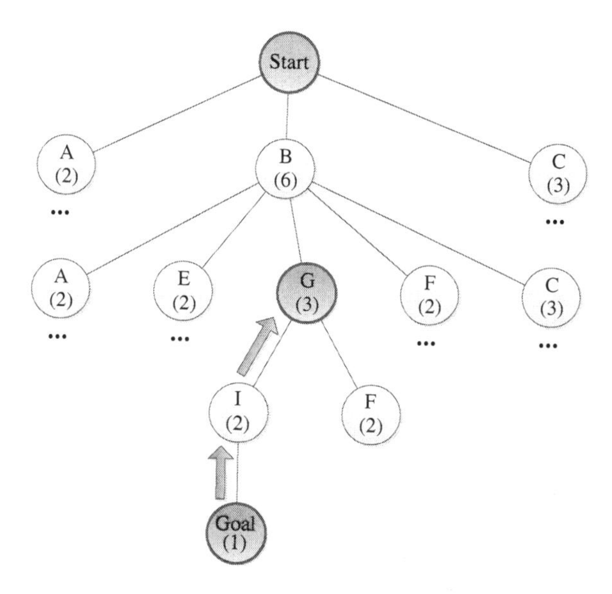

Fig. 1 Selecting a sub goal

leaf node of the skewed tree. Next, the best-first search with the number of degrees as evaluation function is used for finding the shortest path, as depicted in Fig. 2.

Table 1 is the proposed algorithm for searching the shortest path. The path can be detected quickly by preferential search of a hub user in large SN, when using the proposed shortest path search method. However, our method differs from complete enumeration, in terms of accuracy.

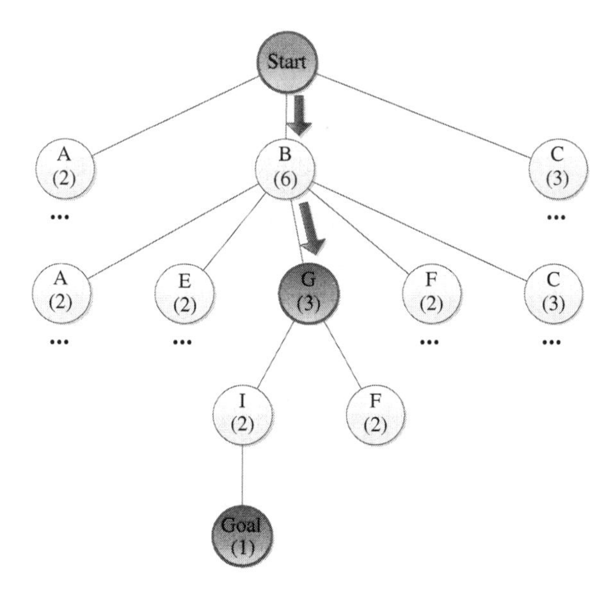

Fig. 2 Finding the shortest path

Table 1 An improved shortest path finding algorithm

```
Algorithm: Modified_Best_First_Search
Input: start_Node, goal_Node
Output: path

1. call module preprocessing(goal_Node)
2. open := [start_Node]
3. closed := [ ]
4. while open != [ ] do
5.  if there is no list in open then return fail
6.  move first node of open to closed, call it X
7.  if X == goal_Node then return path
8.  else if X == subgoal_Node then
9.      path = path + subPath
10.         return path
11.     else
12.         generate children of X
13.         for each children of X do
14.         case
15.             child is not on open or closed
16.                 evaluate the child by heuristic function // in Equation (4)
17.                 add the child to open
18.             child is already on open
19.                 if the child was reached by a shorter path
                    then give the state on open the shorter path
20.             child is already on closed
21.                 if the child was reached by a shorter path
                    then remove the state from closed and add
                    the child to open
22.                  put X on closed
23.                  reorder states on open by evaluated value

24. module preprocessing(goal_Node) // Search and select a subgoal_Node if the
goal_Node is located in a terminal node of a skewed tree, as depicted in Fig. 1
25.         subPath := [ ]
26.         if the degree of parent == 2 then
27.             add the parent to subPath
28.             goal_Node = parent
29.             call module preprocessing(goal_Node)
30.                 else
31.                     subgoal_Node = parent
32.                     return subgoal_Node
```

Evaluation

We conducted experiments to validate the SNA efficiency of the proposed shortest path search method. A computer with Intel Core2duo CPU, 2 GB RAM, and Java as the development toolkit, with a MySql database server, are used for the implementation.

First, we collected 164,910 SNS users for a dataset from Twitter,[1] as shown in Fig. 3. Most users have from one to five links; a few of them have more than five links. We converted the collected datasets into a relational database table, as

[1] http://www.twitter.com

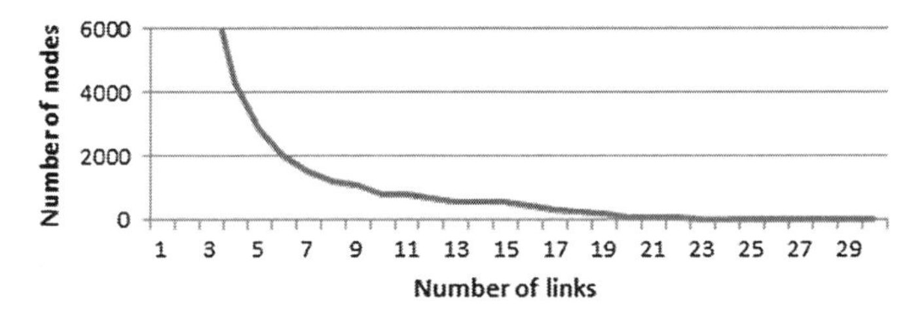

Fig. 3 Distribution of collected social network

demonstrated in Table 2. In Table 2, 'Id' is a unique identification number of each row, 'User' is a twitter user, and 'Following' is a user who follows the 'User'. For instance, if a user follows another ten users, then ten rows are inserted. Next, we designed a java program to accomplish the BFS algorithm and the proposed method, and then selected a pair of 100 arbitrary nodes, to compare the average search performance time.

We compared the proposed shortest path search method and BFS method on the SN, as described in Table 3. As the BFS method gradually searches from the first node to near-level node, it demonstrates the most accurate shortest path. In addition, as the suggested method searches nodes that have a lot of edges, it extracts the same or larger paths better, than those of the BFS method.

The BFS method always calculates the optimal shortest path, because it enumerates every possible path, to find the shortest one. On the other hand, the proposed method lists several paths selectively, using the heuristic evaluation function, so that it cannot always assure optimal values. Our proposed method can compute an average shortest path of about 80 %, compared to that of the BFS method. The method searches nodes less than those of the BFS method, so the average searching time is much faster, by about 7 times.

In addition, we calculated the betweenness centrality and closeness centrality, using performance comparison of the SN datasets, as depicted in Table 4. This shows the search results for both the BFS method and the proposed searching

Table 2 Social network database table

Id	User	Following
1	1,142,682	384,752
2	1,142,682	957,841
3	384,752	248,571
…	… omitted…	…

Table 3 Performance comparison of finding the shortest path

Algorithm	Mean path	Mean search	Mean time
BFS	2.45	83,069	1,781.69
Proposed	2.97	38,062	240.86

Table 4 Performance measurement of both the BFS and the proposed method

Algorithm	Criteria	BFS	Proposed	± Ratio
Betweenness centrality	Average time (s)	153,474	22,331	6.8 times
	Average number of searched nodes	255.33	181.36	−1.4 times
Closeness centrality	Average time (s)	109,336	59,418	1.8 times
	Average number of searched nodes	217.92	169.34	−1.3 times

method, to calculate the betweenness centrality and closeness centrality. In Table 4, the time to compute the betweenness centrality improved by about 6.8 times, but the number of searched nodes decreased by about 1.4 times. Besides, the time to compute closeness centrality improved by about 1.8 times, but the number of searched nodes decreased by about 1.3 times.

Conclusion

As web services are based on Web 2.0 and Social Web, online SNS users are gradually increased. At the same time, SNS posts share information between themselves, and impact on various fields. Then, many researchers in a variety of areas, such as Sociology, Economics, Politics, etc. have attempted to analyze SNSs. Since most real online SNSs consist of huge amounts of nodes, however, it is difficult for diverse SNA methods to be applied to real SNSs.

In this paper, we suggested a shortest path search method to improve the time performance for SNA, such as betweenness centrality, closeness centrality, and degree centrality. In other words, we presented a modified heuristic method, which is appropriate to online SNS. Though the method has less accuracy than BFS by about 80 %, it can compute closeness and betweenness centralities more than 7 times faster. Moreover, we can improve betweenness and closeness centrality analysis efficiency by 6.8 times and 1.8 times respectively.

Using our heuristic method, we can apply various SNA methods to large datasets. The method can also be used to determine social phenomena, user trends, and political traits in various fields, such as Politics, Sociology, Economics, etc.

References

1. Ellison NB, Steinfield C, Lampe C (2007) The benefits of facebook "friends:" social capital and college students' use of online social network sites. J Comput Mediated Commun 12(4):1143–1168
2. Kwak H, Lee C, Park HS, Moon S (2010) What is twitter, a social network or a news media? In: Proceedings of the 19th international conference on world wide web, North Carolina, USA, ACM, pp 591–600
3. Sohn JS, Chung IJ (2013) Dynamic FOAF management method for social network in the social web environment. J Supercomput

4. Otte E, Rousseau R (2002) Social network analysis: a powerful strategy, also for the information sciences. J Inf Sci 28(6):441–453
5. Huh MH, Lee YG (2011) Applying Monte-Carlo method in social network analysis. Appl Stat Res 24(2):401–409
6. Cho ID, Kim NK (2011) Recommending core and connecting keywords of research area using social network and data mining techniques. J Intell Inf Syst 17(1):127–138
7. Kim HK, Choi IY, Ha KM, Kim JK (2010) Development of user based recommender system using social network for u-healthcare. J Intell Inf Syst 16(3):181–199
8. Newman MEJ, Girvan M (2004) Finding and evaluating community structure in networks. Phys Rev E 69(2):26–113
9. Hummon NP, Doreian P (1990) Computational methods for social network analysis. Soc Netw 12(4):273–288
10. Shen Y, Pei WJ, Wang K, Wang SP (2009) A self-organizing shortest path finding strategy on complex networks. Chin Phys B 18(9):3783
11. Nan D, Wu B, Pei X, Wang B, Xu L (2007) Community detection in large-scale social networks. In: Proceedings of joint 9th WEBKDD and 1st SNA-KDD
12. Ahmet ES, Kamer K, Erik S, Umit VC (2013) Incremental algorithms for network management and analysis based on closeness centrality. In: Proceedings of arXiv:1303.0422v1
13. Heer J, Boyd D (2005) Vizster: visualizing online social networks. Information visualization. In: Proceedings of IEEE symposium on INFOVIS

Part II
Special Session

Power-Saving Scheduling Algorithm for Wireless Sensor Networks

Ting-Chu Chi, Pin-Jui Chen, Wei-Yuan Chang, Kai-Chien Yang and Der-Jiunn Deng

Abstract Due to advances in communications technology in recent years, prompting more extensive application of wireless sensor networks. Such sensors are limited in battery energy supply and produce the energy hole problem, this even causes paralysis of part of the system. In the cluster architecture, burden of the cluster head is bound to become the energy consumption of the maximum point, so our method is focused on reducing energy consumption of the cluster head. To this end, we propose a new scheduling mechanism based on cluster architecture. In this mechanism, we use the "polling" method to make the cluster head have an absolutely effective data receiving. In addition, we also introduced the "sleeping" mechanism to ensure that the cluster head can achieve power saving under the premise of the most effective data receiving.

Keywords Power saving · Wireless sensor networks

T.-C. Chi · P.-J. Chen · W.-Y. Chang · K.-C. Yang · D.-J. Deng (✉)
Department of Computer Science and Information Engineering National Changhua, University of Education, Taiwan, China
e-mail: djdeng@cc.ncue.edu.tw

T.-C. Chi
e-mail: a22533884@hotmail.com

P.-J. Chen
e-mail: jeffery2936@hotmail.com

W.-Y. Chang
e-mail: v123582@gmail.com

K.-C. Yang
e-mail: as5374942@hotmail.com

Y.-M. Huang et al. (eds.), *Advanced Technologies, Embedded and Multimedia for Human-centric Computing*, Lecture Notes in Electrical Engineering 260,
DOI: 10.1007/978-94-007-7262-5_14,

Introduction

In wireless sensor networks, a large number of sensor nodes need to sense data sent to the data collection point, therefore, many protocol for data transmission were been developed. And the easiest way is the direct transmission protocol [1]. This protocol need to consider the distance from sensing node and the base station. Energy consumption will be generated when sensor nodes send data directly to the data collection point. The distance between sense node and data collection point will affect the energy for sensing nodes to transmit data, thus limits the scope of transmission. In order to increase the scope of transmission, a way to transfer information by node relaying in multiple-hop was generated, which is called the multi-hop transmission [2]. Although these two transmission purposes are the same to send data to the data collection point, but the latter can increase the effective sensing scope.

Multiple-hop architecture can be subdivided into cluster-based architecture, chain-based architecture and tree-based architecture to collect data. In multiple-hop transmission method, the nodes far from data collection point will send data to the nodes closer to the base station first, this way can avoid a large amount of energy-consuming because of the transmission from distant distance [3]. However, the nodes closer to the base station need to relay the data from the other nodes very frequently, therefore, the energy consumption will be particularly high. When the loss of these nodes happened, the transmission distance of other nodes will be increased, and the energy consumption will also be increased, these will cause the fail of data transmission. As a result, it will lead to the partial sensing node lose effectiveness, even more, the energy hole problem will be generated. As the following diagram shown in Fig. 1, the energy consumption

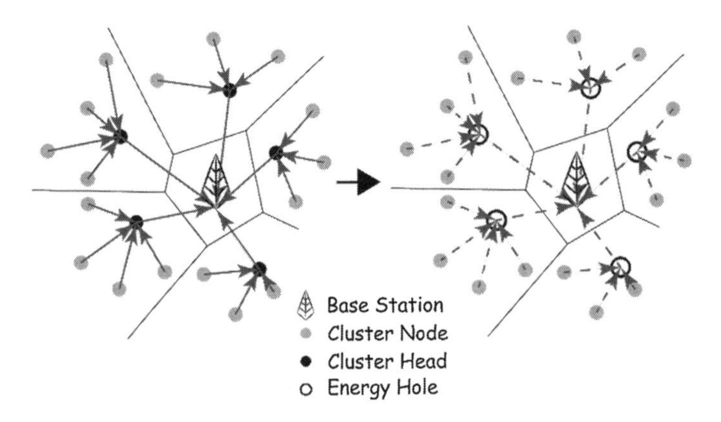

Fig. 1 Energy hole

of the overall system will be extremely uneven, and leading to the shortening of life-cycle [4].

The energy hole problem is difficult to be completely avoided, only to defer the occurrence of this problem. Thus, as much as possible to make the energy consumption of each node be balanced to avoid the energy hole problem casing by the loss of partial sensing nodes. That is, to extend the life cycle of the system by making the energy be balanced [5].

In this paper, we use the cluster's infrastructure architecture. In this architecture, cluster head usually become the point which has the heaviest energy load, and it is easily to cause becoming of the energy hole. So we mainly investigate the problem of the energy consumption of the cluster head. We take a particular scheduling method to make cluster head can achieve the optimal power saving schedule under the premise that satisfying the time tolerance.

Networkmodel

We use cluster-based network architecture, using the homogeneous sensor nodes, but divide them into two classes, a cluster head and several Cluster-Nodes, and doing operation by taking the architecture shown in Fig. 2.

Cluster is a circle, the cluster head as the center, and the distance R is the radius of the circle. The deployment of the nodes is a topology formed by taking random uniform distribution.

Propose Scheme

Initial Phase

The cluster head must tell other nodes in the network that it is the cluster head. Cluster head broadcasts an initial message (Ini-msg) to all nodes in the cluster at

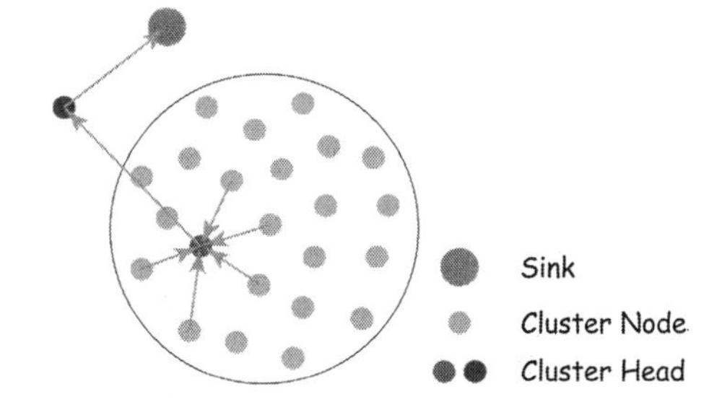

Fig. 2 Cluster-based network architecture

first. The Ini-msg contains an ID of the cluster head and some environmental parameters, such as sample rate, channel rate, etc. After each non-cluster-head node receives the Ini-msg, it transmits a Join-req (join-request) message to ask for joining the cluster by using CSMA/CA protocol. After the cluster head receives the Join-req message, it will base on the relevant environmental parameters to decide whether to accept this request. Once the non-cluster-head node is joined, the cluster head will create a corresponding token buffer.

Running Phase

When the systems are in this stage, it can be divided into four mechanisms:

- **Token Buffer Mechanism**

The cluster head establish token buffer, token buffer products token over sample rate, to let it know that it should receive the data from corresponding node.

- **Detection Mechanism**

Cluster head scans every Token buffer. Once finding the Token, it will add node ID into the Poll list and start scanning again from the first token buffer.

- **Polling Mechanism**

If Poll list is empty, cluster head will keep in sleep mode. If Poll list is not empty, cluster head will switch to active mode, and sent a polling packet to the corresponding cluster node and the node can know it should deliver the data to cluster-head. When cluster node received the polling packet it started to deliver the data, until cluster head reply a power saving poll (PS-Poll) packet and remove the node ID from the Polling list. After the completion of action system will restart Polling Mechanism.

- **Node Processing**

Cluster node keeps in sleep mode and sensing data. When node sensed data, the device will switch into active mode, and receiving the Polling packet from cluster head. Once cluster node receives PS-Polling packet, the node switches to sleep mode again.

Pseudo code

```
Function Polling_List_Creating(){
   WHILE (TRUE){
     head scans all nodes;
     IF (scan == token){
         add nodeID to Polling list;
         pointer move back to the first buffer;}
     ELSE{scan again from the first buffer to the last; }}
 scan again from the first buffer to the last;
}

Function Polling_Sending(){
   WHILE (TRUE){
    IF (Polling list != empty){
       head turns to active mode;
       send polling pkt to node i;}
    ELSE { head turns to sleeping mode; }}
}
```

Simulation

We refer to [6] and [7], our simulation environment is built in a circle. Other parameters are listed in the following (Table 1):

In Fig. 3, we consider that the energy consumption in our method will dynamically increase with the increasing of the node. In this way, we can reduce unnecessary waste by according to the amount of nodes. When there is no node needs to transmit, the head will also turn into sleep mode to conserve battery power. In addition, our approach will no longer receive data in the time of the upper bound of nodes in the system. The reason is that the information are time out, even if it receives is for naught, that is, to reduce energy consumption by avoiding unnecessary receiving. Because of our sleeping mechanism and the filter

Table 1 The variables of simulation

Parameters	
Sample period	30,000 unit time
Poll time/Ack time/data time	5/1/50 unit time
Tolerable time of node	23,000 unit time
Energy consumption of sending/receiving/sleeping	280/204/14 mA
Head power capacity	$1.31072 * 10^{11}$ m Ah
Channel rate	2 Mbps

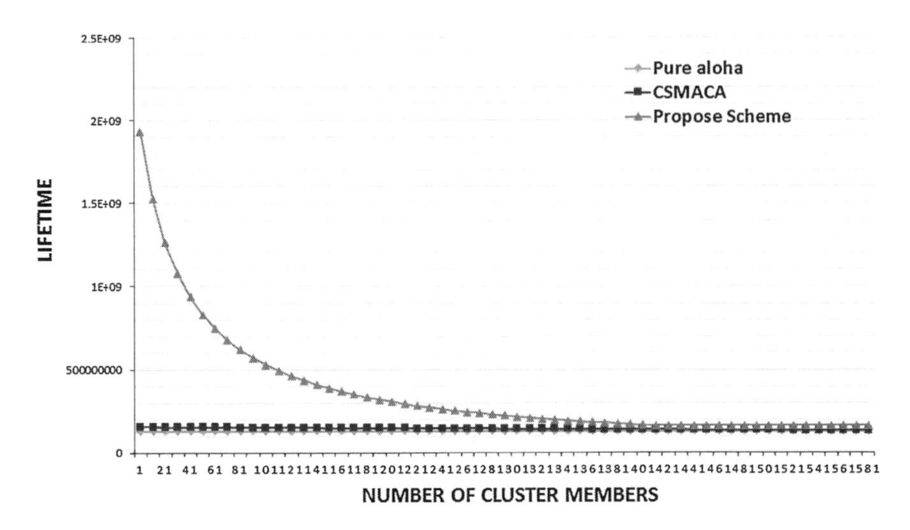

Fig. 3 Simulation of lifetime

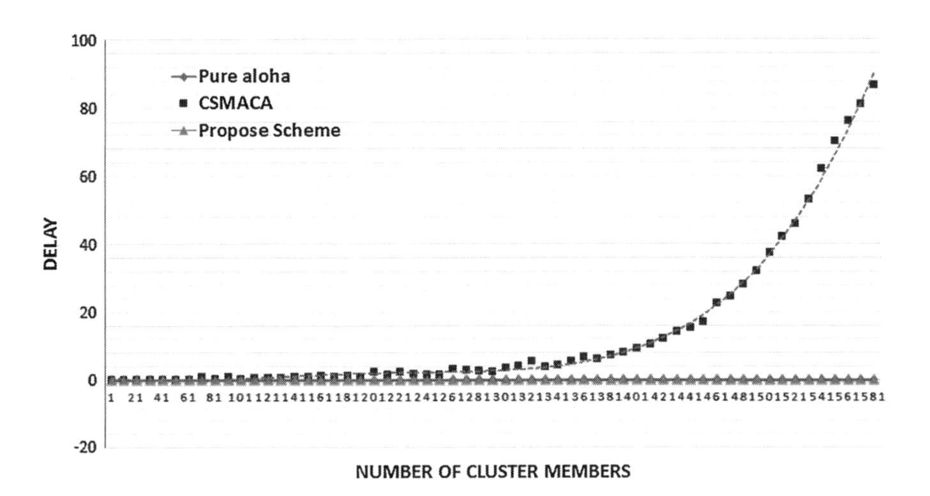

Fig. 4 Simulation of delay

mechanism in the initial step, the power saving is achieved, and the lifetime of the cluster head is also be prolonged.

Figure 4 shows the simulation of delay, the vertical axis represents the delay time of the cluster, and the horizontal axis represents the number of cluster. Because the data transmitting and receiving in cluster in our method will follow the schedule by using our "polling" method to make the cluster head have an absolutely effective data receiving, so that there is no delay in the cluster.

Conclusion

Burden of the cluster head will cause the cluster head more quickly lose energy, and this makes the cluster head naturally become the energy hole location because of its maximum energy consumption in the cluster. Thus, to reduce the energy consumption of the cluster head to be an important issue. And the most important part is to prolong the cluster head's life cycle in which prioritization is key to optimizing the overall performance of the cluster. The "polling" method to make the cluster head has an absolutely effective data receiving. In other hand, the "sleeping" mechanism to ensure that the cluster can provide the most effective data receiving under the premise of saving more power of the cluster.

References

1. Muruganathan SD, Ma DCF, Bhasin RI, Fapojuwo AO (2005) A centralized energy-efficient routing protocol for wireless sensor networks. IEEE Commun Mag 43:S8–S13
2. Noori M, Ardakani M (2008) Characterizing the traffic distribution in linear wireless sensor networks. IEEE Commun Lett 12(8):554–556
3. Guo P, Jiang T, Zhang K, Chen HH (2009) Clustering algorithm in initialization of multi-hop wireless sensor networks. IEEE Trans Wireless Commun 8(12):5713–5717
4. Dietrich I, Dressler F (2009) On the lifetime of wireless sensor networks. ACM Trans Sens Netw 5(1):Article 5
5. Li J, Mohapatra P (2005) An analytical model for the energy hole problem in many-to-one sensor networks. In: Proceedings of IEEE 62nd vehicular technology conference, vol 4, pp 2721–2725
6. Chen B, Jamieson K, Balakrishnan H, Morris R (2002) Span: an energy-efficient coordination algorithm for topology maintenance in ad hoc wireless networks. ACM Wireless Netw 8(5):481–494
7. Bouabdallah F, Bouabdallah N, Boutaba R (2009) On balancing energy consumption in wireless sensor networks. IEEE Trans Veh Technol 58(6):2909–2924

Localization Algorithm for Wireless Sensor Networks

Yin-Chun Chen, Der-Jiunn Deng and Yeong-Sheng Chen

Abstract In recent years, many localization algorithms are proposed for wireless sensor networks because that is crucial to identifying the accurate positions of sensor nodes. This study proposes an analytic localization algorithm by utilizing radical centers. Assume that a target node can measure its distances to four or more anchor nodes. By picking four distance measurements to four anchor nodes, a radical center is computed and treated as the target node location. To further improve and fuse these estimations, effective filtering mechanisms are then proposed to filter out the improper estimations. Afterwards, the remaining radical centers are averaged, and the solution is the final estimation of the target node location. The location errors of the proposed method and the conventional Minimum Mean Square Error method (MMSE) are analytically compared. Extensive computer simulations were carried out and the results verify the advantage of the proposed location algorithm over the MMSE approach.

Keywords: Wireless sensor networks · Localization · Radical centers · Minimum mean square error

Y.-C. Chen · D.-J. Deng (✉)
Department of Computer Science and Information Engineering, National Changhua University of Education, Changhua, Taiwan
e-mail: djdeng@cc.ncue.edu.tw

Y.-C. Chen
e-mail: fpwking32@hotmail.com

Y.-S. Chen
Department of Computer Science, National Taipei University of Education, Changhua, Taiwan
e-mail: yschen@tea.ntue.edu.tw

Y.-M. Huang et al. (eds.), *Advanced Technologies, Embedded and Multimedia for Human-centric Computing*, Lecture Notes in Electrical Engineering 260,
DOI: 10.1007/978-94-007-7262-5_15,

Introduction

Wireless sensor networks (WSNs) have great potential in a lot of control and monitor applications such as data collection, environment observation, and battlefield surveillance, and so on. Since most sensor nodes are randomly deployed without the knowledge of their positions, localization of sensor nodes is an essential issue for the operation, management, and applications of WSNs.

A lot of researchers have proposed many different solutions for the localization problem in WSNs [1–8]. Minimum Mean Square Error (MMSE) estimation method [1, 2] uses least squares solution to estimate the position of the target node. It has been widely studied for 2D localization in wireless sensor networks. However present network environment and technologies demand 3D localization [3–5]. In this paper, a novel range-based localization algorithm in wireless sensor networks is developed. Assume that the target node can measure its distances to four or more anchor nodes. We propose a simple and efficient mechanism for deriving better location accuracy based on the existing technologies for distance measurements using the conventional trilateration approach. The contribution of this study is that based on the existing technologies for distance measurements and without any extra hardware cost, the proposed mechanism provides an efficient algorithm for localization with better accuracy.

Location Estimation Using Trilateration

Trilateration is a common localization algorithm to identify the position of target node by using similar geometric concept of triangulation [9, 10]. Let $A(x_A, y_A, z_A)$, $B(x_B, y_B, z_B)$, $C(x_C, y_C, z_C)$ and $D(x_D, y_D, z_D)$ denote four anchor nodes. The actual distances from the target blind node T(x, y, z) to A, B, C and D are denoted as d_A, d_B, d_C and d_D; whereas, the estimated distances from T to A, B, C and D are denoted as e_A, e_B, e_C and e_D.

Consider node A. Equation $(x - x_A)^2 + (y - y_A)^2 + (z - z_A)^2 = d_A^2$ represents the sphere with center (x_A, y_A, z_A) and radius d_A. Ideally, the target node T(x, y, z) must be a point on the spherical surface. However, in practical implementation, due to the measurement errors, only the estimated distances (e_A, e_B, e_C and e_D) can be derived. Similarly, after also considering nodes B, C and D, we have the following system of equations.

$$\begin{cases} (x - x_A)^2 + (y - y_A)^2 + (z - z_A)^2 = e_A^2 \\ (x - x_B)^2 + (y - y_B)^2 + (z - z_B)^2 = e_B^2 \\ (x - x_C)^2 + (y - y_C)^2 + (z - z_C)^2 = e_C^2 \\ (x - x_D)^2 + (y - y_D)^2 + (z - z_D)^2 = e_D^2 \end{cases} \tag{1}$$

In the realistic case, there will be no pair of coordinates (x, y, z) satisfying (**1**). To tackle this problem, based on the theory of the radical centers of four spheres, an efficient analytic solution to the localization problem that can be formulated as (**1**) is proposed as described as follows.

Localization Utilizing Radical Centers

Related Definitions and Theorems

Definition 1 (*Power of a point*) The power of a point X with respect to a sphere with center O and radius r is defined as $(XO)^2 - r^2$ [12].

Definition 2 (*Radical Plane*) The radical plane of two spheres is the locus of points that have the same power with respect to both spheres [12].

Theorem 1 *For two spheres* A *and* B *whose centers and radii are* (x_A, y_A, z_A), (x_B, y_B, z_B), d_A *and* d_B, *respectively, the equation of the radical plane of spheres* A *and* B *is*:

$$x_A^2 - x_B^2 + y_A^2 - y_B^2 + z_A^2 - z_B^2 - x(2x_A - 2x_B) - y(2y_A - 2y_B) - z(2z_A - 2z_B) = d_A^2 - d_B^2$$

Theorem 2 *The radical planes of four spheres (no two of them are concentric) are concurrent or parallel or coincident* [13, 14]. *The point of concurrence is called the 'radical center' of the four spheres. (For brevity's sake, the proof is omitted.)*

Computation and Selection of the Radical Center

In practical implementation, the coordinates of the radical center can be easily derived by using Cramer's Rule [11], the solution [i.e., the coordinates of the radical center (x', y', z')], is

$$\begin{cases}
x' = \frac{1}{2} \times \frac{\begin{vmatrix} (e_A^2 - e_B^2 - x_A^2 + x_B^2 - y_A^2 + y_B^2 - z_A^2 + z_B^2) & (y_B - y_A) & (z_B - z_A) \\ (e_B^2 - e_C^2 - x_B^2 + x_C^2 - y_B^2 + y_C^2 - z_B^2 + z_C^2) & (y_C - y_B) & (z_C - z_B) \\ (e_C^2 - e_D^2 - x_C^2 + x_D^2 - y_C^2 + y_D^2 - z_C^2 + z_D^2) & (y_D - y_C) & (z_D - z_C) \end{vmatrix}}{K} \\
y' = \frac{1}{2} \times \frac{\begin{vmatrix} (x_B - x_A) & (e_A^2 - e_B^2 - x_A^2 + x_B^2 - y_A^2 + y_B^2 - z_A^2 + z_B^2) & (z_B - z_A) \\ (x_C - x_B) & (e_B^2 - e_C^2 - x_B^2 + x_C^2 - y_B^2 + y_C^2 - z_B^2 + z_C^2) & (z_C - z_B) \\ (x_D - x_C) & (e_C^2 - e_D^2 - x_C^2 + x_D^2 - y_C^2 + y_D^2 - z_C^2 + z_D^2) & (z_D - z_C) \end{vmatrix}}{K} \\
z' = \frac{1}{2} \times \frac{\begin{vmatrix} (x_B - x_A) & (y_B - y_A) & (e_A^2 - e_B^2 - x_A^2 + x_B^2 - y_A^2 + y_B^2 - z_A^2 + z_B^2) \\ (x_C - x_B) & (y_C - y_B) & (e_B^2 - e_C^2 - x_B^2 + x_C^2 - y_B^2 + y_C^2 - z_B^2 + z_C^2) \\ (x_D - x_C) & (y_D - y_C) & (e_C^2 - e_D^2 - x_C^2 + x_D^2 - y_C^2 + y_D^2 - z_C^2 + z_D^2) \end{vmatrix}}{K}
\end{cases} \tag{2}$$

Definition 3 (*Valid set of four anchor nodes*) For four anchor nodes, if their radical center is not at a point of infinity or no intersection, the set of these four anchor nodes is called *valid.*

Theorem 3 *Four anchor nodes* $A(x_A, y_A, z_A)$, $B(x_B, y_B, z_B)$, $C(x_C, y_C, z_C)$ *and* $D(x_D, y_D, z_D)$ *form a valid set iff* $K = \begin{vmatrix} (x_B - x_A) & (y_B - y_A) & (z_B - z_A) \\ (x_C - x_B) & (y_C - y_B) & (z_C - z_B) \\ (x_D - x_C) & (y_D - y_C) & (z_D - z_C) \end{vmatrix} \neq 0$

(For brevity's sake, the proof is omitted.) Theorem 3 provides simple and useful rules to pick proper anchor nodes for computing the radical center.

Further Investigation of the Errors of the Radical Center

To analyze the error between the radical center and actual location of the target node, we can check the values of $x - x'$, $y - y'$ and $z - z'$. Let e and d respectively denote the estimated and actual distance from the target node to an anchor node. Thus, we have $e = d + \varepsilon$, where ε is the measurement error. In this study, it is assumed that $\varepsilon = \rho d$, where ρ is called the *measurement error coefficient*. That is, the inherent location error of the radical center is formulated as (3), (4) and (5).

$$\begin{aligned} x - x' = {} & \frac{\left(\frac{1}{(1+\rho)^2} - 1\right)\left(e_A^2 - e_B^2\right)\left[(y_C - y_B)(z_D - z_C) - (y_D - y_C)(z_C - z_B)\right]}{2K} \\ & + \frac{\left(\frac{1}{(1+\rho)^2} - 1\right)\left(e_B^2 - e_C^2\right)\left[(y_D - y_C)(z_B - z_A) - (y_B - y_A)(z_D - z_C)\right]}{2K} \\ & + \frac{\left(\frac{1}{(1+\rho)^2} - 1\right)\left(e_C^2 - e_D^2\right)\left[(y_B - y_A)(z_C - z_B) - (y_C - y_B)(z_B - z_A)\right]}{2K} \end{aligned} \tag{3}$$

$$\begin{aligned} y - y' = {} & \frac{\left(\frac{1}{(1+\rho)^2} - 1\right)\left(e_A^2 - e_B^2\right)\left[(z_C - z_B)(x_D - x_C) - (z_D - z_C)(x_C - x_B)\right]}{2K} \\ & + \frac{\left(\frac{1}{(1+\rho)^2} - 1\right)\left(e_B^2 - e_C^2\right)\left[(z_D - z_C)(x_B - x_A) - (z_B - z_A)(x_D - x_C)\right]}{2K} \\ & + \frac{\left(\frac{1}{(1+\rho)^2} - 1\right)\left(e_C^2 - e_D^2\right)\left[(z_B - z_A)(x_C - x_B) - (z_C - z_B)(x_B - x_A)\right]}{2K} \end{aligned} \tag{4}$$

$$z - z' = \frac{\left(\frac{1}{(1+\rho)^2} - 1\right)\left(e_A^2 - e_B^2\right)[(y_C - y_B)(x_D - x_C) - (y_D - y_C)(x_C - x_B)]}{2K} + \frac{\left(\frac{1}{(1+\rho)^2} - 1\right)\left(e_B^2 - e_C^2\right)[(y_D - y_C)(x_B - x_A) - (y_B - y_A)(x_D - x_C)]}{2K} + \frac{\left(\frac{1}{(1+\rho)^2} - 1\right)\left(e_C^2 - e_D^2\right)[(y_B - y_A)(x_C - x_B) - (y_C - y_B)(x_B - x_A)]}{2K} \quad (5)$$

Filtering Mechanism

In (3), (4) and (5), it is likely that the smaller the value of the denominator K is, the larger the error ($x - x'$, $y - y'$ and $z - z'$) will be. Thus, K can be taken as a metric for filtering out the improper selections of anchor nodes. We utilize K by sorting all the estimated points according to its K value, and then some certain percentage of the estimated points with small K values is filtered out. This mechanism is called K-filtering.

Comparisons with MMSE and Simulation Results

With the MMSE method, we also check the values of ($x - x'$, $y - y'$ and $z - z'$) as follows. Let $P = (x_1 - x_n)^2 + \cdots + (x_{n-1} - x_n)^2$, $Q = (x_1 - x_n)(y_1 - y_n) + \cdots + (x_{n-1} - x_n)(y_{n-1} - y_n)$, $R = (x_1 - x_n)(z_1 - z_n) + \cdots + (x_{n-1} - x_n)(z_{n-1} - z_n)$, $S = (y_1 - y_n)^2 + \cdots + (y_{n-1} - y_n)^2$,
$T = (y_1 - y_n)(z_1 - z_n) + \cdots + (y_{n-1} - y_n)(z_{n-1} - z_n)$,
$U = (z_1 - z_n)^2 + \cdots + (z_{n-1} - z_n)^2$, we have

$$\begin{bmatrix} x - x' \\ y - y' \\ z - z' \end{bmatrix} = \frac{-1 * \left(\frac{1}{(1+\rho)^2} - 1\right)}{2 * (PSU + 2 * QTR - R^2S - Q^2U - T^2P)} * \begin{bmatrix} [(SU - T^2)(x_1 - x_n) + (RT - QU)(y_1 - y_n) + (QT - RS)(z_1 - z_n)](e_1^2 - e_n^2) + \cdots + [(SU - T^2)(x_{n-1} - x_n) + (RT - QU)(y_{n-1} - y_n) + (QT - RS)(z_{n-1} - z_n)](e_{n-1}^2 - e_n^2) \\ [(RT - QU)(x_1 - x_n) + (PU - R^2)(y_1 - y_n) + (RQ - PT)(z_1 - z_n)](e_1^2 - e_n^2) + \cdots + [(RT - QU)(x_{n-1} - x_n) + (PU - R^2)(y_{n-1} - y_n) + (RQ - PT)(z_{n-1} - z_n)](e_{n-1}^2 - e_n^2) \\ [(QT - RS)(x_1 - x_n) + (RQ - PT)(y_1 - y_n) + (PS - Q^2)(z_1 - z_n)](e_1^2 - e_n^2) + \cdots + [(QT - RS)(x_{n-1} - x_n) + (RQ - PT)(y_{n-1} - y_n) + (PS - Q^2)(z_{n-1} - z_n)](e_{n-1}^2 - e_n^2) \end{bmatrix} \quad (6)$$

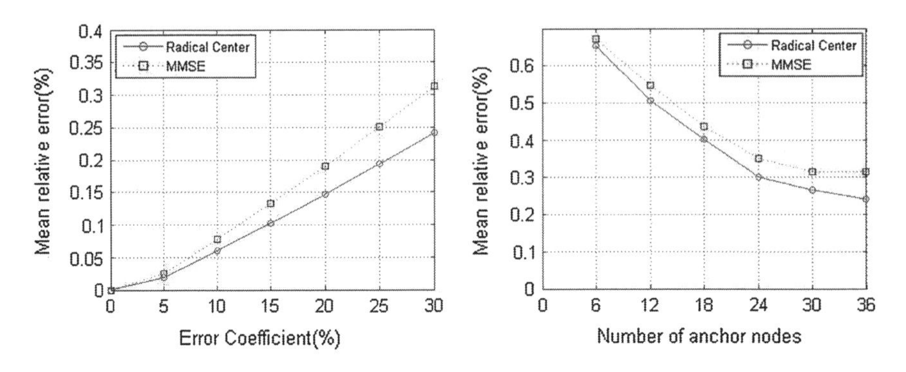

Fig. 1 Filtering percentage = 10 %

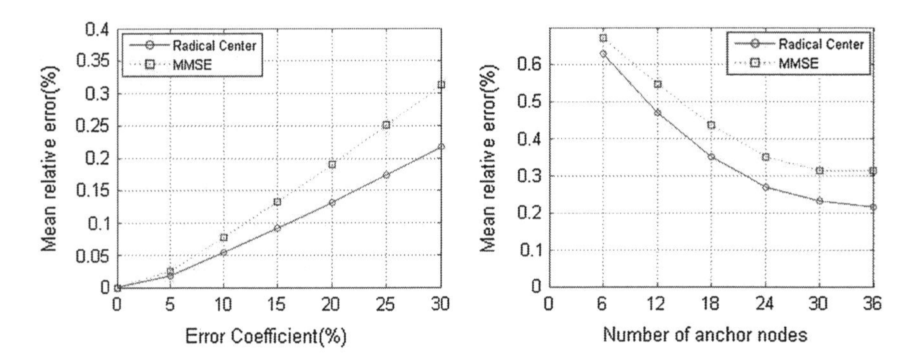

Fig. 2 Filtering percentage = 30 %

From (6), we can see that with MMSE, the location error is the summation of n-1 items. However, with the proposed algorithm, the estimated error, is composed of only three items. Thus, it is easy to reason that the resulting location error from the MMSE approach will be larger than the proposed algorithm. In the simulation environment, there is a target node and 36 anchor nodes and they are randomly placed. The distance measurement error coefficients are set to be 5, 10, 15, 20, 25 or 30 %. The filtering percentages in the filtering process are set to be 10, 20, or 30 %. We have conducted extensive simulations to look into the effectiveness of the proposed approach. For brevity's sake, parts of the simulation results are described here (Figs. 1 and 2).

As shown in Figures, the proposed approach always outperforms the MMSE method in accuracy and efficiency and also has better location accuracy than the MMSE in different number of anchor nodes.

Conclusions

Localization is one of the key issues in WSNs. We propose a novel range-based localization algorithm by utilizing radical centers. With our proposed algorithms, the target node computes the radical centers for location estimation with any four anchor nodes. Then, it can effectively filter out the improper estimations (radical centers) with the proposed filtering mechanisms so as to enhance the location accuracy. The advantages of the proposed algorithm over the conventional MMSE approach have been analytically analyzed and compared; and the conduced simulations demonstrated that the proposed algorithm effectively derive better location accuracy than the MMSE approach.

References

1. Wan J, Yu N, Feng R, Wu Y, Su C (2009) Localization refinement for wireless sensor networks. Comput Commun 32:1515–1524
2. Qi Y, Kobayashi H, Suda H (2006) Analysis of wireless geolocation in a non-line-of-sight environment. IEEE Trans Wireless Commun 5:672–681
3. Kuruoglu S, Erol M, Oktug S (2009) Three dimensional localization in wireless sensor networks using the adapted multi-lateration technique considering range measurement errors. In: Proceedings of IEEE GLOBECOM, pp 1–5
4. Zhang Y, Liu S, Jia Z (2012) Localization using joint distance and angle information for 3D wireless sensor networks. IEEE Commun Lett 16:809–811
5. Davis JG, Sloan R, Peyton AJ (2011) A three-dimensional positioning algorithm for networked wireless sensors. IEEE Trans Instrum Meas 60:1423–1432
6. Barsocchi P, Lenzi S, Chessa S, Giunta G (2009) A novel approach to indoor RSSI localization by automatic calibration of the wireless propagation model. In: Proceedings of IEEE 69th vehicular technology conference, pp 1–5
7. Gracioli A, Fröhlich A, Pires RP, Wanner L (2011) Evaluation of an RSSI-based location algorithm for wireless sensor networks. IEEE Latin Am Trans 9:830–835
8. Niculescu D, Nath B (2003) Ad hoc positioning system (APS) using AOA. In: Proceedings of 22nd annual joint conference of the IEEE computer and communications societies (INFOCOM 2003), vol 22, pp 1734–1743
9. Evrendilek C, Akcan H (2011) On the complexity of trilateration with noisy range measurements. IEEE Commun Lett 15:1097–1099
10. Yang Z, Liu Y, Li XY (2010) Beyond trilateration: on the localizability of wireless ad hoc networks. IEEE/ACM Trans Netw 18:1806–1814
11. Anton H, Rorres C (2010) Elementary linear algebra: applications version, 10th edn. Wiley, New York
12. Coxeter HSM, Greitzer SL (1967) Geometry revisited. The Mathematical Association of America, Washington, DC, pp 27–34
13. Dorrie H (1965) 100 great problems of elementary mathematics. Dover Publications, New York, p 153
14. Weisstein EW (2003) CRC concise encyclopedia of mathematics, 2nd edn. CRC Press, USA, p 502

Exploring Community Structures by Comparing Group Characteristics

Guanling Lee, Chia-Jung Chang and Sheng-Lung Peng

Abstract In recent years, more and more researchers devoted to identifying community structure in social networks. The characteristics of the social network are analyzed by clustering the social network users according to user's relationships. However, the users of current popular social networks such as LiveJournal and Flickr, can join to or create the communities according to their interests. Instead of grouping the users according to the cluster strategies which are wildly used in previous works, the purpose of the paper is to explore the structures and characteristics of the social networks according to the community the users actually joined. Moreover, we experiment on four real datasets, LiveJournal, Flickr, Orkut and Youtube, to analyze the characteristics hidden behind the social networks.

Keywords Social network · Community structure · Cluster

Introduction

Accompanying the growth of social networks, such as Facebook, more and more users rely on the social networks to communicate with their friends and share their daily life. In order to understand the users' behavior, identifying the characteristics behind the connections of the social network users become an important research topic. In [1, 2], the problem of how to measure the effect of information dissemination in a social network is discussed. By analyzing the number of users that will be affected by a certain user, the roles of *leader* and *follower* in a social network are defined in [3]. Moreover, in [4–6], the problem of how to partition the

G. Lee (✉) · C.-J. Chang · S.-L. Peng
Department of Computer Science and Information Engineering, National Dong Hwa University, Hualien 974, Taiwan, Republic of China
e-mail: guanling@mail.ndhu.edu.tw

Y.-M. Huang et al. (eds.), *Advanced Technologies, Embedded and Multimedia for Human-centric Computing*, Lecture Notes in Electrical Engineering 260,
DOI: 10.1007/978-94-007-7262-5_16,

users into the clusters is discussed. In previous works, the users are categorized into the same cluster if the degrees of intra connection is high and inter connection is low. However, in current popular social network, the users have the right to join to or create his/her own *communities*. Therefore, the purpose of the paper is to explore the structures and characteristics of the communities that the users actually joined, and use the characteristics to represent the relationship among the social network users.

The datasets we used to analyze the community structure are collected from four popular social networks, LiveJournal, Flickr, Orkut and Youtube. Moreover, we propose several measurements to model the characteristics and structures of the communities. The paper is organized as follows. The problem and three measurements are presented in section Problem Definition. We experiment on four real datasets and discuss the results in section Experimental Results. And finally, Conclusion concludes the work.

Problem Definition

As discussed in [7], a social network is defined as an *interaction Graph* $G = (V, E)$, where V denotes the vertex set of G and represents the social network users, E denotes the edge set and e_{ij} is contained in E if nodes i and j are friends in the network. In previous works, the users (nodes in G) are partitioned into clusters according to the connection degree among them. That is, the nodes are grouped into a cluster if the intra similarity is much larger than inter similarity. And similarity is usually measured by a function of the number of connections among the nodes. However, as mentioned above, in current popular social network, the users have the right to join to or create his/her own *communities*. Moreover, a user can join many communities according to his/her interests. That is, differ to the cluster concept proposed in previous works, in the real community model, a node can belong to many communities and a community is not necessary to have a tight connectivity. Therefore, the measurements for modeling a cluster are not suitable to measure the characteristics of the community. In the following, we propose three ideas for measuring the characteristics of a community.

Center of the community: If a user has many friends in the community, then the information posed by him would be noticed by many users in the same community. Therefore, we define the center of a community is the node whose number of connections in the community is much larger than that of other nodes in the same community. Therefore, the *center degree* of $node_i$ in community A, denoted as c_i^A, can be measured by the following equation.

$$c_i^A = \frac{\text{degree}_i^A}{|N_A| - 1} \tag{1}$$

In the equation, $degree_i^A$ denotes the number of connections (friends) of $node_i$ in community A and $|N_A|$ is the number of members in community A. Therefore, $|N_A| - 1$ is the maximum connections of a node in community A. When c_i^A is larger than a predefined threshold α, $node_i$ is said to be a center of community A. Moreover, the center set of community A, denoted as C_A, is the collection of centers in community A.

Intra connection degree: We propose the idea of *intra connection degree* to measure the inner structure of a community. And the complete connections concept is adapted to measure it. The intra connection degree of community A which is denoted by $Intra_A$, is measured by the following equation.

$$Intra_A = \frac{|E_A|}{(|N_A| \times (|N_A| - 1))/2} \tag{2}$$

In the equation, $|E_A|$ denotes the number of edges contain in A. The denominator indicates the maximum number of edges that community A can have. Therefore, the larger the value of $Intra_A$, the tighter the members in community A is.

Inter connection degree: We use the idea of centers of community to measure the inter connection degree. The basic concept is that if a center of community A is also the center of other community, then community A has a strong connection to the other community. And we denote the node which is the center of at least two communities as OC. The inter connection degree of community A, denoted as $Inter_A$, is measured by

$$Inter_A = \frac{|\{n|\, n \in C_A \, and \, n \, is \, OC\}|}{|C_A|} \tag{3}$$

The numerator of the equation is the number of centers which is also an OC in community A. A large $Inter_A$ indicates the proportion of OC in C_A is high. And therefore, community A has a strong connection to other communities.

In next section, a set of experiment is performed on real datasets. And by comparing the proposed measurements to the characteristics of the dataset, a thorough discussion is made.

Experimental Results

Four real datasets of online social network platforms, LiveJournal, Orkut, Flickr and Youtube, are collected from [8] to perform the experiment. LiveJournal was built by Brad and Fitzpatrick in 1999, and is a vibrant global social media platform where users share common passions and interests. Orkut is a social network service provided by Google. Users can build their own virtual social links in the internet by using the platform. Flickr was developed by Ludicorp company. In the platform, users can upload and share their pictures. Moreover, users can create *tags* for the pictures to ease the browsing process. YouTube is a video-sharing website, on

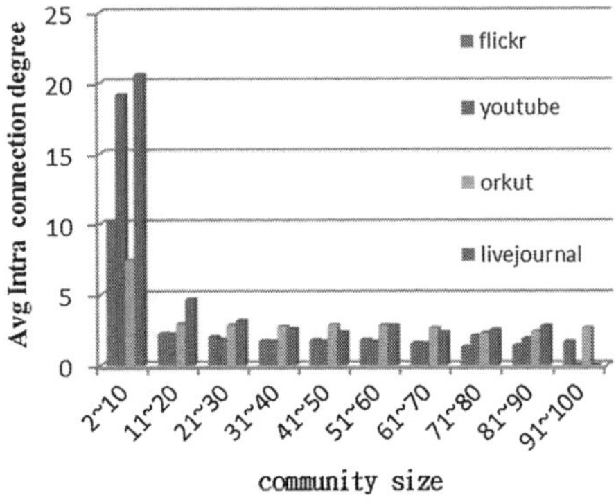

Fig. 1 The average intra connection degree of small size communities

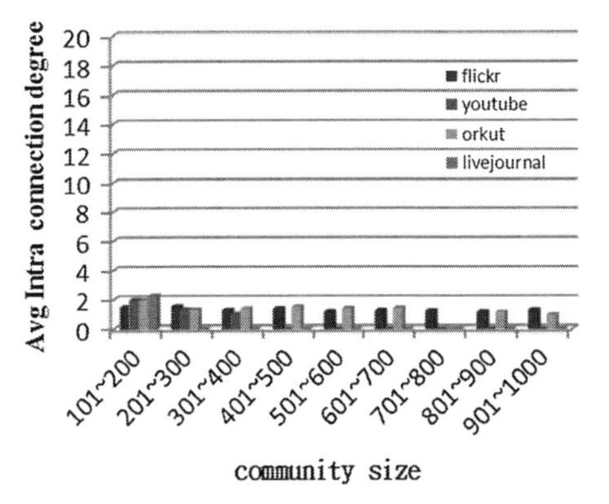

Fig. 2 The average intra connection degree of large size communities

which users can upload, view and share videos. Most of the content on YouTube has been uploaded by individuals, although media corporations including CBS, and other organizations offer some of their material via the site. In the platform, unregistered users can watch videos, while registered users can upload an unlimited number of videos.

The first experiment shows the relationship between the number of community members and the intra connection degree of the community. The results for small size and large size communities are shown in Figs. 1 and 2, respectively.

As shown in the results, the average intra connection degree of the communities whose sizes are within 2–10 is much larger than that of other group of communities. It indicates that the relationship between the members in a small community is very tight. Moreover, when the community size exceeds 10, the intra connection degree becomes quite small, which means the communities have a loose connection structure. This is because the users join the community according to their interests, which means the members belong to the same community have the similar interests and they are not necessary to know each other.

The second experiment shows the relationship between the number of community members and the inter connection degree of the community. The results

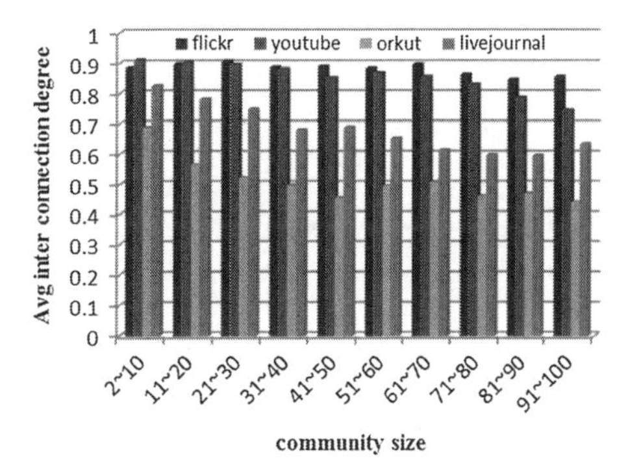

Fig. 3 The average inter connection degree of small-size communities

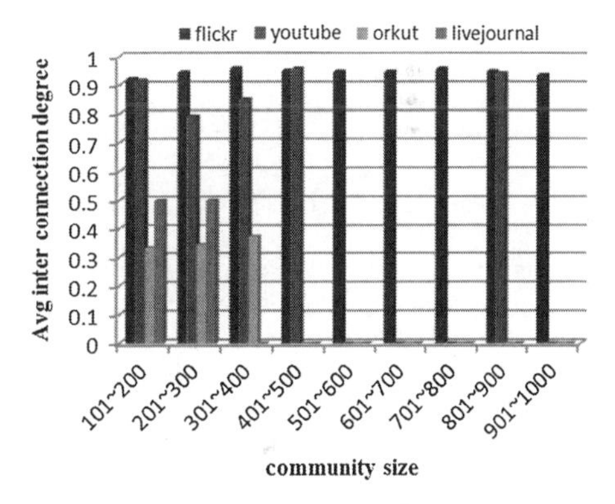

Fig. 4 The average inter connection degree of small-size communities

for small-size and large-size communities are shown in Figs. 3 and 4, respectively. As indicated in the results, the inter connection degrees of Flickr and Youtube are quite large, which means the centers of a community are also likely to be the centers of other community. The simulation result shows that the connection between real communities is high, and is a very different result comparing to the idea of cluster analysis proposed in previous works.

Conclusions

In this paper, by exploring the structures and characteristics of the social networks according to the community which the users actually joined, the characteristics of social networks are discussed. We propose several measurement methods to model the characteristics of the social network based on the community structure. To measure the inter connection degree of a community, we introduce the concept of centers and define what is a strong connection between the communities. Moreover, we experiment on four real datasets, LiveJournal, Flickr, Orkut and Youtube, to analyze the characteristics hidden behind the social networks. According to the experiment results, we find that the community has a loose connection structure when its size exceeds 10. Moreover, the inter connection between communities is high especially in Flickr and Youtube. It is a very different result by comparing to the concept of cluster analysis proposed in previous works.

Acknowledgments This work was partially supported by the National Science Council of Taiwan, under contracts NSC 101-2221-E-259-002 and NSC 101-2221-E-259-004.

References

1. Chen W, Wang Y, Yang S (2009) Efficient Influence maximization in social networks. In: Proceedings of the 15th ACM SIGKDD international conference on knowledge discovery and data mining, pp 199–208
2. Kempe D, Kleinberg J, Tardos E (2003) Maximizing the spread of influence through a social network. In: Proceedings of SIGKDD 2003, pp 3–9
3. Goyal A, Bonchi F, Lakshmanan L (2008) Discovering leaders from community actions. In: Proceedings of ACM conference on Information and knowledge management, pp 3–7
4. Biryukov M (2008) Co-author network analysis in DBLP: classifying personal names. Springer, Berlin Heidelberg, pp 403–407
5. Lancichinetti A, Fortunato S, Kertesz J (2009) Detecting the overlapping and hierarchical community structure of complex networks. Physics 11(3):4–14 (arxiv.org)
6. Backstrom L, Huttenlocher D, Kleinberg J, Lan X (2006) Group formation in large social networks: membership, growth, and evolution. In: Proceedings of ACM KDD 2006, pp 44–53
7. Licamele L, Getoor L (2006) Social capital in friendship-event networks. In: Proceedings of the sixth IEEE international conference on data mining, ICDM 2006, pp 1–12
8. http://socialnetworks.mpi-sws.org/data-imc2007.html

Emergency Broadcast in VANET by Considering Human Satisfaction

Yu-Shou Chang, Shou-Chih Lo and Sheng-Lung Peng

Abstract The emergency broadcast is an important service in Vehicular Ad Hoc Network (VANET) for the safety of vehicle drivers. The design of a broadcast scheme in a city environment becomes challenging. In this paper, we extend our previous work of using the concept of water wave propagation by further considering human satisfaction. As long as the drivers have sufficient time to react to an emergency event, a lazy rebroadcast approach is applied to significantly reduce rebroadcast times and network traffic.

Keywords VANET · Emergency broadcast · Human satisfaction

Introduction

Vehicular Ad Hoc Network (VANET) [1] is a type of mobile wireless network which supports for multi-hop wireless communications between vehicles. The most important application of VANET is to disseminate emergency messages to drivers in case of dangerous events [2]. A warning message needs to be delivered with low delay and high reliability to those vehicles that are located within a warning area. Also, this warning message needs to be delivered to those vehicles that newly enter the warning area within a warning time. This kind of service relies on emergency broadcast.

The broadcast design in VANETs suffers from new challenges beside the well-known broadcast storm problem [3], and these challenges are the connection hole problem, the building shadow problem, and the intersection problem [4]. These problems become more serious in a city environment with many street roads.

Y.-S. Chang · S.-C. Lo (✉) · S.-L. Peng
Department of Computer Science and Information Engineering, National Dong Hwa University Hualien, Hualien 974, Taiwan, Republic of China
e-mail: sclo@mail.ndhu.edu.tw

Y.-M. Huang et al. (eds.), *Advanced Technologies, Embedded and Multimedia for Human-centric Computing*, Lecture Notes in Electrical Engineering 260,
DOI: 10.1007/978-94-007-7262-5_17, © Springer Science+Business Media Dordrecht 2014

The core technique of any broadcast scheme is to select next rebroadcast nodes. That is, when a node listens to a broadcast packet, this node should follow a certain criterion to verify whether to rebroadcast this packet or not. Since there might have several nodes listening to the same broadcast, a contention or selection mechanism is applied to these nodes. Based on the different philosophies of selection mechanisms, a comprehensive survey of emergency broadcast schemes was given in [5].

Some existing broadcast schemes such as WPP [6], RBM [7], and UVCAST [8] cannot fully and efficiently solve the new challenges. In this paper, we propose an emergency broadcast scheme that is suitable for city environments. This scheme is extended from our previous one called Water-Wave Broadcast (WWB) [5] by further considering human satisfaction at the warning service. Unlike other broadcast schemes that always perform rebroadcasting immediately, our proposed scheme would perform rebroadcasting lazily as long as the drivers are satisfied with the warning service.

The remainder of this paper is organized as follows. The proposed scheme is illustrated in section Emergency Broadcast, followed by the performance evaluation in section Performance Evaluation. A brief conclusion is given in section Conclusions.

Emergency Broadcast

WWB is our previously proposed scheme that follows the concept of water wave propagation. This work further improves WWB by considering human satisfaction. Basically, we divide a warning area with circle shape into three ranges (WA_1, WA_2, and WA_3). Denote $A_{x,y}$ to be the area with the distance to the center of the warning area between x and y. For example, if the radius of the warning area (R) is 200 m, we have $WA_1 = A_{0,100}$, $WA_2 = A_{100,150}$, and $WA_3 = A_{150,200}$. The allowable waiting time to listen to the emergency warning is t_i for WA_i. In our setting, $t_1 = 1$ s, $t_2 = 7$ s, and $t_3 = 11$ s. A person (vehicle, or node) is satisfied with the warning service if one of the following three cases is true:

Case 1: A node that has already located in WA_i can receive the warning within t_i after the beginning of the emergency broadcast.

Case 2: A node that is newly entering into WA_i can receive the warning within t_i after the entering.

Case 3: A node that is newly entering into WA_i encounters the ending of the emergency broadcast (i.e., warning time is over) within t_i after the entering, no matter whether the warning is received or not.

A warning wave is simulated as an emergency event is detected by a vehicle, and the propagation medium is vehicles on the road. The head wave will spread the event to the whole warning area. Vehicles hold the head wave as they move and forward it to other encounter vehicles. The warning area keeps rippling until the end of the warning time. Any vehicle entering into a ripple area will be notified of this event through other encounter vehicles. Accordingly, we identify the following types of tasks that are performed by a vehicle (or node).

- Normal-node task: Regular jobs to each node.
- Source-node task: Dedicated jobs to a node triggering an emergency event.
- Head-node task: Dedicated jobs to a node promoting a head wave.
- Receiver-node task: Dedicated jobs to a node receiving a broadcast packet.

Normal-Node Task

Each node periodically announces its status to all its one-hop neighbors by broadcasting a hello packet. The cycle time is called a *hello interval*. Hello packets have two formats: basic and extended. These packets in both formats carry the current location of a node. An extended hello packet additionally carries an event list that summaries what emergency events have been received and are still valid for a node. This extended hello packet is broadcast only when any new neighbor is found.

Moreover, each node maintains two tables: neighbor table and event table. The neighbor table records the information of its one-hop neighbors by listening to hello packets from them. The successive location data received from a neighbor are used to estimate the motion vector of this neighboring node. The event table records valid emergency events that have been received or generated. An event list is generated by listing the identification numbers of all entries in the event table.

Source-Node Task

If an emergency event is detected by a node, this node additionally performs this type of task. An emergency broadcast packet which specifies a warning area and a warning time is generated and broadcast then. An emergency event is *valid* for a node if the warning time is not expired and this node is currently located within the warning area. Next, this node ends the source-node task and starts performing the head-node task.

Head-Node Task

The mission of this task is to carry an emergency broadcast packet and forward it by broadcasting to the other neighbors. After the broadcast, a *reliability check* is performed to confirm that any neighbors continue rebroadcasting the packet. Otherwise, this node rebroadcasts the packet once again. Then, this node decides whether to give up this head-node task or not according to a *head-node-selection* criterion.

A node S satisfies the head-node-selection criterion, if there is no neighbor of S that is ahead of or behind S and driving the same direction as S. This node S usually locates on the border of current event propagation and cannot temporally spread the event forward or backward. Therefore, this node S carries the emergency broadcast packet and starts rebroadcasting the packet when encountering a new neighbor.

Receiver-Node Task

If any hello or emergency broadcast packets are received, a node starts performing this type of task. The mission of this task is to parse a received packet and performs jobs accordingly. The receiver node updates its neighbor table when receiving a hello packet. If an extended hello packet is received, the receiver node checks whether the node sending this packet misses any emergency events by comparing its own event list with the received one. The closest node to this sending node will handle any event missing by locally rebroadcasting the missing events. By considering human satisfaction, this local rebroadcast need not be performed immediately, or in other words, need not be performed each time. This is called a *lazy rebroadcast* approach. The lazy rebroadcast is based on a *local rebroadcast probability* as computed in (1).

$$\text{Local rebroadcast probability} = \begin{cases} 1, & \text{if } T_R gt;\, \alpha \cdot T_W \\ (1 - D/R) \times 0.5 + (T_R/(\alpha \cdot T_w)) \times 0.5, & \text{otherwise} \end{cases} \quad (1)$$

T_R is the remaining warning time and T_W is the warning time. When T_R is less than and equal to a certain portion of T_W (controlled by parameter α, $0 \leq \alpha \leq 1$), the local rebroadcast probability becomes small, since the warning time is almost over. Moreover, we consider the distance between the node of missing an event to the center of the warning area (denoted as D). If D is large, the local rebroadcast probability becomes small too, since this warning event is not urgent to this node.

If a non-locally broadcast packet is received, a node i waits for a certain time period as computed in (2) and then decides whether to rebroadcast the packet or not based on a *global rebroadcast probability*.

$$\text{Waiting time} = WT_{\max} \times \mathit{random}\,(0, NB_i)/NB_i \quad (2)$$

$WT_{\max}$ is set to be twice the average one-hop communication delay. NB_i is the number of neighbors of node i (including node i itself). During the waiting period, node i may listen to ongoing rebroadcasts of the same packet from its neighbors. After the waiting time is over, node i records these rebroadcast nodes including the original broadcast node in set RS_i (Rebroadcast Set). This node i then need not rebroadcast the packet if all its neighbors are under the communication ranges of (in other words, are covered by) these nodes in RS_i. Here, we make an assumption that a node can receive any packets with high probability from another node within the communication range. This coverage situation is used to compute a global rebroadcast probability in (3). Denote $|RS_i|$ to be the number of nodes in the set. CV_i is the number of neighbors of node i that are also covered by nodes in RS_i.

$$\text{Global rebroadcast probability} = \begin{cases} 1, & \text{if } |RS_i| = 1 \\ 0, & \text{if } NB_i = 1 \\ (1 - CV_i/(NB_i - 1))^{|RS_i|}, & \text{otherwise} \end{cases} \tag{3}$$

For every packet rebroadcast, a reliability check is performed always to increase delivery reliability. Finally, we check whether the receiver node is suitable to be a head node by checking the head-node-selection criterion.

Performance Evaluation

To evaluate the performance, we carried out simulations using NS-2. We consider a real street environment which is imported from the TIGER [9]. A city street map of size 2,000 × 2,000 m is used as in the paper [5]. Under the street model, vehicles are generated and their moving patterns are controlled by the tool VanetMobiSim [10]. For each simulation run, one vehicle is randomly selected as an event source node. The warning area is a circle centered at the current location of the source node. The default parameter settings in our simulation are listed in Table 1.

We compare our proposed scheme (WWB) with WPP, RBM, and UVCAST. The cost metrics are *satisfy ratio* (percentage of the number of nodes satisfying the warning service in the warning area to the number of nodes entering into the warning area), *rebroadcast times* (number of times that an emergency packet is

Table 1 Parameter settings

Parameter	Value
Transmission radius	100 m
MAC protocol	IEEE 802.11p
Propagation model	Two-ray ground
Number of nodes	50–250
Vehicle speed	20–60 km/hr
Warning radius (R)	200 m
Warning time	180 s
Hello interval	1 s
WT_{max}	4 ms
Simulation time	240 s

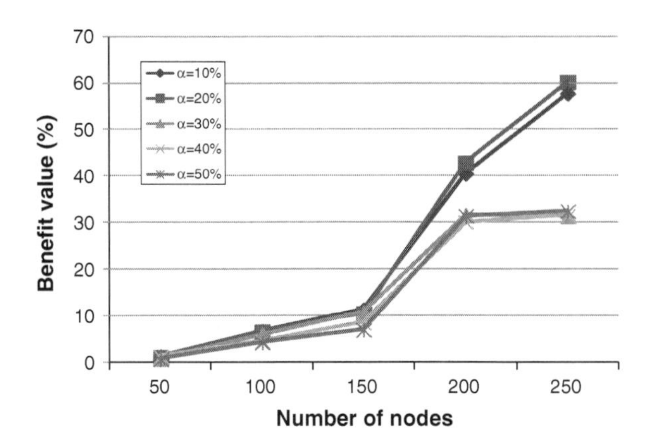

Fig. 1 The setting of parameter α

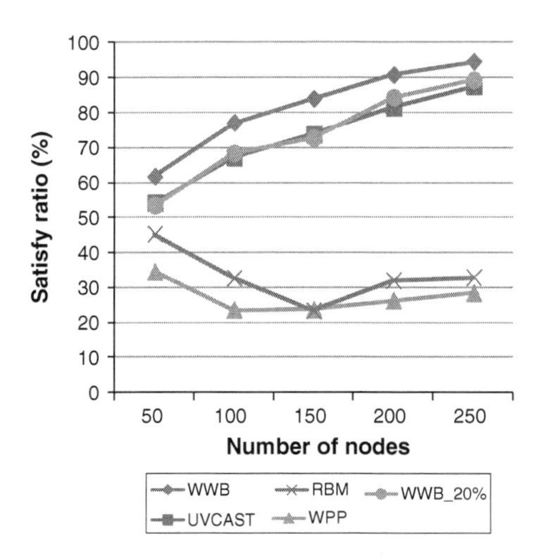

Fig. 2 Comparison of satisfy ratio

rebroadcast), and *delay time* (average elapsed time from the moment when a node enters into a warning area to the moment when the node receives the emergency packet).

At first, we evaluate the setting of α in (1) by observing the benefit value (the percentage of the amount of reduced rebroadcast times to the amount of decreased satisfy ratios). The benefit value is the best as $\alpha = 20$ % (Fig. 1), which implies to use the lazy rebroadcast when the remaining warning time is less than 36 s. If the lazy rebroadcast is applied, we sacrifice a certain amount of satisfy ratios, but the proposed scheme can still compete with other schemes (Fig. 2). Also, we greatly reduce the rebroadcast times (Fig. 3), and the delay time is still acceptable (Fig. 4).

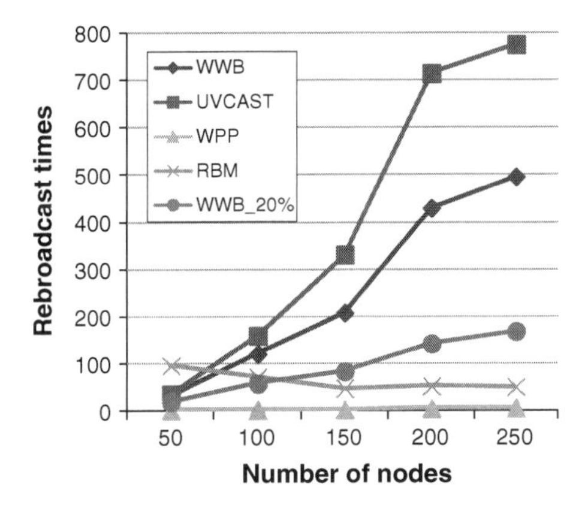

Fig. 3 Comparison of rebroadcast time

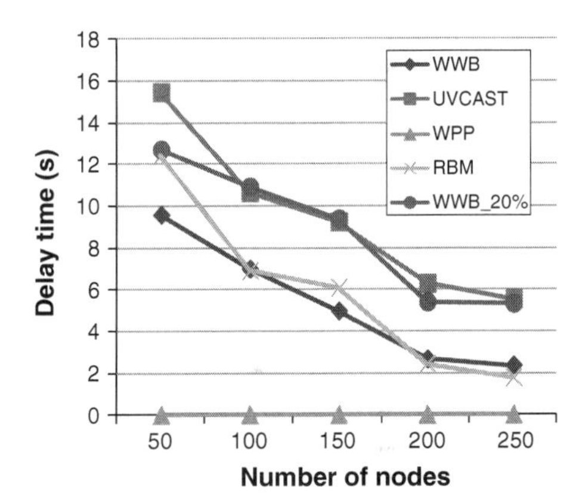

Fig. 4 Comparison of delay time

Conclusions

The dissemination of safety-related messages is an important application in a vehicular network environment. Our proposed scheme follows the water wave propagation to disseminate emergency warning messages along the street. To reduce the rebroadcast times in the whole network, we propose a lazy rebroadcast scheme by considering human satisfaction at the warning service.

References

1. Yousefi S, Mousavi MS, Fathy M (2006) Vehicular ad hoc networks (VANETs): challenges and perspectives. In: Proceedings of international conference on ITS telecommunications 2006, pp 761–766
2. Toor Y, Muhlethaler P, Laouiti A (2008) Vehicle ad hoc networks: applications and related technical issues. IEEE Commun Surv Tutorials 10:74–88

3. Ni SY, Tseng YC, Chen YS, Sheu JP (1999) The broadcast storm problem in a mobile ad hoc network. In: Proceedings of ACM MOBICOM 1999, pp 151–162
4. Yang CY, Lo SC (2010) Street broadcast with smart relay for emergency messages in VANET. In: Proceedings of IEEE AINA workshop 2010, pp 323–328
5. Lo SC, Gao JS, Tseng CC (2012) A water-wave broadcast scheme for emergency messages in VANET. Wireless Pers Commun. doi:10.1007/s11277-012-0812-2
6. Wisitpongphan N, Tonguz OK, Parikh JS, Mudalige P, Bai F, Sadekar V (2007) Broadcast storm mitigation techniques in vehicular ad hoc networks. IEEE Wirel Commun 14:84–94
7. Khakbaz S, Fathy M (2008) A reliable method for disseminating safety information in vehicular ad hoc networks considering fragmentation problem. In: Proceedings of international conference on wireless and mobile communications 2008, pp 25–30
8. Viriyasitavat W, Tonguz OK, Bai F (2011) UV-CAST: an urban vehicular broadcast protocol. IEEE Commun Mag 49:116–124
9. TIGER: topologically integrated geographic encoding and referencing http://www.census.gov/geo/www/tiger
10. VanetMobiSim http://vanet.eurecom.fr/

The Comparative Study for Cloud-Game-Based Learning from Primary and Secondary School Education Between Taiwan and America

Hsing-Wen Wang and Claudia Pong

Abstract A new way of learning has been introduced and it is here to stay: e-learning. E-learning stands for electronic learning and includes m-learning (mobile learning), u-learning (ubiquitous learning), computer-based learning, web-based learning and cloud-game-based Learning. The aim of this paper is to study cloud-game-based Learning in Taiwan and the United States of America through primary and secondary education and explore the fundamental differences between these countries. The most important ambition is to reach collaborative learning, or providing an environment where both teachers and students can debate and achieve synergy so that it is necessary to stimulate interest, improve children's performance and increase efficiency through games, problem solving and team-work. This new learning paradigm seems to be vastly popular because it completely integrates people into learning, no matter what socioeconomic status or urban/rural area; the only two things that are significant are the willingness to learn and the teacher's effort. The paper is divided into the follows sections, including the section covers the purpose and relevance of this study supported by a review of recent e-learning and cloud-game-based Learning research; the section contains the analysis of cloud-game-based Learning in primary and secondary schools in Taiwan; the section examines cloud-game-based Learning in primary and secondary schools in the USA; the section compares cloud-game-based Learning in primary and secondary schools in Taiwan and USA. Finally, the last section covers the conclusion and suggestions.

H.-W. Wang (✉)
Department of Business Administration, Changhua University of Education, Changhua, Taiwan, Republic of China
e-mail: shinwen@cc.ncue.edu.tw

C. Pong
Department of Licenciatura en Economía, Universidad Nacional del Sur, Bahía Blanca, Argentina

Y.-M. Huang et al. (eds.), *Advanced Technologies, Embedded and Multimedia for Human-centric Computing*, Lecture Notes in Electrical Engineering 260,
DOI: 10.1007/978-94-007-7262-5_18,

Keywords E-learning · Ubiquitous learning · Cloud-game-based learning · Comparative study · Primary and secondary schools

Introduction

A new learning paradigm has emerged, it is called e-learning and it is student centered. The main objectives are increase motivation, effectiveness and fun in learning activities. The purpose of this paper is to study Cloud-game-based Learning in Taiwan and the United States of America through primary and secondary education and explore the fundamental differences between these countries. The research methodology will be a review of a set of e-learning and Cloud-game-based Learning in the US and Taiwan. The first step is to define a game, its qualities and elements, relation to learning theory and advantages of Cloud-game-based Learning.

In the second step, it is going to be explained four games designed in Taiwan for primary and secondary schools: (1) 3D role play for learning anti-Japanese war during Qing dynasty and geography of Southern Taiwan; (2) Communicative language teaching for English learning; (3) Chinese language learning and (4) Gjun system. In third place, five games are famous in the USA for teaching primary and secondary students: (1) Making history about World War II; (2) Massive multiplayer online game (MMOG) for Mathematics, Language Arts, Science and Social Studies; (3) Dimention MTM for Mathematics; (4) Immune Attack for Science and (5) Survival Master for STEM (Science, Technology, Engineering and Mathematics).

In section Comparison Between Cloud-Game-Based Learning in Primary and Secondary Schools in Taiwan and USA, a summary is presented through a table where it is compared the target, design, hardware and software, objectives and improvements obtained by using Cloud-game-based Learning in Taiwan and USA. Finally, in section Conclusions and Suggestions, the conclusion and suggestions for Taiwan.

Literatures Reviews of Joyful Learning

In 2003, a movement was started for using video games in teaching and training. This initiative, known as serious games, has changed the way that educators viewed instruction to meet the needs of the Net generation. The perceived change in learning needs of the ‘Games Generation’ (Prensky 2001) or ‘Net Generation’ (Oblinger 2004) coupled with the ongoing growth in use and acceptability of a range of communications technology that has precipitated a growing interest in the potential of games and computer games for learning.

There are some elements that define an activity as a game: (1) Competition: the score-keeping element and/or winning conditions which motivate the players and provide an assessment of their performance. (2) Engagement: or intrinsic motivation means that once the learner starts, he or she does not want to stop before the game is over and the four sources are challenge, curiosity, control and fantasy (Beck and Wade 2004; Prensky 2006). (3) Immediate Rewards: Players receive victory, points or descriptive feedback, as soon as goals are accomplished.

Games fulfill a number of educational purposes. Some games are explicitly designed with educational purposes, while others may have incidental or secondary educational value. All types of games might be used in an educational environment. Educational games are games that are designed to teach people about certain subjects, expand concepts, reinforce development, understand an historical event or culture or assist them in learning a skill as they play (Aldrich 2004; Foreman et al. 2004; Prensky 2001; Quinn 2005).

However, a game is educational when it makes learning integral to scoring and winning. It is not enough to simply incorporate course material into a game and if it is possible to score and win without learning, students are likely to do so. There are different kinds of games [1]: (1) Video Games: These are played over the Internet, on personal computers or on specific game consoles hooked up to televisions. (2) Role-Playing Games: These are generally cooperative and highly engaging with a subtle way of handling scoring. (3) Board and Card Games: These tend to emphasize strategy elements rather than being completely random games of chance. Some of board and miniature games take hours or even days to play. (4) Sports: Students do not need to be physically fit to enjoy running around chasing things. (5) Scavenger Hunts, Raffles, etc.: When these events are organized as fundraisers for students and they tend to be quite popular with students.

Not only does the integration of learning with gaming make science more fun; it also motivates students to learn through doing, immerses them in the material so they learn more effectively and encourages them to learn from their mistakes. Games are such a great escape from the real world because bad consequences are rarely serious or lasting, they are only a game and if students lose, they can start the game over and try again.

These findings frame the three key aspects to Cloud-game-based Learning: motivation, skill development and immersive learning environments. The very nature of games provides three main factors for motivation: fantasy, challenge and curiosity (Malone 1981). Fantasy relates to the use of imagination and the child's inherent inclination towards play (Opie and Opie 1969). There is freedom to fail, experiment, fashion identities, freedom of effort and interpretation that create a learning space where new ideas and problem-solutions can emerge (Klopfer et al. 2009).

Beyond increased motivation, teachers using games in classroom have also noted improvement in several key skills areas (Joyce et al. 2009): personal skills (such as initiative, persistence, planning and data-handling), spatial and motor skills (such as coordination and speed of reflexes), social (such as teamwork, communication, negotiating skills and group decision-making) and intellectual (such as problem-solving, strategic thinking and application of numbers).

About learning environments, games allow players to enter environments that would be impossible to access in any other way; for instance, going back in history, understanding the complexity of running a major city, managing entire civilizations or nurturing families. They require engagement with complex decisions like exploring the effects of different choices and a multiplicity of variables offering ongoing and responsive feedback on choices. They also stimulate conversation and discussion; players share ideas, hints and tips in what increasingly tend to be lively and supportive learning communities (ELSPA 2006).

According to James Paul Gee (2003), digital games create 'semiotic domains' which are any set of practices that recruits one or more modalities (for example, oral or written language, images, equations, symbols, sounds, gestures, graphs, artifacts, etc.) to communicate distinctive types of meanings. The semiotic domain for a game is the world or culture it creates and is shared by those participating in the game together where they share knowledge, skills, experiences and resources. Active and successful participation in a semiotic domains demonstrated by 'active learning', where group members gain there sources and skills to solve problems within and perhaps beyond the domain as well as 'critical learning', which includes thinking about the game at a 'meta' level so that they cannot only operate within the game but within the social structure that surrounds the game as well (Williamson 2003).

However, teachers are also consistently found to be critical components ineffective Cloud-game-based Learning. Where the game is just the tool, the teacher is essential to effective implementation of the game through direction of the learning approach, discussion, debrief and support in construction of the social learning culture that surrounds the game-play. Numerous researchers have stated that learning with educational video games is not likely to be effective without additional instructional support and effective strategies for implementation (Leemkuil et al. 2003; O'Neil et al. 2005; Wolfe 1997).

Research Methodology and Issues

The paper aims to identify key issues and themes arising from the literature reviewed, the case studies produced and the consultation undertaken. The review comprises a meta-review that is a review of literature reviews, and literature has been grouped in relevant categories according to selected themes or issues. Literature was sourced from keyword searches of electronic databases, key journals in the field and a general search of the internet. Selected criteria include significant meta-reviews, relevance to Cloud-game-based Learning and empirical studies of the use of games. The criteria were used to identify relevant literature for inclusion

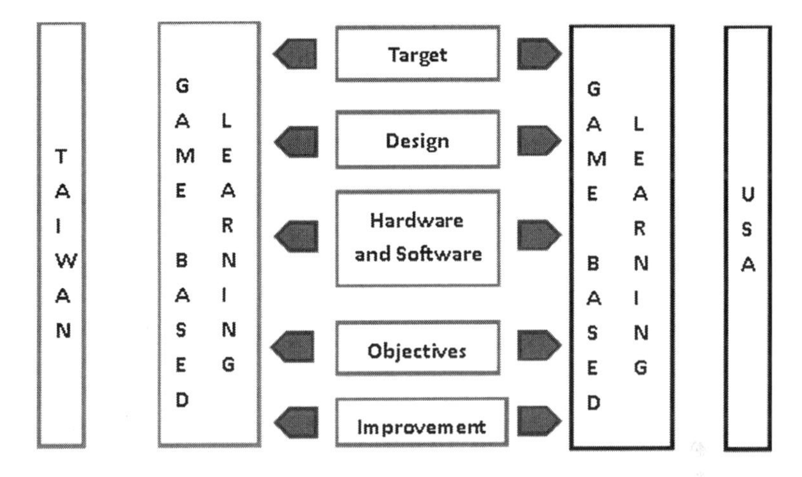

Fig. 1 The research structure of our framework

in the paper. Recommendations from experts in the field were also used to identify key articles and texts relating to examples from the practice.

In the introduction and literature review, there were a definition of game, its qualities, types and elements, relation to learning theory and advantages of Cloud-game-based Learning. In section Literatures Reviews of Joyful Learning, it is developed four games designed in Taiwan for primary and secondary schools: (1) 3D role play for learning anti-Japanese war during Qing dynasty and geography of Southern Taiwan; (2) Communicative language teaching for English learning; (3) Chinese language learning and (4) Gjun system.

In section Research Methodology and Issues, it is explained five games that are famous in the USA for teaching primary and secondary students: (1) Making history about World War II; (2) Massive multiplayer online game (MMOG) for Mathematics, Language Arts, Science and Social Studies; (3) Dimention MTM for Mathematics; (4) Immune Attack for Science and (5) Survival Master for STEM (Science, Technology, Engineering and Mathematics).

In section Comparison Between Cloud-Game-Based Learning in Primary and Secondary Schools in Taiwan and USA, a summary is presented through a table where it is compared the target, design, hardware and software, objectives and improvements obtained by using Cloud-game-based Learning in Taiwan and USA. Finally, in section Conclusions and Suggestions, the conclusion and suggestions for Taiwan. The following figure shows briefly section Comparison Between Cloud-Game-Based Learning in Primary and Secondary Schools in Taiwan and USA mentioned above: Fig. 1

Comparison Between Cloud-Game-Based Learning in Primary and Secondary Schools in Taiwan and USA

When looking for papers for the USA about Cloud-game-based Learning in primary and secondary school, most of them were orientated to college and university applications. While in Taiwan, most of the research papers are based on Cloud-game-based Learning for primary and secondary education. One reason for this is that Taiwan wants their student to get accustomed to information technology and computing devices so that it is increased the competitiveness of the students and their clerical skills. However, as part of its Connected Educator Month, the U.S. Department of Education notes that Cloud-game-based Learning is gaining considerable attention as more and more young people are learning from games outside of school, and more and more teachers are leveraging the power of games to engage students in school. Well-designed games can motivate students to actively engage in meaningful and challenging tasks, and through this process to learn content and sharpen critical-thinking and problem-solving skills.

Most of the games found for the USA were focus on one specific lesson instead of the complete subject in primary and secondary education. In Taiwan, the games that were analyzed include exercises and answers for the students as well as test and record graphics. However, these functions need more equipment and requirements i.e. while teaching only one lesson needs only one software (that can be saved with other programs), in order to teach a complete subject in primary and secondary schools, it is necessary a hardware or electronic device for the software or video game.

On one hand, Cloud-game-based Learning in the USA mainly pays attention to the achievement of higher grades in the students' subjects and the improvement of skills like problem solving, team working and strategic planning. On the other hand, in Taiwan is important to increase motivation in students and develop a deep understanding, an enjoyable experience and cultural immersion inside and outside the classrooms Table 1.

Table 1 Comparison between cloud-game-based learning in primary and secondary schools in Taiwan and the USA

	USA	Taiwan
Target	Universities	Primary and secondary schools
Design	Only one lesson	Subject and exams
Hardware and software	Software	Software and hardware
Objectives	Achievement	Motivation
Improvement	Problem solving	Deep understanding
	Team working	Enjoyable experience
	Strategic planning	Cultural immersion

Conclusions and Suggestions

It is clear from the data that Cloud-game-based Learning presents an opportunity to engage students in activities, which can enhance their learning. Like any successful pedagogy, outcomes need to be well planned and classrooms carefully organized to enable all students to engage in learning. What is notable about using games for learning is the potential they have for allowing many children to bring their existing interests, skills and knowledge into the classroom and then use games as a hook or stimulus to build the activities for learning around them. In many ways these findings reflect those of earlier media education programs, which sought to capitalize on children's own interest in television and film and build activities around them.

Although, it is good to have Cloud-game-based Learning in primary and secondary schools in Taiwan, it would be convenient to extend Cloud-game-based Learning to college and kindergarten too as well as develop games in different languages so that different countries can take advantage of them because students perceived a range of educational benefits as a result of participating in the Cloud-game-based Learning approaches, including increased collaboration, creativity and communication.

For future research, it would be valuable to investigate how to cultivate more interpersonal relationships, how to improve privacy and security in using online video games and integrate Cloud-game-based Learning with u-learning and m-learning.

Acknowledgments This work was supported in part by the National Science Council (NSC), Taiwan, ROC, under Grant NSC Taiwan Tech Trek 2012, NSC 101-2511-S-018-016.

References

1. Blunt RD (2005) Knowledge area module V: a framework for the pedagogical evaluation of video cloud-game-based learning environments, Applied Management and Decision Sciences. Walden University, Minneapolis
2. Carleton College http://serc.carleton.edu/introgeo/games/index.html
3. Groff J et al. (2010) The impact of console games in the classroom: evidence from schools in Scotland. Futurelab, Innovation in Education. www.futurelab.com.uk
4. Watson WR et al. (2010) A case study of the in-class use of a video game for teaching high school history. Elsevier, computers and education, Purdue University, Department of curriculum and instruction, United States, contents lists available at Science Direct. www.elsevier.com/locate/compedu
5. Swearingen DK (2011) Effect of digital game based learning on ninth grade students' mathematics achievement. University of Oklahoma, ProQuest LLC, Oklahoma
6. Hacker M, Kiggens J (2011) Gaming to learn: a promising approach using educational games to stimulate STEM learning. In: Barak M, Hacker M (eds) Sense publishers, fostering human development through engineering and technology education, pp 257–279

7. Leonard AA (2008) Video games in education: why they should be used and how they are being used. Theor Pract, N C State Univ 47(3):229–239
8. Whitton N (2007) Motivation and computer game based learning. Education and Social Research Institute, Manchester Metropolitan University, Singapore
9. de Freitas S (2006). Learning in immersive worlds. A review of cloud-game-based learning. JISC e-learning programme
10. Shih JL et al. (2010) Designing a role-play game for learning Taiwan history and geography. IEEE computer society, 2010 IEEE international conference on digital game and intelligent toy enhanced learning
11. Wang YH (2010) Using communicative language games in teaching and learning english in Taiwanese primary schools. J Eng Technol Educ, Kainan Univ 7(1):126–142
12. National Changhua University of Education–NCUE (2012) Chinese language e-learning site content, effectiveness and relationship management for user satisfaction
13. Hwang WY et al (2011) Effects of reviewing annotations and homework solutions on math learning achievement. Br J Educ Technol 42(6):1016–1028
14. Huang YM et al (2012) A ubiquitous English vocabulary learning system: evidence of active/passive attitudes vs. usefulness/ease-of-use. Comput Educ 58(1):273–282 Elsevier, Science Direct
15. Huang YM et al (2011) The design and implementation of a meaningful learning-based evaluation method for ubiquitous learning. Comput Educ 57(4):2291–2302 Elsevier, Science Direct
16. Huang YM et al (2011) Development of a diagnostic system using a testing-based approach for strengthening student prior knowledge. Comput Educ 57:1557–1570 Elsevier, Science Direct
17. Huang YM et al (2010) An automatic group composition system for composing collaborative learning groups using enhanced particle swarm optimization. Comput Educ 55:1483–1493 Elsevier, Science Direct
18. Huang YM et al (2011) Applying adaptive swarm intelligence technology with structuration in web-based collaborative learning. Comput Educ 52:789–799 Elsevier, Science Direct
19. Huang YM et al (2009) An adaptive testing system for supporting versatile educational assessment. Comput Educ 52:53–67 Elsevier, Science Direct
20. Huang YM et al (2008) Toward interactive mobile synchronous learning environment with context-awareness service. Comput Educ 51:1205–1226 Elsevier, Science Direct
21. Huang YM et al (2010) Effectiveness of a mobile plant learning system in a science curriculum in Taiwanese elementary education. Comput Educ 54:47–58 Elsevier, Science Direct
22. Huang YM et al (2008) A blog-based dynamic learning map. Comput Educ 51:262–278 Elsevier, Science Direct

The New Imperative for Creating and Profiting from Network Technology Under Open Innovation: Evidences from Taiwan and USA

Hsing-Wen Wang, Raymond Liu, Natalie Chang and Jason Chang

Abstract In the rapidly changing Web 2.0 era innovation has become a key focus in organizations around the world. Innovation can be described as the creation of new products that take advantage of changing markets and improved technology, but innovation also means the ability to adapt to new technologies and create new networks. In this paper we focus on organizations in different fields that attempt to adapt to advances in network technology. The fields we examine are online web stores. With regards to social media, we will examine how the design and allowance of user feedback affects the success of E-stores. In addition we also compare the studies done on these fields from Taiwan and America in order to understand how regional cultural biases affect research style and direction. This paper will study the effects of being innovative in the three fields described above as well as the different views of being innovative in America and Taiwan.

Keywords Web 2.0 · Open innovation · E-stores · Network technology

H.-W. Wang (✉)
Department of Business Administration, Changhua University of Education, Changhua, Taiwan, Republic of China
e-mail: shinwen@cc.ncue.edu.tw

R. Liu
Department of Economics and International Relations, University of California Davis, Davis, CA, USA

N. Chang
Department of Economics, Simon Fraser University, Burnaby, Canada

J. Chang
Department of Business, University of Iowa, Iowa, USA

Y.-M. Huang et al. (eds.), *Advanced Technologies, Embedded and Multimedia for Human-centric Computing*, Lecture Notes in Electrical Engineering 260,
DOI: 10.1007/978-94-007-7262-5_19,

Introduction

The amount of people using the internet to communicate, search for information, and create new web pages has grown tremendously in the past 20 years. Not only has the user base for the internet grown, its capabilities and purposes have expanded greatly as well. With such an explosive growth in users over the past twenty, no organization in the world can ignore the importance of integrating themselves to the online network. Older functions of the internet included looking up information and communication via IRC channels. Now in the Web 2.0 the internet provides cloud storage capabilities, e-commerce, and gigantic social media community sites such as Facebook. Companies must be adaptable to the growth of the internet user base and to the creation of new platforms of internet usage. Organizations achieve this by being innovative and taking the initiative to understand each new internet application before it becomes widespread.

The rapid improvement in network technologies has made it possible for over two billion users to access the internet. In addition the internet now processes over 21 exabytes of information per month. Organizations are eager to integrate themselves with the digital world but there is still a lack of research regarding the most efficient ways to use web 2.0 network technologies. Companies are looking for the best ways to design their online store websites; they want to understand which features promote user connectivity and make communication easier. Successful E-stores need to quick and simple, as well as having the platforms necessary to form user communities that will help promote brand products (Fig. 1).

Organizations have also begun focusing their attention on popular social media sites such as Facebook to serve as a new advertising and marketing platform. With over 873 million users (http://www.checkfacebook.com/), Facebook is by far the largest social media site. Companies and other independent organizations see great potential in the Facebook fan page application, a page which allows Facebook users to show they "like" a brand name or cause. Companies can upload pictures and videos, provide updates, and create special Facebook apps in order to build rapport with customers and improve their brand image. Fan pages are free and represent an efficient minimal cost advertising technique. Fan pages save companies money and raises brand name visibility, increasing profits.

The ability to store data online via cloud storage is a boon to institutions such as hospitals. With thousands of patient files, hospitals can benefit tremendously from online data storage, as long as they are able to keep that information secure. One such application by hospitals is the use of balance score cards as a feedback tool to inspect the efficiency level of internal operation. These score cards are costly and time consuming evaluations that could become more efficient if the process of collecting and sorting data became online-based. If these score cards could be shared online, low scoring hospitals could study higher rated hospitals and adapt their management techniques and fund allocations.

Adverse drug event (ADE) is an injury resulting from the use of a drug. Hospitals need to be careful of the drug being use in the institution. Although the medications

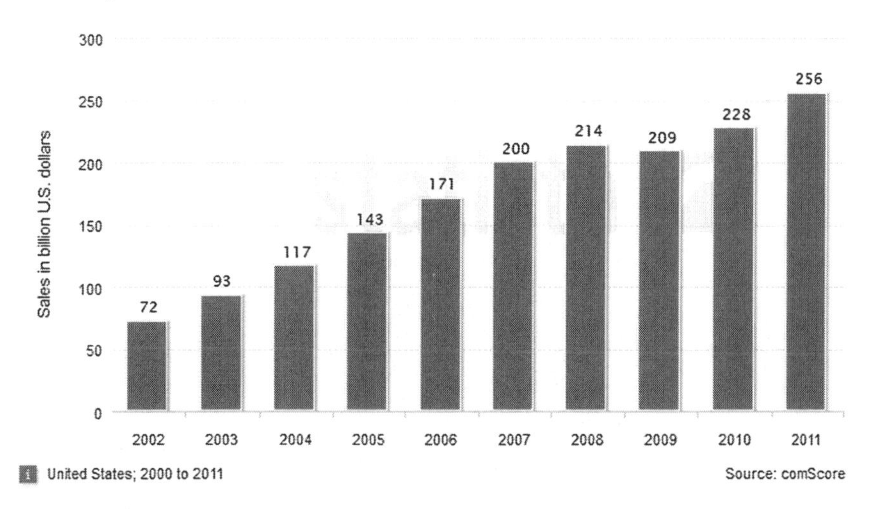

Fig. 1 The trend toward e-commerce in America

being used are tested, constant monitoring to insure safety is still required. Hence, hospitals are also dependent on online database collect and monitor patients with drug usage. Prior to 2007, adverse drug event is under recognized and underreported within the U.S. Department of Veterans Affair. Two methods for extracting and collating ADEs into national databases where developed in years 2007. They are the Adverse Reaction Tracking package and Veteran Affairs Adverse Drug Event Reporting System. It is necessary to conduct multiple studies on how to achieve the most efficient usage of all this Web 2.0 technology. We intend to examine research done in Taiwan and the United States in order to gain two different perspectives on how innovation can be best achieved. Researchers from each country will have their own cultural backgrounds that affect how their study will be designed and what conclusions they will draw from their datasets. By examining the work of two different countries, we hope that our conclusions will show how to profit from the global future of network technologies.

Literatures Reviews

E-commerce refers to a broad range of online business activities for products and services. Using private network, Electronic Data Interchange (EDI), to transact business between companies was the early form of e-commerce. As defined by

Kestenbaum and Straight [1], e-mail, electronic fund transfer, electronic data interchange, and any related technological integration could be identified as elements of a business system. However, different definitions to technological network has brought up by Kalakota and Whinston [2]. Consolidating the principles of e-commerce can be described as four types: business to business, business to consumer, consumer to consumer, and consumer to business, which are all usually associated with conducting any commercial transaction through the Internet.

With a rapid change in network technology, firms are vigorously seeking innovative strategies to advance performance and it has aroused numerous researchers' interests to study this phenomenon. Many researchers focus their studying on how corporations' existing business model influences by innovation and how they adapt these changes and integrate with old business model to form new marketing strategies. According to lots of research papers, utilizing Internet technology to expand business has more pros than cons. And also most of reports have pointed out the common reasons that corporations have chosen to launch online service are based on the factors of transaction costs, consumer behaviors, and services improvement.

The marketing strategies that the majority of companies utilize for e-business have listed out in the studies as well; such strategies are as auction, point collection, and lottery. Auction is the process of sale in which goods are sold to the highest bidder. Reward point is a program for customers to redeem points they accumulated from purchasing. And lottery involves the drawing of lots for a prize. The reward programs are powerful mechanisms for raising sales or brand loyalty [3] and the two mechanisms that make positive sales impact are "points pressure" and "rewarded behavior" [4]. Effort to earn a reward by increasing purchase is a short term impact, described as points pressure, and the long term impact whereby customers rise their purchase rate after obtaining the reward, can be defined as rewarded behavior.

Opening an e-business market can acquire profitability by offering businesses the opportunity to reduce their costs dramatically. Companies are realized that their online systems are more valuable than the companies themselves. And this could be happened is because of the innovative technology assisting corporations to maximize the benefit of transactional cost, which it always refers to the six transactional sigma identified by Downes and Mui [5]. The six types of transaction costs are search costs, information costs, bargaining costs, decision costs, policing costs, enforcement costs and IT costs. Firms can less depend on broker dealer to sell products and have fully control on the quality of products and services. Additionally, it trims down the cost on time and money to find responsible suppliers and developing new customers. Via the Web, companies can directly interact with their customers to strengthen consumer relationship and also attract more new customers who are relying on online shopping by sharing digital content. Virtual stores can provide fastest and recent updates of new products and user experiences than physical stores as well. Widespread Internet marketing has certainly opened many various unique possibilities for firms and helps to expand the corporate image. Innovative technologies can improve different aspects for a corporation and has presented an optimistic outlook of the e-business market.

The Research Framework

The topics of innovative web design, targeted Facebook fan page marketing and maximizing hospital efficiency are all centered about the usage of network technology and the importance of organizations being innovative. However, each topic deigns its own research methodology as a study designed to research hospital efficiency lacks the tools necessary to understand why consumers prefer Starbuck's fan page over Subway's fan page. In this section we examine our three different research methodologies. Hypothesis: H1: User interaction affects purchase intention; H2: Brand image affects purchase intention; H3: Event creation affects user participation; H4: Consumer attitude affects purchase intention; H5: Heuristic Information Processing is a driving force behind purchase intention; H6: Network technology usage increases profits.

Comparative methodology will be using in this research to compare and analyze the companies in manufacturing sector, 3 M and Procter and Gamble (P&G), between US and Taiwan. Both of 3 M and Procter and Gamble are multinational manufacturers and sell from home and leisure products to health care products. As attributions of markets in diverse regions, these two companies in US and Taiwan would operate on different tactics of online marketing to compete domestically. Since the online order center of 3 M in US is only open for the channel partners and other business customers, the virtual store for daily consumers have no longer existed, Procter and Gamble as a supportive sample in the comparison.

3 M in Taiwan has opened an online shopping site to serve its customers directly and conveniently which customers can purchase various 3 M products from displays and graphics, health care, to home and leisure. Since virtual store has helped 3 M sell products without broker, 3 M have fully control to promote its products in low prices irregularly to raise sells. There are different tactics can be seen on online web site of 3 M Taiwan, (1) Auction: 3 M sets a $1 in New Taiwan dollar reserve price by offering consumers up for bid. (2) Reward point: Customers can obtain points when they purchase products from 3 M online and which customers can use these collected points to exchange 3 M products; (3) Lottery: Once customers have become members online, they will receive a lottery number while logging into the Web. Customers will win a prize when his/her numbers are drawn (Fig. 2).

The listed marketing strategies are as allures to attract more shoppers and also can help 3 M to understand consumer behavior. From studying consumer behavior, 3 M can implement dynamic product adjustment to correspond with market trend and can acquire immediate responses from customers when new products are released.

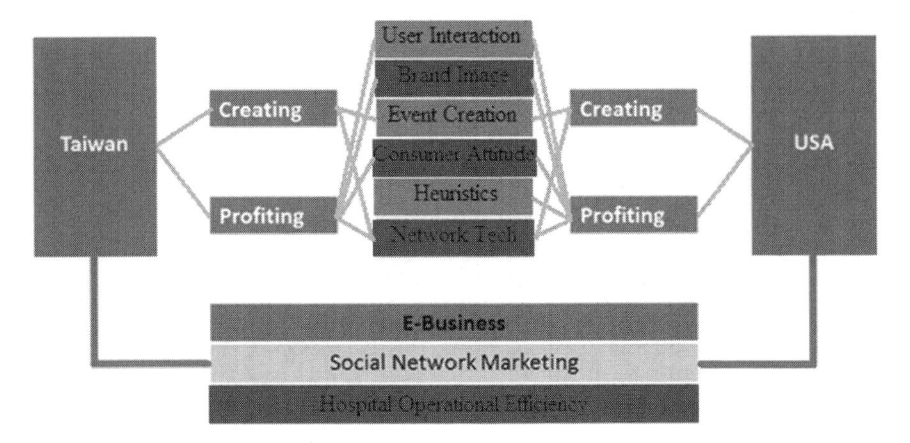

Fig. 2 The framework of this research

Additionally, 3 M has created a forum and fan page on Facebook to have live interaction with consumers. This has given a platform for consumers to share their experiences using 3 M products and for 3 M to share and update its recent news and new products vividly and instantly by uploading audios or pictures. Either social media or virtual store, innovations exponentially grow consumer affinity for the brand and mutually raise sales for 3 M and its retail partners.

As shown from the comparisons, 3 M and Procter and Gamble are vigorously seeking advance network technologies to strengthen their business. Since consumers move based on the development of the market and their shopping habits, e-business continues to grow faster than traditional retail. Percentage of e-business revenue is expected to more than double in next few years even now e-business total revenue is only 6–12 % of a traditional retailer's total revenue. Figure 1 indicates the future outlook of online shopping market is optimistic, therefore; 3 M and Procter and Gamble should gradually shift their concentration from traditional retail to e-business. 3 M and Procter and Gamble must adjust their marketing strategies as attributions in diverse regions and operate on different ideals of network technology to compete domestically.

Results and Discussions

Posts on 3 M Taiwan Facebook and individual brands of Procter and Gamble US show different atmospheres to online users. The tactic that Procter and Gamble used in social media is to partner with celebrities and it would lead to a creation of commercial polished environment which shopping decisions of some customers will influence by the celebrities. 23 % of the respondents strongly agree celebrity endorsement is a method of persuasion. Sales and brand visibility will tend to react with media exposures and reputation of celebrities positively because a high-status endorser can facilitate reassure the true quality of a product to consumers [6]. 50.7 % of the respondents agree that using celebrity endorsement will make the brand stand out in the clutter. Economic value of celebrity endorsement is profitable. As the statement from P&G India, the effect of celebrity endorsement helped one of the major brands of P&G to rank as the market leader with over 45 % market share. Different than what Procter and Gamble US approach, 3 M Taiwan provides a simple platform for consumers to share experiences and reviews. Certainly, the simplicity builds an intimate friendship between 3 M Taiwan and consumers.

Conversely, 3 M Taiwan shows creativity on its online shopping store than Procter and Gamble US. 3 M Taiwan offers a loyalty program including reward point, lottery and auction, to optimize its sales and also establish an unique way to interact with shoppers. Reward program is a promotional tool to incentivize consumers on basis of cumulative purchases from a firm [7]. From a case study, a consumer will spend between 20 and 25 % more per visit in a well-constructed loyalty program. As Lottery, shoppers require to log into 3 M virtual store in order to receive a daily lottery number. Lottery acts as a temptation to consumers for visiting 3 M virtual store everyday and it successfully increases chances of customers to purchase products even customers are not planning to. E-Store, name of Procter and Gamble's virtual store which is partnership with PFS Web, does not offer any reward programs and it makes E-Store less competitive to the market (Fig. 3).

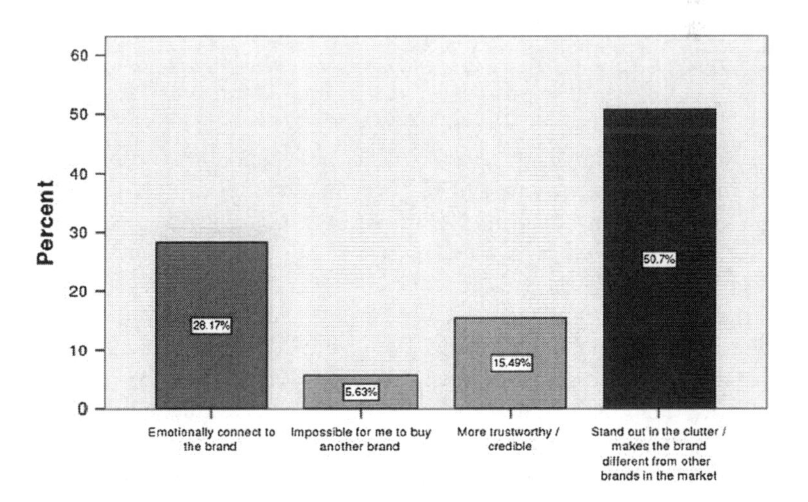

Fig. 3 The result of this research

Conclusions and Suggestions

More similar characteristics of companies will start an online shopping business to emulate and for advancing their services and the existed virtual stores should be out with the old to make way for the new in order to strive for success. The suggestions will be: (1) Strengthen relationship with online consumers. Companies must divide up the market serves to prioritize the primary target markets they will focus on, which the segments will facilitate companies to profoundly comprehend market characteristics and types of consumers. From the analysis, companies can release different kinds of discounts for distinct customers and update to keep websites continuously fresh. (2) Partnership with other firms in different industries. Companies can cooperate with other types of reward programs to reinforce services. For example, if Procter and Gamble signed an agreement with Air Miles (Air Miles is a Canada based reward program offering flight mileage on a multiplicity of products and services, which was launched in the United States in 1992): (a) Members can also earn Air Mile points when they purchase products from any brands of Procter and Gamble; (b) Members can transfer their points or flight miles between both reward programs; (c) Members can redeem Air Miles points for Procter and Gamble products.

Advantages of partnership are attracting more prospective customers and customers can enjoy a variety services from two or more companies. Either 3 M or Procter and Gamble can consider forming partnership with other firms in different industries to expand their businesses. Additionally, 3 M or Procter and Gamble can complement with its partners to better allocate their business. Besides, firms can design customer satisfaction surveys and questionnaire for existing and prospective customers to obtain information about: (1) Appealing of web pages atmosphere; (2) How good selection of products was present; (3) How satisfied consumers are with their purchases via virtual store; (4) Age range; (5) What kind of goods consumers usually buy online; (6) Online purchase intention.

Surveys and questionnaire quantify firms' strengths in growth opportunities, area for improvement, and competitive threats. Furthermore, firms are able to view their performances objectively and adjust business model and target market on the right track. As attributions of markets in diverse regions, 3 M and Procter and Gamble in US and Taiwan would operate different ideals of network technology to compete domestically and internationally. If 3 M or Procter and Gamble and other loyalty programs collaborate on rewarding, it will enhance brand visibility in multiple countries and consumers will acquire diversifies services. Besides, they can mutually help one another in business. Network technology brings professional and creativity to the business market and convenience to either firms or consumers. Innovation is inevitable.

Acknowledgments This work was supported in part by the National Science Council (NSC), Taiwan, ROC, under Grant NSC Taiwan Tech Trek 2012, NSC 101-2511-S-018-016.

References

1. Kestenbaum MI, Straight RL (1996) Paperless grants via the internet. Public Adm Rev 56(1):114–120 Washington, D.C
2. Kalakota R, Whinston AB (1996) Frontiers of Electronic Commerce. Addison Wesley Longman Publishing Co., Inc., Redwood City, pp 1–3
3. Kopalle PK, Neslin SA (2003) The economic viability of frequency reward programs in a strategic competitive environment. Rev Mark Sci 1(1):1–39
4. Taylor GA, Neslin SA (2004) The current and future sales impact of a retail frequency reward program. http://dbs.ncue.edu.tw:2057/science/article/pii/S0022435905000692
5. Downes L, Mui C (2012) Unleashing the killer app: digital strategies for market dominance. Amazon.com
6. Chung KYC, Derdenger TP, Srinivasan K (2012) Economic value of celebrity endorsements. http://www.andrew.cmu.edu/user/derdenge/CelebrityEndorsements.pdf
7. Kim BD, Shi M, Srinivasan K (2001) Reward programs and tacit collusion. http://dbs.ncue.edu.tw:2105/stable/pdfplus/3181632.pdf?acceptTC=true
8. Je-How Y (2009) A study of applying internet marketing strategy to achieve clicks-and-mortar business
9. Liu SH, Huang JX (2011) FB million fans operating technique. Common Wealth Magazine p 464
10. Venkatesh V, Morris MG, Davis GB, Davis FD (2003) User acceptance of information technology: toward a unified view. MIS Q 27(3):425–478
11. VerticalNet (1999). http://www.verticalnet.com
12. Webopedia (what is database management system) (2012) http://www.webopedia.com/TERM/D/database_management_system_DBMS.html
13. Tseng WH (2010) A study of the effects of brand image and purchase intention on e-learning facebook fan page with moderator of cloud service. National Changhua University of Education, Changhua

Part III
Exploration of Scientific Evidence on Affective Learning

What is Affective Learning?

Wen-Yen Wang, Ling-Chin Ko, Yueh-Min Huang, Yao-Ren Liu and Shen-Mao Lin

Abstract Affective computing can be used to evaluate human psychological reactions after using affective equipment. And, learners produce intrinsic and extrinsic affective reactions in the learning process called as affective learning. However, some work evaluates affective learning through non-computing method as compared to computing one. Based on the survey, this work demonstrates two examples that used different methods to evaluate affective learning. In addition, the characteristics of the methods have been discussed and explored to understand affective learning further.

Keywords Affective computing · Affection

Introduction

What is affective learning? Affective learning means that learners generate intrinsic and extrinsic affective reactions in the learning process [1]. The reactions include personal emotion, feeling, fancy, attitude, and so on. Learners may also produce personal emotion performance and extrinsic reactions facing specific courses, teaching materials and subjects in the learning. These affective reactions

W.-Y. Wang (✉) · Y.-R. Liu · S.-M. Lin
Department of Information Engineering, Kun Shan University, No. 949, Da-Wan Road, Yung-Kang, Tainan city 71003, Taiwan, Republic of China
e-mail: wwang@mail.ksu.edu.tw

L.-C. Ko
Information and Communication, Kun Shan University, No. 949, Da-Wan Road, Yung-Kang, Tainan city 71003, Taiwan, Republic of China

Y.-M. Huang
Department of Engineering Science, National Cheng Kung University, Taiwan, No. 1, University Road, Tainan city 701, Taiwan, Republic of China

Y.-M. Huang et al. (eds.), *Advanced Technologies, Embedded and Multimedia for Human-centric Computing*, Lecture Notes in Electrical Engineering 260,
DOI: 10.1007/978-94-007-7262-5_20,

have cognitive changes with personal favor degree, referring to the learners' cognitive thoughts and behavioral performance [1]. The changes in the emotion performance can be regarded as a class of emotion. The extrinsic affective reactions of the learners in studying various things can generate different emotion combinations, which are regarded as a learning reaction of affective learning.

As the equipment for affective computing is complicated and expensive, instead of using affective computing for discussion, some studies observed the learners' intrinsic and extrinsic emotion in learning, and used adaptive strategy for discussion, so as to evaluate the learner's affective reactions.

Related Studies

Christian (2010) indicated the affective computing technology could be used to simulate the human cognitive inference capability, and implemented the test environment for simulation [2]. It was a 3D emotional space displaying the continuous development combination of human inner thoughts and somatic reactions. The facial expression of emotional change could be calculated by affective computing, so as to know how the affective reactions generated continuous combination process of a person's emotional response to body change through the effect of cognition. Kiavash et al. (2012) proposed an improved learning framework, using web camera and microphone for learning [3]. The camera and microphone were used to collect the learners' facial expressions, operating conditions and learning reactions. The learners' learning data were analyzed by affective computing, so as to help the learners observing their learning behaviors, and improve their learning effectiveness, flexibility and expandability. Nik and Tanya (2012) indicated that the facial expression of learners could be observed by computational analysis of affective learning and computer arithmetic, so as to identify the present emotion type of the learners [4]. The results could be helpful in evaluating the different emotional reactions in various environments. The learning changes and states of the learners could be analyzed effectively. Guhe et al. [5] designed an emotion mouse. When the users used the mouse, the mouse could sense the users' physiological data, including heart rate, hand temperature, conductivity of skin, and so on. The present physiological state of the users was recorded to analyze the learners' physiological state in various learning environments. Arindam and Amlan (2012) found that the affective computing technology could improve learning [6]. The learners' facial expressions were analyzed and classified by collecting the biological signal of emotion detection. The learners' response was known from the classified facial expression, and the suitable learning style was found as detailed classification for adjusting learning.

The affective learning state can calculate and measure the affective state of the learner in learning. Some scholars have observed the intrinsic and extrinsic emotion of the learner in learning, and used adaptive strategy for discussion for evaluating the learners' affective reactions. This is called affective learning, but is seldom discussed. Anderson and Krathwohl [7] proposed the theoretical five-hierarchy architecture of affective learning, including receiving, responding, valuing, organization and internalization from bottom to top. At the low hierarchy of the primary structure of emotion theory, the behavior of emotion hierarchy is more specific and apparent; at the high hierarchy, the hierarchy of emotion is more abstract and complex.

Although affective learning implements different strategies by using or not using affective computing, they aim to discuss the emotions of learners in learning, and to use these emotions to improve the learning state of learners.

Example of Affective Computing Strategy

Kwok et al. [8] suggested helping students to understand and utilize Six Thinking Hats in SAMAL to generate creative solution, instead of asking them to wear six colors of hats, and using the six interactions and emotions to evoke their mental states for problem solving. The SAMAL (emotional atmosphere learning) provides an unique integrated environment, using cognitive and environmental emotions to improve learning. Bono's Six Thinking Hats approach in learning process proposed a research model to check the learners' affective experience, learning participation and creativity in positive SAMAL environment. In the research model, compared with physical setting, SAMAL setting uses learners' vision, hearing, sense of smell and interactive feeling to evoke appropriate affective and psychological conditions, thus stimulating learners to participate in the process of Six Thinking Hats, and generating creative solutions in SAMAL. Therefore, the ambient stimulation or message is delivered to learner. The learner produces the personal perception after receiving the delivery, then express emotion that the perception affects his or her thinking. The entire procedure is named as ambient affective computing as shown in Fig. 1.

Example of Strategy not Using Affective Computing

Besides measuring biological features of learners, some scholars have discussed another aspect of affective learning, which is the basic concept of emotion. First, the learner receives learning triggering action, so that the internal learner responds and evaluates the value and experience obtained from learning. The value and experience of learning are reorganized, and the combined value and personal value

Fig. 1 Ambient affective leaning

Fig. 2 Affective learning with diverse affection

are combined. The architecture of affective learning theory is formed systematically to attain the deep level objective of educational psychology aspect. Krathwohl et al. [9] proposed the research direction of using five learning phases to observe the learners' emotional change and external response in learning. The extrinsic learning of learners is converted into personal psychological features, including intrinsic interest, attitude and value. Finally, American educationalists Anderson and Krathwohl (2001) proposed the overall theory of affective learning [7]. The theory of affective learning, from bottom to top, is divided into five hierarchies, including receiving, responding, valuing, organization, and internalization as Fig. 2. At the low hierarchy of the primary structure of the emotion theory, the behavior of emotion hierarchy is more specific and apparent; at the high hierarchy, the hierarchy of emotion is more abstract and complex. The architecture of affective learning is formed systematically to attain the deep level objective of educational psychology aspect. This is another part discussed in this study. The five hierarchies are introduced as follows. Receiving means the learner is willing or active to receive or participate in learning activities with some kind of stimulation. Responding indicates the learner is willing to participate in learning activities, and participates in learning as interested in learning. Thereafter, the affection is passed to valuing. On this stage, in terms of the learner's cognition or impression of persons and objects in an environment, the objects are judged and measured by personal inner assessment standard. It can be regarded as personal intrinsic value measurement criteria. Then, the organization stage expresses the learner conceptualizes the learned or referenced values, and then classifies and integrates them into a new value by systematization. Finally, the learner characterizes various exotic values, integrated with personal value by personal judgment into personal shared value belief system as the basis of behaving and dealing with matters in the future. This system can integrate belief, concept and value structures into consistent intrinsic system.

Conclusion and Future Research

Using auxiliary system to measure the features of learners in learning can reach very high accuracy, however, such measurement and computing are very costly and complex, and cannot be used extensively. Thus, the affective learning computing has not yet been extensively used in learning so far [1].

In terms of whether affective learning has used affective computing, there has not yet been questionnaire for affective learning in the affective assessment of affective learning in studies. Future studies can design a questionnaire for affective learning factors.

References

1. Picard R, Papert S, Bender W, Blumberg B, Breazeal C, Cavallo D et al (2004) Affective learning: a manifesto. BT Technol J 22(4):253–269
2. Becker-Asano C, Wachsmuth I (2010) Affective computing with primary and secondary emotions in a virtual human. Auton Agent Multi-Agent Syst 20(1):32–49
3. Bahreini K, Nadolski R, Westera W (2012) FILTWAM-a framework for online affective computing in serious games. Procedia Comput Sci 15:45–52
4. Thompson N, McGill TJ (2012) Affective tutoring systems: enhancing e-learning with the emotional awareness of a human tutor. Int J Inf Commun Technol Educ 8(4):75–89
5. Guhe M, Gray WD, Schoelles MJ, Liao W, Zhu Z, Ji Q (2005) Non-intrusive measurement of workload in real-time. In: Proceedings of the human factors and ergonomics society annual meeting, 2005. SAGE Publications, pp 1157–1161
6. MacLean EL, Matthews LJ, Hare BA, Nunn CL, Anderson RC, Aureli F et al (2012) How does cognition evolve? phylogenetic comparative psychology. Anim Cogn 15(2):223–238
7. Anderson LW, Krathwohl DR, Airiasian W, Cruikshank K, Mayer R, Pintrich P (2001) A taxonomy for learning, teaching and assessing: a revision of Bloom's taxonomy of educational outcomes, Complete edition. Longman, New York
8. Kwok R, Cheng SH, Ho-Shing Ip H, Kong J (2011) Design of affectively evocative smart ambient media for learning. Comput Educ 56(1):101–111
9. Krathwohl DR, Bloom BS, Masia BB (1964) II: handbook II: affective domain. David McKay, New York

The Influences of Emotional Reactions on Learning Gains During a Computerized Self-Assessment Test

Yueh-Min Huang, Chin-Fei Huang, Ming-Chi Liu and Chang-Tzuoh Wu

Abstract This study aims to examine learning gains and emotional reactions by receiving applause during computerized self-assessment testing for elementary school students. The participants were asked to solve mathematics problems in a computer-assisted self-assessment system with or without pre-recorded applause as emotional feedback while EEG measurements were taken. The results of this study provide support for the belief that it is useful to improve students' learning achievement by using emotional reactions such as applause during computer-assisted self-assessment testing, especially for male students. It is suggested that teachers may create such a positive emotional self-assessment learning environment as encourage students to learn by themselves more efficiently.

Keywords Applause · EEG · Emotion · Computer-assisted self-assessment

Introduction

Assessment is one of the useful instruments in the education domain since it can be frequently carried out by teachers to evaluate students' learning gains. However, from the students' perspective, assessment normally turns into an invisible source of stress and anxiety if they cannot achieve the expected grades [1]. A self-assessment test system is typically considered as an effective instructional strategy for training students to evaluate their own learning progress and helping them

Y.-M. Huang (✉) · M.-C. Liu
Department of Engineering Science, National Cheng Kung University,
Tainan City, Taiwan, Republic of China
e-mail: chiaju1105@gmail.com

C.-F. Huang · C.-T. Wu
Graduate Institute of Science Education, National Kaohsiung Normal University,
Kaohsiung, Taiwan, Republic of China

Y.-M. Huang et al. (eds.), *Advanced Technologies, Embedded and Multimedia for Human-centric Computing*, Lecture Notes in Electrical Engineering 260,
DOI: 10.1007/978-94-007-7262-5_21,

prepare to face anxiety and other emotional states during tests [2]. Self-assessment provides a practice chance for students to rehearse the course content and discover unfamiliar content, so they will be prepared for in-class tests [3]. Incidentally, an increase in students' test achievement and a decrease in their anxiety will occur through the training of self-assessment (Snooks 2004).

Although self-assessment is beneficial for students to prepare for tests and to maintain learning motivation, paper-based assessment always creates the stressful experience of a typical exam [4]. The evidence from past studies indicates that computer-based assessments (CBA) could be more user friendly for students [2]. Besides, the advantages of CBA include security, reducing the time and cost of assessment, recording students' test results automatically, and providing instant feedback [5]. To sum up, it is both important and suitable for learners to integrate self-assessment and computer-based assessment [2], (Nicol and Macfarlane-Dick 2006). In this study, we therefore combined self-assessment and computer-based assessment into a computerized self-assessment test.

Affect is the basis of human experience. There are two types of affect, positive and negative [6], where positive affect (e.g. acceptance, joy, confidence, etc.) has a positive impact on learning, memory, and thinking. On the contrary, negative affect (e.g. anxiety, anger, fear, sadness, etc.) has a negative impact on motivation, and leads to inattention. Moridis and Economides [2] indicated that negative affect impedes learning. More negative affect is caused when the learner replies to a question with the wrong answer. Therefore, affect has a tremendous impact on learners' learning. However, in paper-based assessment, it is difficult to understand learners' affect and to give them immediate feedback [7].

Issues of immediate affective feedback for computer-assisted self-assessment have been increasingly discussed in recent studies [2]. These studies all suggest that immediate affective feedback for computer-assisted self-assessment could help to promote students' self-confidence and performance. In this study, we adopted neuroscience technology to explain the reasons for the gender differences and why the rewarding feelings influence students' test performance. The results of this study could provide psychological evidence to interpret the underlying reasons for the findings and could provide useful suggestions for the design of further systems.

Methodolgy

This study adopted the Hot Potatoes System to design the computer-self-assessment tests. The experimental task and the controlled task designs are shown in Fig. 1. In the first step, the single choice question type was chosen to compose the tests for both the experimental and controlled tasks. Second, fifteen single choice type mathematics questions were individually imported into the computer-assisted test for both the experimental and controlled tasks. In other words, there are 30 single choice test questions in this study allotted to the two parallel tasks. Third, it was designed that the participants in the experimental task would receive applause when they got the right answers, whereas the participants in the controlled task would not. Fourth, all of the participants needed to complete the computer-assisted self-assessment test. The correlation of these two tests for the two tasks is 0.96. This high correlation indicates that the two tests are highly consistent with each other.

The details of the third step are that a pre-recorded sound of applause was imported into the Hot Potatoes System for the experimental task. If the students get the correct answer, the applause sound is played, accompanied by the word "correct" shown on the computer screen. If the students get the wrong answer, no sound is played and the word "wrong" is shown on the screen. In the controlled task, there is no applause played if the answer is correct; however, the word "correct" is shown on the screen. If the students get the wrong answer, the word "wrong" is shown. The computer-assisted self-assessment tests for the experimental and controlled tasks were exported to create the achievement tests for this study.

This study was conducted at an urban university in Taiwan. A total of 30 students (n = 30, 15 males, 15 females; mean age ± S.D. = 19.2 ± 2.0 years) participated in this study, divided into two groups, one male group and one female group. All participants were asked to complete both the experimental and controlled task tests using neuroscience technology throughout the whole process to collect the psychological data. This study conformed to The Code of Ethics of the World Medical Association (Declaration of Helsinki) and was approved by the ethics committee of National Kaohsiung Normal University.

In this study, students' anxiety levels were measured using the State-Trait Anxiety Inventory (STAI) [8] which was the same instrument used in the research of [2]. The self-report inventory included 20 items to assess the participants' state of anxiety. Responses to the items ranged from 1 to 4, as follows: (1) not at all; (2) somewhat; (3) moderately so, and (4) very much, according to the students' feelings. Scores range from a minimum of 20 (highest anxiety) to a maximum of 80 (lowest anxiety). The validity of the contents of the Chinese version was assessed by three professional psychologists, and the Cronbach's alpha was 0.91 for the state subscale.

Two mathematics tests on the addition of two-digit numbers were administered in this study, consisting of 15 single choice type questions with a perfect score of 30. The correlation of these two tests is 0.96, and the Cronbach's alpha values were 0.84 and 0.87 for the two tests respectively.

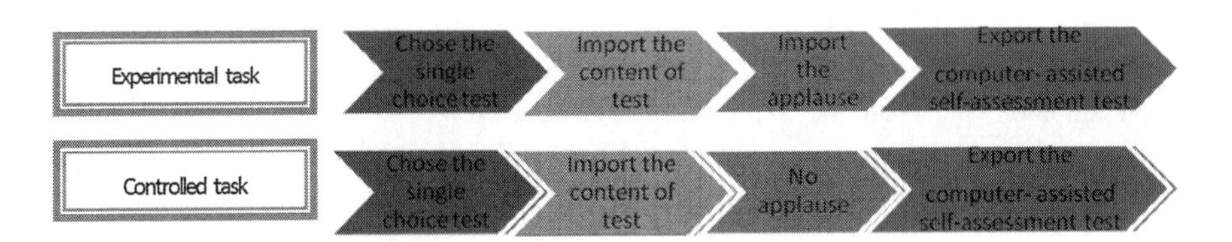

Fig. 1 The hot potatoes system

In this study, the EEG supplied neuroscience data indicative of the effect of gender differences on emotional reactions during a computer-assisted self-assessment test, and to the best of our knowledge, this is the first study to provide concrete evidence regarding this issue. The neuroscience technology adopted in this study is EEG (Electroencephalography) which is a procedure to measure the electrical activity of the brain through the skull and scalp [9]. When participants recognize specific emotional reflections, the corresponding electrical activities in the brain are induced [10].

All students were advised that they needed to take a computer-assisted final mathematics exam in this study. They were also told that they could participate in a computer-assisted self-assessment test twice as practice before the exam. All students in this study took the computer-assisted self-assessment test twice, the controlled test first followed by the experimental test. The two tests were separated by an interval of one week. The duration of each test was approximately 20 min.

The state of anxiety questionnaire was distributed both before and after the controlled and the experimental tests. In the process of completing the tests, all participants had to wear the electrode cap in order to collect the EEG data. At the beginning of the EEG experiment, all students were asked to sit down and relax. The EEG of their rest state was collected to be the individual brain wave baseline.

We recorded all participants' EEG signals when they were completing the experiments. Frequency analysis was performed in the delta (1–4 Hz), theta (4–7 Hz), alpha (8–12 Hz), beta (13–30 Hz) and gamma (>30 Hz) frequency bands. The evidence of brain activities from these frequencies and power values could help to verify the hypothesis that the part of the brain that generates feelings of reward is more active in males than in females during the computer-assisted self-assessment test. The extracted data were analyzed using a *t* test and ANCOVA analysis (SPSS version 17.0).

Results and Discussion

Computer-assisted self-assessment tests with and without applause were administered in this study and the statistical data from the state of anxiety questionnaire were collected before and after each test. The first to be administered was the controlled test without applause as emotional feedback. The results of this test are discussed first. The results show that the male group scored 12.4 ± 1.5 points compared with 11.3 ± 1.8 points for the female group in the controlled task mathematics test. There is no significant difference between the two groups' performance on the test. However, for the male group, the state of anxiety after the controlled test is significantly higher than before the test.

Then, t-test analysis was used to assess the differences in the scores of the state of anxiety of the male and female groups before the controlled computer-assisted self-assessment test. The results show that the females had a significantly higher state of anxiety before the test ($t = -2.05$, $p < 0.05$) than the males. Hence, analysis of covariance (ANCOVA) was used to assess the differences in the scores of the state of anxiety of the two groups after the test. The result shows that there are no significant differences ($F = 2.96$, $p > 0.05$, $\eta 2 = 0.099$) in the scores of state of anxiety between gender after the controlled computer-assisted self-assessment test without applause. This finding is consistent with the results of the research of [2] who suggested that the main effect of gender on state of anxiety is not significant.

Our findings by anxiety questionnaire mostly in line with those found in literatures. However, we argue that anxiety should be a continuous state rather than an outcome of a period. Apparently, the data from questionnaire in the aftermath of the test presumably regarded as the emotion state needs to be confirmed. In short, a continuous observation is a substantial step to conclude the questionnaire. For the purpose, the EEG data of the males and females who performed the experimental computer-assisted self-assessment test with applause were analyzed. The topographical map of the brain is shown in Fig. 2. For the male group, the result reveals that the power values of the alpha 1 and alpha 2 frequencies are more active on the two sides of the frontal lobe than they are for the female group. Blackhart et al. [11] mentioned that the power values of the alpha 1 and alpha 2 frequencies from the two sides of the frontal lobe are often induced by the appearance of positive emotions [11]. In other words, the higher power values of the alpha 1 and alpha 2 frequencies mean more positive emotion activities. Therefore, the findings indicate that, in the experimental computer-assisted self-assessment test with applause, the part of the brain that generates feelings of reward is more active in males than in females during computer-assisted self-assessment testing (Table 1).

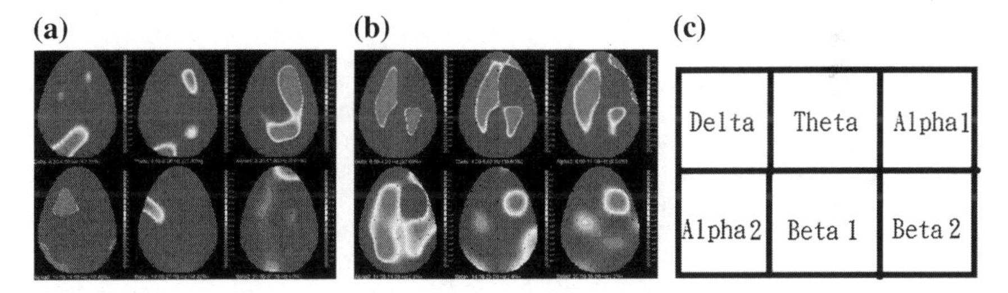

Fig. 2 The hot potatoes system. **a** Females. **b** Males. **c** Frequency table

Table 1 The scores of state of anxiety before and after the computer-assisted self-assessment test for males and females not receiving applause (controlled task)

Task	Gender group	Variables	Mean ± S.D.	*t*	Cohen's *d*
Controlled task	Male group	State of anxiety (pretest)	63.87 ± 3.70	−2.41*	0.030
		State of anxiety (posttest)	61.93 ± 3.13		
	Female group	State of anxiety (pretest)	59.73 ± 6.90	−0.65(n.s.)	0.049
		State of anxiety (posttest)	60.07 ± 6.94		

Huang and Liu [12] mentioned that the frontal lobe of the human brain dominates humans' high-level thinking, mental rotation and math calculations. Ho et al. [10] also indicated that the delta frequency in the frontal lobe of the human brain is related to humans' high-level cognitive processing. For this reason, this study further analyzed the delta frequency power values in the frontal lobe (F4 electrode). The results show that both the male and female groups had higher delta frequency power values when completing the controlled computer-assisted self-assessment test without applause feedback than for the test with applause feedback (see Fig. 2).

References

1. Caraway K, Tucker CM, Reinke WM, Hall C (2003) Self-efficacy, goal orientation, and fear of failure as predictors of school engagement in high school students. Psychol Schools 40(4):417–427
2. Moridis CN, Economides AA (2012) Applause as an achievement-based reward during a computerised self-assessment test. Br J Educ Technol 43(3):489–504
3. Kulik JA, Kulik CLC, Bangert RL (1984) Effects of practice on aptitude and achievement-test scores. Am Educ Res J 21(2):435–447
4. Ricketts C, Wilks SJ (2002) Improving student performance through computer-based assessment: insights from recent research. Assess Eval High Educ 27(5):475–479
5. Terzis V, Economides AA (2011) The acceptance and use of computer based assessment. Comput Educ 56(4):1032–1044
6. Moridis CN, Economides AA (2008) Toward computer-aided affective learning systems: a literature review. J Educ Comput Res 39(4):313–337
7. Sung YT, Chang KE, Chiou SK, Hou HT (2005) The design and application of a web-based self- and peer-assessment system. Comput Educ 45(2):187–202
8. Spielberger CD (2005) State-trait anxiety inventory for adults. Mind Garden, Redwood City
9. Coles MGH, Rugg MD (1995) Event-related brain potentials: an introduction. In: Rugg MD, Coles MGH (eds) Electrophysiology of mind: event-related brain potentials and cognition. Oxford University Press, New York, pp 1–26
10. Ho MC, Chou CY, Huang CF, Lin YT, Shih CS, Han SY, Shen MH, Chen TC, Liang CL, Lu MC, Liu C-J (2012) Age-related changes of task-specific brain activity in normal aging. Neurosci Lett 507:78–83
11. Blackhart GC, Kline JP, Donohue KF, LaRowe SD, Joiner TE (2002) Affective responses to EEG preparation and their link to resting anterior EEG symmetry. Pers Individ Differ 32:167–174
12. Huang CF, Liu CJ (2012) An event-related potentials study of mental rotation in identifying chemical structural formulas. Eur J Educ Res 1(1):37–54

A Conceptual Framework for Using the Affective Computing Techniques to Evaluate the Outcome of Digital Game-Based Learning

Chih-Hung Wu, Yi-Lin Tzeng and Ray Yueh Min Huang

Abstract That's an interesting issue for how the outcome that educators use angry bird to teach projectile motion physics theorem. For verifying the possibility of playing angry bird to learn the projectile motion physics problem, this study design an experiment that include two different learning methods. One is the tradition learning method, and the other one is to learn the projectile motion using Angry Bird. When student learning, their eye movement data, brain wave and heart beat will be measured for analyzing their attention, emotion and the strategy of solving problem. After learning, they take a posttest to prove the digital game-based learning method can help student learning.

Keywords Affective computing · Eye movement · Brain wave · Heart rhythm coherence · Digital game-based learning

Introduction

The motivation of games could be combined with curricular contents into "Digital Game-Based Learning (DGBL)" [1]. Games that encompass educational objectives and subject matter are believed to hold the potential to render learning of academic subjects more learner-centered, easier, more enjoyable, and more

C.-H. Wu (✉) · Y.-L. Tzeng
National Taichung University of Education, Taichung, Taiwan, Republic of China
e-mail: chwu@ntcu.edu.tw

Y.-L. Tzeng
e-mail: bit099101@gm.ntcu.edu.tw

R. Y. M. Huang
National Cheng Kung University, Tainan, Taiwan, Republic of China
e-mail: huang@mail.ncku.edu.tw

Y.-M. Huang et al. (eds.), *Advanced Technologies, Embedded and Multimedia for Human-centric Computing*, Lecture Notes in Electrical Engineering 260,
DOI: 10.1007/978-94-007-7262-5_22,

interesting. Although games are believed to be motivational and educationally effective, the empirical evidence to support this assumption is still limited and contradictory [2]. The mobile game "Angry Bird" is a very famous game. Because "Angry Bird" has relations with projectile motion physics theorem. Some educators combine this game and physics course to enhance the student's motive. But the educators use angry bird to teach that is just for fun, or really can promote student's grade in real course. That's an interesting issue.

For verifying the possibility of playing angry bird to learn the projectile motion physics problem, this study design an experiment that include two different learning methods. One is the tradition learning method, and the other one is to learn the projectile motion using Angry Bird. When student learning, their eye movement data, brain wave and heart beat will be measured for analyzing their attention, emotion and the strategy of solving problem. After learning, they take a posttest to prove the digital game-based learning method can help student learning.

Literature Review

Digital Game-Based Learning

Several studies found the digital game can change the players' emotion. Ravaja et al. [3] looked at facial EMG activity and skin conductance responses as well as assessments of mood during game-play (joy, pleasant relaxation, fear, anger, and depressed feeling) in response to short duration emotional game events. They found the game events did in fact lead to emotion state [3].

Affective Computing in Learning

Since Affective Computing was proposed, there has been a burst of research that focuses on creating technologies that can monitor and appropriately respond to the affective states of the user [4]. Because this new Artificial Intelligence area, computers able to recognize human emotions in different ways. Why human emotion is an important research area? The latest scientific findings indicate that emotions play an essential role in decision-making, perception, learning and more [5].

The Physiological Input Signals this Study Selected

The physiological input signals of eye movement, EEG and ECG were selected to input our learning affective recognition system. According the past studies, several techniques need to be combined to estimate the state of attention and emotion. Eye movements provides information about location of attention and the nature, sequence and timing of cognitive operations [6]. With the emergence of Electroencephalography (EEG) technology, learner's brain characteristics could be accessed directly and the outcome may well hand-in-hand supported the conventional test recognize a learner's Learning Style [7]. And the arousal state of the brain [8], alertness, cognition, and memory [9, 10] also can be measure. Heart rate variability from ECG, has gained widespread acceptance as a sensitive indicator of mental workload [6]. And positive emotions may change the HF components of HRV [11].

Method

Research Hypotheses

To examine the relationships among different learning methods, learner attention, emotion, strategy and learning outcome, this study utilized the following two different learning methods: digital game-based learning, study the projectile motion physics problem by Angry Bird; tradition learning, study by text description, coordinates and formula. To fairly compare how digital game-based learning affect learning attention, emotions, strategy and learning outcome, the two learning methods in this study have the same learning content and learning objectives; that is, the same learning materials are presented in different methods. Figure 1 shows the relationship framework of the discussed research variables in this study (Tables 1, 2).

Research Variables

The input and output variables in this study are shown in Table 3. Learner attention is recognized by the Neurosky system, it was used to detect neuron electric triggering activity, and it has the earphone appearance. According the NeuroSky proprietary Attention and Meditation eSense algorithms, NeuroSky can report the attention score each second. The range of attention score is 1–100 (1 = very low attention level and 100 = very high attention level). Learner emotion is recognized by the emWave system, which uses human pulse physiological signals to identify Coherence score every 5 s. Coherence score have 0, 1 and 2 (0 = negative emotion, 1 = peaceful and 2 = positive emotion). Strategy includes visual attention and sequential analysis. To successfully solve this problem, participants need to distinguish the key factors. Fixation duration on options and factors provide the data to show that learners' thinking will pay much

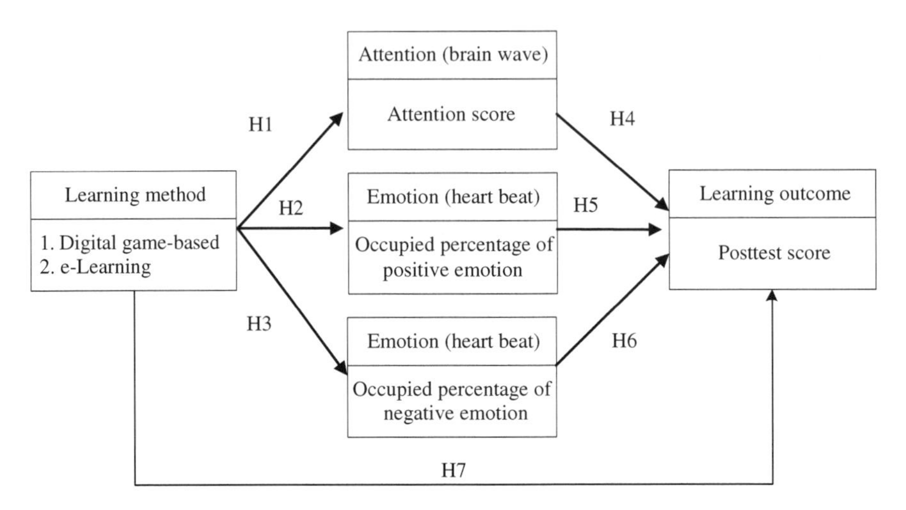

Fig. 1 Relationship framework of the research variables discussed in this study

Table 1 Multi physiological feature system review

		Physiological input signals					
Research object	Reference	Eye	EEG	ECG	Facial	Speech	SCR
Emotion recognition	[12]			×			×
Neonatal seizures	[13]		×	×			
Emotion recognition	[14]				×	×	
Emotion recognition	[6]	×		×	×		
Visual search task	[15]	×	×				
Emotion recognition	[16]		×		×		
Emotional distractors	[17]				×		
Emotion recognition	[18]					×	
Emotion recognition	[19]			×			
Brain computer interface	[20]						
Reading process	[21]	×	×				
Emotion recognition	[22]	×			×		
Learning state	[23]			×			
Driver fatigue	[24]			×			
Driver fatigue	[25]		×	×			
Emotion recognition	[8]		×				
Epilepsy state	[26]		×	×			
Learning state	*My research*	×	×	×			

According to the table, we found the physiological signals of eye movement, EEG and ECG have become the research trends. But it not exist a system combined these signals to recognize the affective of human

attention to refer which key factors. In addition, sequential analysis can effectively infer the overall behavioral path patterns during learners' thinking. We can observe the problem solving logic of learners. The assessment of learning outcome is based on pretest and posttest results.

Table 2 Hypotheses and reference in this study

Hypotheses	Reference
H1. Learners use digital game-based learning method that have better attention score, and have significant difference	[27]
H2. Learners use digital game-based learning method that have more occupied percentage of positive emotion, and have significant difference	[3, 28]
H3. Learners use digital game-based learning method that have less occupied percentage of negative emotion, and have significant difference	[3, 28]
H4. There is a significant positive correlation between attention and learning outcome	[29, 30]
H5. There is a significant positive correlation between positive emotion and learning outcome	[23, 31]
H6. There is a significant negative correlation between negative emotion and learning outcome	[23, 31]
H7. Learners use digital game-based learning method that have better learning outcome, and have significant difference	[2]

Table 3 Input and output variables in this study

Input device	Input variables	Output
Neurosky	Attention score will be calculated each second. (The range of attention score is 1 to 100; 1 = very low attention level and 100 = very high attention level.)	Attention
emWave	Coherence score will be calculated every 5 s. (Coherence score have 0, 1 and 2; 0 = negative emotion, 1 = peaceful and 2 = positive emotion)	Emotion
Eye tracker	1. Visual attention: fixation duration on options and factors 2. Scan paths for sequential analysis	Strategy

Participants and Design

Thirty university students will participate in this study. All of them took the fundamental projectile motion course in the past. Therefore, they already possessed some prior knowledge for solving the physics problem. All participants had good visions and passed the eye-tracking calibrations. For verifying the possibility of playing angry bird to learn the projectile motion physics problem, this study refers a multiple-choice science problem solving study [32]. Four images in which four factors (velocity in slingshot, degree in slingshot and pigs from a variety of structures) were included were designed to be inspected by each participant during the problem solving task. These four factors correspond with the velocity, degree, texture and fall point in projectile formula. In projectile motion course, the projectile formula is shown in following:

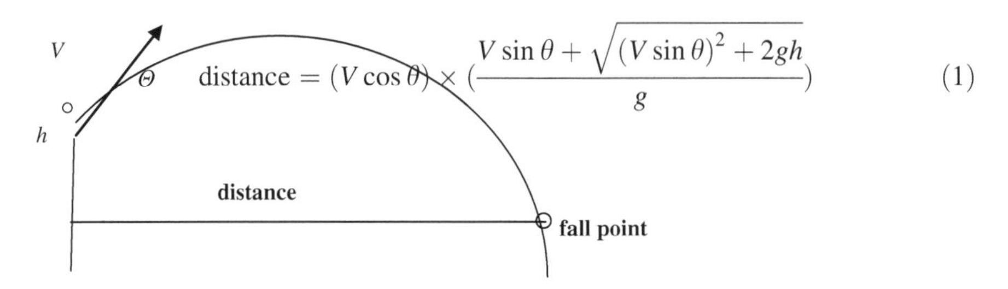

$$\text{distance} = (V\cos\theta) \times \left(\frac{V\sin\theta + \sqrt{(V\sin\theta)^2 + 2gh}}{g}\right) \tag{1}$$

where V is velocity, θ is degree, h is the distance between ball and ground, distance is the length of the ball from slingshot flying to fall point.

Procedure

Participants will test individually. On arrival and after attaching physiological sensors, the participants will be asked to read an introduction. Participants will be told that four images in which four factors (velocity in slingshot, degree in slingshot, birds and pigs from a variety of structures) will be displayed for 5 m on a screen in front of them. On the top of the screen, the problem will be initiated by a statement describing, "According the first shot, please select an image inferring a best combination you will choose and justify your selection." Before the experiment begins, all of the participants pass the calibrations by eye tracker and Ware have signal input. Next, the experiment begin, the emotion and eye movement data will be recorded when participant solving the problem. In addition, participants

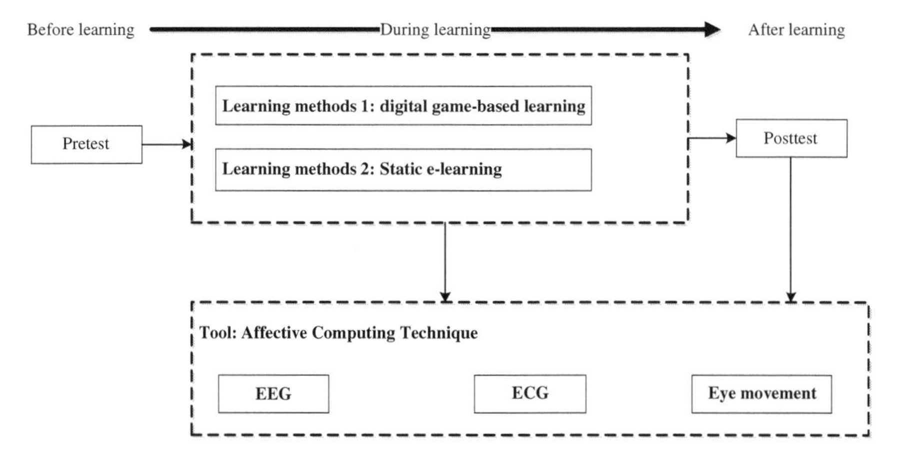

Fig. 2 The flow of experiment

will be asked to think aloud while solving the problem. By doing so, we can accumulate participants' justifications which are used to check against their selection. A think aloud training will be conducted before the experiment start. The entire experiment will lasting approximately 10 min. Participants' eye movement data, emotion and attention state and question responses will be recorded. The flow of experiment is shown in Fig. 2.

References

1. Prensky M (2003) Digital game-based learning. ACM Comput Entertainment 1:1–4
2. Marina P (2009) Digital game-based learning in high school computer science education: impact on educational effectiveness and student motivation. Comput Educ 52:1–12
3. Ravaja N, Turpeinen M, Saari T, Puttonen S, Keltikangas-Jarvinen L (2008) The psychophysiology of James Bond: Phasic emotional responses to violent video game events. Emotion 8:114–120
4. Picard R (1997) Affective computing. MIT Press, Cambridge
5. Ben Ammar M, Neji M, Alimi AM, Gouardères G (2010) The affective tutoring system. Expert Syst Appl 37:3013–3023
6. Lin T, Imamiya A, Mao X (2008) Using multiple data sources to get closer insights into user cost and task performance. Interact Comput 20:364–374
7. Rashid NA, Taib MN, Lias S, Sulaiman N, Murat ZH, Kadir RSSA (2011) Learners' learning style classification related to IQ and stress based on EEG. Procedia Soc Behav Sci 29:1061–1070
8. Zhang Q, Lee M (2012) Emotion development system by interacting with human EEG and natural scene understanding. Cogn Syst Res 14:37–49
9. Berka C, Levendowski DJ, Cvetinovic MM, Petrovic MM, Davis G, Lumicao MN, Zivkovic VT, Popovic MV, Olmstead R (2004) Real-Time Analysis of EEG indexes of alertness, cognition, and memory acquired with a wireless EEG headset. Int J Hum Comput Interact 17:151–170
10. Berka C, Levendowski DJ, Lumicao MN, Yau A, Davis G, Zivkovic VT, Olmstead RE, Tremoulet PD, Craven PL (2007) EEG correlates of task engagement and mental workload in vigilance, learning, and memory tasks. Aviat Space Environ Med 78:B231–244
11. von Borell E, Langbein J, Després G, Hansen S, Leterrier C, Marchant-Forde J, Marchant-Forde R, Minero M, Mohr E, Prunier A, Valance D, Veissier I (2007) Heart rate variability as a measure of autonomic regulation of cardiac activity for assessing stress and welfare in farm animals—A review. Physiol Behav 92:293–316
12. Kim KH, Ban SW, Kim SR (2004) Emotion recognition system using short-term monitoring of physiological signals. Med Biol Eng Comput 42:419–427
13. Greene BR, Boylan GB, Reilly RB, de Chazal P, Connolly S (2007) Combination of EEG and ECG for improved automatic neonatal seizure detection. Clin Neurophysiol 118:1348–1359
14. Ruffman T, Henry JD, Livingstone V, Phillips LH (2008) A meta-analytic review of emotion recognition and aging: implications for neuropsychological models of aging. Neurosci Biobehav Rev 32:863–881
15. Latanov AV, Konovalova NS, Yermachenko AA (2008) EEG and EYE tracking for visual search task investigation in humans. Int J Psychophysiol 69:140
16. Zhang Q, Lee M (2010) A hierarchical positive and negative emotion understanding system based on integrated analysis of visual and brain signals. Neurocomputing 73:3264–3272
17. Srinivasan N, Gupta R (2010) Emotion-attention interactions in recognition memory for distractor faces. Emotion 10:207–215

18. Yang B, Lugger M (2010) Emotion recognition from speech signals using New Harmony features. Signal Process 90:1415–1423
19. Murugappan M, Ramachandran N, Sazali Y (2010) Classification of human emotion from EEG using discrete wavelet transform. J Biomed Sci Eng 3:390–396
20. Lee EC, Woo JC, Kim JH, Whang M, Park KR (2010) A brain–computer interface method combined with eye tracking for 3D interaction. J Neurosci Methods 190:289–298
21. Dimigen O, Sommer W, Hohlfeld A, Jacobs AM, Kliegl R (2011) Coregistration of eye movements and EEG in natural reading: analyses and review. J Exp Psychol Gen 140:552–572
22. Schmid PC, Schmid Mast M, Bombari D, Mast FW, Lobmaier JS (2011) How mood states affect information processing during facial emotion recognition: an eye tracking study. Swiss J Psychol 70:223–231
23. Chen C-M, Wang H-P (2011) Using emotion recognition technology to assess the effects of different multimedia materials on learning emotion and performance. Libr Inf Sci Res 33:244–255
24. Patel M, Lal SKL, Kavanagh D, Rossiter P (2011) Applying neural network analysis on heart rate variability data to assess driver fatigue. Expert Syst Appl 38:7235–7242
25. Zhao C, Zhao M, Liu J, Zheng C (2012) Electroencephalogram and electrocardiograph assessment of mental fatigue in a driving simulator. Accid Anal Prev 45:83–90
26. Valderrama M, Alvarado C, Nikolopoulos S, Martinerie J, Adam C, Navarro V, Le Van Quyen M (2012) Identifying an increased risk of epileptic seizures using a multi-feature EEG–ECG classification. Biomed Signal Process Control
27. Wen-Hao H (2011) Evaluating learners' motivational and cognitive processing in an online game-based learning environment. Comput Hum Behav 27:694–704
28. Baldaro B, Tuozzi G, Codispoti M, Montebarocci O, Barbagli F, Trombini E, Rossi N (2004) Aggressive and non-violent videogames: short-term psychological and cardiovascular effects on habitual players. Stress Health 20:203–208
29. Annette MS (2004) Attention performance in young adults with learning disabilities. Learn Individ Differ 14:125–133
30. Robinson K, Winner D (1998) Rehabilitation of attentional deficits following brain injury. J Cogn Rehabil 16:8–15
31. Goleman D (1995) Emotional intelligence. Bantam Books, New York
32. Tsai M-J, Hou H-T, Lai M-L, Liu W-Y, Yang F-Y (2012) Visual attention for solving multiple-choice science problem: an eye-tracking analysis. Comput Educ 58:375–385

Adopt Technology Acceptance Model to Analyze Factors Influencing Students' Intention on Using a Disaster Prevention Education System

Yong-Ming Huang, Chien-Hung Liu, Yueh-Min Huang and Yung-Hsin Yeh

Abstract This paper explores the potential of geographic information system (GIS) in disaster prevention education. Open source GIS is applied to build a disaster prevention education system used to assist students in strengthening their knowledge of typhoon prevention and enhancing awareness of typhoon disaster. An experiment which the technology acceptance model was applied as the theoretical fundamental was designed to investigate students' intention on using the system. A total of 34 university students participated in using the proposed system. Results show that (1) perceived ease of use has a positive and significant influence on attitude toward use and perceived usefulness; (2) perceived usefulness has a positive and significant influence on attitude toward use and behavioral intentions; (3) attitude toward usage does not have a significant influence on the students' intention to use the system.

Keywords GIS · Disaster prevention education · Technology acceptance model

Y.-M. Huang
Department of Applied Informatics and Multimedia, Chia Nan University of Pharmacy and Science, Tainan, Taiwan, Republic of China
e-mail: ym.huang.tw@gmail.com

C.-H. Liu (✉)
Department of Network Multimedia Design, Hsing Kuo University of Management, Tainan, Taiwan, Republic of China
e-mail: chliu@mail.hku.edu.tw

Y.-M. Huang · Y.-H. Yeh
Department of Engineering Science, National Cheng Kung University, Tainan, Taiwan, Republic of China
e-mail: huang@mail.ncku.edu.tw

Y.-H. Yeh
e-mail: nimo.tw@gmail.com

Y.-M. Huang et al. (eds.), *Advanced Technologies, Embedded and Multimedia for Human-centric Computing*, Lecture Notes in Electrical Engineering 260,
DOI: 10.1007/978-94-007-7262-5_23,

Introduction

Disasters always caused the losses of life, property damage as well as social and economic disruption. Disaster is a serious disruption of functions of a community/society, which includes human, material, economic or environmental losses [1]. Disasters involve natural disasters, technological disasters, and man-made disasters [2]. Earthquake, flood, landslide, windstorm, drought and wildfire are a type of natural disaster [2]. Industrial and transport accident as well as bomb explosion are a type of technological disaster [2]. Man-made disaster is an event brought extensive damage and social disruption through complex technological, organizational and social process such as terrorist activity [2, 3]. Once disasters occur, it always led to huge loss no matter what caused it. Thus, it is a vital issue to develop a sound approach to mitigate the effect of disasters.

Disaster education is one of useful ways to mitigate the effects of disasters [4–7]. Early on, Vitek and Berta proposed that education is the most reliable means of gaining information about disasters and learning how to react during emergencies [7]. Later, Becker reported a technological disaster education project which was used to foster the ability of students to collaboratively deal with a chemical or nuclear disaster [4]. Ronan and Johnston examined the role of disaster education for increasing youths' resilience to disasters, and their findings revealed that disaster education was helpful to increase youths' resilience to disasters [5]. Recently, Tanaka explored whether disaster education can enhance people's readiness for disaster. His results showed people with disaster education are more prepared than people without disaster education [6]. Overall, disaster education can assist people in realizing the seriousness of disasters and promote people's capacity for handling disasters further.

Among the studies of disaster education, natural disaster education has been regarded as the most important issues in some countries such as Taiwan, because such disasters such as typhoons and torrential rains often caused huge property damage. In these countries, typhoons easily brought severe wind, floods, landslides, and debris flows from Kalmaegi (July 2007), Sinlaku (September 2007), and Jangmi (September 2008). A famous case is that Typhoon Morakot hit Taiwan on 8 August 2009 which caused the second highest damages of school facility in history [8]. Consequently, it is a vital issue to assist the people in these countries in strengthening their knowledge of disaster prevention and promoting disaster awareness for reducing the loss of life and property damage.

In this study, we used an open source geographic information system to develop a disaster prevention education system, and help students strengthen their knowledge of natural disaster prevention and promote natural disaster awareness. To explore the perspectives of students on the system, an experiment based on the technology acceptance model (TAM) was constructed [9, 10]. Specifically, we implemented the system and deployed it at a university. A questionnaire was designed to explore students' perspectives on the system. Finally, a series of analyses were conducted to examine the model and draw a conclusion about the analyses.

Research Design

Research Tool

In this work, we aimed to develop a disaster prevention education system and intended to support student engagement in a typhoon prevention education curriculum. To this end, MapGuide Open Source was used to develop the system. MapGuide Open Source is a web-based platform that enables researchers to develop and deploy web mapping applications. More importantly, it provides users with interactive system design that includes support for feature selection, property inspection, map tips, and operations. Figure 1a shows the user interface which A area shows the name of the system, B area shows the menu of system that supports students in choosing the learning topic, C area shows the description of system, and D area shows the usage of the system. Furthermore, in order to support student engagement in learning more realistically, the historical data such as photo and video is included to ensure that students can be fully immersed in the learning and achieve further meaningful learning. Figure 1b shows an example. System will play the then disaster video when students view the historical disaster event of a certain region.

Research Model and Hypotheses

TAM is regarded as one of significant roles in the successful development of e-learning system [11, 12]. It is one of famous means to evaluate users' perspective on acceptance of technology [9, 10], which was developed by Davis and his colleagues. Davis et al. proposed four main perceived constructs to develop TAM, that is, perceived ease of use (PEU), perceived usefulness (PU), attitude toward use (AT), and behavioral intentions (BI). PEU refers to a person believes that using a technology would be free of effort [9]. PU refers to a person believes that using a technology would enhance his/her job performance [9]. AT refers to a person's

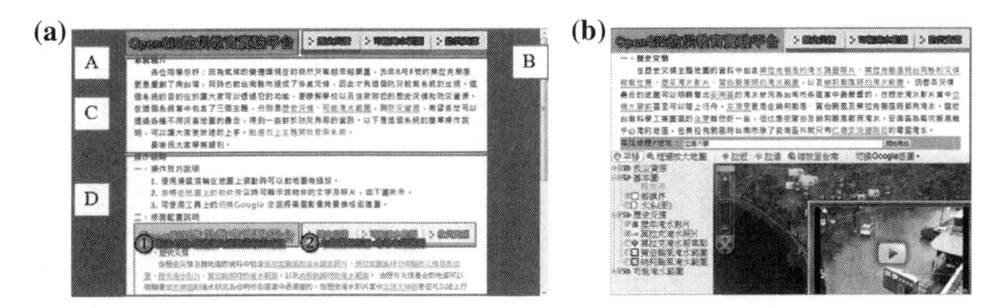

Fig. 1 The disaster prevention education system interface. **a** User interface. **b** Student views the historical disaster event of a certain region

general feeling of a favorableness or unfavorableness toward some stimulus object [13]. BI refers to a person's subjective probability that he/she will perform a specified behavior [14]. Through TAM, researchers can understand whether the system meets users' requirements and demonstrate the systems' value further. Consequently, TAM is adopted to investigate students' perspectives on the disaster prevention education system. Figure 2 shows the research model, which originates from TAM theory. The model consists of five hypotheses, which are described as follows:

From the studies of TAM [9, 10], PEU was hypothesized to influence PU and AT, and subsequently PU was hypothesized to influence AT and BI, and AT was hypothesized to influence BI. Consequently, the third to the seventh hypothesis are shown as follows:

H1. PEU is positively related to PU.
H2. PEU is positively related to AT.
H3. PU is positively related to AT.
H4. PU is positively related to BI.
H5. AT is positively related to BI.

Participants, Questionnaire, and Procedure

The participants were students from a university in Tainan City, Taiwan. A total of 34 students enrolled in the experiment. The framework for questionnaire design based on a review of prior studies [9–12, 14] as well as feedback from two experts. The questionnaire included four constructs, that is, PEU, PU, AT, and BI. At the start of the experimental procedure, all the participants executed a learning activity through the disaster prevention education system. In the activity, the participants used the system to strengthen their knowledge of typhoon prevention. When the activity was completed, the participants were asked to fill out the questionnaire that examined the proposed research model.

Results

In this study, the partial least squares (PLS) approach was used to analyze the questionnaire data, due to the small sample size. In this paper, SmartPLS 2.0 was used to assess the measurement and structural models [15]. The measurement

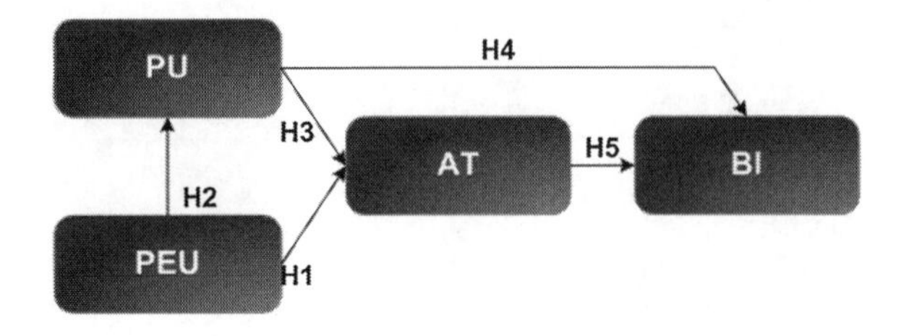

Fig. 2 Research model

Table 1 The convergent validity, reliability of measure, discriminant validity for the measurement model

	Convergent validity	Reliability of measure		Discriminant validity			
	AVE	Composite reliability	Cronbach's alpha	Latent variable correlations			
				PEU	PU	AT	BI
PEU	0.83	0.93	0.90	0.91			
PU	0.91	0.96	0.95	0.70	0.95		
AT	0.88	0.95	0.93	0.66	0.69	0.93	
BI	0.77	0.91	0.85	0.67	0.70	0.58	0.87

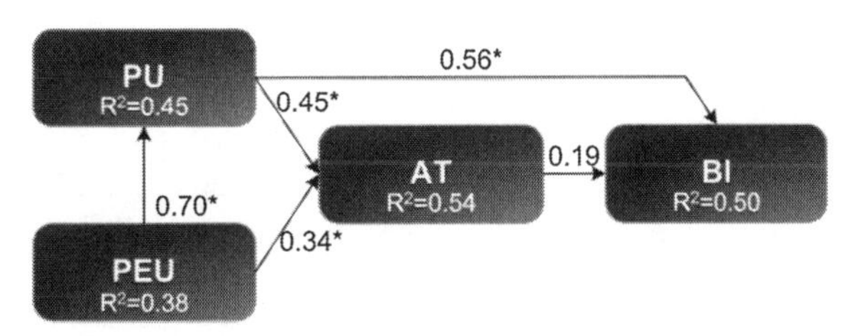

Fig. 3 The results of the structural model

model was assessed by convergent validity, reliability of measure, and discriminant validity. Table 1 shows the results of the measurement model is acceptable, since all the values meet the standard levels.

The structural model was used to verify the hypotheses by using path coefficients and R^2 value. The R^2 was used to assess the ability of the model to explain the variance in the dependent variables. The path coefficients were used to assess the statistical significance of the hypotheses. These results indicated that one hypothesis refuted the predictions, that is, H5; while the others confirmed the predictions. One reason for rejecting H5 is that the participants in this study did not perceive their use of technology to be mandatory, so that AT is not a significant predictor of intention to use technology. This result is consistent with previous research suggesting that AT is a significant predictor of intention to use technology mainly under mandatory conditions of technology use [16] (Fig. 3).

Conclusions

This study used an open source geographic information system to develop a disaster prevention education system to help students strengthen their knowledge of typhoon disaster prevention. To explore students' perspectives of the system, TAM was applied to build the research model, and PLS was used to assess the model. The results revealed that the system was successfully accepted by the students in the sample, but attitude toward use does not have a significant influence on the students' intention to use the system.

Limitations of this study include the type of the measurements, and the relatively small sample size. In this study, all of the measurements of this study are limited to students' self-reported perceptions. In future work, we will introduce additional measurements to explore the effects of the proposed system on disaster prevention education. Furthermore, increasing the sample size to obtain stronger evidence for the proposed system will be expected because the small sample size might limit the power of this study.

Acknowledgments The authors would like to thank the National Science Council of the Republic of China for financially supporting this research under Contract No. NSC 102-2511-S-041-001, and NSC 101-2511-S-432-001.

References

1. ISDR (2011) International strategy for disaster reduction. Retrieved 3 Apr 2011 from http://www.unisdr.org/eng/terminology/terminology-2009-eng.html
2. Mansourian A, Rajabifard A, Valadan Zoej MJ, Williamson I (2006) Using SDI and web-based system to facilitate disaster management. Comput Geosci 32(3):303–315
3. Shaluf IM, Ahmadun F, Said AM, Mustapha S, Sharif R (2002) Technological man-made disaster precondition phase model for major accidents. Disaster Prev Manage 11(5):380–388
4. Becker SM (2000) Environmental disaster education at the university level: an integrative approach. Saf Sci 35(1–3):95–104
5. Ronan KR, Johnston DM (2001) Correlates of hazard education programs for youth. Risk Anal 21(6):1055–1064
6. Tanaka K (2005) The impact of disaster education on public preparation and mitigation for earthquakes: a cross-country comparison between Fukui, Japan and the San Francisco Bay Area, California, USA. Appl Geogr 25(3):201–225
7. Vitek JD, Berta SM (1982) Improving perception of and response to natural hazards: the need for local education. J Geogr 81(6):225–228
8. Chen CY, Lee WC (2012) Damages to school infrastructure and development to disaster prevention education strategy after Typhoon Morakot in Taiwan. Disaster Prev Manage 21(5):541–555
9. Davis FD (1989) Perceived usefulness, perceived ease of use and user acceptance of information technology. MIS Quart 13(3):319–340
10. Davis FD, Bagozzi RP, Warshaw PR (1989) User acceptance of computer technology: a comparison of two theoretical models. Manage Sci 35(8):982–1003
11. Liu IF, Chen MC, Sun YS, Wible D, Kuo CH (2010) Extending the TAM model to explore the factors that affect intention to use an online learning community. Comput Educ 54(2):600–610
12. Sanchez-Franco MJ (2010) WebCT—the quasimoderating effect of perceived affective quality on an extending technology acceptance model. Comput Educ 54(1):37–46
13. Fishbein M, Azjen I (1975) Belief, attitude, intention and behavior: an introduction to theory and research. Addison-Wesley, Reading
14. Chatzoglou PD, Sarigiannidis L, Vraimaki E, Diamantidis A (2009) Investigating Greek employees' intention to use web-based training. Comput Educ 53(3):877–889
15. Ringle CM, Wende S, Will A (2005) SmartPLS 2.0 (beta). Retrieved 22 Oct 2010 from http://www.smartpls.de
16. Teo T, Noyes J (2011) An assessment of the influence of perceived enjoyment and attitude on the intention to use technology among pre-service teachers: a structural equation modeling approach. Comput Educ 57(2):1645–1653

Designing an Interactive RFID Game System for Improving Students' Motivation in Mathematical Learning

Ho-Yuan Chen, Ding-Chau Wang, Chao-Chun Chen and Chien-Hung Liu

Abstract Game-based learning becomes a critical issue in the e-learning field. Many instructors want to make their students can do the studying with fun and their interest activities. For this reason, this paper designs an interactive RFID learning system for improving learners' motivation and performance by adopting game-based learning. There are several advantages for using this RFID learning system while learners behave well, such as learners will not feel they are engaging in a traditional learning environment when playing the RFID learning system. The purpose of this research was to develop an interactive RFID learning device and competitive learning environment for enhancing students' learning motivation and number sense in mathematics subject. The research is to investigate whether and how this RFID learning system can be developed to help users learn mathematics with enjoyment. This study considers ideas in game design, motivation issues, and mathematics learning to develop a strategy to engage users with the interactive RFID system. In this research, the learners can do the synchronic learning with other classmates outside of the classroom. Moreover, instructors can

H.-Y. Chen
Graduate School of Education, Chung Yuan Christian University, Jhongli, Taiwan, Republic of China
e-mail: huc140@cycu.edu.tw

D.-C. Wang (✉)
Department of Information Management, Southern Tainan University, Tainan, Taiwan, Republic of China
e-mail: dcwang@mail.stust.edu.tw

C.-C. Chen
Institute of Manufacturing Information and Systems, National Cheng-Kung University, Tainan, Taiwan, Republic of China
e-mail: chaochun@mail.ncku.edu.tw

C.-H. Liu
Department of Network Multimedia Design, HsingKuo University of Management, Tainan, Taiwan, Republic of China
e-mail: chliu@mail.hku.edu.tw

Y.-M. Huang et al. (eds.), *Advanced Technologies, Embedded and Multimedia for Human-centric Computing*, Lecture Notes in Electrical Engineering 260,
DOI: 10.1007/978-94-007-7262-5_24,

evaluate the individual's learning levels in mathematics in this research by analyzing the database of the game-based RFID learning system.

Keywords Interactive RFID applications · Game-based learning · Mathematics education · Number sense

Introduction

Clearly, instructional technology changes have influenced many educational activities, especially in the field of game-based learning and e-learning. In fact, instructors face a complex task in designing, developing, and evaluating e-learning courses, which include many different factors [7, 10, 11]. For this reason, program planners must consider several factors as they provide their learners with effective learning activities by using technology. Many researchers have pointed out that computer games as one kind of math learning tools with considerable potential for students to learn mathematics in game-based context. In reference to these concepts, we are developing the interactive RFID learning system which can be designed to assist students to learn mathematics.

We have to understand the composition of the interactive RFID learning system in order to design a game-based learning module first. Then we will discuss some concepts related to the computer-based learning. Moreover, we will describe several issues we faced to develop the interactive RFID learning system. Finally, the contributions of the RFID learning system in the game-based learning will be introduced. The research designs an interactive RFID learning system for improving learners' motivation and performance by adopting the car racing game-based learning. The RFID learning system meets individual's learning needs and provides various learning situations in the game-based learning activities.

There are several advantages for using a RFID learning system while learners behave well.

First, the interactive RFID learning system is a game-based learning device. Users will not feel they are engaging in a traditional learning environment when playing the RFID learning module.

Second, the competition system in the interactive RFID learning system is based on their prior knowledge in Mathematics. In this research, we designed a competition system as an example to deliver different level task-based questions according to the answer learners choose is right or wrong.

Third, many scholars have pointed out that distance education involves teachers and students separated by geographic and time factors [5]. The learners can do the synchronic learning with other classmates outside of the classroom, because we developed the interactive RFID learning system can be accessed through the internet.

The rest paper is organized as follows. Section System Architecture discusses the system architecture. We proposed the interactive RFID system designs for improving users' learning motivation, number sense, and performance in mathematics in section Designs of Game-Based Learning RFID Module. Then, section Demonstration shows the prototype implemented based on the interactive RFID system. Finally, we conclude our study in section Conclusion.

System Architecture

Figure 1 shows the reference architecture of the interactive RFID system. In the system, the database of interactive RFID learning system includes three factors: the ranking data of scores that users get, the inferential ability of the mathematics system, and the logic ability of the mathematics system. The logic ability of the mathematics system provided various logic questions for learners to figure out the right answers. The inferential ability of the mathematics system assists learners to learn the different concepts related to mathematics. Besides, the system can record users' learning processes for instructors to analyze the different learning behaviors and levels. In refer to this concept, instructors can meet individual's learning needs and situations based on this database.

Figure 2 shows the steering wheel which attached the RFID Tag. Learners can use the steering wheel device to choose the plus, minus, times, or divide when they are playing the game.

Designs of Game-Based Learning RFID Module

The game-based learning RFID system mainly include the RFID communication component (e.g., RFID reader and tag) and the database component. In this

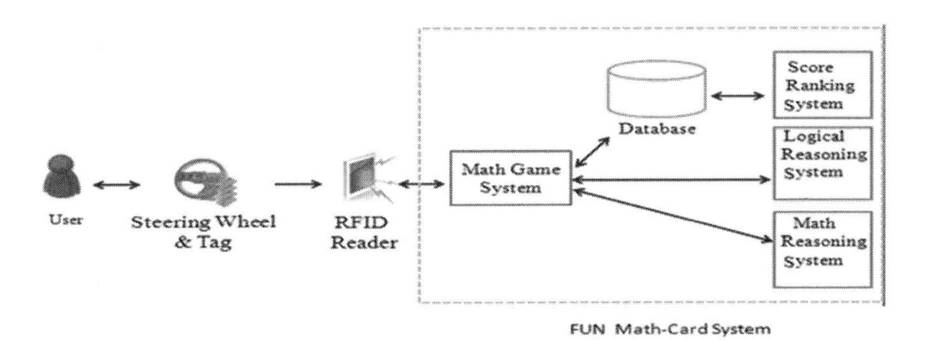

Fig. 1 System architecture diagram

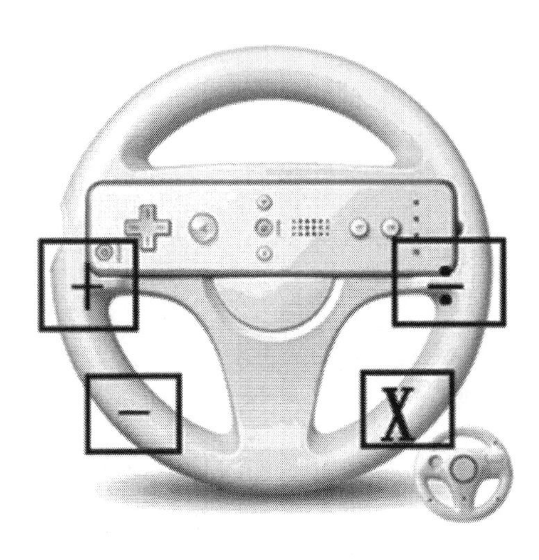

Fig. 2 The steering wheel which attached the RFID tag

section, we design the game-based learning RFID module in order to improve the users' learning motivation and number sense in mathematics field.

The Interactive RFID Learning Module

Figure 3 shows the interactive RFID learning system we designed. The RFID Reader can receive the information from the RFID Tag which is attached with steering wheel. Then, the RFID Reader delivers the learning data to the computer screen.

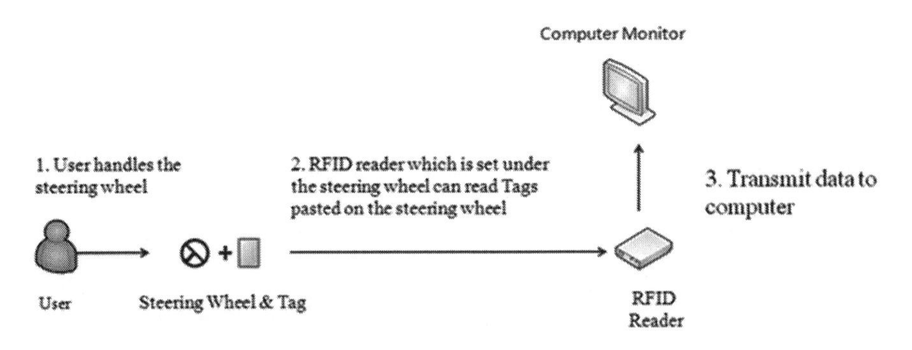

Fig. 3 System operation diagram

The Playing Steps of the Interactive RFID System

There are several important play steps in this interactive RFID system:

Step 1 The screen of the computer will show the questions when users turn on the start button. Also, the car will be started to run in the computer interface when the users start to play the game.

Step 2 The right or wrong answers:

2.1 If the users choose the right answer, the interface will show the score of the question.

2.2 If the users choose the wrong answer, they could not get any score.

Step 3 The system will continue showing the new questions when users finish one question. The learners can have 5 min in each section.

Step 4 The users will receive the ranking of the scores from the computer screen when they finish each section.

Computer-Based Learning Environment

Computer-based learning (CBL) lets learners can learn new knowledge or skills from not only in traditional learning environment but also by computers. In the computer-based learning system, learners can practice themselves in anytime and anywhere. As the technology innovating, many instructors start to adopt computer-based learning in order to improve students' learning motivation [9].

Besides, many researches indicated that number sense of the learners had increased greatly by using computer-based learning tools [1, 4]. Moreover, the internet become more and more popular among out life, learners can do the competitive and collaborative learning activities via internet. The National Council of Teachers of Mathematics (NCTM 2000) indicated that the use of computer software can assist learners to study the mathematics.

Many researchers pointed out that game-base learning can attract learners' attentions and motivation [2, 9]. Also, the game-based learning can promote the math test performance more than the traditional learning approaches. In this research, the design strategy adopts a competitive game as the learning context in order to keep students' attention and motivate students to engage the learning activities.

In reference to this concept, we design the interactive RFID learning module to combine the game activities for learning mathematics. However, the design of effective game-based learning environment and to make learners to engage game-based learning environments is not simple. Moreover, motivation plays a critical role in the learning of mathematics [3]. Many researchers pointed out that the competition environment is an effective way to motivate learners to engage the learning environments [8]. For this reason, this research design a system for users

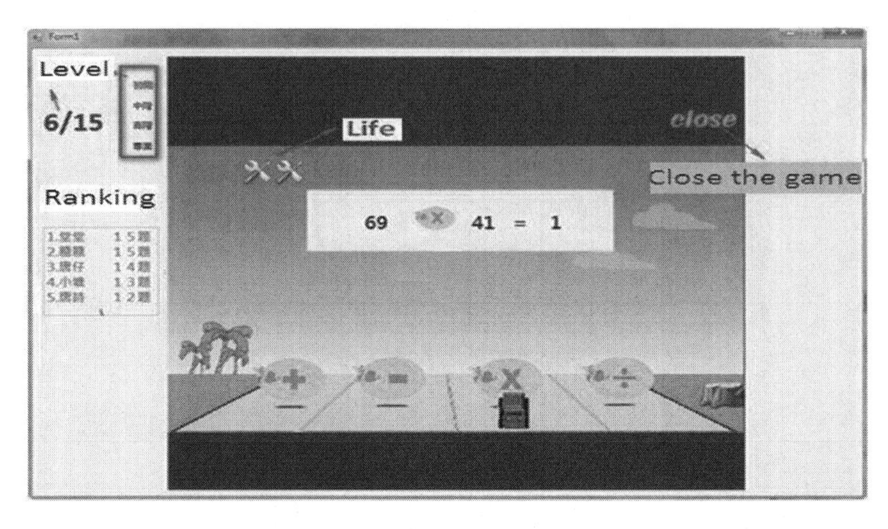

Fig. 4 Practice interface

which can help them receive the information related to the ranking of scores in this race game and who are also login at the same time. The main purpose of this research is to develop an online interactive RFID learning system on computer-based environment for learners competing themselves with other classmates. Learners will keep practicing the mathematical skills when they are trying to receive the higher ranking in this game-based learning environment. Finally, the users' records related to their learning processes of the interactive RFID learning system is recorded in the database of the system. In case good game-based learning quality is adopted in our research experiment. Moreover, we left the issue related to develop the system on mobile devices in future works.

Demonstration

We have developed the interactive RFID learning system for instructors to make students do the learning with fun. Figure 4 shows the main interface of the game-based learning system on computer. The interface will provide several information included questions, different sections, and the ranking. In educational field, games usually are learner-center and have several important characteristics such as rules design, competition, challenging activities, choices, and the ranking of the score points. One of the primary advantages of the interactive RFID learning system is that they can play the game in competition environments. Game-based learning environment has the potential to improve students' number sense in mathematics education. In this research, the interactive RFID learning system can automatically deliver appropriate feedback related to the answer is wrong or right in the interface of the race game. Each user has five chances to finish one question. This interactive RFID learning device can be used with a wide range of users along a continuum of different ability levels of mathematics. Learners can move to the advance level sections when they chance the right answer. If they could not complete the question during the five chances, the interactive RFID learning system would automatically deliver the other same level question for users to practice again. Besides, the users can check their ranking information in this result interface. Moreover, the interactive learning system will increase the speed of the racing car after the user complete one section in order to improve learners' number sense in the game-based learning context.

Conclusions

This research implemented a game-based learning environment by using an interactive RFID learning system to improve learners' learning motivation, number sense, learning performance, and also collected three different kinds of data: the logic ability of the mathematics, the inferential ability of the mathematics, and the ranking data of scores.

For the instructors, instructors usually adopt the exam to evaluate the student's learning level in the ending of the course in the traditional learning environment. However, teachers can analyze the database of the game-based learning system in order to realize the individual's learning levels in mathematics in this research and also realize how the game system can be develop to assist students to practice their number sense. Besides, the instructors can utilize the interactive RFID learning system to develop the self-learning game to increase learners' motivation.

In refer to the learners, the interactive RFID learning system is fun and will not make users feel they are studying or testing. Users can feel they are engaging a racing game and have the opportunity to compete with other classmates. The game-based learning becomes an important issue in e-learning field in order to make learners can do learning with fun and their interest subjects. In this research, learner could be encouraged to contribute more times to do the game-based learning when they get higher ranking of scores more than other classmates. At the same time, learners can improve their number sense and approach to the solution of a given mathematical questions.

Future Work

There are several future works needed to consider to improve this research. First, the mobile device technology has become very popular nowadays. In the e-learning field, the basic use of information communication technologies innovation is to assist both instructors and students to engage in interactive educational opportunities across many barriers.

In regard to these reasons, we should combine the online interactive RFID learning system into mobile device. Learners can play the game with others by using smart phone or e-pad. Besides, the learners can utilize their mobile device to practice the skills they received or to learn more information and knowledge in this field.

The other future work is to develop the system related to the educational feedback and reward system. In traditional web-based learning environment, learners usually receive the feedback from the learning module or instructor, such as the ranking of scores, a summary of the learning performance, or provide the information for them to challenge the next difficult level questions or tasks. However, the reward system is not only the one of feedbacks but also is an efficient way to keep learners' motivation and psychological needs.

Acknowledgments Thanks to Chin-Yin Lin and Yen-Ju Tsai, students at Southern Taiwan University, for writing code to establish the system, and Huei-Err Hsu, Si-Yu Su, and Pin-Juin Chen for designing and drawing the user interface.

References

1. Cavanagh S (2008) Playing games in class helps students grasp math. Educ Digest: Essent Readings Condens Quick Rev 74(3):43–46
2. Ke F (2008) Alternative goal structures for computer game-based learning. Int J Comput Support Collaborative Learn 3(4):429–445
3. Ma X, Kishor N (1997) Attitude toward self, social factors, and achievement in mathematics: a meta-analytic review. Educ Psychol Rev 9(2):89–120
4. Miller D, Brown A, Robinson L (2002) Widgets on the Web: using computer-based learning tools. Teach Except Child 35(2-):24–28
5. Moore MG, Kearsley G (1996) Distance education: a systems view. Wadsworth, Belmont
6. National Research Council (2000) How people learn: Brain, mind, experience, and school. National Academy Press, Washington, DC
7. Pearson J, Trinidad S (2005) An instrument for refining the design of e-learning environments. J Comput Assist Learn 21:396–404
8. Schwabe G, Göth C (2005) Mobile learning with a mobile game: design and motivational effects. J Comput Assist Learn 21(3):204–216. doi:10.1111/j.1365-2729.2005.00128.x
9. Sedig K (2008) From play to thoughtful learning: a design strategy to engage children with mathematical representations. J Comput Math Sci Teach 27(1):65–101
10. Thurmond VA, Wambach K, Connors H, Frey B (2002) Evaluation of student satisfaction: determining the impact of a Web-based environment by controlling for student characteristics. Am J Distance Educ 16(3):169–189
11. Trinidad S, Aldridge J, Fraser B (2005) Development and use of an online learning environment survey. J Educ Technol 21(1):60–81

Part VI
Multimedia Technology for Education

Design and Development of an Innovation Product Engineering Process Curriculum at Peking University

Win-Bin Huang, Junjie Shang, Jiang Chen, Yanyi Huang, Ge Li and Haixia Zhang

Abstract Over the past decade, the elements of innovative wisdom are not only a business organization to survive in the dangerous environment, school organizations to enhance the quality of education to meet the needs of the community. For the succession challenge in the continuous impact, an innovation level of a top university covered the administration, curriculum, teaching, equipment, environment and so on is quite extensive. This paper overviews a novel curriculum at Peking University, called Innovation Product Engineering Process, established by six interdisciplinary teachers for school students in various professional fields. The curriculum aims at inspiring students to break through professional limitations for experiencing the innovation process from idea into product. The students are self-organized as a team and construct a prototype collaboratively. Instructors from industrial give a practical perspective lesson and provide market information, funding and technical support. Students in the course are fostered six expected abilities, including creativity, practical, engineering process, team-working, communication and expressiveness. Ideas from students become the topic of a project after competing in three eliminating rounds. All competitions are graded and ranked by the participators (teachers, students, instructors from academic and

W.-B. Huang (✉)
Department of Information Management, Peking University, Beijing 100871, China
e-mail: sebastian.huangwb@gmail.com

J. Shang
Graduate School of Education, Peking University, Beijing 100871, China

J. Chen
Department of Electronics, Peking University, Beijing 100871, China

Y. Huang
College of Engineering, Peking University, Beijing 100871, China

G. Li
Department of Computer Science, Peking University, Beijing 100871, China

H. Zhang
Institute of Microelectronics, Peking University, Beijing 100871, China

Y.-M. Huang et al. (eds.), *Advanced Technologies, Embedded and Multimedia for Human-centric Computing*, Lecture Notes in Electrical Engineering 260,
DOI: 10.1007/978-94-007-7262-5_25,

industry). Finally, six ideas having the opportunity to become prototypes are developed successfully with various properties. Most of students indicated that the curriculum provided them a new training experience, interesting learning style and useful content of courses.

Keywords Workforce development • Product engineering process • Course development

Introduction

Cultivating and training innovative talents is certainly important in order to building an innovative country and improving international competitiveness. School students having high social resilience and entrepreneurial capacity is one of the aims of the long-term education reformation and development program from 2010 to 2020 in China. It should be achieved through education and training of various fundamental science, research and practice. High level education in engineering, besides, in China mainly fosters people translating science and technology knowledge into productive power [1]. Developing students' innovation through engineering practice, integrating technology, humanities, economics and management knowledge, is also an important part of advance engineering education. The interdisciplinary cooperation, however, is few and far not only between academics but also between colleges and enterprises. Advance engineering education, therefore, should be oriented strenuously towards the practice of engineering process. The integrated engineering activities training a student solving complex, comprehensive and interdisciplinary problem collaboratively with different professionals have become the key point of education around the world.

Currently, the reformation of engineering practice is initialed in US and few academics have set up the courses for inspiring students in product processing engineering. The Conceive Design Implement Operate (CDIO) project [2] in Massachusetts Institute of Technology is regarded as the representative in this teaching model and its teaching manner provides students to experience a life cycle of a product process in the real environment. Project-based learning (PBL) in engineering practice is adopted in Canada [3]. Students as a team to completing the production of the specified projects put the theories and methods what they have learned in use. Different from the traditional curriculum, learning assessment is not only dominated teachers but taking a combination of student self-evaluation and peer evaluation by team members. Evaluation system and the way of the reformation of engineering practice courses, as a result, is an advisable merit of the PBL. The reformation in British perform practice course intending to improve the learning experiences of the science and engineering student to be "engineer" instead of "student." Furthermore, the engineers in industry are invited to give lectures or workshops for importing working experiences and instructing interdisciplinary exercises. Academics also connect a two-way interaction with industries in order to understand the demand for graduates. The courses recently increase the number of knowledge in humanities, economics and social sciences [4].

In recent years, the reformation of engineering education and practice are speeded up in China, and a lot of improvements on student participation and engineering practice are achieved. Comparing with the developed countries, however, two problems as follows are still critical. (1) The course of engineering and practice mostly focus on technical contents without much attention on general educations. (2) Teaching proportion of the arrangement in a course is unreasonable and the importance of a teacher in engineering practice education is over-valued. Consequently, students emphasize technical tools and skills too much during their learning in education. Self-innovation of students is probably disappeared under the situation of education. In order to increase the creativity and innovation of students at Peking University, the co-authors from different fields open a novel and interesting course with the following originality in China.

- Inviting engineers and managers from industry to sharing market information, providing finance, importing experiences and so on.
- Providing comprehensive knowledge including social science, engineering and management to students.
- Fostering students to experience product engineering process through completing a project.
- Allowing students to participate in the evaluation system of the course and scoring an idea or a work with the "press" equipment.

According to the outcome of the course and feedback from students, well interaction among teachers and students is observed, and curiosity about engineering and practice capability of students are improved. Students also understand and respect the opinions and works of others. School members comprehend the needs of industries and the discipline of the projects includes chemistry, computer science, communication, signal processing, the design of application on tablet PC or smart phone and so on. At the end of lesson, a company in China even intends to invest its resource in one of them as a product in the future.

Course Description

The course being team taught by the co-authors of this paper and titled, "Innovation Product Engineering Process," at Peking University has the following purposes in creating this course:

- Design a course integrating topics from interdisciplinary content relevant to product engineering process.

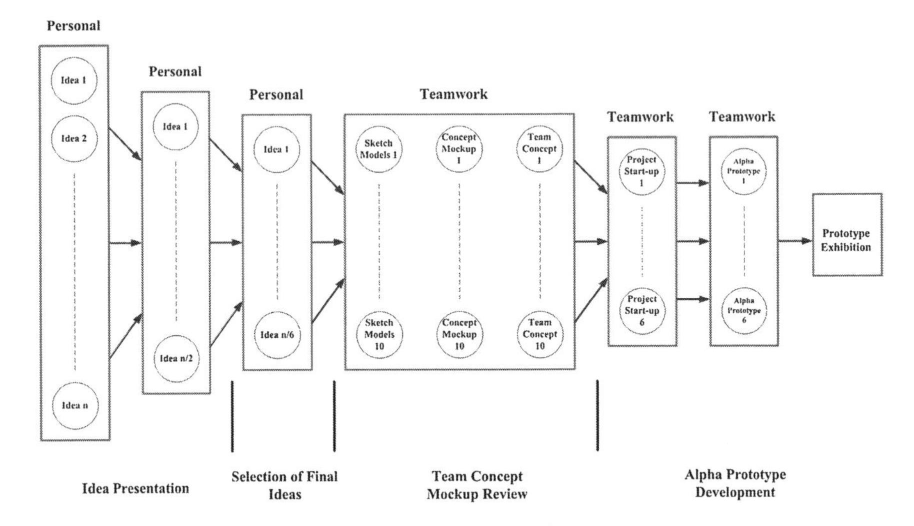

Fig. 1 A high-level schematic of the project workflow in the course is delineated

- Design a course to target audience of undergraduate students of various professional fields.
- Design a course with a teaching team including members in academics and industries.
- Design a course to provide students funding and technical supports to experience from idea creation to prototype making collaboratively.
- Design an evaluation system to allow students participating in.

A high-level schematic of the project workflow is shown in Fig. 1. The process leading to a functional and workable prototype is separated into four major milestones: the ideas presentation; selection of final ideas; team concept mockup review; and prototype development. Every student, at first, in the course creates more than an idea, and these initial ideas are examined in three eliminated round. In the stage, ideas are made under students' brain storming, surveying literature and sharing information. Ten initial ideas then become the major topics of a team project—they work in full cooperation and virtually coordinating resource as appropriate before team concept mockup review. The goal of the team concept mockup review is to inform instructors about the state of the project's functional, the concept of final prototype. This feedback also provides each responsible team prioritize improvements for the final presentation. Course content coverage is described as follows:

- Ideas Presentation
  - Week 1: Overview and introduction to innovation, creativity and entrepreneurship [Lecture].

  - Week 2: All students share personal ideas as more as possible and make a defense to others' interrogation. Here 50 % of the ideas are eliminated.
  - Week 3: Research and design process in an innovated project. [Lecture]
  - Week 4: The students survived in the last competition represent the selected idea in 3 min specifically and also make a defense to others' interrogation. Here 30 % of ideas are collected to next eliminating rounds.
- Selection of Final Ideas
  - Week 5: Fundamentals of proposal writing, and how to make a business plan. [Lecture].
  - Week 6: Each one of the remainder ideas should be represented clearly and specifically in 10 min, including requirement, application and scenario. The students initiated these ideas make a defense to others' interrogation and only ten ideas are selected.
  - Week 7: A special workshop is hold in the class for interaction freely among the original designer and other classmates, and then making a team finally. The designer with his team members, moreover, develops sketch, technical drawing and preliminary plan of the idea as a project.
  - Week 8: Each team has 15 min to make a detail presentation on their preliminary project, and the responses to others' interrogation are also considered. In the end, only six teams are obtained the opportunity to implement their idea with all supports, such as finance, technical, laboratory. Besides, a school teacher is responsible for technical advising, trouble shooting and schedule control of a team selected. Each team also has a budget of near ¥ 3,000, depending on the discussion of all school teachers, to purchase materials, supplies, and resources for the project. The members of the eliminated teams are separated themselves into the succeed teams.
- Team Concept Mockup Review
  - Week 9: The relation between technology and product: why failed? [Lecture].
  - Week 10: A discussion is hold for interaction, exchange and information sharing among all members in the class.
  - Week 11: Management and control of a project. [Lecture].
  - Week 12: A discussion is hold for interaction, exchange and information sharing among all members in the class.
  - Week 13: Leadership, communication and exchange. [Lecture].
- Alpha Prototype Development
  - Week 14: Each team must plan and work to keep their projects on budget and on scheduled. Moreover, they have to report their progress, balance of appropriations and technical detail. The teachers consider budget extensions while a budget overrun for further completing the project well is required.
  - Week 15–18: Each team keeps doing their project and reports to responsible school guidance. Furthermore, a few lectures and discussions of which the speakers are invited from enterprise or industry are held from time to time.

– Week 19: A whole-school exhibition is held for final presentation made by the team's representative and the demonstration of its alpha prototype in interesting, funny, formal or surprising way is necessary.

The lectures with different topics are given by different professional co-authors, engineers, and managers from industry. This course was offered for the first time in spring 2013 and met class once a week with each lecture and workshop being approximately 180-mins. Student's feedback and evaluation of course will be taken into account to improve the course contents and organization in future offerings. Students' performance on idea propagation, teamwork performance, system design, and final presentation is considered as the measure of this course.

Project Budget

Most of financial support in the course is provided by Peking University. In the stage of selection of final ideas, the teachers determine the practical financial needs according to the proposal, scheduling, scenario and development requirements. The final budget of a team is accepted with over 50 % guidance's agreements. On average, each team has a budget of ¥ 3,000 to purchase materials, supplies, and resources for the project. Each team member must participate in planning and keeping their projects on budget and on scheduled. The project budget is not compliant after verdict in the class. The teachers, however, consider budget extensions of a project while a budget overrun for further completion is required. There is only a chance to add their budget up while reviewing the mockup of team concept. Each team member will pay an equal portion of deficit if their budget is overrun. Moreover, each team is allowed to have sponsorship fee from enterprises.

Grading

The overall score of a student is graded based on four parts with its proportion: idea propagation (25 %), teamwork performance (25 %), system design (40 %), final presentation (10 %). Before forming a team, the performance of a student is evaluated as a partial personal grade. All members including teachers and students in the course vote pass or fail to the idea into next eliminated round, and the assistant instructor calculates its score. Here customer needs, thoughtfulness, clarity and quality of the design alternative of the ideas are considered. Once the six ideas are determined as the major content of a team, personal score of a student is graded in the "idea propagation" part. After that, the review contributes to a portion of a shared team-wide grade and members of course or team participate in the rest review process. Key grading critical in teamwork performance are operating, activity, workload and communication in coordination. The score of a student in this part is graded by all team members and guidance. Furthermore, all members in the course participate in grading a team in system design, of which key critical contains mechanical design details, system integration, details of prototype execution and manufacturing. In the final presentation, it takes place in one day of the summer vacation and provides each team with the opportunity to show their works to various audience including academics and industrial visitors. All participators in the exhibition evaluate a team's work based on team's performance in customer data, market information, specifications, or benchmarks for the product.

Prototype Exhibition

The final milestone in the course is a formal presentation which is attended by the overall members in the course, all guidance, sponsors, and guests from academic and industry. A portion of the shared team-wide grade is contributed in the exhibition. Each team has the opportunity to demonstrate their work to all participators. Students may learn how to prepare a complete technical presentation in a life-styled, educated, technical, or business oriented way. A team is also allowed to seek investors for their product to start-up a company. Each team is evaluated and graded by all participators based on its presentation quality, business assessment; technology, the prototype, and overall potential to become a real product.

Summary

A new curriculum combining innovation, learning, co-operation and practice is in development at Peking University in line with the need of interdisciplinary training to product engineering process. This course covers creativity, teamwork, management and practice of a project realization emphasizing pioneering aspects. After taking this course, students are expected to contribute to the start-up aspect of industrial projects related to product engineering. Students will be able to understand vulnerabilities and difficulties to the product engineering process in addition to realizing the basic principles of project workflow. Students are expected to critically analyze the interdependencies of related workflow in product engineering process and apply the interdisciplinary principles that they have learned in starting a practical idea up. After the course, many meaningful and useful prototypes created by the students are impressive.

Acknowledgements Authors would like to thank Peking University for financially supporting the work reported in this paper and the experienced lecturers from academic and industry for a professional speech.

References

1. Chen Jin (2010) Building the innovative country: Theory and practice. Science Press
2. Crawley EF (2001) The CDIO Syllabus. A statement of goals for Undergraduate Engineering Education
3. Michel J (2009) Management of change-implementation of problem-based and project-based learning in engineering. Eur J Eng Educ 34(6):606
4. The Imperial Study Guide [EB/OL] (2011) Imperial, College London pp 7–12

The Design of an Educational Game for Mobile Devices

Daniela Giordano and Francesco Maiorana

Abstract The importance of computing education is well known across different fields from STEM to Computer Science, from Humanities to Social Science. Educating the younger generation to 21st century skills is advocated by many international organizations since these skills can be used across several disciplines. Shifting from educating students to be user of a software tool to be designer of customization of existing tool to their needs or even creators of software artifacts designed around specific needs is deemed the major challenge that educators are facing. This paper describes an educational game that by using modern and appealing technologies such as smartphones and mobile devices presents a game that interleaves ludic and educational aspects. The paper describes the main design goals of an educational game as well as the educational design of the learning path centered around a set of topics organized in different levels. The levels are designed in accordance with the Bloom taxonomy and each level has different stages with increasing grade of difficulties.

Keywords Educational game · Game design · Mobile educational game

Introduction

The importance of acquiring a set of basic competencies and abilities in science, technology, engineering, and mathematics (STEM) education as well as in Computer Science education is internationally recognized. Official documents

---

D. Giordano (✉) · F. Maiorana
Department of Electrical, Electronic and Computer Engineering, University of Catania, Catania, Italy
e-mail: daniela.giordano@dieei.unict.it

F. Maiorana
e-mail: francesco.maiorana@dieei.unict.it

Y.-M. Huang et al. (eds.), *Advanced Technologies, Embedded and Multimedia for Human-centric Computing*, Lecture Notes in Electrical Engineering 260,
DOI: 10.1007/978-94-007-7262-5_26,

such as [1–3] emphasize the importance of acquiring an in-depth knowledge of the fundamental concepts around Computer Science and underlines the necessity of a suitable acquisition of "21st century skills" such as problem solving and critical and creative thinking, as well as communication and cooperation skills.

The "CS principles project" [4], for example, has designed and put into practice, both at school and university level, a curriculum built around seven central ideas:

1. computing is a creative activity,
2. abstraction reduces information and detail to facilitate focus on relevant concepts,
3. data and information facilitate the creation of knowledge,
4. algorithms are used to develop and express solutions to computational problems,
5. programming enables problem solving, human expression and creation of knowledge,
6. the internet pervades modern computing
7. computing has global impacts.

The project aims to teach to a large audience of students, without restricting to small elective courses, the basic concepts of Computer Science in such a way to make these concepts a common basic background that can be used by younger generation in all their further study both in STEM education as well as humanities.

For these reasons there is the necessity to modify both the content of the curricula and the pedagogy and the way of teaching these new materials. These modifications are necessary since the first years of school in such a way to transmit and educate, as soon as possible, not only the ability related to the use of computers and software tools, such as internet, e-mail, text writing or tools to manipulate data, but also creative capacity of designing and implementation of software tools able to resolve a given problem at hand. Once the skill to design and implement a software artifact are acquired, they can be used to customize existing software tools to personal work-related needs and to personal fields of interest.

The experimentation concerning the curriculum is focused on content aiming at developing, through modern instruments, the above mentioned capacities. Modern pedagogies are centered around constructivist theories that put the students at the center, privileging their central and active role. The students should no more be passive listeners of the lessons but have to participate, inside a group of peers, to knowledge creation through the realization of practical projects.

Some pedagogical theories such as the "inverted classroom" [5] banish passive activities of lecturing, by delegating learning to homework and concentrating all class activities on laboratory projects or the development of practical activities.

A necessary tool to customize existing software or to develop new software is represented by a programming language. The knowledge of one or more programming languages and of the techniques to design and implement algorithms, to choose the best and most suitable data structures represents one of the greatest obstacles in the introductory programming course both in high school and in first level University courses.

There is an sample scientific literature dealing with the problems related to teaching an introductory programming course. A recent review can be found in [6] where the authors describe the curricula, the teaching pedagogies, the choice of the programming languages and the tools that can be useful in teaching.

In general, all the introductory programming courses have one of the highest levels of drop out or poor results. In [7] for example, the authors have analyzed the results of 67 international institutions, arriving at the conclusion that, as a general rule, failure percentage is between 30 and 60 % in courses in introductory programming with relatively large classes.

In order to overcome such difficulties one of the possible approaches is to use educational games inside the educational path within a course of an entire curriculum both in school and universities.

As stated in [8] it is natural to combine the content of a learning path with the great motivation and attraction that the educational game have on students and in particular in teenagers. Nevertheless, in a recent literature review the authors point out the need of more quantitative studies on the ability of educational games to foster greater reasoning skills [9]. It is, hence, necessary to collect more quantitative data in order to carry out analytical studies on the validity of educational game as tools supporting education and modern pedagogies.

Another pressure in the direction of technology innovation is represented by the use of smartphones and mobile devices. It is clearly stated in the literature that nowadays the majority of the population in developed countries has access to a mobile network. A recent study [10] reports that since 2010, 90 % of the population has access to a mobile network. The percentage increases if the sample population is restricted to teenagers and younger people, who use mobile devices in many daily activities such e-mail reading, internet access, use of chat, social network access, storing and sharing photos and so on. Younger students, moreover, are used to always bringing a smartphone or a mobile device along with them. These types of technologies have a strong appeal on younger generations

In the light of the above consideration it is natural to use these new technologies in the educational field.

This work has the aim to describe the design and implementation of an educational game that blends entertainment and game play with educational questions. The educational questions have been designed to guide the students in the learning path typical of an introductory programming course both for a major in Computer Science and for non-majors. The game has been designed to be utilized with mobile devices and in particular with smartphones. The game can be adapted for use in STEM courses or in the humanities, or in high school or university, by allowing the personalization of the multiple choice questions. The educational game and the questions have been designed in such a way as to guide the students in their learning path.

This paper is organized as follows: section Game and questions design briefly describes the design considerations both for the game and for the questions, section three describes the game and implementation details, section Game evaluation plan presents an outline of a questionnaire to be administered to the students and the expected outcome from the analysis of both the qualitative and quantitative data gathered from the user answers and from the log data gathered during game play in a client server version of the game that we plan to develop, section Conclusions and further work draws some conclusions and highlights future work.

Game and Questions Design

In designing the game the following aspects were taken into consideration:

- The game should be designed for small mobile devices and in particular for smartphones
- The game should blend recreational aspects with educational activities in the form of questions related to one or more disciplines
- The game should provide, upon request, immediate feedback to the user. In order to avoid abuse of the feedback mechanism and stimulate self-discovery, the feedback should have a cost for the user in terms of scores
- The educational aspects of the game should be embedded in questions carefully designed in order to guide the users, through different level of difficulties, in their learning path.
- The game play should be customizable to the user needs and pace. This customization should account for different levels of initial knowledge and different expectations and goals.
- The game should provide progressive levels of difficulties both in the ludic and educational aspects.

The design of the questions should be organized into topics and each topic in levels, and each level in an increasing grade of difficulty. For example, with an introductory programming course in mind, the topics can be the main part of a procedural language such as: input instructions, variable declarations, arithmetical, relational and logical operators, expressions, input instructions, procedures and functions, conditional instructions, cycles, array, matrices, recursion and linear and non-linear data structure. The level can be, in accordance with the Bloom Taxonomy of cognitive processes [11] organized into the following scale:

1. Find syntactic errors. This type of error is detected by a compiler and is used to test the knowledge of the syntactic structure of the programming language. This type of question represents the lowest level of the Bloom taxonomy.
2. Program understanding: typical question are related to guess the output of a piece of code.
3. Design procedure and functions in terms of their parameters or output value
4. Sort a sequence of instructions in order to obtain a correct algorithm

5. Find semantic errors: given a program description and a piece of code find the semantic errors.
6. Complete a fragment of code with a missing part. This can be framed into the highest level of the Bloom taxonomy.

There are several means to increase the grade of difficulties such as increasing the number of possible answers in a multiple choice game, increasing the number of missing parts or providing more possible answers to choose from, and so on.

Game Implementation

The first prototype of the game was implemented using Eclipse [12], the Android Software Development Kit (SDK) [13] and the Android Development Tools (ADT) plugin, in particular the Android Api level8 and the SqLite [14] database to store the questions.

The game can be framed as a shooting game: two types of characters appear on the screen: good and bad ones, along with other distracting elements. These characters move randomly on the screen. The player has to touch on the screen all the good characters in a fixed amount of time. At increasing levels, the number of characters increases and so the game difficulty. If all the bad characters are not hit in the amount of time the player loses a life.

According to the game level, each time a fixed number of good characters are hit a multiple choice question appears on the screen. The level of the questions is chosen in accordance with the game progression and the above sketched hierarchy.

The game can be played as a complete game from start to end, or the user can choose a level as a starting point in order to customize the learning path and progression. Figure 1 shows a screenshot of the game in the shooting mode.

In Fig. 1 the good and bad characters are also differentiated by a different color of the bounding rectangle.

Figure 2 shows an example of the game in question mode. The game was designed with an object-oriented approach. A database stores the questions. The questions can also be uploaded from a file where the fields of each questions are separated by commas, thus allowing easy customization of the game to different domains.

Game Evaluation Plan

In order to evaluate the game we plan to gather both qualitative and quantitative data. The qualitative data will be gathered through a questionnaire administered to educators and to the game players. The questionnaire will gather data in accordance with the evaluation model presented in Table 1. The table reports the main

Fig. 1 A screenshot of the game in “game mode”

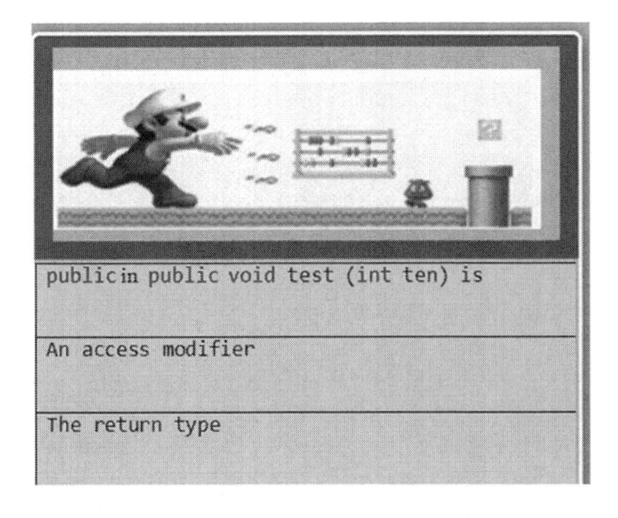

Fig. 2 A screenshot of the game in “question mode”

Table 1 Evaluation model

Dimension	Type of data	Expected results
Game design	Questionnaire aimed at collecting data on the game usability, playability and interactivity.	Evaluation of the usability, playability and interactivity.
Question design	Data gathered from student response and log data. Questionnaire aimed at collecting data about question difficulties and cognitive load and perceived effort by the users. Data gathered on the explanation of the cognitive strategies and process followed by the player in answering the questions	Using classical test theory and item response theory to validate both the single item and the overall quality of the questions
Effects on learning	Pre-test post-test and retention test. Questionnaire asking the perceived effectiveness of the learning path followed during the game	Evaluation of the effects on learning and the learning process.

aspects of the investigation, the type of data gathered and the main results expected from the data analysis process. In particular, the quantitative analysis will be based on the log data, collecting information such as the time spent playing the game, the time spent to answer each question, the number of corrections and so on. Both the quantitative and qualitative data can give an indication of the effects on learner motivation; on the quality of the questions; and on the effect on learning.

Conclusions and Further Work

This work has presented the main design principle of an educational game that by posing questions arranged around topics, with each topic divided into different levels in accordance with the Bloom taxonomy, and each level with a different grade of difficulty allows for a customizable environment that, even it has been designed for initial programming courses, can be used, with a careful design of the questions, in different domains ranging from STEM to languages courses or humanities.

As further work we plan to fully develop the database of questions and to use the game as a study aid and as an assessment tool in an introductory programming course for non-majors. At a more advanced level, not only textual but also graphical questions could be posed to the users, so that they may also develop some skills in designing the human-computer interfaces of modern information systems, in which system functionalities should be controlled by suitable multimedia interfaces, as suggested in [15]. Each questionnaire may be stored in a server with a short abstract and keywords so that the educational material may be clustered to give rise to an organizational memory, as envisaged in [16–19], that allows the users to download the game most suitable for their educational needs of the user, and also may facilitate interaction across peers, as in a social network, to select the more engaging games, and eventually, to create a channel for exchanging answers. This approach would also support a pedagogy oriented to social learning, and could be easily implemented by foreseeing in the game also some open questions with no feedback. Analysis of the game usage and answering paths of the students can be used to obtain a deeper insight on the main student difficulties in getting acquainted with programming skills, which is nowadays an important aspect of education.

References

1. ACM & CSTA (2010) Running on empty: The failure to teach K–12 computer science in the digital age
2. ACM-IEEE (2012) Computer science curricula 2013 Strawman Draft
3. Royal Society (2012) Shut down or restart: the way forward for computing in UK schools, January, 12
4. Astrachan O, Briggs A (2012) The CS principle projects. ACM Inroads 3(2):38–42
5. Lage MJ, Platt GJ, Treglia M (2000) Inverting the classroom: a gateway to creating an inclusive learning environment. J Econ Educ 3(1):30–43
6. Pears A, Seidman S, Malmi L, Mannila L, Adams E, Bennedsen J et al (2007) A survey of literature on the teaching of introductory programming. SIGCSE Bull 39(4):204–223
7. Bennedsen J, Caspersen ME (2007) Failure rates in introductory programming. SIGCSE Bulletin 39(2):32–36
8. Prensky M, Prensky M (2003) Digital game-based learning. Mcgraw Hill Book Co
9. Connolly TM, Boyle EA, MacArthur E, Hainey T, Boyle JM (2012) A systematic literature review of empirical evidence on computer games and serious games. [doi: 10.1016/j.compedu.2012.03.004]. Comput Educ 59(2):661–686
10. Johnson L, Smith R, Willis H, Levine A, Haywood K (2011) The 2011 horizon report. The New Media Consortium, Austin, Texas
11. Anderson LW, Krathwohl DR, Airasian PW, Bloom BS, Cruikshank KA, Pintrich PR, Mayer RE (2001) A taxonomy for learning, teaching and assessing, pp 67–68. Addison Wesley Longman, Inc, complete edn
12. Eclipse available at http://www.eclipse.org/
13. Android SDK available at http://developer.android.com/sdk/index.html
14. SqLite database available at http://www.sqlite.org/download.html
15. Faro A, Giordano D (2000) Ontology, esthetics and creativity at the crossroads in information system design. knowledge-based systems, 13(7–8), (1 December 2000), 515–525
16. Faro A, Giordano D (1998) Concept formation from design cases: why reusing experience and why not. Knowl-Based Syst 11(7):437–448
17. Faro A, Giordano D (1998) StoryNet: an evolving network of cases to learn information systems design. In Software, IEE Proc 145(4):119–127, IET
18. Giordano D (2004) Shared values as anchors of a learning community: a case study in information systems design. J Educ Media 29(3):213–227
19. Giordano D Evolution of interactive graphical representations into a design language: a distributed cognition account. Int J Human-Comput Stud 57(4):317–345
20. Faro A, Giordano D (2003) Design memories as evolutionary systems socio-technical architecture and genetics. ProcIEEE Int Conf Syst Man Cybernetics 5:4334–4339

Activating Natural Science Learning by Augmented Reality and Indoor Positioning Technology

Tien-Chi Huang, Yu-Wen Chou, Yu Shu and Ting-Chieh Yeh

Abstract In recent years, with the rapid development of information technology, educational technologies have been used successfully in enriching learning content and improve learning efficiency. This study attempts to develop an assisted learning system for increasing students' motivation by creating flexible learning path. The purpose of the designed system is to bridging the gap between formal and informal learning on natural science subject. Augmented reality and indoor positioning technologies have been implemented in the system to guide learners to construct their knowledge in the informal learning environment, National Museum of Natural Science. Learners can not only receive virtual information left by other learners in such space, but also leave their own learning experience and share with others. Additionally, the system also analyzes individualized learning subject. By doing so, other learners who have the same interests could find an efficient way to learn.

Keywords Educational technology · Augmented reality · Indoor positioning · Self-regulated learning · Formal and informal learning

T.-C. Huang (✉) · Y.-W. Chou · T.-C. Yeh
Department of Information Management, National Taichung University of Science and Technology, Taichung, Taiwan, Republic of China
e-mail: tchuang@nutc.edu.tw

Y.-W. Chou
e-mail: s1801B103@nutc.edu.tw

T.-C. Yeh
e-mail: s13013048@nutc.edu.tw

Y. Shu
Taichung Shinmin Senior High School, Taichung, Taiwan, Republic of China
e-mail: h1257@shinmin.tc.edu.tw

Y.-M. Huang et al. (eds.), *Advanced Technologies, Embedded and Multimedia for Human-centric Computing*, Lecture Notes in Electrical Engineering 260,
DOI: 10.1007/978-94-007-7262-5_27,

Introduction

Classroom learning at school is a way of formal learning for students; however abundant resources outside of the classroom cannot be ignored. For example, informal learning in the National Museum of Natural Science (NMNS) combines both formal and informal learning to create a diverse learning, enhancing student's ability of self-regulated learning. Nevertheless, visitors may get lost in the spacious museum so guides are required to lead the route and to explain to the visitors. The same applies to students getting educated at this site. The fact is that not every site has a guide to explain to the visitors, or else a guide may be subject to limited time; also a guide may not be suitable for the learning path for everyone.

Usually when students are in the face of new knowledge, they feel at loss because they know nothing about the topics covered and they can only explore by guessing. Consequently, students spend too much effort during discovery process in search of the topic or may even digress from the topic, resulting in ineffective self-regulated learning. If there is one guideline to enable the students to take less pointless routes and put more effort to explore the knowledge more deeply, they will have better performance on learning. It is not the case that these knowledge are never been discovered, but only that current museum of science has not recorded the history study log systematically to share the experience with others. Hence future generations can only rely on self-learning ability. When knowledge sharing is not available, students may struggle and get lost in searching of the topics, eventually losing interest in learning and then giving up learning.

Therefore, how to establish one knowledge-learning platform to help future generations to follow their predecessors' footsteps for learning, and further aid self-regulated learning is the main goal of this study. A mobile carrier is used to assist students' learning at museum of science. Two additional technologies have been adopted with the mobile carrier. Indoor positioning is adopted to calculate the current position of the learner; AR (augmented reality) technology is then adopted to gradually guide the learning route. The learning route is defined from a learning mode analyzed in accordance with previous learning experience, aiming to reduce time cost in groundless searching. The platform records the learning process of each person with added nodes to further monitor learning progress, as well to share with other learning partners. Through such guiding and sharing, we expect that students' self-regulated learning at the NMNS is assisted.

Literature Review

Augmented Reality

Augmented reality (AR) is a technology combining virtual environment and real world. Past studies have shown that AR supplements inadequacies of the real world in the way that cumulative learning experience triggers learner's thinking skill and understanding of concepts, instead of replacing the real environment [1]. In essence, the educational value of AR is not merely on the use of technology but how to implement AR to formal and informal learning environments. Furthermore, AR application has been demonstrated actually enhances learning motivation [2]. Therefore, this study adopts AR technology on mobile carrier, making the screen displaying auxiliary information and combines indoor positioning technology to display learning process so as to guide the direction of learning theme.

Indoor Positioning

In the recent years, the development of indoor positioning technology has become increasingly mature. At the current developmental stage, indoor positioning technology can already be used in indoor navigation. The majority of navigation application requires specific learning tasks in the past [3]; learners can only passively accept the task and cannot select appropriate learning resources according to their respective needs. In order to tackle this problem, this study implements indoor positioning technology to detect students' current position to provide learning routes. Indoor positioning technology can identify users' indoor position and provide effective information to aid users to orientate resources in a limited indoor environment. NMNS is one informal learning environment for learners to gain extracurricular knowledge. Learners are impelled to enroll in active learning in NMNS and receive help in learning, thereby allowing learners to learn effectively in informal educational environment [4].

Self-Regulated Learning

Self-regulated learning has been one of the main objectives of formal education set by researchers in the past. If learners possess this ability, they are able to determine current learning needs and reflect on learning performance. Therefore, self-regulated learning activities contribute to the mediation of individual, learning context and actual performance [5, 6]. This study explores in depth how to develop one learning platform in which students can gain information and aids effective learning in informal learning environment such as NMNS, Botanical Garden and so on. With the effective learning platform, the study further combines it with adequate learning strategies during learning process to develop learning mode suitable for learning in NMNS.

Methodology

Bridging Formal and Informal Learning

This study proposes a learning mode to be applied in National Museum of Natural Science by integrating AR and indoor positioning technology to illustrate how students can be diverted from the original formal learning environment into informal learning environment. The study adopts the concept of self-regulated learning proposed by Zimmerman et al. in 1996 [8] and the concept of resource management strategies proposed by Pintrich et al. [9] to construct the "self-regulated learning mode in NMNS." The constructed learning mode is combined with learning activities to promote students to effectively seek for learning resources in informal learning environment and to link the concept and practice.

The design has two levels: self-regulated learning and resource management strategies, as shown in Table 1. "Self-regulated learning" is a structure rendered by informal learning mode; self-regulated learning theory is adopted to explore in depth, including self-assessment and monitoring, goal setting, strategic planning, implementation and monitoring of strategies, and monitoring and correction of strategic results. The process of adjustment learning emphasizes on combining with learning resources to set up appropriate learning objectives, formulation and implementation of study plan, effective time management, monitoring environment, and seek for human resources for discussion on specific implementation of learning activities. In other words, resource management strategy helps in learning strategic design to well and truly carry out learning activities. The purpose of this study is to develop a learning system for cultivating students' self-regulated ability, thus the self-regulated learning mode is used to aid learning. Depending on students' learning conditions, the corresponding indicator at every stage of self-regulated learning is a reference for students to adjust their learning pace. Also, learning resources are integrated to learning tasks at every stage for learning.

Learning Activities in the Informal Learning Environment

This study sets natural science and technology for third-graders as the learning subject. The subject of natural science and technology emphasizes on experiments; the required knowledge is complete only when the learners grasp the concept and can verify with experiments. AR technology plays a role to provide visual multimedia information to aid in learning so that students can fully understand the purpose of the experiment with existing equipment and instruments in the NMNS. At the same time, the gapbetween experimental results demonstrated on instruments and the cognitive results can be reduced. Also, the learning route is effectively planned so students can accurately hold onto learning resources without getting lost during the learning process and hence greatly enhances learning

Tableô1 Self-regulated learning model in NMNS

Stage for learning	Teaching mode of self-regulated learning [10]	Resource management strategies [9]	Implementation of learning activities [11]
Performance prior to learning	Self-assessment and monitoring	Set up appropriate learning objectives Rightful achievement attributes to attitude	*Goal Setting* Carry out action learning in the NMNS in accordance with progress of experimental course
	Goal setting and strategic planning	Formulation and implementation of study plan	*Strategy Learning* Set up steps for learning
Performance in learning	Implementation and monitoring of strategies	Effective time management	*Self-management* Effective management of learning time, and control of personal learning condition
Post-learning performance	Monitoring and correction of strategic results	Monitoring environment and seek for human resources	*Strengthening Mutual Aid* Seek for human resources to solve any learning problems *Cooperative Learning* Solve problems with cooperation to meet expected progress *Self-feedback* Complete learning progress and gain experience *Correction on Implementation* Adjust learning steps appropriately; strengthening ability of self-regulated learning

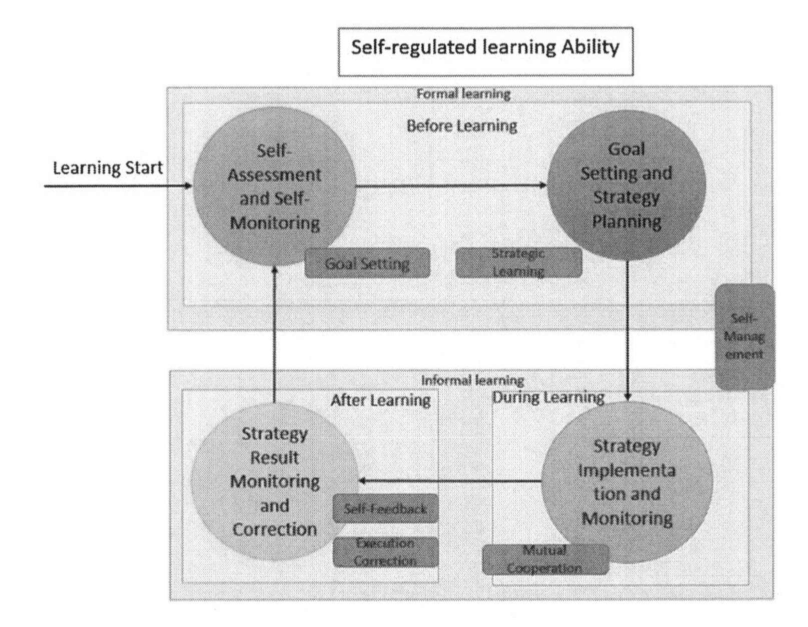

Fig. 1 Self-regulated learning diagram in formal and informal learning contexts

efficiency. This study aims to provide one set of appropriate learning steps (as shown in Fig. 1) designed for learning convergence to guide the students when they encounter the gap between learned knowledge and experimental results when studying natural science and technology subject. For the sake, we provide a solution functioning as bridge between fundamental concept of knowledge and relevant experiments, driving students to have profound experience in learning science and technology as well as to strengthen learning motivation.

Guiding Learning Objectives with AR and Indoor Positioning Technology

As indoor configuration of the NMNS is irregular and with multi-thematic distribution, it is time consuming to look for themes relating to experimental topic corresponding to the curriculum. In order to learn effectively and get the learning resources quickly, this study designs a learning system applicable in the NMNS utilizing the AR and indoor positioning technology.

Augmented Reality

In this study, the use of AR technology assists learners to effectively find the learning resources. The mobile device held by learner will display virtual tag and text prompts to guide the leaner the route to specific learning resources. The problem of time-consuming to find specific learning resource or getting lost along the way during self-regulated learning is thus improved. During the learning process, the system constantly records new learning routes and conditions of usage; such data can be further used in the future to analyze learning effectiveness of the learner.

Indoor Positioning

The usage of indoor positioning technology effectively navigates the learning route and records learning nodes. The technology is implemented to learning routes via learning steps, enabling students to monitor and control their own learning pace and to cultivate self-adjustment learning ability. Students can effectively do self-assessment on their learning performance and the goal of cultivating students' self-adjustment learning ability is therefore achieved.

System Demonstration

This system is designed against the subject of Nature & Life Science for the 3rd-grader. The learning goal, which enables the students to understand the topic learning goal thoroughly, is set up based on the learning requirement of students after ending up with the basic concept program. Moreover, it helps students to find out the learning sources accurately through the assistance of learning platform during the informal learning environment. The virtual arrow and text prompt displayed on the hand-hold device can help the learner not to lose the direction whilst searching for resource indoors and the searching time can also be reduced. Meanwhile, continuous tracking can be executed by using such system and the learning status and usage condition can both be recorded.

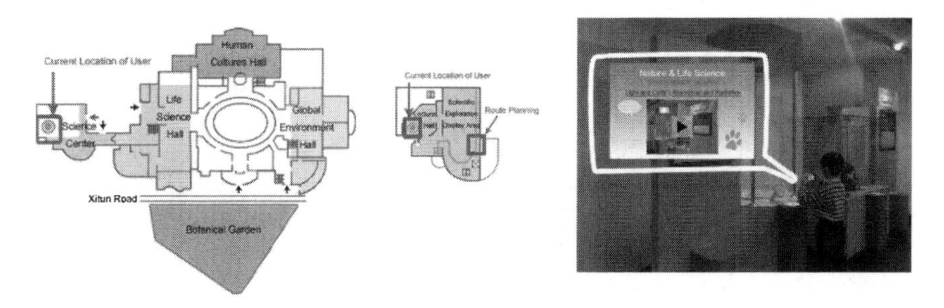

Fig. 2 Simulation screen of interior location / real learning condition of student using such learning platform

The proposed learning system is able to show up the learning path, record the learning nodal point and even present content embedded in learning path effectively by means of the combination of augmented reality technology and indoor positioning technology, as shown in Figs. 2 and 3. In this way, the system not only enriches learning experiences but also draws students' attention and inspire their motivation. During the learning process, self-learning regulation is aroused in order to bring up and regulate the learning habit gradually to achieve the goal of self-regulated learning designed by this study.

Results and Discussion

Augmented Reality is used to Strengthen the Effectiveness of Self-Regulated Learning upon Academic Study

This study will further build up a knowledge learning platform. It will integrate the learning topics in the formal context and learning paths for the learning history of each user through systematic analysis.

When the user is learning in NMNS, the location-based annotation function can be used to hold the learning experience combining with the built-up coordinate system at the current location of learning. Each node can store the learning type of user in NMNS, such as audio visual learning and thinking of some topic. By node learning record, the user can grasp the self-learning progress and share with peers.

Enhance the Learning Efficiency by AR and Indoor Positioning

It is a topic worth concerns if a learner can learn effectively in the environment of informal learning, in which if informal learning is performed in NMNS, the learning status of each learner may not be taken care under the limited human resources. The system are designed to combine the augmented reality and indoor positioning technology to help the learner get the direction whilst searching for resource indoors and reduce searching time. Meanwhile, since the problem of that subject can be provided properly through the system, the learner may not miss the learning point in NMNS. This system can also be used to record the learning history and track the usage condition of system continuously. At last, the learner can not only absorb the knowledge but also manage the self-learning regulation during the learning process and thereupon improve and train the habit of self-learning gradually.

Conclusion and Future Prospect

The purpose to build up this learning system is to combine the environment of formal learning with informal learning. The traditional teaching model of teacher guiding to learn is changed. In a new wave of learning, the students will be guided into a diverse learning environment to enhance the learning experience and knowledge complementary outside class and the self-regulated learning ability in the large-scale environment can be trained. This system provides the assistance of the subject of Nature & Life Science required by the experiment learned by the students, which helps to encourage the students performing expeditionary learning in NMNS, and further strengthen the critical thinking and train the organizational ability of students. The more important point is to be capable of self-regulated learning during the learning process and after it; a set of effective learning model can be obtained. In the future, we will further investigate such system being used in the environment of informal learning and carry out an experiment on the subject of Nature & Life Science learned by the 3rd Grade students of an Elementary School. Meanwhile, we will continuously pay attention to the effect of the self-regulated learning ability developed by the augmented reality and indoor positioning technology.

References

1. Wu HK, Lee SWY, Chang HY, Liang JC (2012) Current status, opportunities and challenges of augmented reality in education. Comput Educ 62, 41–49. doi: 10.1016/j.compedu.2012.10.024
2. DiSerio A, Ibáñez MB, Kloos CD (2012) Impact of an augmented reality system on students' motivation for a visual art course. Comput Educ 1–11. doi:10.1016/j.compedu.2012.03.002
3. Sung YT, Chang KE, Lee YH, Yu WC (2008) Effects of a mobile electronic guidebook on visitors' attention and visiting behaviors. Educ Technol Soc 11(2):67–80
4. Paris SG, Hapgood SE (2002) Children learning with objects in informal learning environments. In: Paris SG (ed) Perspectives on objects-centered learning in museums. Lawrence Erlbaum, Mahwah, NJ, pp 37–54
5. Accurate Mobile Indoor Positioning Industry Alliance, called In-Location, to promote deployment of location-based indoor services and solutions, http://press.nokia.com/2012/08/23/accurate-mobile-indoor-positioning-industry-alliance-called-in-location-to-promote-deployment-of-location-based-indoor-services-and-solutions/ (2012, August 23)
6. Boekaerts M (1997) Self-regulated learning: a new concept embraced by researchers, policy makers, educators, teachers and students. Learning and Instruction. 7(2):161–186. doi: 10.1016/S0959-4752(96)00015-1
7. Pintrich PR (2000) The role of goal orientation in self-regulated learning. In: Boekaert M, Pintrich PR (eds) Handbook of self-regulation Academic Press, San Diego, pp. 13–39
8. Zimmerman BJ (2000) Attaining self-regulation: A social cognitive perspective. In: Boekaerts M, Pintrich PR, Zeidner M (eds) Handbook of self-regulation Academic Press, San Diego, pp. 13–39
9. Pintrich PR, Smith DA, Mckeachie WJ (1989) A manual for the use of the motivated strategies for learning questionnaire (MSLQ). National Center for Research to Improve Postsecondary Teaching and Learning, School of Education, The University Michigan, Mich

10. Zimmerman BJ, Bonner S, Kovach R (1996) Developing self-regulated learners: Beyond achievement to self-efficacy. American Psychological Association, Washington, DC
11. Cherng BL (2001) The Relations Among Motivations, Goal Setting, Action Control, and Learning Strategies: The Construct and Verification of Self-regulated Learning Process Model. J Nat Taiwan Norm Univ 46(1):67–92

Using Particle Swarm Method to Optimize the Proportion of Class Label for Prototype Generation in Nearest Neighbor Classification

Jui-Le Chen, Shih-Pang Tseng and Chu-Sing Yang

Abstract Nearest classification with prototype generation methods would be successful on classification in data mining. In this paper, we modify the encoded form of the individual to combine with the proportion for each class label as the extra attributes in each individual solution, besides the use of the PSO algorithm with the Pittsburgh's encoding method that include the attributes of all of the prototypes and get the perfect accuracy, and then to raise up the rate of prediction accuracy.

Keywords Particle swarm optimization · Prototype generation · Evolutionary algorithms · Classification

Introduction

The nearest neighbor algorithm [1] has a significant effect on classification prediction. To calculate the similarity between the predicted target and the known samples is the way to find the nearest neighbor. This method provides a very high accuracy rate and having the characteristic that the more precise with the more the number of samples. However, there are some drawbacks for the method. The costs of the calculation are too high and the accuracy is susceptible to noise interference.

J.-L. Chen (✉)
Department of Multimedia Design, Tajen University, Tajen, Taiwan, Republic of China
e-mail: reler@mail.tajen.edu.tw

S.-P. Tseng
Department of Computer Science and Information Engineering, National Cheng Kung University, Cheng Kung, Taiwan, Republic of China

J.-L. Chen · C.-S. Yang
The Institute of Computer and Communication Engineering, National Cheng Kung University, Cheng Kung, Taiwan, Republic of China

Y.-M. Huang et al. (eds.), *Advanced Technologies, Embedded and Multimedia for Human-centric Computing*, Lecture Notes in Electrical Engineering 260,
DOI: 10.1007/978-94-007-7262-5_28,

In order to solve the above problem, some method can achieve this goal by thinking about reduction of the number of samples. There are two main proposed methods that try to select the reasonable ones from all of the samples and then to perform the nearest neighbor algorithm. These two methods are named prototype selection (PS) [2–7] and the prototype generation (PG) [8–11].

For the purposes of prototype selection to perform the classification, those prototypes are base on the new selection of suitable samples from the training set. There are two main methods for PS problem. One is the concentration method [2], the main idea is to avoid the proportion of certain types of samples are more large than others that make the error decision. For the reason that those samples are eliminated for the properties may be too similar or unrelated. By the second method, the main purpose is to focus on removing those samples would interfere with decision or cause confusion then the follow-up prediction would be more accurate [3]. For prototype generation (PG), not only choose the appropriate samples but also modify the attributes of individual sample. At the result, the decision of classification would be more obvious and distinguished [12, 13].

The main purpose of PG method is to choose or modify the n samples' data from the training set. After that generates a new set, GS, which contains the r prototypes, in which $n > r$. These newly generated prototype can be used for the classification to accelerate the prediction efficiency because of the fewer number of samples would achieve the better accuracy. Find subsets guaranteeing zero errors with N-prototypes for each class when the original data set which is submitted to the Prototype Generation Classifier.

In general, PS and PG problem can be considered as combination and optimization problem, there are many evolutionary search method applied in this problem. PG is regarded as a continuous space search problem. The evolutionary optimization search methods using particle swarm optimization (PSO) and differential evolution (DE) are suitable for continuous spaces. Many schemes are presented on this topic, such as [12–15].

Most of the methods for prototype generation that gives a suggestion for the proportional to classes label is equal to the average, but does not completely arrive at ideal accuracy. The proportion of class for the prototypes in the AMPSO [12] method is uncertain, which is determined by the execution results of each run. In the SFLSDE [13] method, the proportional to the number of all classes for the prototypes is based on the training set. As a result, this method got a good prediction accuracy.

The main contribution of this paper follows a idea to that presented in [13] that is the use of the PSO algorithm with the Pittsburgh's encoding method that include the attributes of all of the prototypes and get the perfect accuracy, but we modify the encoded form of the individual that add the proportion for each class label as the extra attributes in each individual solution and then to raise up the rate of prediction accuracy.

The remainder of the paper is organized as follows. Section "Proposed Method for Prototype Generation" describes in detail the proposed method. Performance evaluation of the proposed method is presented in section "Experimental Framework and Results". Conclusion is drawn in section "Conclusions".

Proposed Method for Prototype Generation

In this section, the Particle Swarm Optimization (PSO) method is applied for the prototype generation. The PSO follows the general process of the Evolutionary Algorithm. PSO initializes the population with N members as the candidate solutions (NP), for each solution of NP is named a individual.

Encoding of Prototype

The encoding method will be used the Pittsburgh method, all of the prototype will be encoded into each individual, the size for one individual can be denoted D. The individual with dimension D in DE method can be regarded as a target vector. As the result, $D = r \times n + m$, r is the number of prototypes in the same individual. n is the number of attributes in the prototype. m is the number of type of class label. In addition to those prototypes and it's attributes as the part of individual solution, the number of distinct class labels is also as a part of solution.

Table 1 describes the structure of an individual. Each prototype p_i has a corresponding class label. Within entire evolutionary cycle of PSO, the value of this class label remains unchanged. It means that by the operation of the PSO, the class labels assigned but still fixed from the initialization phase of each prototype to the end phase, the class labels are not part of the individual. Typically, it is necessary to normalize the values of each attribute before PSO processing. That prevents the attributes in large ranges to influence some attributes in smaller ranges. The normalization means that to transform a value v of a attribute A to v' in the range $[-1, 1]$ by computing with Equation(1) where $\min_A$ and $\max_A$ are the minimum and maximum values of attribute A.

Table 1 Encoding of a set of prototypes

	Prototype 1	Prototype 2	…	Prototype r	Proportion
Attributes	$p_{11}, p_{12}, \ldots, p_{1n}$	$p_{21}, p_{22}, \ldots, p_{2n}$	…	$p_{r1}, p_{r2}, \ldots, p_{rn}$	$d_0, d_1, \ldots, d_m$
Vectors	$x_1, x_2, \ldots, x_n$	$x_{n+1}, x_{n+2}, \ldots, x_{2n}$	…	$x_{(r-1)n+1}, x_{(r-1)n+2}, \ldots, x_{rn}$	$x_{rn+1}, \ldots, x_{rn+m+1}$
Class	d_0	d_1	…	d_m	

r is the number of prototypes
n is the number of attributes
m is the number of class label which denotes as $d_0, d_1, \ldots, d_m$

$$v' = \left(\frac{v - \min_A}{\max_A - \min_A}\right) \times 2 - 1 \tag{1}$$

Algorithm and Movement

During the initialization process, there is one thing is needed to ensure that each class has at least a prototype to be represented. All of the prototypes are combined and encoded in each individual. These prototypes is proportional to each class with the number of samples in the training set. The individual should include each class with at least one prototype S_i.

Assume an D-dimensional search space S, and a swarm comprising Np particles. The position $X = [x_1, \ldots, x_D] = [x_1, x_2, \ldots, x_{rn}, x_{rn+1}, \ldots, x_{rn+(m+1)}]$ of a particle in the search space S denotes a candidate solution.

The current position of particle i is an D-dimensional vector $X_i = [x_{i1}, x_{i2}, \ldots, x_{iD}]^t$ belong to S in iteration t.

The velocity of this particle is also an D-dimensional vector $V_i = [v_{i1}, v_{i2}, \ldots, v_{iD}]^t$ belong to S in iteration t, which indicates the displacement for updating the position of each particle in the search space.

The best position encountered by particle i is denoted as $P_i = [p_{i1}, p_{i2}, \ldots, p_{iD}]^t$ belong to S. Assume that g is the index of the particle that attained the best position found by all particles in the neighborhood of particle i.

The swarm is manipulated by the following equations:

$$v_{id}^{t+1} = wv_{id}^t + c_1 w_1 \left(p_{id}^t - x_{id}^t\right) + c_2 w_2 \left(p_{gd}^t - x_{gd}^t\right) \tag{2}$$

$$x_{id}^{t+1} = x_{id}^t + v_{id}^{t+1} \tag{3}$$

Where $i = 1,2,\ldots,Np$, is the particles index; $d = 1,2,\ldots,D$, is the dimension index; $t = 1,2,\ldots,T$, is the iteration number; The variable w is a parameter called inertia weight, which balances global and local searches in the PSO. The two positive constants c_1 and c_2 are cognitive and social parameters, respectively. Proper fine tuning of c_1 and c_2 may improve the performance of the PSO. $c_1 = c_2 = 2$ were recommended as default values. The w_1, w_2 generates a random number uniformly distributed within the interval [0, 1].

Finally, if the selected individual obtains the best fitness in the population, and returns the best individual found during the evolutionary process.

Experimental Framework and Results

In this paper, the performance of the proposed algorithm is evaluated by using it to solve the prototype generation in nearest neighbor classification problem. All the experimental results are obtained by running on an IBM X3650 machine with 2.4 GHz Xeon CPU and 16 GB of memory using CentOS 6.0 with Linux 2.6.32. Moreover, all the programs are written in C++ and compiled using GNU C++ compiler.

Parameter Settings and Datasets

We perform experimentation on the problems summarized in Table 2. They are well-known real problems taken from the University of California, Irvine, collection, used for comparison with other classification algorithms.

Table 2 summarizes the properties of the selected data sets. For each data set that include the number of examples, the number of attributes, and the number of classes. For the results of classification, the data sets are using the ten fold cross-validation to perform the prediction.

Experimental Results

In this section, we describe the results of the experiments and perform comparisons between the GA, PSO, AMPSO and proposed method PPGPSO with same population size and iterations. Those parameters for setting. In all experimental results show in this section, we use the following notation: a "(+)" tag to the result means that the average result was significantly better than the result of the other's method. We also use boldface style to highlight the best result (Table 3).

In Table 4, we compare the average success rate of GA, PSO, AMPSO and proposed method PPGPSO. It shows that PPGPSO has more opportunity to do better than others in those problems.

Table 2 Summary description for classification data sets

Data set	#Examples	#Attributes	#Classes
Australian	690	14	2
Breast	286	9	2
German	1000	20	2
Glass	214	9	7
Heart	270	13	2
Iris	150	4	3
Wine	178	13	3

Table 3 Parameter specification for all the methods employed in the experimentation

Algorithm	Parameters
GA	PopulationSize = 50, Iterations = 1,000, reduction rate(r) = 5 %, CR = 0.9, MR = 0.05, CO = TwoPoint
PSO	SwarmSize = 50, Iterations = 1,000, reduction rate(r) = 5 %, C1 = 2, C2 = 2, Vmax = 0.25, Wstart = 1.5, Wend = 0.5

Table 4 Average success rate (in percent) that compared with EA algorithm for prototype generation

Problem	GA	PSO	AMPSO	PPGPSO	Proportion
Australian	64.79	82.43	**87.00** (+)	85.83	(1:1)
Breast	62.31	65.90	**66.16** (+)	65.25	(1:1)
German	70.04	74.54	75.05	**76.49**	(1:1)
Glass	74.43	72.93	82.62	**87.01**	(1:2:1:1:1:3:1)
Heart	95.41	96.28	97.43	**97.54**	(1:1)
Iris	94.31	98.93	**99.99** (+)	99.97	(3:2:1)
Wine	94.43	95.14	96.02	**96.51**	(1:1:1)

Conclusions

In summary, we have proposed differential evolution as a prototype generation for data reduction method. Specifically, it was used to optimize the proportional to the prototypes for the nearest neighbor classification and to perform as a prototype generation method.

The main aim of this paper to modify the encoded form of the individual to plus the proportion for each class label as the extra attributes in each individual solution, besides the use of the DE algorithm with the Pittsburgh's encoding method that include the attributes of all of the prototypes and get the perfect accuracy, and then to raise up the rate of prediction accuracy.

The ongoing work of experimental study would be performed which be allowed us to justify the behavior of DE algorithms when dealing with small and large datasets.

References

1. Cover T, Hart P (1967) Nearest neighbor pattern classification. IEEE Trans Inf Theory 13(1):21–27
2. Gowda K, Krishna G (1979) The condensed nearest neighbor rule using the concept of mutual nearest neighborhood (corresp.). IEEE Trans Inf Theory 25(4):488–490
3. Wilson DL (1972) Asymptotic properties of nearest neighbor rules using edited data. IEEE Trans Syst, Man and Cybern 3:408–421
4. Brighton H, Mellish C (2002) Advances in instance selection for instance-based learning algorithms. Data Min Knowl Disc 6(2):153–172

5. Marchiori E (2008) Hit miss networks with applications to instance selection. J Mach Learn Res 9:997–1017
6. Fayed HA, Atiya A (2009) A novel template reduction approach for the k-nearest neighbor method. IEEE Trans Neural Networks 20(5):890
7. Marchiori E (2010) Class conditional nearest neighbor for large margin instance selec- tion. IEEE Trans Pattern Anal Mach Intell 32(2):364–370
8. Wilson DR, Martinez TR (2000) Reduction techniques for instance-based learning algorithms. Mach Learn 38(3):257–286
9. Fayed HA, Hashem SR, Atiya AF (2007) Self-generating prototypes for pattern classification. Pattern Recogn 40(5):1498–1509
10. Lam W, Keung C-K, Liu D (2002) Discovering useful concept prototypes for classification based on filtering and abstraction. IEEE Trans Pattern Anal Mach Intell 24(8):1075–1090
11. Bezdek JC, Kuncheva LI (2001) Nearest prototype classi_er designs: an exper-imental study. Int J Intell Syst 16(12):1445–1473
12. Cervantes A, Galván IM, Isasi P (2009) Ampso: a new particle swarm method for nearest neighborhood classification. IEEE Trans Syst, Man, Cybern, Part B: Cybern 39(5):1082–1091
13. Triguero I, Garcia S, Herrera F (2011) Differential evolution for optimizing the positioning of prototypes in nearest neighbor classification. Pattern Recog 44(4):901–916
14. Nanni L, Lumini A (2009) Particle swarm optimization for prototype reduction. Neurocomputing 72(4):1092–1097
15. Triguero I, Garcia S, Herrera F (2010) A preliminary study on the use of differfiential evolution for adjusting the position of examples in nearest neighbor classification. In 2010 IEEE congress on evolutionary computation (CEC). IEEE, pp 1–8

A Novel Genetic Algorithm for Test Sheet Assembling Problem in Learning Cloud

Shih-Pang Tseng, Long-Yeu Chung, Po-Lin Huang, Ming-Chao Chiang and Chu-Sing Yang

Abstract The assessment is the most effectively tool for the teachers to realized the learning status of the learners. The test sheet assembling is an important job in the E-learning. In the future learning cloud environment, the large amount of items would be aggregated into the itembank from various sources. The test sheet assembling algorithm should be with the ability of abstract the needed information directly from the items. This paper proposed an effective method based on genetic algorithm to solve the test sheet assembling problem. The experimental result shows the effectiveness of the proposed method.

Keywords E-learning · Genetic algorithm · Test sheet assembling

S.-P. Tseng (✉) · M.-C. Chiang
Department of Computer Science and Engineering, National Sun Yat-sen University, Kaohsiung, Taiwan, Republic of China
e-mail: tsp@mail.tajen.edu.tw

M.-C. Chiang
e-mail: Chungly1@mail.chna.edu.tw

L.-Y. Chung
Department of Applied Informatics and Multimedia, Chia Nan University of Pharmacy and Science, Tainan, Taiwan, Republic of China
e-mail: mcchiang@cse.nsysu.edu.tw

P.-L. Huang
Center of General Education, Kao Yuan University, Kaohsiung, Taiwan, Republic of China
e-mail: tf0155@cc.kyu.edu.tw

C.-S. Yang
Department of Electrical Engineering, National Cheng Kung University, Tainan, Taiwan, Republic of China
e-mail: csyang@ee.ncku.edu.tw

S.-P. Tseng
Department of Computer Science and Information Engineering, Tajen University, Pingtung, Taiwan, Republic of China

Y.-M. Huang et al. (eds.), *Advanced Technologies, Embedded and Multimedia for Human-centric Computing*, Lecture Notes in Electrical Engineering 260,
DOI: 10.1007/978-94-007-7262-5_29,

Introduction

In past 20 years, the development of Internet has changed the modern human living in several domains. The modern learning is also changed by Internet. Learning is a major human activity to accommodate the environment or the society better. Because of the rapid changing of modern society, the learning efficiency becomes more important than the past. The E-learning [1] is proposed and developed in the first decade of the 21st century to enhance the human learning efficiency. In this time, the various kinds of E-learning are widely applied on various types of educations, such as primary education and continuing education.

Cloud computing [2] is the developing trend of information technology. These unprecedented amount of computing and storage resources would be elastically management and organized to support heterogeneous computing needs. The computing applications, such as entertainment and learning, are transformed into services delivered via Internet. The concept of *learning cloud* is proposed to represent the learning service supported by cloud computing. Because of the computing and storage ability, the learning cloud can be used to process and store more learning contents than the traditional learning management system (LMS), and these learning contents can be presented in various different forms and mediums. The learning contents may be from many different sources, such as e-books, wikis and blogs. The integration and organization of learning contents are important issues in learning cloud. Tseng et al. [3] proposed a method to integrate the learning contents from different sources. The large amount of learning content would be prospectively aggregated in the future learning cloud.

Because of the learners with various different knowledge backgrounds and learning experiences, it is almost impossible that the teacher can design a learning plan which can suitable to each student. The individualization [4] of learning tries to provide different learning plans and activities to different students. It is almost impossible in traditional education because of the cost. But it has become an important issue in E-learning domain because the Internet can provide more interactions among the teacher and the students. For the individualization, it is necessary to gather information about the students by assessment process [5]. Computer-based test (CBT), or e-assessment, [6–8] can provide more performance and lower cost than traditional paper–pencil test (PPT) to realize what the students know. The test sheet assembling, which selects the candidate items from the itembank to generate the test sheet, is the basis of E-assessment. The quality of the itembank and the method of assembling are the two main factors which influence the quality of the test sheet. This paper tries to propose a novel method based on genetic algorithm to solve the test sheet assembling problem in learning cloud environment.

The remainder of the paper is organized as follows. The related works are in section Related Works. Section Proposed Method introduced the proposed method. The experimental results are in section Experimental Result. Conclusion is given in section Conclusion.

Related Works

In the last 30 years, the metaheuristics [9] has been successfully applied on various discrete and continuous problems. There are two categories of metaheuristics, single-solution and population based. The single solution based metaheuristics, such as simulating annealing (SA) [10] and tabu search (TS) [11], search the neighbours of the only one solution. The population based metaheuristics, such as genetic algorithm (GA) [12] and particle swarm optimization (PSO) [13], do the parallel searching by a set of solutions and would interchange the information among the solutions.

Genetic algorithm [12] is originally proposed by J. H. Holland. It is inspired by the natural evolutionary process. There are already various variants of GA proposed to different problems, such as travelling salesman problem and numerical function optimization.

Hwang et al. [14] proposed a tabu search method to solve the test sheet assembling problem. Hwangs' work focus on the degree of discrimination and consider the constrains of the expected testing time and the expected ratio of unit concepts. But the weakness of Hwangs' work is that the itembank must be well-organized and can provide all necessary information to the tabu search algorithm for generating the test sheet. It is not usually practical in the learning cloud environment. The items of the itembank may be from several different sources. The organization of the itembank should be usually messy unless a large amount of human work is used to put all items in order. In addition, the information about one item may be incomplete. These situations would restrict the application of Hwangs' work.

Leung et al. [15], propose a personalized genetic algorithm (PGA) for the test sheet assembling problem. The PGA is based on the item response theory and focused on the personalization of assessment. It is the same with the weakness that the itembank must be well-organized.

Proposed Method

In learning cloud environment, the itembank aggregates a large amount of items from different sources. It is not practical to re-organized these items by humans. In one topic, there are many items which contains some redundancy. One item may be identical with another item. Or the item is the variant of another item, the two items do test similar but not the same concepts. Or part of the concepts of one items are in another item. The test sheet assembling in learning cloud environment should reduce the redundancy and maximize the number of concepts in finite items of the test sheet. For this purpose, the Maximum Concepts Genetic Algorithm (MCGA) is proposed in learning cloud environment.

Fig. 1 System architecture of test sheet assembling

Figure 1 shows the system architecture of test sheet assembling. Before the MCGA, there are two stages of pre-processing. At first, all the items of the itembank should be segmented into words. This work focuses on the assessment itembanks on the Taiwan elementary school. Because the characteristic of Chinese, the Chinese item should be segmented into some meaningful words. The Chinese word segmentation is a important research issue in the Chinese natural language processing. We use the Chinese word segmentation service provide by the Chinese Knowledge and Information Processing (CKIP) group [16], Institute of Information Science, Academia Sinica, Taiwan. The accuracy of this Chinese word segmentation is about 96 %. Secondly, the data cleanup stage is responsible for removing all the stopwords in a Chinese item according to the attributes of each word. It is similar to the Stemmers algorithm [17] for English text. But there is no the stop word list which can be used to remove all the useless words. The Chinese word segmentation service provide the attribute of each word. The noun, verb, adjective and adverb is reserved into MCGA, the others is removed.

The solution in MCGAis represented as a bit-string showed as Fig. 2. The length of bit-string, *N,* is the number of total items in the itembank. If the test sheet consists of *n* items; there are only *n* bits of ones in the bit-string, the other bits are zeros.

Algorithm 1 Maximum concepts genetic algorithm

1: Randomly initiate the population
2: **while** The terminate condition is not met **do**
3: Selection()
4: Crossover()
5: Mutation()
6: **end while**
7: Output the result

Algorithm 1 shows the outline of MCGA. At first, all the solutions in the population are initiated randomly. The main loop would be terminated after the pre-defined number of iterations. In the main loop, like the general genetic algorithm, there are three steps: *selection*, *crossover* and *mutation*. The tournament selection [18] is used as the selection operator in this work and the tournament size is set to 3. The selected solutions would be store a temporary population as the parents of crossover.

The simple 2-points crossover is used to reproduced the next generation solutions. Figure 3 shows an example which is used to illustrate the 2-points crossover in MCGA. There are total 10 items in the itembank of the example, and the

0	1	1	0	0	………	0	1	0

Fig. 2 The representation of solutions

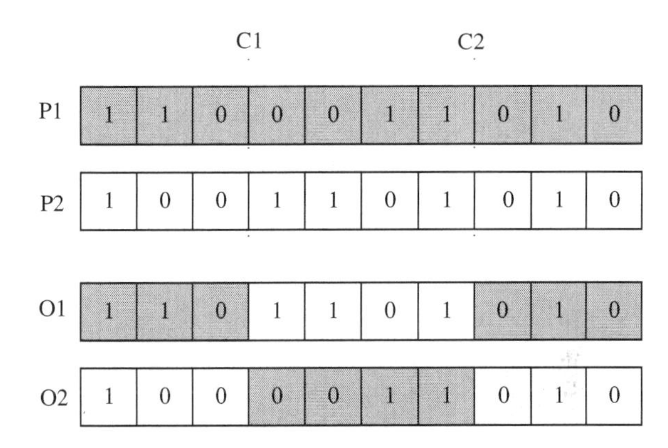

Fig. 3 The example of crossover in MCGA

testsheet consists of 5 items. The parents, $P1$ and $P2$, are legal solution, but the ofspring, $O1$ and $O2$, are both illegal. There are 6 ones in $O1$, and only 4 ones in $O2$. It is necessary to repair the offspring solutions.

In the binary bit-string representation of solutions, the simplest mutation is flipping one arbitrary bit. Because the repairing may be needed after the crossover step, we designed two kinds of mutation, *increment mutation* and *decrement mutation*. Figure 4 shows the example of mutations. If the number of the solution's one bits is greater than n, the increment mutation would be applied. On the other hand, If the number of the solution's one bits is less than n, the decrement mutation would be applied. The increment mutation would choose the zero-bits to flip until the solution become a legal solution. In addition, the decrement mutation would choose the one-bits to flip until the solution become a legal solution. In Fig. 4, the solution $O1$ is applied the decrement mutation, the 2nd bit is changed from one to zero. And the solution $O2$ is applied the increment mutation, the 4th bit is chosen to flip from zero to one. There are two strategies used to choose the flip bits in the

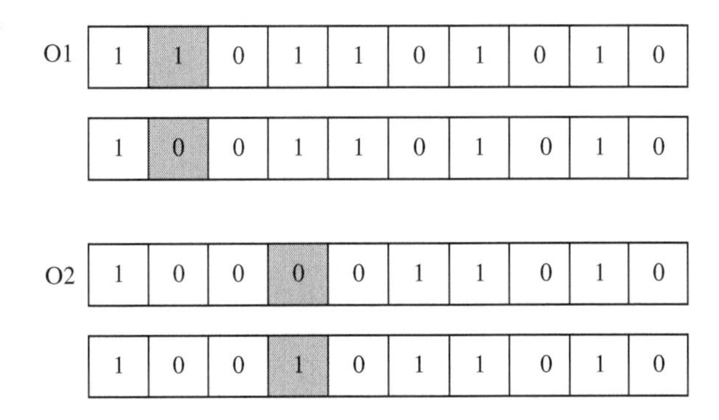

Fig. 4 The representation of solutions

both increment and decrement mutations. The first flip-bit choosing strategy is *random choosing*. This is simple and easy to implement. The second flip-bit choosing strategy is *heuristic choosing*. The heuristic choosing strategy on the increment mutation would prefer the item which can increase the maximum fitness value. The heuristic choosing strategy on the decrement mutation would prefer the item which would decrease the minimum fitness value.

Experimental Result

For illustrating the effectiveness of MCGA, we have implemented the MCGA on a HP DL165 G7 machine with 2.6 GHz AMD Opteron CPU and 12 GB of memory using Ubuntu 12.04. Moreover, all the programs are written in C++ and compiled using g++ (GNU C++ compiler). The two variants of MCGA are implemented. The first one is the MCGA with *random choosing* mutation, denoted by $MCGA_R$. The other one is the MCGA with *heuristic choosing* mutation, denoted by $MCGA_H$. In this paper, the population size is set to 40, and the crossover rate is 0.8. The number of iterations is equal to 100. The initial solutions are generated by randomizing. Each experiment repeated 30 trials and all results shown in this paper are the average of 30 trials.

As shown in Table 1, ten itembanks—denoted DS1-10—are used to measure the performance of the MCGA. The DS1 itembank is from the textbook published by the Han Lin Publishing [19]. And the DS2 itembank is from the textbook published by the Kang Hsuan Publishing [20]. These two books are for the grade 4 Society course of the elementary school in Taiwan. These two books are edited according to the same standard of Society textbook; so the contents of these two books are eventually identical, but with the different schemas. The DS1 itembank has 6 chapters, 13 sections, and 697 items. The DS2 itembank has 3 chapters, 15 sections, and 464 items. The DS3-DS6 itembanks are the subsets of DS1. The DS3 contains the first 3 chapters of DS1, DS4 contains only the 2nd chapter. The items of the last 3 chapters are in DS5, and the DS6 contains only the 5th chapter. The DS7-10 are based on DS3-6 and integrated the items from DS2 by using the method [3]. All of the itembanks are segmented by using the service of the Chinese word segmentation system [16].

In this work, *coverage* are chosen as the fitness value and the performance measure. The coverage evaluation function could be described as Eq. (1). We assume that the concepts in one item can be represented by its keywords. And all identical keywords of the itembank can be as the domain of the itembank. The coverage means that the ratio of the itembank's domain is covered by the test sheet.

$$MSE = \frac{\text{The identical keywords in the test sheet}}{\text{The identical keywords in the itembank}} \tag{1}$$

Table 1 Itemsets

	Item#	Keyword#	Identical keywords
DS1	697	7,268	1,874
DS2	464	3,908	1,255
DS3	340	3,395	1,058
DS4	123	1,230	443
DS5	357	3,873	1,144
DS6	157	1,707	611
DS7	551	5,137	1,481
DS8	182	1,702	616
DS9	516	5,248	1,525
DS10	254	2,555	887

Table 2 Experimental result of MCGA, the test sheet size is 50

	Random	MCGAR		MCGAH	
	Coverage	Coverage	Time	Coverage	Time
DS1	0.19	0.36	0.78	0.39	5.81
DS2	0.23	0.37	0.50	0.39	1.53
DS3	0.30	0.54	0.42	0.57	1.16
DS4	0.58	0.86	0.16	0.88	0.24
DS5	0.31	0.51	0.46	0.54	1.54
DS6	0.50	0.76	0.23	0.78	0.38
DS7	0.22	0.39	0.62	0.42	3.77
DS8	0.44	0.70	0.23	0.72	0.36
DS9	0.23	0.38	0.63	0.41	3.30
DS10	0.34	0.56	0.34	0.58	0.86

Table 3 Experimental result of MCGA, the test sheet size is 100

	Random	MCGAR		MCGAH	
	Coverage	Coverage	Time	Coverage	Time
DS1	0.33	0.54	1.56	0.59	10.02
DS2	0.38	0.58	0.97	0.61	3.23
DS3	0.50	0.76	0.80	0.80	1.83
DS4	0.89	1.00	0.30	1.00	0.45
DS5	0.50	0.74	0.87	0.78	2.19
DS6	0.77	0.96	0.42	0.97	0.67
DS7	0.36	0.58	1.20	0.63	5.46
DS8	0.70	0.93	0.43	0.94	0.46
DS9	0.37	0.58	1.24	0.62	5.48
DS10	0.57	0.80	0.65	0.82	1.43

The size of test sheet is set to 50 and 100. The Tables 2 and 3 show the comparison of *random*, $MCGA_R$ and $MCGA_H$. Both $MCGA_R$ and $MCGA_H$ are dramatically better than the random method in all itembanks. In addition, the coverage of the $MCGA_H$ is better than the $MCGA_R$. But $MCGA_H$ is slower than the $MCGA_R$.

Conclusion

The test sheet assembling is an important job in the E-learning. In the future learning cloud environment, the large amount of items would be aggregated into the itembank from various sources. In this situation, the well-organized itembank is not practical. The test sheet assembling algorithm should be with the ability of abstract the needed information directly from the items. The proposed MCGAis based on genetic algorithm and incorporated with the domain-specific heuristic. The experimental result shows the MCGA can effectively this problem. In the future, we try to use the multi-objective optimization to assemble the test sheet for more different test requirements.

Acknowledgments The authors would also like to thank Kang Hsuan Publishing and Han Lin Publishing for providing their itembanks and the Chinese Knowledge and Information Processing (CKIP) group, Institute of Information Science, Academia Sinica for providing their Chinese Word Segmentation System to support this research.

References

1. Zhang D, Zhao JL, Zhou L, Nunamaker JF Jr (2004) Can e-learning replace classroom learning? Commun ACM 47(5):75–79
2. Armbrust M, Fox A, Griffith R, Joseph AD, Katz R, Konwinski A, Lee G, Patterson D, Rabkin A, Stoica I, Zaharia M (2010) A view of cloud computing. Commun ACM 53(4):50–58
3. Tseng SP, Chiang MC, Yang CS, Tsai CW (2010) An efficient algorithm for integrating heterogeneous itembanks. Int J Innovative Comput, Inf Control 6(10):4319–4334
4. Bork A, Gunnarsdottir S (2001) Individualization and interaction. Tutorial Distance Learning, vol 12 of Innovations in Science Education and Technology. Springer, Netherlands, pp 47–62
5. Pellegrino JW, Chudowsky N, Glaser R (2001) Knowing what students know: the science and design of educational assessment. The National Academies Press, USA
6. Chua YP (2012) Effects of computer-based testing on test performance and testing motivation. Comput Hum Behav 28(5):1580–1586
7. Llamas-Nistal M, Fernndez-Iglesias MJ, Gonzlez-Tato J, Mikic-Fonte FA (2013) Blended e-assessment: migrating classical exams to the digital world. Comput Educ 62:72–87
8. JISC (2007) Effective practice with e-assessment: an overview of technologies, policies and practice in further and higher education
9. Blum C, Roli A (2003) Metaheuristics in combinatorial optimization: overview and conceptual comparison. ACM Comput Surv 35(3):268–308
10. Kirkpatrick S, Gelatt CD, Vecchi MP (1983) Optimization by simulated annealing. Science 220(4598):671–680
11. Glover F, Laguna M (1997) Tabu search. Kluwer Academic Publishers, Heidelberg

12. Holland JH (1992) Adaptation in Natural and Artificial Systems. MIT Press, Boston
13. Kennedy J, Eberhart RC (1995) Particle swarm optimization. In: Proceedings of the IEEE international conference on neural networks. pp 1942–1948
14. Hwang GJ, Yin PY, Yeh SH (2006) A tabu search approach to generating test sheets for multiple assessment criteria. Education, IEEE Transactions on 49(1):88–97
15. Gu P, Niu Z, Chen X, Chen W (2011) A personalized genetic algorithm approach for test sheet assembling. In Leung H, Popescu E, Cao Y, Lau R, Nejdl W (eds) Advances in web-based learning—ICWL 2011. Vol 7048 of Lecture notes in computer science, Springer, Berlin, pp 164–173
16. Chinese Document Segmentation (2008). http://ckipsvr.iis.sinica.edu.tw/
17. Porter MF (1997) Readings in information retrieval. Morgan Kaufmann Publishers Inc., San Francisco, pp 313–316
18. Miller BL, Miller BL, Goldberg DE, Goldberg DE (1995) Genetic algorithms, tournament selection, and the effects of noise. Complex Systems 9:193–212
19. Han Lin Publishing (2002). http://www.hle.com.tw/
20. Kang Hsuan Publishing (2002). http://www.knsh.com.tw/

Part V
Modern Learning Technologies and Applications with Smart Mobile Devices

Investigation of Google+ with Mobile Device Implement into Public Health Nursing Practice Course

Ting-Ting Wu and Shu-Hsien Huang

Abstract In recent years, mobile device assisted clinical practice learning is popular for the nursing school students. The introduction of mobile devices not only saves manpower and avoids the errors, but also enhances nursing school students' professional knowledge and skills. In order to respond to the demands of different learning strategies and reinforce the maintenance of learning system, new Cloud Learning is gradually introduced to instructional environment. This study introduces mobile devices and Cloud Learning in public health nursing practice with their advantages and adopts Google+ integrating different application tools as the learning platform. The users can save and use the data by all devices with wireless internet. According to findings of this study, learning effectiveness of the learners adopting Google+ is higher than that in traditional learning. Most of the students and nursing educator have positive attitude and are satisfied with the innovative learning method.

Keywords Mobile device · Public health nursing practice · Google+

Introduction

Rise of internet and information technology has changed the learning style and learning becomes more diverse [1, 2]. In highly flexible medical and nursing environment, the immediateness and continuity of mobile devices allow the users

---

T.-T. Wu (✉)
Department of Information Management, Chia-Nan University of Pharmacy and Science, Tainan, Taiwan, Republic of China
e-mail: wutt0331@mail.chna.edu.tw

S.-H. Huang
Department of Engineering Science, National Cheng Kung University, Tainan, Taiwan, Republic of China

Y.-M. Huang et al. (eds.), *Advanced Technologies, Embedded and Multimedia for Human-centric Computing,* Lecture Notes in Electrical Engineering 260,
DOI: 10.1007/978-94-007-7262-5_30,

to immediately acquire the resources needed and solve the problems [3]. Besides general hospitals, nursing related courses gradually introduce mobile devices in nursing school students' practices, such as clinical practice [4], tracking medication administration [5], outpatient clinics [6] and nursing homes [7], etc. Introduction of mobile devices in nursing practice saves the manpower, avoids the errors and provides rapid and immediate information searching [8]. In this study treats mobile devices as main learning tools and introduces them in public health care nursing practice. By the immediate support and service of mobile devices, the learners can acquire critical learning guidance and content at the right time and construct theoretical and practical connection in learning process to achieve the purposes of nursing practice.

Besides, in order to enhance the maintenance of learning system, reduce the problems of program upgrading and increase the reuse of instructional materials [9], Cloud Computing is used to design and plan the system. On Cloud platform, according to the users' different needs, the system can dynamically and immediately arrange and reorganize the support and functions to provide substantial or virtual services [10, 11]. Cloud Learning resembles a virtual office or campus. The learners and the nursing educator can acquire the resources, instructional materials, programs and services needed freely or by low cost. The function of high calculation and saving capacities allows the users to save all learning data and instructional materials on Cloud. By internet, the users can access and operate the data through internet. High flow of Cloud can enhance the reuse of instructional materials and avoid the incompatibility between data or programs [12]. In the highly flexible and convenient Cloud Learning environment, the learners having different types of mobile devices can acquire the learning content and instructional materials needed through internet at any time and in any place. Thus, this study suggests constructing learning management system on Cloud and the users can upload and access learning content and materials by mobile devices and internet. With the attributes and characteristics of Cloud Learning, instructional strategy and application of collaborative learning can be accomplished. Google+ is the learning management system of this study and it is currently integrated with the programs related to Google. With Google+, the learners can experience the effectiveness of Cloud Learning by varied application tools.

The purpose of this study is to help nursing learners connect the theories and practices by the introduction of new information and transform the abstract concept into specific operation, enhance the learners' knowledge structure and increase the learners' application of public health care theories. In addition, with the assistance of technology, the learners' obstacles in home visit will be reduced to enhance the interest in public health care practice. Besides, learning system can record all learning portfolio and the data can be the criteria for the teachers' instruction after class and adjustment of learning content.

Cloud Learning System for Public Health Practice

This study treats Google+ as the learning management system of Cloud environment (Fig. 1). Google+ integrates the application tools related to Google (Google Picasa、Google Docs、Google Map、Google Location、Hangouts…etc.). The learners can apply for the accounts of Google to use varied services of tools through internet. They can rapidly post, access and use the information, avoid the incompatibility of programs and increase the flow and use of information. Through the setting of high degree of privacy on Google+, the learners can classify different types of social circles and they can construct the public groups in information sharing. It increases the authority of personal privacy. In addition, besides the basic operation of words, pictures and videos, Google+ can position the locations of the data uploaded and offer the map. Besides, Google+ also functions as multi-user word chat room and video conference. In any places, the users can exchange the information. All behaviors on Google+ will be saved and recorded on Cloud and the nursing educator can download all activities of the learners, analyze and recognize the learners' learning situation and provide proper feedback and assistance.

Introduction of Cloud Learning allows nursing practice students to access and browse Google+ by wireless computers and mobile devices. The learners can use the functions on the platform for the practice of home visit. Use interface of the

Fig. 1 Interface of cloud learning system (Google+)

Fig. 2 User's interface showing on mobile device

learners is shown in Fig. 2. Before home visit, the learners of different groups can plan the visit routes by Google Map and post the planned routes and maps on Google+. In the process of home visit, they can access learning platform by mobile devices and read the planned routes of home visit. The learners can even search for the locations of the cases with the guidance of Google Map to avoid the confusion with the routes. In addition, Google+ learning platform is the supporter in the practice. Through the platform, the nursing educator can control the home visit practice of the students in different groups at the same time and enhance the interactive learning between the nursing educator and the students. The nursing educator can actively care about and guide the passive learners and passively assist with active learners. Thus, in the home visit, the guidance information will support the practice. The learners will realize the contribution of basic medical care institutions to overall medical care system and it will effectively fulfill the application of knowledge related to health education.

Research Design

Participants

Two classes participated in this study and they were the grade 4 students in five-year nursing college. The students have taken the course of public health care. There were totally 66 subjects and they voluntarily participated in the experiment. The researcher randomly divided the subjects into experimental group (34 students) and control group (32 students). In the practice of home visit, experimental group students treated Google as learning management system and used smart mobile phones as the mobile devices of home visit. Through Cloud Learning, the learners could operate, use and save the functions and information on learning platform and rapidly share and exchange the information. Control group students received traditional instruction and in home visit, they recorded the process and exchanged with each other by paper. The researcher tried to find the difference of learning effectiveness of the students in two groups after the introduction of Cloud Learning.

The nursing educator of two classes was the same person who had rich instructional experience and was considerably interested in the course of nursing practice upon technology. Thus, the said nursing educator voluntarily participated in the experiment. In addition, in order to make sure that each subject had positive execution ability and learning process in experiment, the scores would be included in the subjects' academic performance in this semester.

Experiment Procedure

Public health care nursing practice of the experiment lasted for 4 weeks and the course took 7 h every day. One day was designed for the practice of home visit every week. When the students did not go out for home visit, the nursing educator would introduce basic public health care in health office to allow the learners to approach the functions and jobs in basic medical health care institutions and recognize the roles and contribution of public health care nurses. Before the practice of home visit, the nursing educator introduced the activity, divided the students into groups (two students in a group) and distributed the areas of home visit; in addition, for experimental group students, the nursing educator would arrange one day for explaining the learning system and allow the students to use it. In the process, the technical personnel would help the learners be familiar with the interface. Besides making the nursing educator and the learners to approach the system, this study also distributed the scale of computer use to the learners to recognize the learners' past experience and cognition of information use.

In the practice of home visit, experimental group learners could share, save and discuss the related information on Google+ by any devices. In home visit, through smart mobile phones, the learners could immediately check the information, position the locations, upload data, pose the questions and discuss the related information. With the platform, the nursing educator could control the practice of the students in different groups and offer feedback and assistance at the right time. Control group students received traditional instruction and they recorded and planned pre-planning, practice and even discussion after class by paper. The nursing educator controlled the students' practice by phone. Overall experiment process is shown in Fig. 3.

Result and Discussion

As to evaluation of learning effectiveness, *t* test is used to analyze the scores of experimental group and control group, and descriptive statistics is used to explain the effect of immediate and convenient Cloud Learning in public health care nursing practice on learning effectiveness.

Statistical result of scores of experimental group and control group is shown in Table 1. Experimental group and control group are treated as independent variables and the scores of practice are dependent variables for t-test of independent

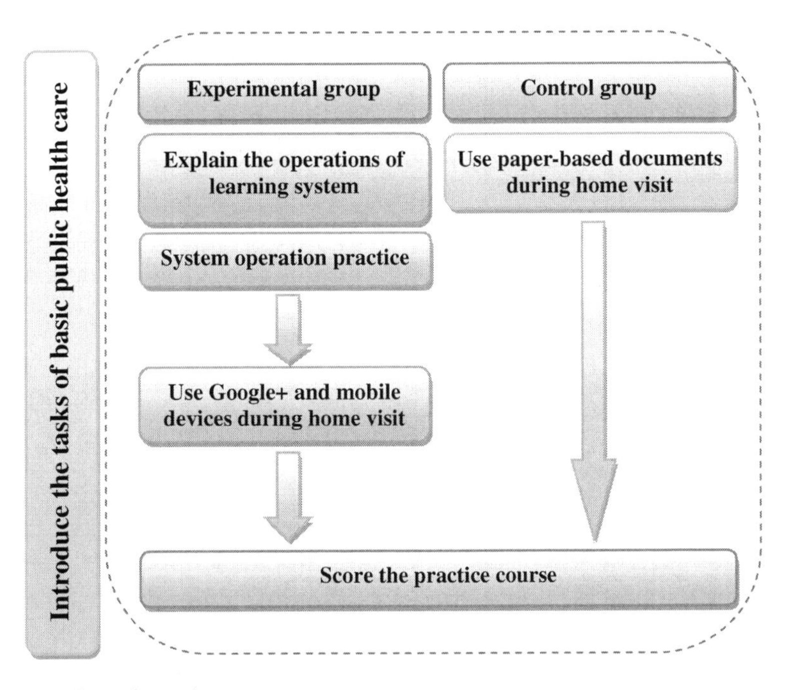

Fig. 3 The flow of experiment process

Table 1 t-test result of the learning effectiveness

	Mean	N	Std. Deviation	*t*	df	*p*
Experimental	85.63	34	2.34	4.26	64	.02*
Control	79.84	32	3.17			

*$p < 0.05$

samples in order to analyze the difference of learning effectiveness of experimental group learners with Google+ from control group with traditional learning.

Before judging the statistical result, it is necessary to test the homogeneity of two groups of variables. According to analytical result, the value is more than 0.05. It means that homogeneity of two groups of variables is supported. Besides, according to statistical result, p value is below 0.05. It means that the practice scores of the learners in two groups are significantly different. According to mean of the practice scores of two groups, experimental group learners' learning effectiveness of home visit with Google+ and Cloud Learning is significantly higher than that with traditional paper instructional strategy. In addition, standard deviation of experimental group is significantly lower than control group. Therefore, Cloud and information system can effectively reduce the learners' gap. The statistical result shows that on Google+ as the platform of Cloud Learning, the learners with mobile devices connect with the practice content and situations at any time. It can effectively reduce the abstract concept and increase practical experience. The varied functions on Cloud allow the learners to effectively internalize the learning content and develop comprehension, operation and integration abilities of home visit.

Conclusion and Future Work

This study tries to introduce Google+ and Cloud Learning in nursing practice course; thus, it only probes into learning effectiveness The application tools related to Google+ are diverse and rich. Future studies can analyze different practice courses and probe into the tools suitable for different practices. In addition, community function of Google+ can be analyzed by collaborative learning or explored by different learning methods in the courses, such as PBL, Concept Map, Jigsaw, etc. In addition, future studies can explore the learners' perceived loading in learning process. This study is still ongoing and the researcher expects that the findings can serve as references for related researches and provide more complete and convenient learning environment and strategy for nursing learners and nursing educator.

Acknowledgments This work was supported in part by the National Science Council (NSC), Taiwan, and ROC, under Grant NSC 102-2511-S-041-002, NSC 101-3113-P-006 023 and NSC 100-2511-S-006-015-MY 3.

References

1. Welsh ET, Wanberg CR, Brown KG, Simmering MJ (2003) E-learning: emerging uses, empirical results and future directions. Int J Training Dev 7(4):245–258
2. Eklund J, Kay M, Lynch HM (2003) E-learning: emerging issues and key trends: a discussion paper. Australian National Training Authority
3. Hwang GJ, Wu TT, Chen YJ (2007) Ubiquitous computing technologies in education. J Distance Educ Technol 5(4):1–4
4. Wu PH, Hwang GJ, Tsai CC, Chen YC, Huang YM (2011) A pilot study on conducting mobile learning activities for clinical nursing courses based on the repertory grid approach. Nurse Educ Today 31:8–15
5. Brian J, Jamieson S (2002) Post-surgical cardiac patients receive new level of care. Caring 21 (3):28–29
6. Edwards DJ (2001) Help is at hand (held): meet the new personal digital assistants that will keep you informed. Nurs Homes Long Term Care Manage 50(4):52–54
7. Hassett M (2002) PDAs: leading the information revolution. Nursing 32(2):66
8. Miller J, Shaw-Kokot JR, Arnold MS, Boggin T, Crowell KE, Allegri F et al (2005) A study of personal digital assistants to enhance undergraduate clinical nursing education. J Nurs Educ 44:19–26
9. Wang Y, Hong A (2010) The construction of virtual learning community of teachers in the times of cloud learning. Chin Educ Technol 1:118–122
10. Raman T (2008) Cloud computing and equal access for all. W4A2008 Keynote, Beijing, China, pp 21–22
11. Sedayao J (2008) Implementing and operating an internet scale distributed application using service oriented architecture principles and cloud computing infrastructure. iiWAS2008, Austria, pp 417–421
12. Al-Zoube M (2009) E-Learning on the cloud. Int Arab J e-Technol 1(2):58

Cognitive Diffusion Model with User-Oriented Context-to-Text Recognition for Learning to Promote High Level Cognitive Processes

Wu-Yuin Hwang, Rustam Shadiev and Yueh-Min Huang

Abstract This study proposed Cognitive Diffusion Model to investigate the diffusion and transition of students' cognitive processes in different learning periods (i.e. pre-schooling, after-schooling, crossing the chasm, and high cognitive processes). In order to enable majority of students crossing the chasm, i.e. bridge lower and higher levels of cognitive processes such as from understanding the knowledge that students learn in class to applying it to solve daily-life problems, this study proposes User-Oriented Context-to-Text Recognition for Learning (U-CTRL). Students participating at learning activities can capture learning objects and then recognize them into text by using U-CTRL. Finally, this study presents a case that shows how to facilitate students' cognition in English through applying the knowledge to solve daily-life problems with U-CTRL and how to evaluate the case.

Keywords Cognitive diffusion model · User-oriented context-to-text recognition for learning · Cognitive processes · EFL learning

W.-Y. Hwang
Graduate Institute of Network Learning Technology, National Central University, No. 300, Jhongda Road, Jhongli 32001, Taiwan, Republic of China
e-mail: wyhwang@cc.ncu.edu.tw

R. Shadiev (✉) · Y.-M. Huang
Department of Engineering Science, National Cheng Kung University, No. 1, University Road, Tainan 70101, Taiwan, Republic of China
e-mail: rustamsh@gmail.com

Y.-M. Huang
e-mail: huang@mail.ncku.edu.tw

Y.-M. Huang et al. (eds.), *Advanced Technologies, Embedded and Multimedia for Human-centric Computing*, Lecture Notes in Electrical Engineering 260,
DOI: 10.1007/978-94-007-7262-5_31,

Introduction

After learning at school, most students usually remember and understand knowledge taught by the teacher and just few of them can apply it to solve daily-life problems. Remember and understand are low level and apply, analyze, evaluate, and create are high-level cognitive processes [1]. Apply level of cognitive processes plays an important role as it separates cognitive domain into high and low level processes. After students remember and understand the knowledge, a very important goal for an instruction becomes enabling students to apply that knowledge to solve daily-life problems. Furthermore, cognitive processes, such as Apply, need to be cultivated in students as they promote other cognitive processes of higher level.

Accordingly, it is important that students learn the knowledge at school as well as apply it outside of school, therefore, learning is not confined to the classroom anymore but takes place in a range of situations. Students will be able to apply the knowledge to solve daily-life problems in context of school district environment (i.e. outside of school or at home) after class time, through exploration and verification of knowledge. In this way, students will learn really useful knowledge and utilize it in different daily-life situations, e.g. paper-based PISA assessment [2].

Cognitive Diffusion Model

Diffusion is the process by which an innovation is communicated through certain channels over time among the members of a social system [3]. There are a total of five categories of the members who adopt an innovation: innovators, early adopters, early majority, late majority, and laggards. To achieve popularity and acceptance of a technology, first, it needs to be utilized by 16 % of members, such as innovators and early adapters. Following early adopters are early majority (about 34 %) who usually want to be sure a technology works and is useful before adopting it. According to Moore [4], there is a chasm between the early adopters and the early majority as they have very different expectations. Crossing the chasm is a very difficult task that any innovation or innovative company must successfully accomplish to reach wide market success.

In order to promote students' cognitive processes from low to higher level, this study proposed the cognitive diffusion model. In the proposed model (Fig. 1), students' cognitive processes distributed into six levels [1] based on data reported by Kocakaya and Gönen [5]. The first (and the highest) level of the model is Create and the last (and the lowest) level is Remember or do not remember.

Crossing the chasm concept [4] was adopted in Cognitive diffusion model. It locates between Apply and Understand levels (Fig. 1); that is crossing the chasm imply that students' cognitive processes reach higher level from low level. It is very important for educators to find a way to cross the chasm. However, it cannot

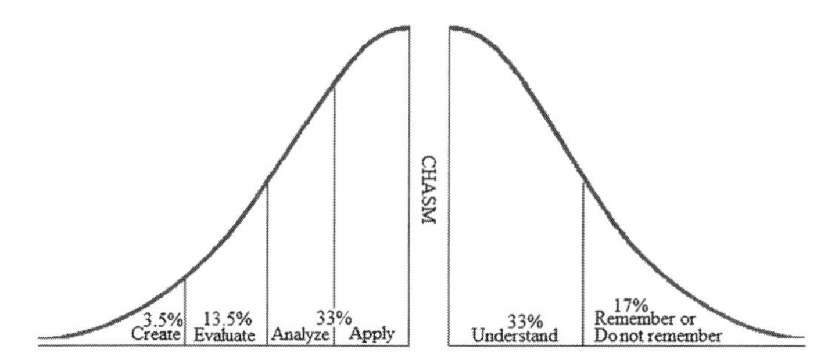

Fig. 1 Cognitive diffusion model chasm

be achieved through paper and pencil tests or exercises. Thus, some effort from educators is required to assist students to reach at least Apply level. The figure shows that the half of students (i.e. 50 %) has crossed the chasm. Cognitive processes of 3.5 % of these students are on Create level, of 13.5 % on Evaluate level, and of 33 % on Apply and Analyze levels. Cognitive processes of the other 33 % of students are on Understand level and of the left 17 % of students are on Remember or do not remember level. However, the percentage of distribution can be different and skewed for different domain knowledge.

Furthermore, this study explores the distribution of students' cognitive processes in cognitive diffusion model according to four learning periods (i.e. pre-schooling, after schooling, crossing the chasm, and high cognitive processes).

In the first period (pre-schooling), most students do or do not remember certain knowledge and only a small number of students can understand it. Therefore, cognitive processes of students in this period are only on the lowest level. In the second period (after schooling) students were instructed about the knowledge and they carried out some related exercises and assignments. In this period, level of cognitive processes of students increased so that most students not only remember the knowledge but also understand it. With further practice, few students even can apply knowledge to solve daily-life problems. The third period (crossing the chasm) is a critical period as it enables most students (at least 50 %) cross from the low level to the high level of cognitive processes (i.e. at least apply). The fourth period is high cognitive processes and cognitive processes of most students (70–80 %) reach highest level, i.e. equal or higher then apply.

In order to better understand why there are four learning stages, an example is given related to learning English as a foreign language (EFL). Most students know English words, how to spell them and their phonetic, but only a small part of them can apply these words to solve daily-life problems, e.g. communicate with someone else. Therefore, even after school learning, most students are still in the second stage (after school learning) and level of their cognitive processes cannot be high. On a contrary, most students have knowledge of native language and they can easily use it for daily conversation. Thus, we may conclude that instruction of native language in elementary school can enable crossing the chasm so that most students reach high level of cognitive processes.

Why there is such a big difference between learning native language and English? In fact, current educational system puts too much emphasis on concept learning and acquisition of knowledge. A little attention is paid on application of knowledge to solve daily-life problems. In most classes, students are requested to do assignments or answer test questions to test their knowledge. No focus is made to enable students to apply knowledge of English to solve daily-life problems. As for native language, obviously students got used to apply it in daily life conversation.

User-Oriented Context-to-Text Recognition for Learning

This study proposes User-Oriented Context-to-Text Recognition for Learning (U-CTRL) mechanism. Students capture objects in the real-life situation then technology converts information into learning text. To enhance cognition students this study suggests extending boundaries of learning environment to outside of classroom so that learning activities are no longer confined to the classroom teaching, instead, students learn anytime and anywhere. U-CTRL provides students with the context of real-life learning situations; that is, students can construct meaningful knowledge and apply it to solve daily-life problems. Students actively capture various learning objects they are interested in (active learning) by using U-CTRL (e.g. tablet computers) and accumulate information in cloud computing database. Therefore, this approach promotes students motivation and interest, and makes learning context richer. Portability of tablet computers will help students to carry out various learning activities outdoor. Information can be inputted through screen keyboard, voice or handwriting, digital camera and etc. in order to carry out individual and collaborative learning. Global positioning system (GPS) will obtain and display students' location in school district. Finally, students will share learning information with peers and it will enhance students' capabilities to search for information, process, and analyzing it. Moreover, it will demonstrate students' ability to apply knowledge to solve daily-life problems. U-CTRL approach is superior comparing to use of RFID or QR code technologies, as the instructor needs to set up learning environment for the latter in advance (e.g. to locate barcode labels or tags). Furthermore, U-CTRL is more context-aware richer comparing to information provided by experts to students.

How to assist students to cross the chasm with U-CTRL? This study proposes four phases (Fig. 2), each of them should have incentive that encourages students to become familiar with U-CTRL and then use it for learning. Phase 1: Training potential students (around 3.5 %) about how to apply knowledge to solve daily-life problems by using U-CTRL. Phase 2: Students (around 3.5 %) with high level (at least apply) tutor students (13.5 %) with lower level of cognitive processes (at least Understand) about how to apply knowledge to solve daily-life problems by using U-CTRL; approximate proportion will be 1 student with higher level to 4 students with lower level. Phase 3: Students (17 %) with at least Apply level tutor students with at least Understand level (33 %) about how to apply knowledge to solve daily-life problems by using U-CTRL; approximate proportion will be 1 student with at least Apply level to 2 students with Understand level. After completion of the three phases, proportion of students to cross the chasm will reach 50 %. That is, it can be said that after schooling reached crossing the chasm, as shown in Fig. 3. Phase four: Students (50 %) with at least Apply level tutor the rest students with Remember level (50 %) about how to apply knowledge to solve daily-life problems by using U-CTRL to promote level of their cognitive processes to, at least, Apply level. After crossing the chasm, to further enhance students' high-level cognitive processes, learning activity and challenges that requires students' higher level of cognitive processes need to be introduced, but slowly.

Applications of U-CTRL for Learning English as a Foreign Language

This study will design English learning activity to provide students with opportunity to learn and practice English in real-life situation. It will include parent-teacher interaction and parental involvement in order to improve students learning achievement [6]. The design will be tested against its potential to contribute to academic achievement of students.

The first step of learning activity includes remember and understand new vocabulary. Students freely explore surrounding context in school district. They learn new vocabulary through capturing learning objects in authentic contexts then the system identify objects and provide their names in English and in native language. In order to better understand pronunciation and meaning of new vocabulary, students will interaction with the teacher and other students about new vocabulary.

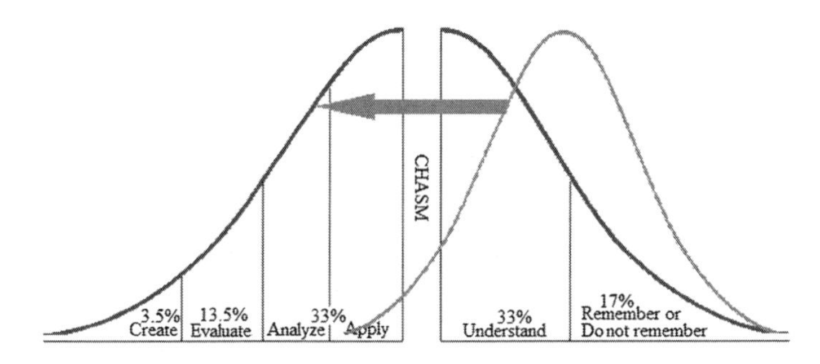

Fig. 2 Crossing the chasm

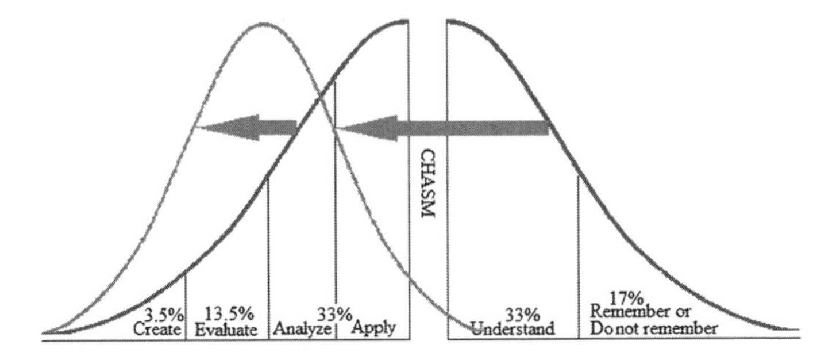

Fig. 3 After crossing the chasm

Capturing learning objects can be combined with specific strategies (meaning discovery strategy and memory reinforcement strategy). Students will collect different objects in school district show how each object relate to a specific situation in the district (meaning discovery strategy). Besides, students will be engaged in vocabulary learning and memorizing new words (memory reinforcement strategy). System identifies captured learning objects and provides their names in English, meaning, pronunciation, different sentence patterns including these words. Students practice new vocabulary repeatedly to facilitate their recall and understanding of new words and the system provides students with self-assessment feature so that students can monitor their learning process and progress. Words that students could not remember or understand during the test will be reminded by the system later so that students can practice them more. The following are some potential sub-activities: (1) Words exercise: Provide students with pictures, text, pronunciation, and meaning of words so that students can exercise words; (2) Vocabulary test: Provide students with pictures and name of words and then students need to give correct meaning of words; (3) Listening test: Provide students with pictures and pronunciation of words and then students need to give correct meaning of words; (4) Pronunciation test: Provide students with pictures of words and name recognized by the system and then students need to speak them out; (5) Spelling test: Provide students with pictures and pronunciation of words and then students need to spell out words; (6) Matching: Provide students with five pictures and five names of words and then then students need to match pictures and their names; (7) Using name of objects in sentences: Students capture objects in learning environment first, the system recognize name of these objects as text, and then students need to use name of objects in sentences.

The second step includes application of new vocabulary to solve daily-life problems. This study suggests using e-books with multimedia annotation tool [7, 8]; it can be carried everywhere to achieve seamless learning (e.g. take home so that parents may engage in learning process by helping their children). This study designed two sub-activities: (1) Introducing family members and (2) Introducing a menu of the dinner today. In the activities students expected to use simple English sentences to introduce family members and a menu of the dinner. Students can use e-books to record voices of family members, describe the dinner at home as well as to interact with their family members during the process of introduction. Teachers will still teach in traditional way and assign regular homework.

Evaluation

This study will carry out an empirical research to evaluate effects of U-CTRL on cognitive processes, analyze learning behavior of students to study with U-CTRL, and investigate students' perceptions and acceptance of the innovative approach.

A quasi-experimental design will be used in this study. Two classes of high grade elementary school students will be invited to participate in the experiment. One class with around thirty randomly assigned students will be the control group (with no treatment) and the other class with around thirty randomly assigned students will be the experimental group (with treatment). This study will administer a pre-test at the beginning of the experiment to assess a prior knowledge and prior cognitive processes of students. At the end of the experiment, after the treatment, this study will administer post-test to assess learning achievement and cognitive processes of students. A pretest–intervention–posttest design will allow evaluating effects of U-CTRL on learning achievement and cognitive processes of students. The targets of pre-test and post-test will focus on evaluating students' cognitive level, rather than scores, by designing test items based on the six levels of Bloom cognition. Therefore, the transition of students' cognitive processes could be analyzed to validate whether U-CTRL could facilitate them to cross the chasm and reach higher cognitive levels.

Meanwhile, regarding evaluating learning behaviors during U-CTRL activities, students will be motivated to capture learning objects, recognize them into text, and use both images and text for learning. All such learning behavior will be recorded and accumulated by students in learning portfolio. This study will explore students' learning behavior and analyze their portfolios to evaluate the transition of cognitive processes and their learning performance throughout the experiment.

Finally, this study will conduct a questionnaire survey and interviews with students to investigate students' perceptions, acceptance, and potential effectiveness of the innovative approach for learning.

Conclusion

This study proposed Cognitive Diffusion Model that distinguishes students' distribution of cognitive processes according pre-schooling, after-schooling, crossing the chasm, and high cognitive processes learning periods. Crossing the chasm is a very critical period as it promotes cognitive processes of most students from low to higher level. To facilitate crossing the chasm period, this study proposed four phases supported by User-Oriented Context-to-Text Recognition for Learning (U-CTRL). Students participate at learning activities by applying the knowledge they learnt in class to solve daily-life problems through capturing learning context, recognizing it into text, and employing for learning by using U-CTRL. Furthermore, one case related to learning English was proposed by using U-CTRL, and how to evaluate effectiveness of U-CTRL approach on cognitive processes of students was discussed.

References

1. Anderson LW, Krathwohl DR (eds) (2001) A taxonomy for learning, teaching and assessing: a revision of Bloom's Taxonomy of educational objectives, Complete edn. Longman, New York
2. PISA (2013) OECD program for international student assessment. Retrieved from http://www.oecd.org/pisa/aboutpisa/ on May 29, 2013
3. Rogers EM (2003) Diffusion of innovations, 5th edn. Free Press, New York
4. Moore GA (1999) Crossing the chasm: marketing and selling high-tech products to mainstream customers, Rev edn. HarperBusiness, New York
5. Kocakaya S, Gönen S (2010) Analysis of Turkish high-school physics-examination questions according to Bloom's taxonomy. Asia-Pacific Forum Sci Learn Teach 11(1)
6. Ho ESC, Kwong WM (2013) Effects of parental involvement and investment on student learning. In Parental Involvement on Children's Education, Springer, Singapore, pp 131–148
7. Hwang WY, Chen NS, Shadiev R, Li JS (2011) Effects of reviewing annotations and homework solutions on math learning achievement. British J Educ Technol 42(6):1016–1028
8. Hwang WY, Shadiev R, Huang SM (2011) A study of a multimedia web annotation system and its effect on the EFL writing and speaking performance of junior high school students. ReCALL 23(2):160–180

A Study of the Reader Recommendation Service Based on Learning Commons for Satisfaction of ePortfolio Users

Yu-Qing Huang, Cheng-Hsu Huang, Jen-Hua Yang and Tien-Wen Sung

Abstract This study implements co-interest recommending service embedded in ePortfolio for managing and sharing reader reading book experience from library records. We also propose five hypotheses based on proposed research model. Two hundred eleven Taiwanese graduate and undergraduate students participated in this research. The experimental results have shown that five hypotheses were supported.

Keywords ePortfolio · Recommendation · Virtual community

Introduction

The learning commons provides a range of programs and services to support learners' learning tasks. The learning commons not only assisting students in managing information, but also helping users manage learning. Recently, constructivism education has become new trend of teaching in universities [6]. On the constructivism learning theory, learning is a process of cooperative construction of the

---

Y.-Q. Huang (✉) · J.-H. Yang
Department of Computer Science and Information Engineering, National Central University, Jhongli, Taoyuan, Taiwan, Republic of China
e-mail: u9115903@ccms.nkfust.edu.tw

J.-H. Yang
e-mail: jhyang@csie.ncu.edu.tw

C.-H. Huang
Department of Computer Science and Information Engineering, Hwa Hsia Institute of Technology, Taipei, Taiwan, Republic of China
e-mail: Jeff@cc.hwh.edu.tw

T.-W. Sung
Department of Network Multimedia Design, Hsing Kuo University, Taipei, Taiwan, Republic of China
e-mail: kevin@mail.hku.edu.tw

Y.-M. Huang et al. (eds.), *Advanced Technologies, Embedded and Multimedia for Human-centric Computing*, Lecture Notes in Electrical Engineering 260,
DOI: 10.1007/978-94-007-7262-5_32,

knowledge by teachers and students. The learning commons can powerful support collaborative learning and learning activities through integrating of various resources and smart services. Learners in constructivism education acquire knowledge by a social process of learners' involvement into knowledge community [7].

Virtual learning community (VLC) is a virtual space of interactive collaboration which limited to network environments. According to constructivism learning theory, the learners of learning commons and VLC must actively probe into meaning of information in a social collaborative learning process. VLC is a new network learning organization established by web technologies to acquire knowledge and perform learning task through communicating with each other. The main distinguishing characteristic of VLC is its virtual space which limited to network environments. Li [6] evaluated VLC as an integral part of the learning commons and constructed a VLC in the learning commons. With the rapid development of web technology, ePortfolio has gradually developed become as a web-based information system for demonstrating the evidence of students' learning process over time in school education [10]. In order to provide more advanced and enhanced learning services, ePortfolio is used to manage the virtual learning community in grid environment [3].

One disadvantage of conventional virtual learning community is that learners are responsible for finding the right people to interact with. Therefore, a mechanism which can seamlessly integrate learning services, collaborators, and learning content is needed to enhance knowledge sharing in VLC and improve learning performance. As a result, finding the right collaborators whom they can derive tacit knowledge in an efficient manner is the most concerned issue of researches at present. The recommendation mechanism in the network is to recommend the information or products to users. Recommendation mechanisms based on social relations among people have become emerging fields. Recommendation people service may be considered an effective mechanism for finding common interest partners. In this study, a school-wide ePortfolios was chose as a platform to develop the co-intertest reader recommendation service. The main goal of recommendation co-interest reader service is to help students to find co-interest partners quickly to construct virtual learning community in ePortfolio.

Literatures Review

With the techniques of Web 2.0 initiated a new era of interactive cooperation and sharing knowledge, the concept of learning commons with the characteristic of cooperation learning was proposed. Li [6] asserted that the learning commons and virtual learning community (VLC) share the same basic concept, which emphasize that learners must not only actively explore the social meaning of information but also regard learning as a social collaborative process. In order to share knowledge in learning commons, finding common interest readers was the important task. Huang et al. [4] have demonstrated the recommending co-interest people service was a useful mechanism to build mobile learning networks. Therefore, recommendation people service may be considered an effective mechanism for finding co-interest partners. An ePortfolio is a digital electronic platform that stores visual and auditory content to demonstrate competencies and reflections in a field of knowledge to a teacher, a colleague, a professional, or a community. With the advent of new web technologies, the ePortfolio is also useful as a personal online virtual learning environment [1]. This study embedded recommending co-interest reader service to construct virtual learning community in ePortfolio. The goal of proposed recommending co-interest people service is to find right partners according to borrowing books from library's database.

Methodology

System Architecture

Finding common interest people was the important task in virtual learning community. Therefore, this study has proposed co-interest reader recommending service to construct virtual community in ePortfolio. The proposed co-intertest reader recommendation service finds co-interest readers according to the library resources of students. The library resources include reading statistics and the reader portfolio of reading authorized students, based on the functionalities provided by open source software called Mahara ePortfolio, which includes student portfolios, virtual community, and Web2.0 tools. The database contains library records, book categories, tables, and student portfolios of borrowing books. Figure 1 illustrates the proposed system architecture.

Instrument and Participants

An ePortfolio served as a virtual learning environment or knowledge-management system which is devised for students to collect, store and manage their learning

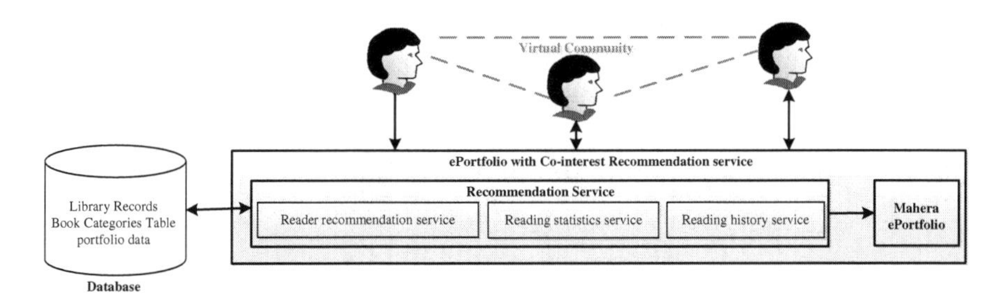

Fig. 1 System architecture

artifacts to demonstrate their competency [1, 5]. The proposed reader recommendation service provides an innovative service to find co-interest reader to construct virtual community in the Mehara ePortfolio system. Thus, this study not only considered ePortfolio as a platform for readers to share their reading experience in efficient manner but also applied recommendation co-interest reader service to help readers find co-interest readers quickly in virtual community. According to [12] KMS success model, the proposed reader recommendation service was used to enhance Mehara ePortfolio platform's system quality, information quality, system use, and learner satisfaction.

This study uses paper-based questionnaires to collect data. A seven-point Likert scale was developed for the measurement. System quality was measured with a 5-item developed scale consistent with [8]. Information quality was measured with a 5-item scale developed by [2]. Perceived system benefits were adapted from [12] with a 4-item scale. Fanally, student satisfaction was measured with a 4-item scale developed by Seddon and Kiew [9] and Wu and Wang [12].

Eighteen items relevant to the four factors of the proposed research model were adopted from existing literature. Data for this study were collected using an online questionnaire to survey universities in Taiwan. All 289 prospective participants were university graduate and undergraduate students. Of the 211 responses received, 189 were considered useful and used for analysis. The rate of useful response was 96.43 %.

Research Model and Hypotheses

Based on the results of a review in existing literature, this study included four factors in the research model (Fig. 2), which are system quality (SQ), information quality (IQ), perceived benefit (PB) and student satisfaction (SAT). The arrows in the research model are causal paths that represent the causal relationships between factors. Five hypotheses are proposed as follow:

H1: System quality positively influences perceived benefits
H2: Information quality positively influences perceived benefits
H3: System quality positively influences student satisfaction

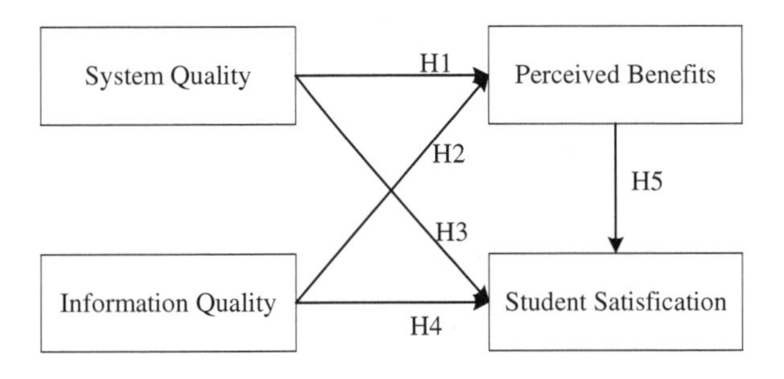

Fig. 2 Research model

Table 1 Results of multiple regression analysis of perceived system benefits (n = 189)

Independent variable	Dependent variable: perceived system benefits		
	β	t-value	Sig.
System quality	0.37	5.59	0.000***
Information quality	0.53	8.22	0.000***
F (2,186)	213.79***		
R^2	0.70		
Adjusted R^2	0.69		

*$p < 0.05$; **$p < 0.01$; ***$p < 0.001$

H4: Information quality positively influences student satisfaction
H5: Perceived benefits positively influences student satisfaction

Results

A multiple regression analysis was conducted to test the hypotheses, shown in Tables 1 and 2. Hypotheses 1–2 examined the relationship between system quality, information quality and perceived system benefits. The result of regression analysis is shown in Table 1. From Table 1, we can know that system quality was a predictor of perceived system benefits ($\beta = .37, p < 0.001$), and that information quality was also a predictor of perceived system benefits ($\beta = .53$, $p < 0.001$). Therefore, system quality and information quality have positive impact on perceived system benefits. The results supported Hypothesis 1 and 2.

Hypotheses 3–5 examined the relationship between system quality, information quality, perceived system benefits and student satisfaction. The result of regression analysis is shown in Table 2. The results of Table 2 showed that system quality was a predictor of student satisfaction ($\beta = .11, p < 0.05$), and that information quality was also a predictor of student satisfaction ($\beta = .21, p < 0.01$), and that perceived system benefits was also a predictor of student satisfaction ($\beta = .57$, $p < 0.001$).

Table 2 Results of multiple regression analysis of student satisfaction (n = 189)

Independent variable	Dependent variable: student satisfaction		
	β	t-value	Sig.
System quality	0.11	2.08	0.039*
Information quality	0.21	3.48	0.001**
Perceived system benefits	0.57	8.89	0.000***
F (5,183)	192.32***		
R^2	0.84		
Adjusted R^2	0.84		

*$p < 0.05$; **$p < 0.01$; ***$p < 0.001$

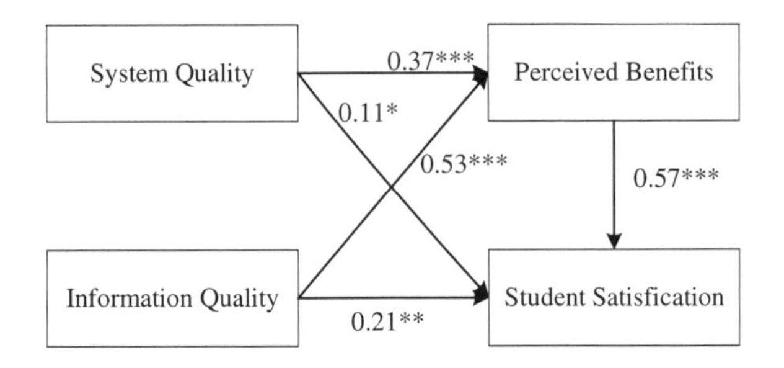

Fig. 3 Final model of multiple regression analysis (* $p < 0.05$; ** $p < 0.01$; *** $p < 0.001$)

Therefore, system quality, information quality and perceived system benefits have positive impact on student satisfaction. The results supported Hypothesis 3–5.

Based on the results of previous studies, system quality, information quality and perceived benefits have positive affect on student satisfaction [2, 8, 11]. From a management information system perspective, perceived benefits have positive impact on satisfaction [9]. Wu and Wang [12] pointed out that perceived benefits have positive impact on satisfaction in view of knowledge information systems. In our study, system quality, information quality, and perceived benefits are all supported, similar to existing studies. (Fig. 3).

Conclusion

This study developed a research model for explaining and predicting factors influencing reader satisfaction in the context of higher education. Based on the concept of learning commons, the co-interest reader recommendation service was implemented to recommend co-interest partners according to library reading records to construct virtual community in ePortfolio. The experimental results have shown that system quality and information quality have positive impact on perceived system benefits. Readers also feel satisfy with system depend on system quality, information quality and perceived benefits. From these experimental results we can know that the co-interest reader recommendation service was benefit to improve the opportunity for sharing knowledge or experience on virtual community of ePortfolio.

Acknowledgments This work is supported by National Science Council, Taiwan under grants NSC 101-2511-S-146 -001.

References

1. Barrett HC (2009) Online personal learning environments: Structuring electronic portfolios for lifelong and life wide learning. On Horiz 17(2):142–152
2. Bliemel B, Hassanein K (2007) Consumer satisfaction with online health information retrieval: a model and empirical study. E-Serv J 5(2):53–81

3. Gouardères G, Conté E (2006) E-Portfolio to promote the virtual learning group communities on the grid. Int J Inf Technol Web Eng 1(2):25–42
4. Huang JJS, Yang SJH, Huang Y-M, Hsiao IYT (2010) Social learning networks: build mobile learning networks based on collaborative services. Educ Technol Soc 13(3):78–92
5. Kim P, Olaciregui C (2008) The effects of electronic portfolio-based learning space on science education. Br J Educ Technol 39(4):700–714
6. Li MJ (2009) Construction of virtual learning community in learning commons of academic libraries. 2009 IEEE international symposium on IT in medicine and education (ITME 2009) vol 1, pp 565–569
7. McMahon M (1997) Social constructivism and the world wide web—a paradigm for learning. 1997 ASCILITE Conference, Curtin University, Perth
8. McKinney V, Kanghyun Y, Fatemeh Z (2002) The measurement of web-customer satisfaction: an expectation and disconfirmation approach. Inf Syst Rev 13(3):296–315
9. Seddon PB, Kiew MY (1996) A partial test and development of the DeLone and McLean's model of IS success. Aust J Inf Syst 4(1):90–109
10. Tosh D, Penny Light T, Fleming K, Haywood J (2005) Engagement with electronic portfolios: challenges from the student perspective. Can J Learn Technol 31:89–110
11. Wixom BH, Todd PA (2005) A theoretical integration of user satisfaction and technology acceptance. Inf Syst Res 16(1):85–102
12. Wu JH, Wang YM (2006) Measuring KMS success: a respecification of the DeLone and McLean's model. Inf Manage 43(2):728–739

A Study of the Wikipedia Knowledge Recommendation Service for Satisfaction of ePortfolio Users

Cheng-Hsu Huang, Yu-Qing Huang, Jen-Hua Yang and Wen-Yen Wang

Abstract This study extended a conventional ePortfolio by proposed Wikipedia knowledge recommendation service (WKRS). Participants included 100 students taking courses at National Central University which were divided into experimental group and control group. The control group students and experimental group students have created their learning portfilios by using ePortfolio with WKRS and conventional ePortfolio without WKRS, respectively. The data for this study was collected over 3 months. The experimental results have shown that the learners' satisfaction, system use, system quality and information/knowledge quality of experimental group students have significant progress than control group students.

Keywords ePortfolio · Wikipedia knowledge · Satisfaction

C.-H. Huang (✉)
Department of Computer Science and Information Engineering, Hwa Hsia Institute of Technology, New Taipei, Taiwan, Republic of China
e-mail: Jeff@cc.hwh.edu.tw

Y.-Q. Huang · J.-H. Yang
Department of Computer Science & Information Engineering, National Central University, Jhongli, Taiwan, Republic of China
e-mail: u9115903@ccms.nkfust.edu.tw

J.-H. Yang
e-mail: jhyang@csie.ncu.edu.tw

W.-Y. Wang
Department of Information Science, Kun Shan University, Tainan, Taiwan, Republic of China
e-mail: wwang@mail.ksu.edu.tw

Y.-M. Huang et al. (eds.), *Advanced Technologies, Embedded and Multimedia for Human-centric Computing*, Lecture Notes in Electrical Engineering 260,
DOI: 10.1007/978-94-007-7262-5_33,

Introduction

An ePortfolio serves as a personal virtual space in which students can construct, collect and organize digital artifacts as well as present their effort, progress and achievements in learning process. An ePortfolio not only provides students with opportunities for self-reflection, but also allows teachers to assess and understand their students' current learning status [1, 2]. Many schools and universities have developed ePortfolios to improve the quality of education and training. During the learning process of school education, students collect, self-reflect, evaluate and connect knowledge artifacts in ePortfolios to achieve their learning goals.

In recent years, many teachers have used ePortfolio to guide their students to deeper learning. However, the benefits of using ePortfolio are weakened because most students do not spend time and effort to maintain their learning portfolios unless they are forced to do so. Jafari [3] has proposed that the "advanced feature", an interactive intelligent service for helping students to maintain their learning portfolios, is one of seven essential attributes for successful ePortfolios. For enhancing the quality of students' learning, advanced features offer a convenient environment for students' reflections in ePortfolio. From this viewpoint, one may say that advanced features can contribute to the stickiness and success of ePortfolios.

Research on ePortfolios in education has been conducted for more than a decade. Gorbunovs [4] classified ePortfolios into three levels: the simplest ePortfolio, higher-level ePortfolio, and modern ePortfolio. The simplest ePortfolio works with MS Office to present students' achievements. Higher-level ePortfolio offer some interactivity, allowing students to communicate with others by using GoogleDocs and Web 2.0. Modern ePortfolio provide interactivity, data management, and reporting systems for assessment. An ePortfolio with more functions and tools can better facilitate students' learning process. In the future, ePortfolios are expected to be capable of offering advanced features.

During learning process for constructing knowledge, the relevant information of knowledge can help students improve the quality and efficiency of learned knowledge. In the recent years, there are many studies have developed searching and recommendation methods for helping knowledge workers to acquire relevant information [5]; (Chen and Chung 2008). For helping students to efficiently construct knowledge in conventional ePortfolios, this study proposes a WKRS based on searching and recommendation technologies to provide relevant information of knowledge. The experimental results have shown that the students' satisfaction, system use, system quality and knowledge/information quality in the experimental group were significantly higher than the control group. This is meaning that the embedded WKRS in ePortfolio can provide suitable knowledge to help students to organize their learning portfolios in specific knowledge domains without any additional system burden.

Literatures Review

Improving Students' Learning in ePortfolio by Searching and Recommendation Technologies

To investigate how to successfully develop ePortfolios, Jafari [3] proposed seven essential attributes that affect their success. These seven essential attributes include ease of use, sustainable business plan, advanced features, robust integrated technology architecture, lifelong support, standards and transportability, and the x-or other-attribute. For the advanced features attribute, ePortfolio should provide attractive, unique, flexible, and interactive intelligent services to help students to complete and maintain their learning portfolios. Sequentially, students can be guide to deeper learning through the advanced intelligent features embedded in ePortfolio. It will be seen from this that the value of ePortfolio is closely with embedded intelligent features. Therefore, this study embedded an intelligent service WKRS in conventional ePortfolio to provide relevant knowledge during students' learning process.

ePortfolios have recently been viewed as knowledge management systems that provide new opportunities for learning [6]. Promoting the knowledge creation and sharing can enhance the effectiveness of individual lifelong learning [7]. Many studies have developed searching methods for improving the effectiveness of sharing and creating knowledge. For example, Liaw [5] indicated that search engines are important knowledge construction tools for creating individual knowledge. The recommendation function also plays an important role for helping students to construct personal knowledge by reading recommended relevant information. The main purpose of existing recommendation systems is to determine the learning materials to be recommended by analyzing the students' learning portfolios [8–10]. For example, Chen and Hsu [9] have implemented a Personalized Intelligent Mobile Learning System for recommending English news articles to students. Therefore, this study develops the WKRS based on searching and recommendation technologies to assist students' knowledge construction in ePortfolio.

Measuring Success in ePortfolios

In e-learning field, the information systems success model can be utilized in examining e-learning systems success [11]. In this model, the effect factors of system success include of system quality, information quality, system use, and user satisfaction. In the KMS success model, the information quality is replaced by knowledge/information quality. Kim and Olaciregui [6] indicated that ePortfolios can be viewed as a knowledge management system (KMS) for students to demonstrate their learning. This study uses the effect factors of system quality, knowledge/information quality, system use, and learners' satisfaction to measure the successful of the proposed ePortfolio with WKRS.

Procedure, Instruments and Hypotheses

To investigate the influences of provided ePortfolio features on students' learning satisfaction, we have develop a conventional ePortfolios platform and extended the conventional ePortfolios platform by proposed intelligent service. This study proposes a WKRS intelligent service that collects Wikipedia content in certain domains (more than 20,000 articles) to provide relevant knowledge in students' learning process. The WKRS provides two functions: (1) Keyword annotation (2) Article recommendation. Figure 1 shows the system architecture of embedding WKRS ePortfolio. The "Keyword annotation" function would annotate keywords of the students' learning articles and sent the relevant knowledge of annotated keywords to students. The relevant knowledge was acquired from google or yahoo searching engine according to students' query decisions. When the learning portfolio was created by students, the "Article recommendation" function would annotate keywords of the created learning portfolio and recommend the relevant Wikipedia knowledge articles of annotated keywords to students.

One hundred students from two classes at National Central University participated in this experiment from February 2010 to April 2010. One class, consisting of 49 students using the embedded WKRS ePortfolio, served as the experimental group. Another class, consisting of 51 students using the conventional ePortfolio, served as the control group. Students received ePortfolio training in March 2009. Every week, students spent three hours defining the goal of their portfolio, collecting information related to the course, recording their reflection and learning experience, and organizing curriculum files on ePortfolio after class. Teachers did not participate in students' creation of ePortfolios or give any comment during the experiment.

This study uses paper-based questionnaires to collect data. A seven-point Likert scale was developed for the measurement. This study included system quality scale, knowledge/information quality scale, ePortfolio system use scale, and ePortfolio learner satisfaction scale. The system quality (7 items, $\alpha = 0.90$) was assessed with scales adopted from Wang et al. [12]. The knowledge/information quality (11 items, $\alpha = 0.88$) and system use (5 items, $\alpha = 0.91$) were assessed

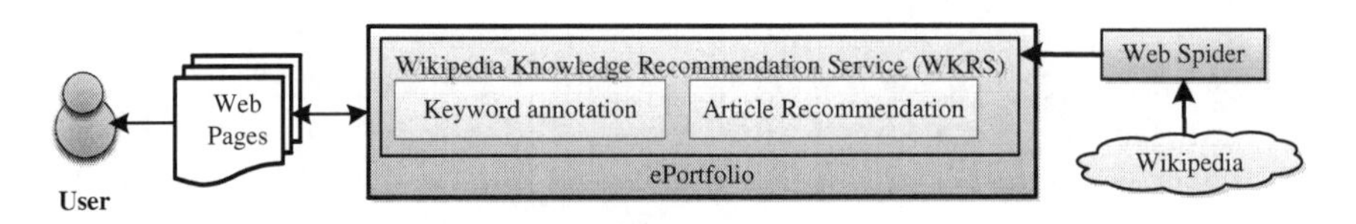

Fig. 1 System architecture of embedded WKRS ePortfolio

with scales adopted from Wu and Wang [13]. Learner satisfaction (4 items, $\alpha = 0.92$) was assessed with scales adopted from Seddon and Kiew [14]. The research question of this study is to investigate the influence of proposed WKRS intelligent service on ePortfolio system quality, knowledge/information quality, system use, and learner satisfaction. So, this study proposes the following hypotheses:

- Hypothesis 1. The WKRS has a positive effect on the system quality of the ePortfolio.
- Hypothesis 2. The WKRS has a positive effect on the knowledge/information quality of the ePortfolio.
- Hypothesis 3. The WKRS has a positive effect on the system use of the ePortfolio.
- Hypothesis 4. The WKRS has a positive effect on the learner satisfaction of the ePortfolio.

Results

Hypotheses 1–4 examine the effects of the WKRS on ePortfolio system quality, knowledge/information quality, system use, and learners' satisfaction. The Independent Samples t test was used to measure the difference in each variable between the experimental and control groups. Table 1 shows that students' satisfaction in the experimental group was significantly higher than the control group ($t = 12.56$, $p < 0.01$). The knowledge/information quality of the experimental group was significantly higher than that of the control group ($t = 9.91$, $p < 0.01$). Learners' satisfaction in the experimental group was also significantly higher than that of the control group ($t = 12.56$, $p < 0.01$), and system use was significantly higher in the experimental group than the control group ($t = 9.96$, $p < 0.01$). These results support Hypotheses 1–4.

Table 1 Independent t-test of prerequisites between experimental and control groups

Prerequisites	Experimental group		Control group		t	Sig.
	Mean	SD	Mean	SD		
System quality	5.65	0.61	4.32	0.59	11.12	0.00^{**}
Knowledge/information Quality	5.55	0.58	4.19	0.77	9.91	0.00^{**}
Students' satisfaction	5.59	0.64	4.00	0.62	12.56	0.00^{**}
System use	5.02	0.81	3.30	0.95	9.96	0.00^{**}

Note $^{*}p < 0.05$, $^{**}p < 0.01$

Conclusion

Intelligent services for managing knowledge can promote the personal core competitiveness of students through deeper learning in ePortfolio. To assist students construct knowledge in ePortfolio, this study proposes the WKRS intelligent service embedded in ePortfolio to automatically provide annotated keywords and recommended articles from Wikipedia after information is collected, chose, and reflected on by students for enhancing knowledge or information quality. Compared with the conventional ePortfolio without WKRS, the extended ePortfolio with WKRS provides more knowledge relevance and better search efficiency. The results of a series of experiments on this ePortfolio show that the WKRS can enhance the system quality, knowledge/information quality, system use, and learner satisfaction. These experimental results show that the WKRS can provide suitable knowledge to help learners organize their ePortfolio in specific knowledge domains with no additional system burden. Students also feel satisfied with the system and effectively use it to meet their needs.

In recent years, the applications of mobile devices have gradually increased in the field of education. The proposed WKRS intelligent service can strengthen the learning feedback of an ePortfolio and enhance learner satisfaction. In the future, we hope to apply the WKRS to other devices, such as smart phones or e-books, to facilitate students' learning in ubiquitous learning (u-learning) environments.

Acknowledgments This work is supported by National Science Council, Taiwan under grants NSC 101-2511-S-146 -001.

References

1. Barrett H (2000) Create your own electronic portfolio. Learn Lead Technol 27(7):14–21
2. Barker KC (2006) Environmental scan: overview of the ePortfolio in general and in the workplace specifically. Retrieved on 20 Oct 2006 from http://www.FuturEd.com
3. Jafari A (2004) The "sticky" ePortfolio system: tackling challenges and identifying attributes. J EDUCAUSE Rev 39(4):38–49
4. Gorbunovs A (2011) Prospective propulsions to embed artificial intelligence into the e-portfolio systems, In Al-Dahoud A et al (eds) Advances in information technology from artificial intelligence to virtual reality. UbiCC Publishers, pp 44–59
5. Liaw S–S (2005) Developing a web assisted knowledge construction system based on the approach of constructivist knowledge analysis of tasks. J Comput Human Behav 21(1):29–44
6. Kim P, Olaciregui C (2008) The effects of electronic portfolio-based learning space on science education. Br J Educ Technol 39(4):700–714
7. Li W, Liu Y (2008) Personal knowledge management in e-learning era. Technol e-learning Digital Entertainment Lect Notes Comput Sci 5093:200–205
8. Aleksandra K-M, Boban V, Mirjana I, Zoran B (2011) E-Learning personalization based on hybrid recommendation strategy and learning style identification. Comput Educ 56(3):885–899
9. Chen CM, Hsu SH (2008) Personalized intelligent mobile learning system for supporting effective english learning. J Educ Technol Soc 11(3):153–180

10. Hsu C-K, Hwang G-J, Chang C-K (2010) Development of a reading material recommendation system based on a knowledge engineering approach. Comput Educ 55(1):76–83
11. Freeze RD, Alshare KA, Lane PL, Wen HJ (2010) IS success model in e-learning context based on students' perceptions. J Inf Syst Educ 21(2):173–184
12. Wang Y-S, Wang H-Y, Shee DY (2007) Measuring e-learning systems success in an organizational context: scale development and validation. Comput Hum Behav 23(1):1792–1808
13. Wu J-H, Wang Y-M (2006) Measuring KMS success: a respecification of the DeLone and McLean's model. J Inf Manage 43(6):728–739
14. Seddon PB, Kiew MY (1994) A partial test and development of the DeLone and McLean model of IS success. In: International conference on information systems. Vancouver, Canada, pp 99–110

Using Personal Smart Devices as User Clients in a Classroom Response System

Tien-Wen Sung, Chu-Sing Yang and Ting-Ting Wu

Abstract This study proposed a classroom rcsponse system (CRS) different from existing commercial product. Modern and widespread used personal smart devices are utilized as the teacher-side controller and student-side response devices in the CRS instead of early infrared or radio frequency-based remote control. A prototype was developed for the proposed CRS, and it will be kept developing for further full functionality in CRS with the advantages and features of smart devices and modern network technologies.

Keywords Classroom response system · Smart device · E-learning

Introduction

A Classroom Response System (CRS) [1] is a technology-enabled learning system that allows instructors to project questions onto the screen, collects students' responses immediately and reports the feedback results in a class. It is also called an Interactive Response System (IRS) [2], an Audience Response System (ARS) [3], or a Student Response System (SRS) [4]. A CRS consists of three major parts: hardware, software, and communications. As mobile technology advances, there are various smartphone or pad products appeared on the consumer electronics

---

T.-W. Sung (✉) · C.-S. Yang
Institute of Computer and Communication Engineering, Department of Electrical Engineering, National Cheng Kung University, Tainan, Taiwan, Republic of China
e-mail: tienwen.sung@gmail.com

C.-S. Yang
e-mail: csyang@ee.ncku.edu.tw

T.-T. Wu
Department of Information Management, Chia-Nan University of Pharmacy and Science, Tainan, Taiwan, Republic of China
e-mail: wutt0331@mail.chna.edu.tw

Y.-M. Huang et al. (eds.), *Advanced Technologies, Embedded and Multimedia for Human-centric Computing*, Lecture Notes in Electrical Engineering 260,
DOI: 10.1007/978-94-007-7262-5_34,

market and used in people's daily life in recent years. These smart mobile devices are fully capable of serving as all-in-one computing devices and can offer a lot of potential applications, especially educational ones. In this paper, a prototype of smart device-based CRS is proposed. The objective is to develop a CRS with modern technologies and personal smart devices instead of earlier ones, and to make CRS more convenient by using smartphones or pads, now in widespread use, as user clients in the classroom.

CRS-Related Works

Earlier CRS systems used specific controllers and infrared signals for user response transmission. For example, EduClick II [2] consists of infrared signal transmitters and a corresponding receiver. The transmission is one-way and works with direct line of sight. A radio frequency-based CRS [5] uses RF signals instead of infrared for response transmission. It has less limitations and disadvantages but still has limited signal transmission range. A web-based CRS [6] utilizes Internet infrastructures to transmit user response data, and uses web technologies instead of specialist software and hardware devices of IR and RF-based CRS. A mobile CRS is another type. In [7], a non-smartphone Java application was presented for enhancing lecture interaction between a teacher and the students. In addition, a short message service (SMS) based classroom interaction system was proposed in [8]. Regarding to the evaluation, Kay et al. [9] have examined and summarized the benefits and challenges when using a CRS by a review of the literatures, and provided suggestions for further investigation. And Lin et al. [10] proposed results of evaluating students' perceptions of an SMS-based e-learning CRS before and after its implementation by applying a postexposure model called the information technology (IT) continuance model.

Proposed Smart Device-Based CRS

Learning Environment

As shown in the Fig. 1, this study proposed a smart device-based classroom response system, SD-CRS for short, which works on a WiFi or 3G enabled campus that provides a wireless data communication network. In the classroom, the teacher uses a smart device (smartphone or pad) held in the hand to control the learning content presentation. Students use hand-held smart devices to make responses. The CRS server can collect the responses and immediately generate a feedback report for showing on the screen and providing important information about learning effects and interaction outcomes.

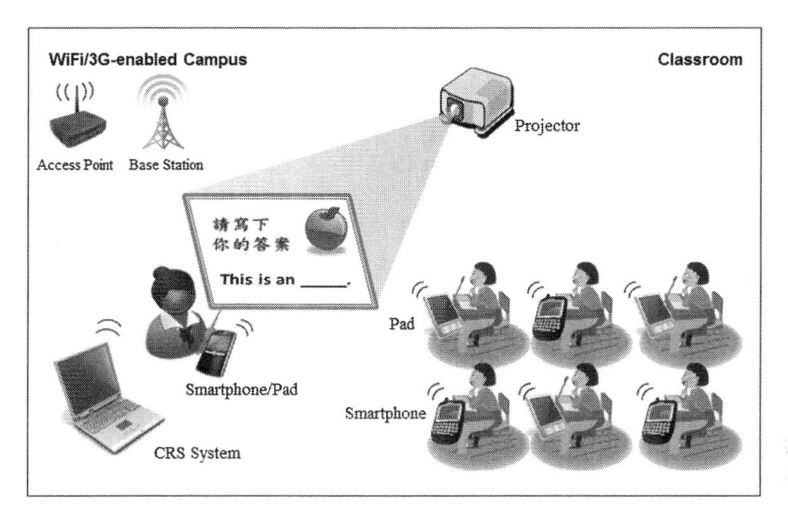

Fig. 1 SD-CRS learning environment and scenario

Design

Although there are some existing commercial CRS systems and applications used in practice, the system still can be promoted to a more convenient level. The key point is the type of the user device. The existing CRS products use specific remote controller with infrared (IR) or radio frequency (RF) signal transmission technology as user client devices. In this study, widespread used smartphones and pads are used instead. The comparison and differences between these two types of CRS are shown as below (Table 1).

Using modern smart devices as client devices has many advantages. A smart device is not merely used as a controller but also able to develop more special applications. For example, the feature of full keyboard can be utilized to input

Table 1 Comparison between general commercial CRS and the proposed SD-CRS

	General commercial CRS	Proposed SD-CRS
Client device	Specific remote controller	Smartphone or pad
Receiver	Specific receiver device	Infrastructure
Communication	Infrared (IR) or radio frequency	WiFi/3G network
Transmission angle	Limited/directional (using IR)	Omni-directional
Transmission range	Shorter	Farther
Obstacle effect	Significant (using IR)	Not significant
Keyboard	Simple buttons	Full keyboard
Screen	No	Yes
Music sound	No	Yes
Computing power	No	Yes
Consumer electronics	Not	Yes

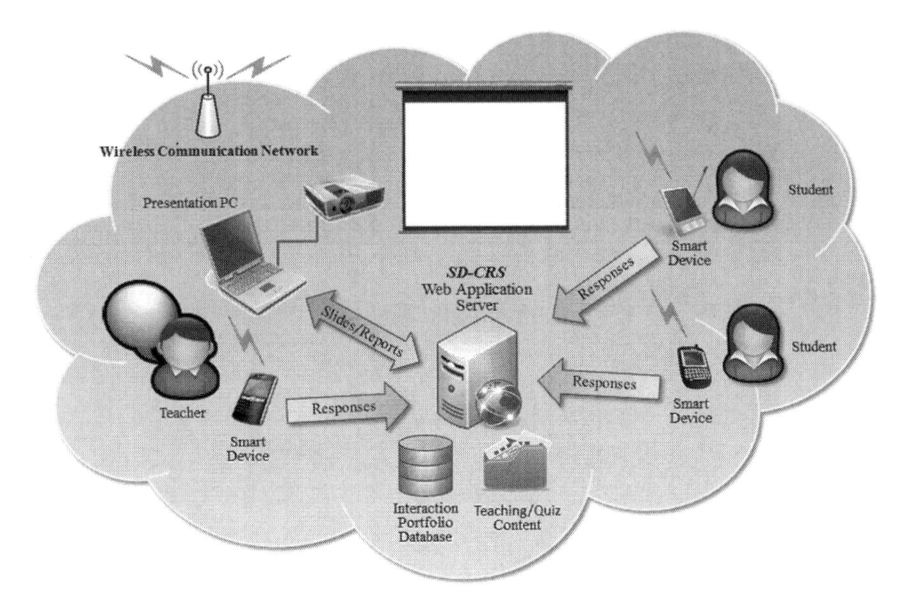

Fig. 2 SD-CRS system architecture

words as a response while simple buttons of traditional controller only can be pressed to select an option. For another example, the screen of a smart device also can be utilized to present additional information or contents within a CRS application.

Figure 2 shows the system architecture of the proposed SD-CRS prototype. The system uses a web-based design, which has three major advantages: (1) The system can work well on various personal smart devices with different OS platforms such as Android, iOS, and Windows Phone. (2) It has no need to install or update SD-CRS APP on the smart devices. The application is on a web server and any modification of the web application should be only made on the SD-CRS server. (3) Even a laptop or desktop PC can be used as a client device to make classroom responses. As shown in the figure, there are teaching and quiz contents stored in the SD-CRS sever, the teacher uses a presentation PC to show the contents by browsing SD-CRS pages and projecting to the large screen. The presentation can be controlled by the teacher's personal smart device. Students can make responses to the contents by the personal smart devices, too. Both teacher-side and student-side devices use web-based application to complete the functions. The interaction portfolio database (IPD) shown in the figure is used to record users' actions and response results in the SD-CRS. The IPD is further used to gather statistics and generate feedback reports regarding the students' responses. These reports are shown on the projection screen immediately. Then, it may cause students a valuable discussion and introspection in the classroom.

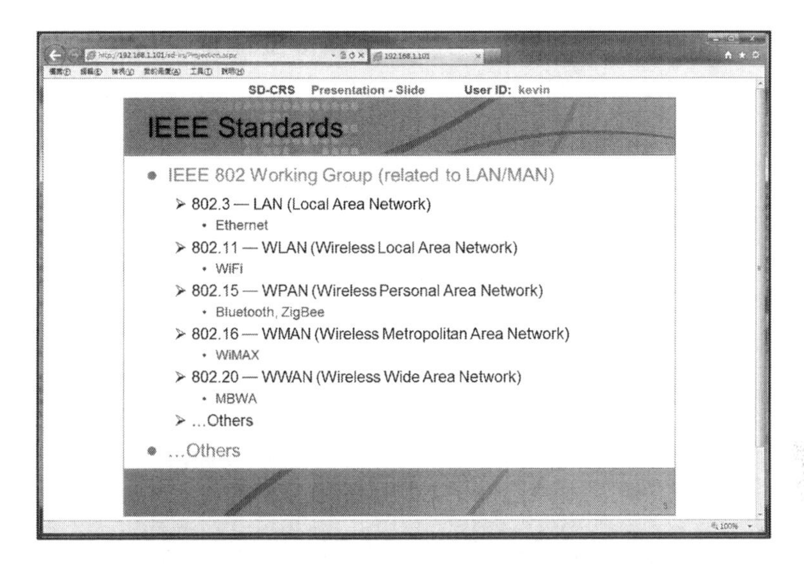

Fig. 3 Web-based teacher-side presentation screen (Slide)

Implementation

The prototype of SD-CRS was implemented by ASP.NET with C# and AJAX. Teaching slides and corresponding quizzes will be shown on the projection screen (Figs. 3 and 4). They can be switched and controlled by the function buttons on the interface of teacher-side smart-device (Fig. 5). A student can response to a quiz question by clicking a button or sending words (Fig. 5). The real-time feedback report then will be shown (Fig. 6) when the teacher clicks 'Report' button.

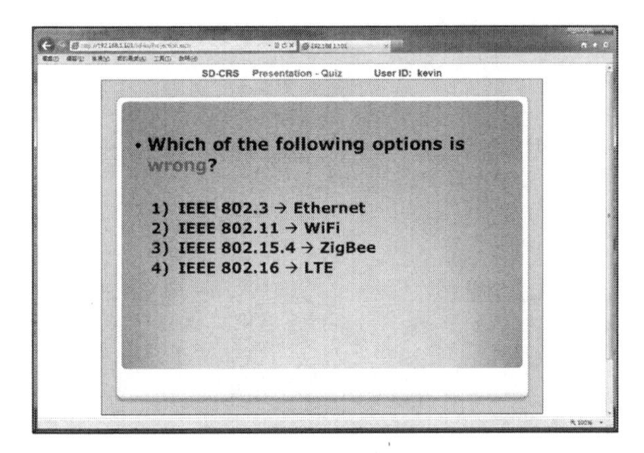

Fig. 4 Web-based teacher-side presentation screen (quiz)

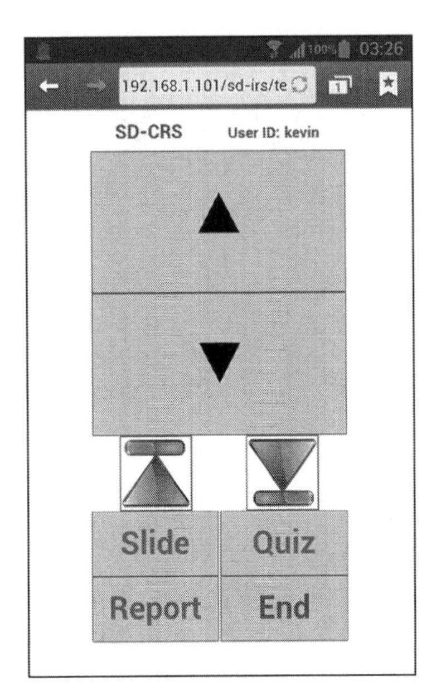

Fig. 5 Smartphone interfaces of teacher-side control (*left*) and student-side response (*right*)

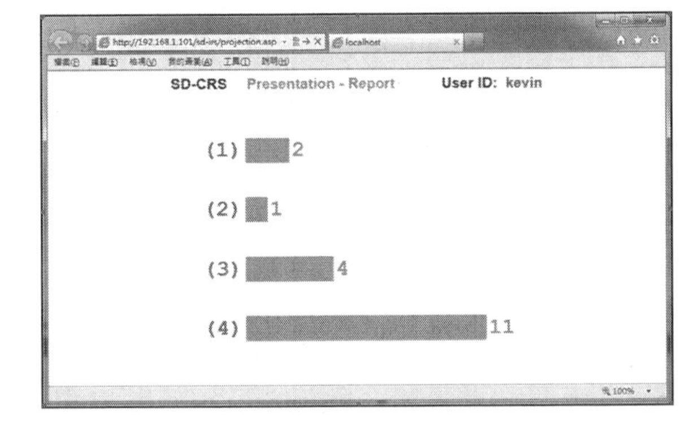

Fig. 6 Web-based teacher-side presentation screen (report)

Conclusion and Future Works

Various smartphone and pad products of consumer electronics are more and more popular and in widespread use. And the infrastructure of wireless communication networks such as WiFi and HSPA + make people easy to use Internet-based applications. This study utilizes the features and advantages of these modern devices and technologies to develop a classroom response system for further assistance in learning.

A prototype is proposed in this paper, and further full functionality of the smart device-based CRS will be kept developing. A learning experiment and evaluation of the completed system will be made in the near future.

Acknowledgments This work is sponsored by National Science Council under grant number NSC 102-2511-S-041-002, Taiwan.

References

1. Siau K, Sheng H, Nah FF (2006) Use of a classroom response system to enhance classroom interactivity. IEEE Trans Educ 49(3):398–403
2. Liu TC, Liang JK, Wang HY, Chan TW, Wei LH (2003) Embedding EduClick in classroom to enhance interaction. In: 11th international conference on computers in education, Hong Kong, pp. 117–125
3. Petr DW (2005) Experience with a multiple-choice audience response system in an engineering classroom. In: 35th annual conference on frontiers in education, Indianapolis, pp S3G-1–S3G-6
4. Hall RH, Collier HL, Thomas ML, Hilgers MG (2005) A student response system for increasing engagement, motivation, and learning in high enrollment lectures. In: 7th Americas conference on information systems, Omaha, pp 1–7
5. Nelson M, Hauck RV (2008) Clicking to learn: a case study of embedding radio-frequency based clickers in an introductory management information systems course. J Inf Syst Educ 19(1):55–64
6. Mantoro T, Ayu MA, Habul E, Khasanah AU (2010) Survnvote: a free web based audience response system to support interactivity in the classroom. In: IEEE conference on open systems, Malaysia, pp 34–39
7. Costa JC, Ojala T, Korhonen J (2008) Mobile lecture interaction: making technology and learning click. In: IADIS international conference mobile learning, Portugal, pp 119–124
8. Scornavacca E, Marshall S (2007) TXT-2-LRN: improving students' learning experience in the classroom through interactive SMS. In: The 40th annual Hawaii international conference on system sciences, Hawaiipp, pp 1–8
9. Kay RH, LeSage A (2009) Examining the benefits and challenges of using audience response systems: a review of the literature. Comput Educ 53:819–827
10. Lin J, Rivera-Sanchez M (2012) Testing the information technology continuance model on a mandatory sms-based student response system. Commun Educ 61(2):89–110

Part VI
Embedded Computing

3D Bidirectional-Channel Routing Algorithm for Network-Based Many-Core Embedded Systems

Wen-Chung Tsai, Yi-Yao Weng, Chun-Jen Wei, Sao-Jie Chen and Yu-Hen Hu

Abstract Network-on-Chip (NoC) is an emerging technology designed for the communication of IPs in an embedded system. This paper proposes a 3D (Three-Dimensional) model for a Bi-directional NoC (BiNoC). This three-dimensional model inspires the development of a new routing algorithm for BiNoC, called Bidirectional Routing (Bi-Routing). Bi-Routing is a fully adaptive routing algorithm using different layers in the proposed three-dimensional model to avoid deadlock without prohibiting the use of any path. As such, Bi-Routing can improve the load balance and reduce the packet latency of an NoC. Experimental simulation results demonstrated superior performance compared with existing routing methods.

Keywords Three-dimensional (3D) · Network-on-chip (NoC) · Bidirectional channel · Routing algorithm

---

W.-C. Tsai (✉)
Department of Information and Communication Engineering, Chaoyang University of Technology, Taichung, Taiwan, Republic of China
e-mail: azongtsai@gmail.com

Y.-Y. Weng · S.-J. Chen
Graduate Institute of Electronics Engineering, National Taiwan University, Taipei, Taiwan, ROC
e-mail: niles90221@gmail.com

S.-J. Chen
e-mail: csj@cc.ee.ntu.edu.tw

C.-J. Wei · S.-J. Chen
Department of Electrical Engineering, National Taiwan University, Taipei, Taiwan, Republic of China
e-mail: d92921022@ntu.edu.tw

Y.-H. Hu
Department of Electrical and Computer Engineering, University of Wisconsin, Madison, WI, USA
e-mail: hu@engr.wisc.edu

Y.-M. Huang et al. (eds.), *Advanced Technologies, Embedded and Multimedia for Human-centric Computing*, Lecture Notes in Electrical Engineering 260,
DOI: 10.1007/978-94-007-7262-5_35,

Introduction

System-on-Chip (SoC) uses numerous kinds of Intellectual Properties (IPs) and interconnections to form an embedded system in a single chip. As the technology progresses, the number and operating frequency of IPs are increasing. The bottleneck has transferred from IPs to interconnections. For example, with the deep sub-micron integrated circuit technology, crossing a chip with a highly optimized interconnects takes between six to ten clock cycles and only one set of IPs can use the traditional bus-based interconnection to transact data, such that the rest numerous IPs are waiting for the using right. Therefore, a new approach to designing the communication subsystem between IPs, Network-on-Chip (NoC), has been proposed in the past years to meet the design productivity and signal integrity challenges of next-generation system designs [1–3].

Routing is to decide which path a packet is to deliver. In other words, given a source and a destination, routing directs a packet where to go. A bad routing algorithm will let numerous packets pass through the same path or choose a longer path. We realize that the same route will lead to the lack of path diversity, and it creates a large load imbalance in the network. So path diversity provided by the adopted routing algorithm determines the performance of an NoC greatly. Most important of all, a deadlock will cause the on-chip interconnection crashed. Thus, routing algorithms must be deadlock-free.

A Bidirectional-channel Network-on-Chip (BiNoC) architecture was proposed to enhance the performance, quality-of-service, and fault-tolerance of on-chip communications [4–7]. BiNoC allows each communication channel to be dynamically self-configured to transmit flits in either direction in order to better utilize on-chip hardware resources. However, the conventional routing methods adopted in these BiNoC studies cannot fully exploit the path diversity of the BiNoC architectures. Accordingly, we present a three-dimensional model of BiNoC and a new routing algorithm for BiNoC called Bidirectional Routing (Bi-Routing). Bi-Routing can reduce packet latency and achieve higher bandwidth utilization due to its high path diversity by conditionally making channel bidirectional. Moreover, deadlock-freedom is provided with Bi-Routing which will be introduced in detail in this paper.

The rest of this paper is organized as follows. In Sect. “Background”, we will first introduce the background about BiNoC and some deadlock-free routing algorithms. Section “Methodology” will describe a 3-dimensional model of BiNoC and a routing algorithm based on the 3-dimensional model will be presented. In Sect. “Experimental Results”, we will show the experimental simulation results. Finally, Sect. “Conclusion” will draw a conclusion.

Background

First, we will introduce Bidirectional-channel Network-on-Chip (BiNoC) in Sect. “Bidirectional-Channel Network-on-Chip”. Next, Sect. “Related Routing Algorithms” will compare several deadlock-free routing algorithms.

Bidirectional-Channel Network-on-Chip

In a conventional router, all channels are unidirectional, thus it may lead to the following scenario where one output channel is busy or in congestion and another channel is idle because the direction of the idle channel is an input channel. BiNoC was proposed to overcome this problem by make all channels bidirectional and to allow each communication channel to be dynamically self-configured to transmit flits in either direction. For example, as shown in Fig. 1a, every vertex represents a task with a value t_j of its computation time, and every edge represents the computing dependence with a value of communication volume. A mesh NoC with most optimized mapping is shown in Fig. 1b. We can find that the NoC only use three unidirectional channels with the other directional channels being idle. However, if we make the same mapping solution on BiNoC which can dynamically change the direction of each channel between each pair of routers as shown in Fig. 1c, the bandwidth utilization will be improved and the total execution time can be reduced.

Related Routing Algorithms

Considering load balance, we prefer to use an adaptive routing algorithm that has more paths to choose for a packet routing delivery. Glass and Ni presented an

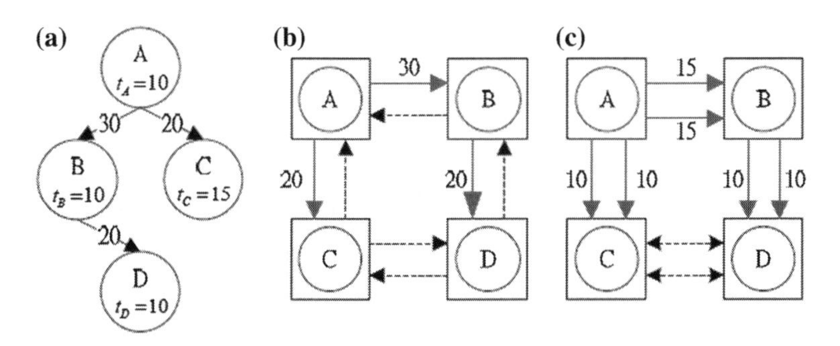

Fig. 1 Example of (**a**) task graph mapping to (**b**) a conventional NoC, and (**c**) a BiNoC

elegant concept of *turn model* [8]. The basic idea of turn model is to prohibit the minimum number of turns that break all of the deadlock cycles such that routing algorithms based on turn model can be deadlock-free. Three adaptive routing algorithms, namely west-first, north-last, and negative-first were designed based on turn model. We show that the four cases of prohibited turns of the three routing algorithms in Fig. 2. Note that the solid lines indicate the allowed turns and the dash line indicate the prohibited turns. For example, Case two uses the turn model that prohibits S–W turn and N–W turn. According to this turn model, west-first routing delivers all the packets to west first if packets need to be delivered to west. Similar with the west-first routing, negative-first routing and north-last routing were designed according to their own turn models. Turn model provides a simple way to design a deadlock-free adaptive routing. Nevertheless, there is a highly uneven routing path use problem in a global view. That is at least half of the source–destination pairs are limited to having only one minimal path, while full adaptive is provided for the rest of the pairs.

To solve the uneven routing path use problem, an odd–even turn model was presented by Chiu in [9]. With the odd–even turn model, any packet is not allowed to take an E-N turn or an E-S turn at any nodes located in an even column, and any packet is not allowed to take an N–W turn or an S–W turn at any nodes located in an odd column. This odd–even turn model restricts certain turns based on the locations such that none of the turns are eliminated in an NoC. Although the odd–even still restricts some turns for a packet to use, these restricted turns are unobvious in a global view. Therefore, the odd–even turn model has higher path diversity than other turn models. Based on the odd–even turn model, we can design an OE-Routing algorithm, which will compare with our proposed Bi-Routing algorithm in Sect. "Experimental Results".

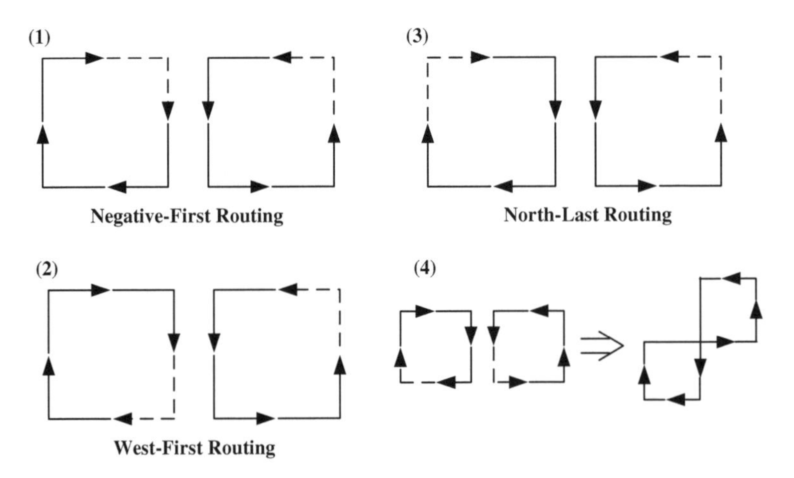

Fig. 2 Four cases of turn models

Methodology

In this section, we present the design methodology of our proposed Bi-Routing routing algorithm to exploit the characteristics of bidirectional channels and provide higher path diversity.

Three-Dimensional Model of BiNoC

Since the original model of mesh NoC cannot show the behavior of BiNoC, we have to represent these four kinds of bidirectional channel patterns as a three-dimensional model in Fig. 3a. The new Z-dimension is time related, which shows the channel diversity during time changed. The three-dimensional graph as shown in Fig. 3a is not a physical three-dimensional IC, but a conceptual model to represent the behavior of a BiNoC. Moreover, as shown in Fig. 3a, the odd–even turn model in BiNoC can also be represented in our three-dimensional model.

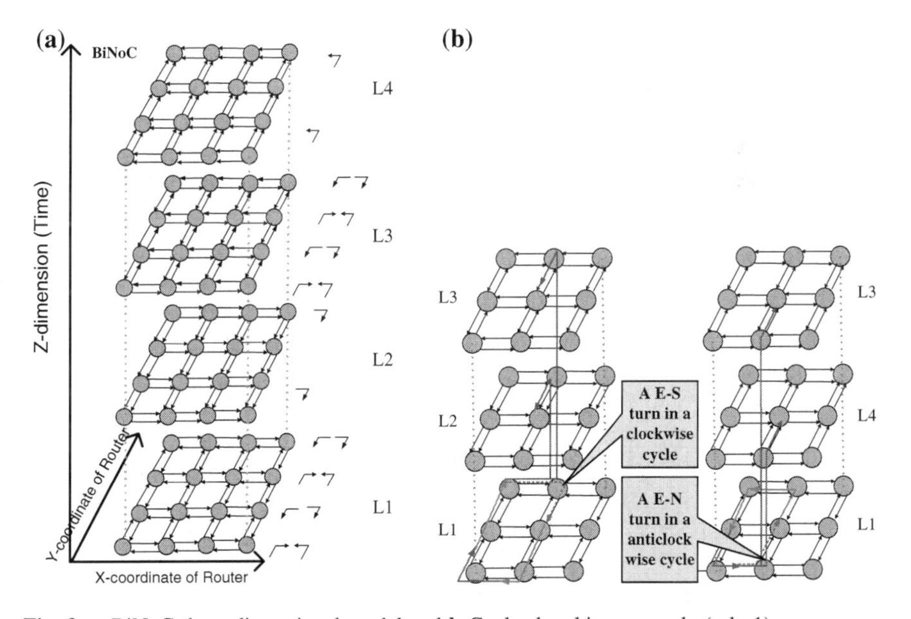

Fig. 3 **a** BiNoC three-dimensional model and **b** Cycles breaking example (rule 1)

Bidirectional Routing Algorithm

The three-dimensional model of BiNoC mentioned in Sect. "Three-Dimensional Model of BiNoC" indicates that BiNoC has higher path diversity than the original unidirectional NoC. We use this path diversity to develop a Bi-Routing algorithm for BiNoC in this section. The Bi-Routing idea is shown in Fig. 3b. On a unidirectional NoC, a deadlock cycle formed by the paths on the same layer can be broken by using another layer of channel (in the Z-dimension). Therefore, we need not prohibit any turn and all paths can be included in the feasible routing set of Bi-Routing. We develop the Bi-Routing based on Theorem 1 brought up in [10].

Theorem 1 A connected and adaptive routing function R for an interconnection network I is deadlock-free, if there are no cycles in its channel dependency graph.

A channel dependency graph D for a given interconnection network I and routing function R is a directed graph, D = G(C, E). The vertices of D are the channels of I. An arcs in D is a pair of channels (c_i, c_j) where there exists a direct dependency from c_i to c_j. The meaning of connected routing function is that for any packet, the connected routing function can find a path to deliver the packet to the destination. Therefore, from Theorem 1, if we can break the cycle in a channel dependency graph, the routing algorithm is deadlock-free. Hence, three rules are brought up for our Bi-Routing algorithm.

Rule 1: Packets use reverse channel at the E-S turn and the E-N turn.

To escape from deadlock in a BiNoC by using another layer to route, as shown in Fig. 3b, we choose E-S turn and E-N turn as a breaking position (using reverse channel in another layer) in clockwise and counter-clockwise cycles.

Rule 2:. Packets from south (north) reverse channel and delivered to north (south) must use reverse channel.

An inter-layer deadlock will appear without Rule 2. Rule 2 indicates that packets should keep using a reserve channel in south or north such that an inter-

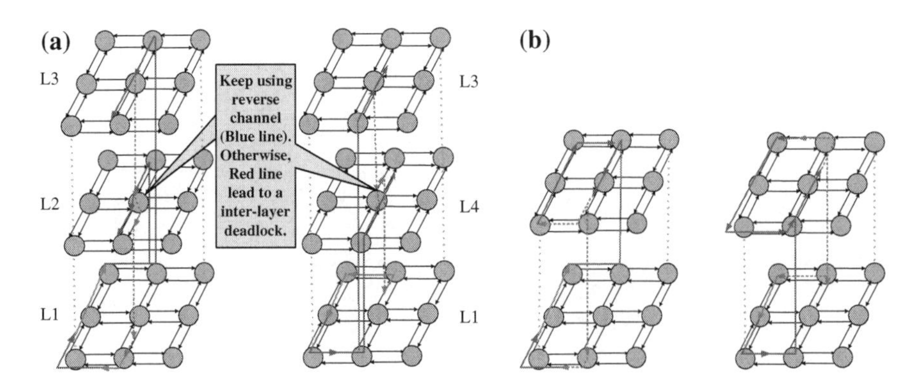

Fig. 4 **a** Example of rule 2 and **b** Example of rule 3

layer deadlock can be removed as shown in Fig. 4a. In which, red dotted lines represent paths violating Rule 2 and lead to inter-layer deadlock conditions.

Rule 3: Packet form reverse channel cannot take S–W or N–W turn.

The essence of Rule 3 is similar to the conventional turn model; it eliminates a turn in just one layer. In other words, reverse channels will make up a cycle, if we do not prohibit packet be routed to a lower layer. With Rule 3, in three-dimensional model, packets cannot take S–W turn and N–W turn when packets are not in the L1 layer as shown in Fig. 4b. Where the red dotted lines represent prohibiting turns in a higher layer and lead to inter-layer deadlock conditions.

With the three rules, Bi-Routing can we provide a fully adaptive routing, which can spread traffic loads to the whole network instead of keeping some parts of network in heavy congestion.

Experimental Results

Our simulation environment comprised an 8 × 8 mesh. Three traffic patterns were used, including uniform, transpose, and hotspot traffics. In the uniform traffic, a node receives a packet from any other node with equal probability. Every node transmits packets to a randomized destination with a probability based on the injection rate. In the transpose traffic, a node at a source with coordinate (i, j) will sent a packet to a destination with coordinate (j,i). In the hotspot traffic, 20 % of the packets change their destination to some selected hotspots while the remaining 80 % of the traffic keep uniform. In this work, we chose (3, 3), (3, 2), (3, 1), (3, 0) as hotspots.

We simulated XY-Routing, west-first routing algorithm (WF-Routing), odd–even routing (OE-Routing), and our proposed Bi-Routing algorithm. The packets in our experiments were composed of 16 flits with one header flit and one tail flit. The capacity of the buffer in each of the 5 directions of channels was 8 flits using wormhole switching. We simulated our network by injecting loads, from 20 flits per clock cycle to 500 flits per clock, at each node. For each injection rate, the simulation time was 25,000 clock cycles. The results of latency in three traffic patterns are shown in Fig. 5a, and the results of throughput are shown in Fig. 5b.

The simulation results show that our bidirectional routing, Bi-Routing, has the best performance among the four algorithms. XY-Routing outperforms OE-Routing and WF-Routing because XY-Routing can distribute packets evenly in the uniform traffic condition. This part of results is the same as in [8, 9]. Our bidirectional routing algorithm still had better saturation throughput than XY-Routing, about a 6.9 % improvement, as shown in Fig. 5a, b. However, the throughput of bidirectional routing decreases much more than XY-Routing in high injection rate.

The transpose and hotspot traffic patterns are close to the real-case embedded system traffics, because in a SoC most IPs have communications with the main CPU core. Adaptive routing algorithms perform better than XY-Routing in transpose and hotspot traffic patterns, because adaptive routing algorithms have

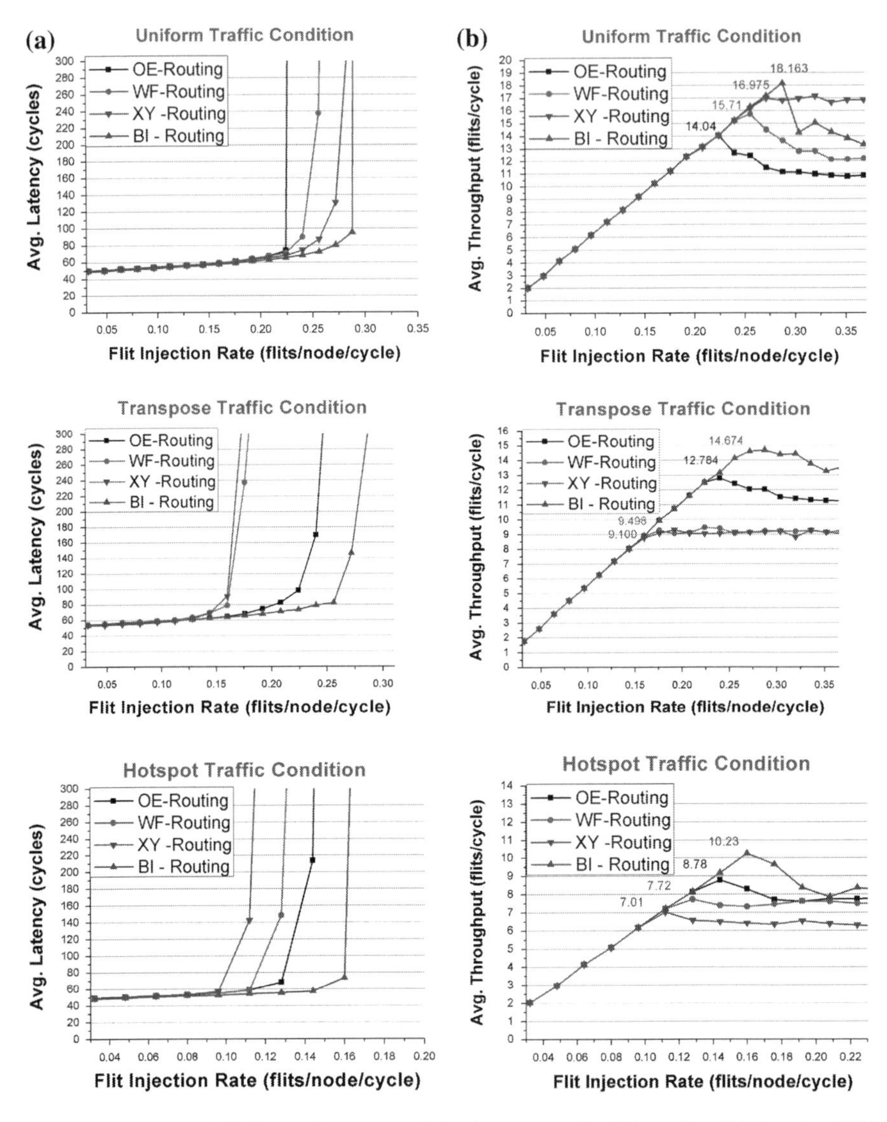

Fig. 5 **a** Latency and **b** Throughput versus injection rate under OE-routing, WF-routing, XY-routing, and Bi-routing

more paths to route. The Bi-Routing method had 14.78 and 16.51 % improvements over the OE-Routing one, in transpose traffic and hotspot traffic, respectively. The reason is our proposed Bi-Routing algorithm can spread traffic loads to relief local traffic congestion.

Conclusion

In this paper, we proposed a three-dimensional (3D) model of BiNoC. Based on this 3D model, we developed a new routing algorithm for BiNoC called Bi-Routing. Bi-Routing used the reversed channel to break the deadlock cycle in BiNoC, if any. Experimental results showed that our proposed Bi-Routing delivers better performance over the original BiNoC with OE-Routing because of the increased path diversity and the enhanced load balance provided by Bi-Routing.

Acknowledgments This work was partially supported by National Science Council, ROC, under grant NSC-101-2220-E-002-008.

References

1. Dally WJ, Towles B (2011) Route packets, not wires: on-chip interconnection networks. In: Proceedings of the design automation conference, pp 684–689
2. Benini L, DeMicheli G (2002) Networks in chips: a new SoC paradigm. IEEE Comput 35(1):70–78
3. Jantsch A, Tenhunen H, Ebrary I (2003) Networks on chip. Kluwer Academic Publishers, Dordrecht
4. Lan YC, Lo SH, Hu YH, Chen SJ (2009) BiNoC: a bidirectional NoC architecture with dynamic self-reconfigurable channel. In: Proceedings of the 3rd ACM/IEEE international symposium on network-on-chip, San Diego, pp 266–275
5. Lan YC, Lin HA, Lo SH, Hu YH, Chen SJ (2011) A bidirectional NoC (BiNoC) architecture with dynamic self-reconfigurable channel. IEEE Trans Comput Aided Des Integr Circuits Syst 20(3):427–440
6. Lo SH, Lan YC, Yeh HH, Tsai WC, Hu YH, Chen SJ (2010) QoS aware BiNoC architecture. In: Proceedings of the 24th IEEE international parallel & distributed processing symposium, Atlanta, pp 1–10
7. Tsai WC, Zheng DY, Chen SJ, Hu YH (2001) A fault-tolerant NoC scheme using bidirectional channel. In: Proceedings of the 48th design automation conference, San Diego, pp 918–923
8. Glass CJ, Ni LM (1994) The turn model for adaptive routing. J ACM 41(5):874–902
9. Chiu GM (2000) The odd–even turn model for adaptive routing. IEEE Trans Parallel Distrib Syst 11(7):729–738
10. Dally WJ, Seitz CL (1987) Deadlock-free message routing in multiprocessor interconnection networks. IEEE Trans Comput C-36(5):547–553

An Energy-Aware Routing Protocol Using Cat Swarm Optimization for Wireless Sensor Networks

Lingping Kong, Chien-Ming Chen, Hong-Chi Shih, Chun-Wei Lin, Bing-Zhe He and Jeng-Shyang Pan

Abstract In this paper, we propose an energy-aware routing protocol for wireless sensor networks. Our design is based on the ladder diffusion algorithm and cat swarm optimization algorithm. With the properties of ladder diffusion algorithm, our protocol can avoid the generation of circle routes and provide the backup routes. Besides, integrating cat swarm optimization can effectively provide better efficiency than previous works. Experimental results demonstrate that our design reduces the execution time for finding the routing path by 57.88 % compared with a very recent research named LD.

Keywords WSN · Routing · Cat swarm optimization

L. Kong · C.-M. Chen (✉) · C.-W. Lin · J.-S. Pan
Innovative Information Industry Research Center, Harbin Institute of Technology Shenzhen Graduate School, Shenzhen, China
e-mail: chienming.taiwan@gmail.com

L. Kong
e-mail: konglingping@utsz.edu.cn

C.-W. Lin
e-mail: jerrylin@ieee.org

J.-S. Pan
e-mail: jengshyangpan@gmail.com

C.-M. Chen · C.-W. Lin · J.-S. Pan
Shenzhen Key Laboratory of Internet Information Collaboration, Shenzhen 518055, China

H.-C. Shih
Department of Electronics Engineering, National Kaohsiung University of Applied Sciences, Kaohsiung, Taiwan, Republic of China
e-mail: hqshi@bit.kuas.edu.tw

B.-Z. He
Department of Computer Science National, Tsing Hua University, Hsinchu, Taiwan, Republic of China
e-mail: ckshjerho@is.cs.nthu.edu.tw

Y.-M. Huang et al. (eds.), *Advanced Technologies, Embedded and Multimedia for Human-centric Computing*, Lecture Notes in Electrical Engineering 260,
DOI: 10.1007/978-94-007-7262-5_36,

Introduction

Recently, wireless sensor networks (WSN) attract considerable research attention since they have been deployed in various applications, such as military, environmental monitor, industry automation and smart space. A WSN is composed of a large number of sensor nodes which collaborate with each other. Each sensor node detects a target within its detection range, gathers useful date, performs simple computations and sends the package to the sink node. In fact, sensor nodes are constrained in battery power and energy capability; therefore, energy saving is necessary to be considered carefully when constructing WSN.

Several routing protocols [1–8] for WSN have been proposed. One well-known protocol named DD (Directed Diffusion) was proposed by C. Intanagonwiwat et al. [2] in 1999. DD attempts to achieve better power consumption by reducing the data relay. More specifically, in DD, the collected data is transmitted only when sensor nodes fit the query from the sink node. However, in a large-sized WSN, the waste of power consumption and storage becomes worse. Besides, the circle route problem also becomes more serious. In 2012, Ho et al. proposed a protocol named LD [3]. LD describes a ladder diffusion algorithm to solve the circle route problem in DD and utilizes ant colony optimization [9–13] to select the most suitable routing path. According to their simulation results, LD indeed reduces power consumption and increases data forwarding efficiency compared with DD. However, LD algorithm still can be improved. In order to obtain the most suitable path, LD requires to perform various iteration operations. It will cause slow performance.

In this paper, we propose an energy-aware routing protocol based on ladder diffusion [3] algorithm and cat swarm optimization [14–17] for WSN. With the properties of ladder diffusion algorithm, our protocol can avoid the generation of circle routes and provide the backup routes if some sensor nodes are captured or unavailable. On the other hand, integrating cat swarm optimization can effectively overcome the weakness of LD that performs various iteration operations. Experiment results demonstrate that our protocol reduces the execution time for finding the routing path by 57.88 % compared with LD. Besides, the energy consumption of LD is about three times greater than our protocol. As a result, the lifetime of whole WSN can effectively be extended.

The remainder of the paper is organized as follows. Section “Related Works” describes a brief review of some closely related works. The proposed routing protocol is illustrated in Sect. “The Proposed Routing Protocol”. Simulation results are shown in Sect. “Simulations”. Finally, we conclude this paper in Sect. “Conclusion”.

Related Works

In this section, we first review the routing protocol LD proposed by Ho et al. [3]. Then, Cat Swarm Optimization (CSO) is described.

LD Algorithm

In 2011, Ho et al. [3] proposed a routing algorithm named LD based on ladder diffusion and ant colony optimization [9–13] for WSN. LD shows better performance in reducing energy consumption and increasing data forwarding efficiency compared with several previous well-known protocols [1, 2].

LD contains two phases, ladder diffusion and ant colony optimization. The purpose of the ladder diffusion phase is to identify routes from sensor nodes to the sink node and avoid the generation of circle routes. This phase also generates a *ladder table* for each sensor node. With the *ladder table*, LD can provide the back-up path and avoid redundant relay.

In the ant colony optimization phase, each sensor node can construct the most suitable routing path from itself to the sink node with the ladder table.

CSO

Chu et al. [14] proposed a new optimization algorithm named cat swarm optimization (CSO) in 2006. CSO imitates the natural behaviors of cats. Although cats spend lots of time in resting, they always remain alert. Once cats sense the presence of a prey, they chase it very quickly spending high energy. This two major behaviors of cats can be modeled into two modes, seeking mode and tracing mode. The seeking mode represents the behavior of cats which is resting and seeking the next position to move to. One the other hand, the tracing mode represents the behavior of cats which trace the target while spending large amount of energy. The detailed steps of CSO can be found in [14–17].

The Proposed Routing Protocol

In this section, we propose an energy-aware routing protocol. Our design has two phases, ladder diffusion phase and cat swarm optimization phase.

Ladder Diffusion Phase

In our protocol, we utilize the ladder diffusion algorithm described in [3]. Here we use a simple example to explain how the ladder diffusion works. Take Fig. 1 as an example. The sink node *a* first broadcasts a ladder-creating package with the grade value of one. Sensor nodes which receive this package are identified as grade one. It means that these grade one sensor nodes transmit data to the sink node require only one hop. In Fig. 1, sensor nodes with green color are categorized into grade one. Then, grade one sensor nodes, for example, *b* and *c*, broadcast another ladder-creating package with the grade value of two. Sensor nodes that receive this package (colored yellow in Fig. 1) are categorized into grade two. These grade two sensor nodes send data to the sink node require two hops count. Similarly, sensor nodes colored blue are categorized into grade three. After finishing the above procedures, all sensor nodes have been classified into a grade.

When a sensor node desires to send data to the sink node, the route is dynamically created by starting with nodes of high grade value and ending with nodes of low grade value. Normally, each sensor node will have more than one route to the sink node. As an example shown in Fig. 1, sensor node *g* can transmit data to the sink node *a* through two candidate nodes *d* and *e*.

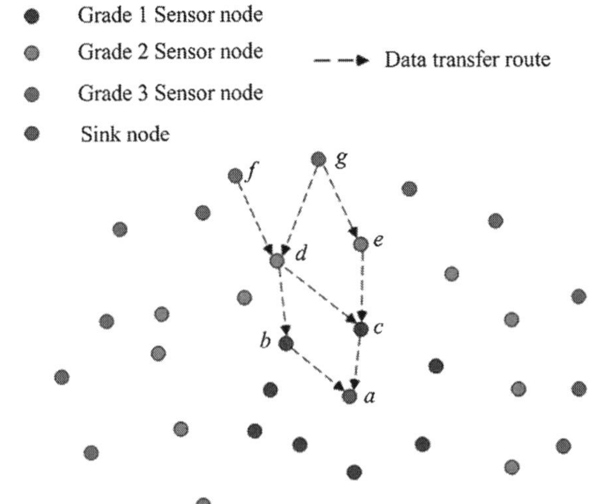

Fig. 1 An example of ladder diffusion algorithm

Cat Swarm Optimization Phase

Since each sensor node has more than one route path to the sink node, in this phase, we integrate the cat swarm optimization (CSO) to obtain the most suitable route path.

Fitness function. In our design, each cat means a routing path from the source node to the sink node. In order to evaluating the energy consumption of each cat, we define a fitness function F_m for Cat_m which is shown in Eq. (1).

$$\mathrm{F}_m = r(\alpha_1 + \alpha_2 \times D_\mathrm{m}) + E_m \tag{1}$$

In Eq. (1), D_m means the summation of the Euclidean distance between any two adjacent nodes for Cat_m; E_m denotes the summation of the remainder energy of sensor nodes for Cat_m; r is the data transfer rate of sensor nodes; α_1 is the distance-independent parameter; α_2 the distance-dependent parameter.

Initialization. Here we use Fig. 2 as an example to describe this phase. Assume that sensor node *tar* (grade six) desires to send a package to sink node *sink*. The configuration of the initialization parameters is as follows.

1. *tar* generates k cats. In this example, we assume that *tar* generates 10 cats. Among these 10 cats, 7 cats are in seeking mode and 3 cats are in tracing mode.
2. Each cat has its own position composed of 5 dimensions and each dimension has its velocity. In Fig. 2, the position of Cat_1 (the path colored red) is $(\chi_{1,1}, \chi_{1,2}, \chi_{1,3}, \chi_{1,4}, \chi_{1,5})$ and the initial velocity for these dimension are $(v_{1,1}, v_{1,2}, v_{1,3}, v_{1,4}, v_{1,5})$. Note that these initial velocities are randomly selected.
3. *tar* utilizes the fitness function (Eq. 1) to calculate the fitness value. Then, *tar* sets the best fitness value as the local best.

Seeking mode. In seeking mode, we first define three essential factors, seeking range of selected dimension (SRD), counts of dimension to change (CDC) and seeking memory pool (SMP). SRD and CDC denote the mutative ratio for the selected dimensions and how many dimensions will be varied respectively. Besides, SMP means the number of copies of a cat in seeking mode.

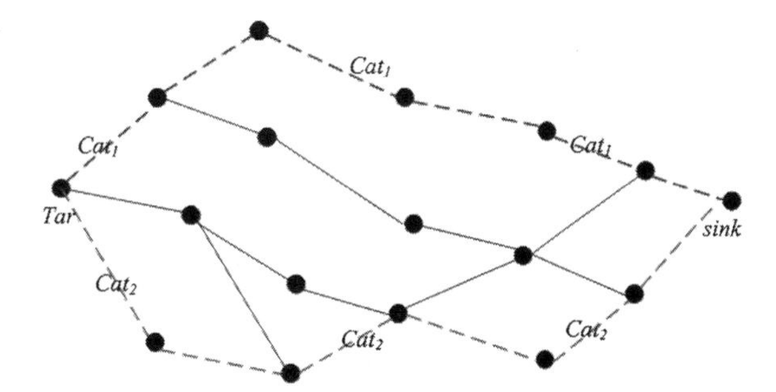

Fig. 2 Routing path from *tar* to *sink*

Assume Cat_2 in Fig. 3 is categorized into seeking mode. Besides, SRD, CDC and SMP are set 0.2, 0.4 and 6 respectively. The procedure of seeking mode is listed as follows

1. Make 6 copies of Cat_2, notes $cat_{2:1}$, $cat_{2:2}$, $cat_{2:3}$, $cat_{2:4}$, $cat_{2:5}$, $cat_{2:6}$.
2. For $Cat_{2:1}$, suppose the mutative nodes are *b* and *d*. According to SRD (20 %), *b* is varied to *g* and *d* is varied to *s*. Consequently, $Cat_{2:1}$ is varied from *(a; b; c; d; e)* to *(a; g; c; s; e)*.

Similarly, other five copies of Cat_2 perform the same procedure as $Cat_{2:1}$.After that, tar calculates the fitness values of these 6 copies of Cat_2 through Eq.1 and then replaces the Cat_2 with the best fitness value copy cat.

Of course, other cats classified into seeking mode do the same procedure as Cat_2. Finally, we get 7 updated cats.

Tracing mode. Assume that Cat_1 shown in Fig. 3 is categorized into tracing mode. It first updates the velocities according to Eq. (2).

$$v_{1,d} = v_{1,d} + r_1 \times c_1 \times (x_{best,d} - x_{1,d}) \quad where\ d = 1, 2, \ldots, 5 \tag{2}$$

$x_{best;d}$ is the *dth* dimension position of Cat_1 who has the best fitness value; $x_{1;d}$ is the *dth* dimension position of Cat_1; c_1 is a constant and r_1 is a random value in the range of [0,1].

Then, $x_{1;d}$ is updated according to $v_{1;d}$ (see Eq. 3).

$$x_{1,d} = x_{1,d} + v_{1,d} \quad where\ d = 1, 2, \ldots, 5 \tag{3}$$

Similarly, other two cats classified into tracing mode also move to a new place. Thus, we get three new cats.

Global Updating

tar evaluates the fitness values of these ten new cats and sets the cat of best fitness value as Cat_{11}. After that, it re-picks 3 cats and sets them into tracing mode and

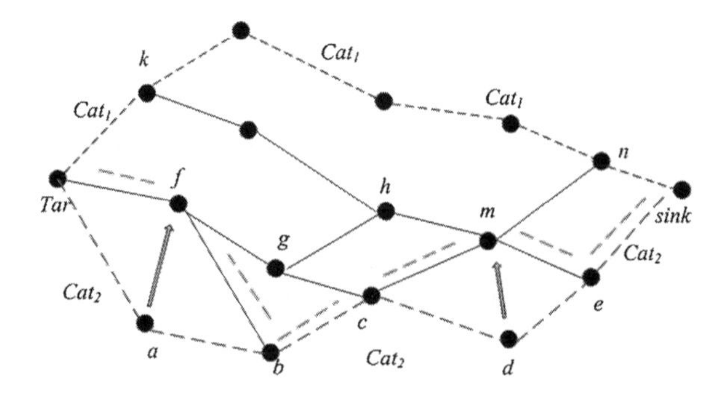

Fig. 3 Seeking mode and tracking mode

Table 1 Average time consumption

	Average time consumption (ms)
LD	25.313
Our design	10.683

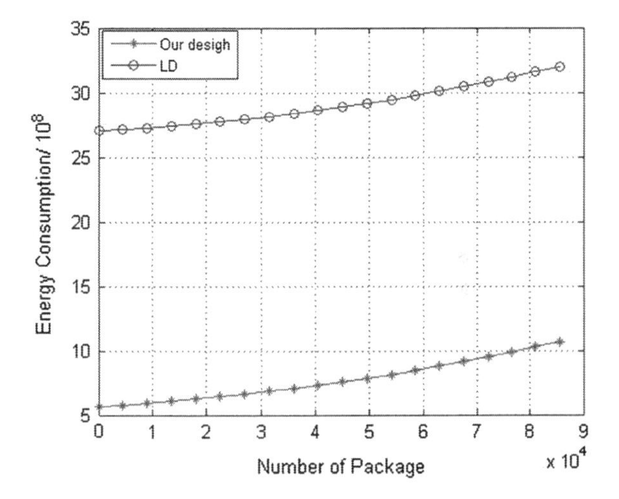

Fig. 4 Energy consumption

sets the rest 7 cats into seeking mode. Then, *tar* performs the above procedures of seeking mode and tracing mode repeatedly. The final path would be the best one since CSO keeps the best path position till it reaches the end of iterations.

Simulations

The simulation results are shown in this section. We implemented LD [3] and our protocol and then compared the performance of these two protocols. In our simulation environment, 3,000 sensor nodes are deployed in an $100 \times 100 \times 100$ unit of three-dimensional space. The transmission range of the sensor nodes are set to 15 units and the Euclidean distance is at least 2 units between any two sensor nodes. The sink node is distributed in the center of the area.

In our simulation, the simulator randomly selects one sensor node and calculates the execution time for this sensor node to find the most suitable routing path. The simulator performs this calculation 90,000 times and then obtains the average execution time. The results are shown in Table 1. Obviously, our work reduces the execution time by 75.84 %.

We also evaluated the energy consumption of our design and LD. The result is shown if Fig. 4. As shown in Fig. 4, the energy consumption of LD is about three times greater than our design.

Conclusion

In this paper, we present an energy-aware routing protocol based on ladder diffusion and cat swarm optimization. Similar to LD, our work can avoid the generation of circle route, reduce the energy consumption and provide back-up routes. Moreover, the proposed protocol further reduces the execution time and energy consumption compared with LD.

References

1. Perkins CE, Royer EM (1999) Ad-hoc on-demand distance vector routing. In: Proceedings of 2nd IEEE workshop on mobile computing systems and applications, pp 90–100
2. Intanagonwiwat C, Govindan R, Estrin D, Heidemann J, Silva F (2003) Directed diffusion for wireless sensor networking. IEEE/ACM Trans Network 11(1):2–16
3. Ho JH, Shih HC, Liao BY, Chu SC (2012) A ladder diffusion algorithm using ant colony optimization for wireless sensor networks. Inf Sci 192:204–212
4. Carballido JA, Ponzoni I, Brignole NB (2007) Cgd-ga: a graph-based genetic algorithm for sensor network design. Inf Sci 177(22):5091–5102
5. He S, Dai Y, Zhou R, Zhao S (2012) A clustering routing protocol for energy balance of wsn based on genetic clustering algorithm. IERI Procedia 2:788–793
6. Nayak P, Ramamurthy G, et al (2012) A novel approach to an energy aware routing protocol for mobile wsn: Qos provision. In: Proceedings of international conference on advances in computing and communications, IEEE, pp 38–41
7. Chen CM, Lin YH, Chen YH, Sun HM (2013) SASHIMI: secure aggregation via successively hierarchical inspecting of message integrity on WSN. J Inf Hiding Multimedia Signal Process 4(1):57–72
8. Chen CM, Lin YH, Lin YC, Sun HM (2012) RCDA: recoverable concealed data aggregation for data integrity in wireless sensor networks. IEEE Trans Parallel Distrib Syst 23(4):727–734
9. Chu SC, Huang HC, Shi Y, Wu SY, Shieh CS (2008) Genetic watermarking for zerotree-based applications. Circuits Syst Signal Process 27(2):171–182
10. Chu SC, Roddick JF, Pan JS (2004) Ant colony system with communication strategies. Inf Sci 167(1):63–76
11. Dorigo M, Gambardella LM (1997) Ant colony system: a cooperative learning approach to the traveling salesman problem. IEEE Trans Evol Comput 1(1):53–66
12. Dorigo M, Maniezzo V, Colorni A (1996) Ant system: optimization by a colony of cooperating agents. IEEE Trans Syst Man Cybern B Cybern 26(1):29–41
13. Misra R, Mandal C (2006) Ant-aggregation: ant colony algorithm for optimal data aggregation in wireless sensor networks. In: In Proceedings of IFIP international conference on wireless and optical communications networks, IEEE, p. 5
14. Chu SC, Tsai PW, Pan JS (2006) Cat swarm optimization. In: PRICAI 2006: Trends in artificial intelligence, pp 854–858
15. Wang ZH, Chang CC, Li MC (2012) Optimizing least-significant-bit substitution using cat swarm optimization strategy. Inf Sci 192:98–108
16. Panda G, Pradhan PM, Majhi B (2011) Iir system identification using cat swarm optimization. Expert Syst Appl 38(10):12671–12683
17. Pradhan PM, Panda G (2012) Solving multi-objective problems using cat swarm optimization. Expert Syst Appl 39(3):2956–2964

Software Baseband Optimization Except Channel Decoding for PC-Based DVB-T Software Radio Receiver

Shu-Ming Tseng, Yao-Teng Hsu, Yen-Yu Chang and Tseng-Chun Lee

Abstract The software radio has the advantages of flexibility, low cost and multimode ability, but the major disadvantage is the slower speed. To increase the speed of PC-based software DVB-T receiver, we need implement the baseband signal processing algorithms much faster. In this thesis, we discuss faster implementation of (1) 16QAM de-mapper to bits, (2) expanding_channel coefficient value in Viterbi decoder, and (3) inner de-interleaver and depuncturer. The speed of these three blocks is 6.64x, 0.57x and 0.88x faster.

Keywords Software radio · DVB-T · SIMD

Introduction

Recently, PC-based software radio systems have been developed, such as software global position system (GPS) [1]. The software GPS reduce the cost and time to modify the hardware. Another PC-based software GPS is in [2]. The new

S.-M. Tseng (✉) · T.-C. Lee
Department of Electronic Engineering, National Taipei University of Technology, Taipei 106, Taiwan, Republic of China
e-mail: shuming@ntut.edu.tw

T.-C. Lee
e-mail: jerryli0527@gmail.com

Y.-T. Hsu
Liteon Corp, Taipei, Taiwan, Republic of China
e-mail: Matthew.Hsu@liteon.com

Y.-Y. Chang
College of Electrical Engineering and Computer Science, National Taipei University of Technology, Taipei 106, Taiwan, Republic of China
e-mail: yeychang6614@gmail.com

Y.-M. Huang et al. (eds.), *Advanced Technologies, Embedded and Multimedia for Human-centric Computing*, Lecture Notes in Electrical Engineering 260,
DOI: 10.1007/978-94-007-7262-5_37, © Springer Science+Business Media Dordrecht 2014

acquisition and tracking method is used to increase operation speed and it can deal with the 12 channels computing requirement. A software digital audio broadcasting (DAB) receiver is implemented in [3]. It is implemented in a notebook PC platform. Besides, a PC-based multimode digital audio broadcasting system includes digital radio mondiale (DRM), DAB in Europe, and HD radio in United States is implemented in [4].

The PC-based software radio has the following advantage. It's easy to modify the signal processing algorithms to enhance the channel capacity and reduce the impact of multipath fading. In addition, we are more familiar with the PC platform than digital signal processing (DSP) or field programmable gate array (FPGA) platforms. Nowadays, the biggest challenge of the software radio is to process massive radio data fast enough. In order to make the speed of the software digital video broadcasting-terrestrial (DVB-T) receiver be fast enough to achieve real-time playing of TV programs, we need to optimize the time-consuming functional blocks of the software DVB-T receiver.

Currently, we have completed the optimization of Viterbi decoder [5], Reed Solomon (RS) decoder [6] and sampling frequency offset compensation [7]. To achieve real-time reception, we also propose a hard iterative channel decoding to enhance the error-correcting ability [8]. We still have to optimize the remaining functional blocks.

The single instruction multiple data stream (SIMD) instructions is often used in H.264 decoder [9] and wavelet transform [10], but not used in communication systems. Now we used SIMD instructions in the PC-based software DVB-T receiver. We also remove branches (IF command, for example), data re-arrangement, and rewrite the program by Assembly to improve performance.

The remaining of the paper is organized as follows: In section "Software and Hardware Environment", we describe the software and hardware environment. In section "Optimization of 16QAM De-mapper", we discuss the optimization of 16 quadrature amplitude modulation (16QAM) de-mapper. In "Optimization of Expanding_Channel Coefficient Value", we discuss the optimization of expanding_channel coefficient value in Viterbi decoder. In section "Optimization of Inner De-interleaver and Depuncturer", we discuss how to speed up the inner de-interleaver and depuncturer operations. And the conclusion of the paper is presented in section "Conclusion".

Software and Hardware Environment

We use SIMD instructions to improve performance, so we choose the operating system and CPU of using 128 bit registers. And the hardware platform is listed in Table 1. Our program takes 215.5 Mbytes in memory, so PC memory can be as low as 1 GB or 2 GB. We use the performance explorer in Microsoft Visual Studio 2010 to measure the performance (speed). The field test data comes from the DVB-T specification currently used in Taiwan (Parameters are shown in Table 2),

Table 1 Hardware list

Item	Model
CPU	Intel Core 2 Quad Q8200(2.33 GHz)
memory	DDR2 800 8 GB
main board	GIGABYTE GA-EP43-DS3L
Graphic card	NVIDIA GeForce 6200 TurboCache(TM)

Table 2 Parameter of Taiwan DVB-T

Arameter	Value
Frame	68 OFDM symbols
Super frame	4 frames
Mmodulation	16 QAM
Transmission modes	8 K
Subcarriers	6,817
Subcarriers (without pilots)	6,048
Guard interval(GI)	1/4
Code rate	2/3

we use a customized USB 2.0 dongle (including RF front end and A/D converter) to record data. Performance calculation is based on the requiring time of decoding 28 frames.

The functional blocks are shown in Fig. 1. In this paper, we discuss faster implementation of the following three blocks: (1) 16QAM de-mapper to bits, (2) expanding_channel coefficient value in Viterbi decoder, and (3) inner de-interleaver and depuncturer.

Optimization of 16QAM De-mapper

We consider 16QAM de-mapper (to 4 bits) in the non-hierarchical mode. For each DVB-T frame, we have 68 orthogonal frequency division multiplexing (OFDM) symbols and each symbol has 6,048 sub-carriers. There are 411,264 de-mapper operations for each frame. This computation is large and speedup is necessary.

For the 16 QAM in Fig. 2, the 4 soft bits ($i1$, $q1$, $i2$, $q2$) are as follows:

$$\begin{aligned} i1 &= -I \\ q1 &= -Q \\ i2 &= \begin{cases} 2D - I, I > 0 \\ 2D + I, I < 0 \end{cases} \\ q2 &= \begin{cases} 2D - Q, Q > 0 \\ 2D + Q, Q < 0 \end{cases} \end{aligned} \tag{1}$$

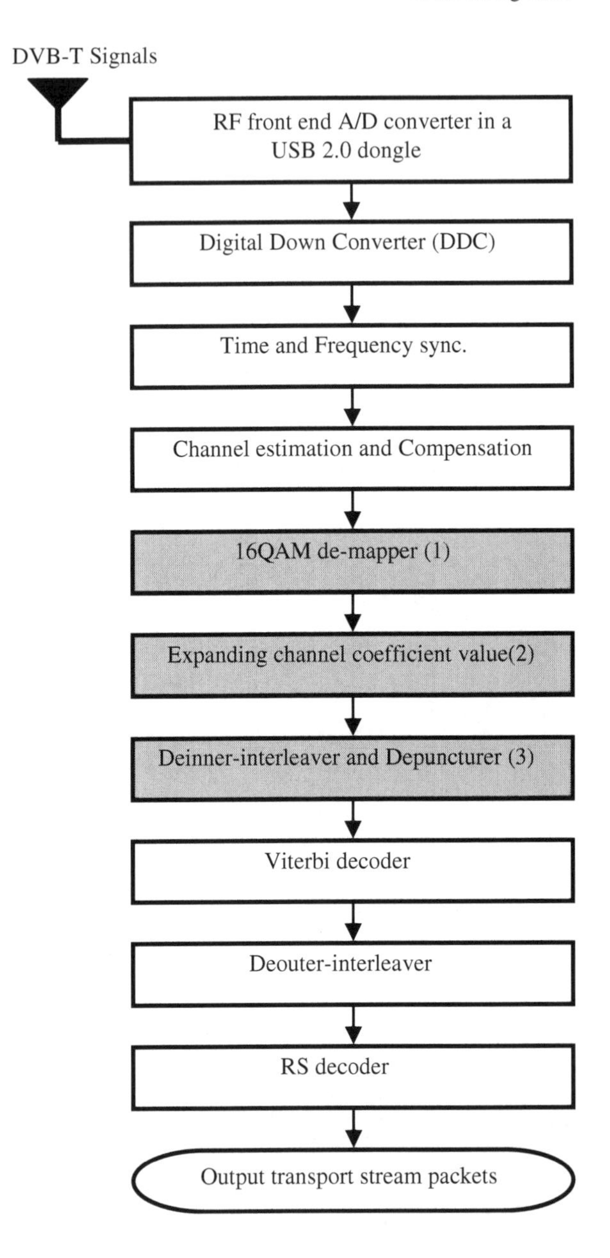

Fig. 1 The block diagram of the software DVB-T receiver

Where I is the magnitude of the in-phase carrier. Q is the magnitude of the quadrature phase carrier. $D = 1$ and is shown in ETSI EN 300 744 [11].

For i2 and q2, we need to check if the input value (I, Q) is positive or negative. The old scheme to do so is shown in Fig. 2a. It can be seen that we have a branch (IF) in this scheme. We use two steps to improve the performance of 16QAM de-mapper, as shown in Fig. 2b. First, we use SIMD instruction set of packed

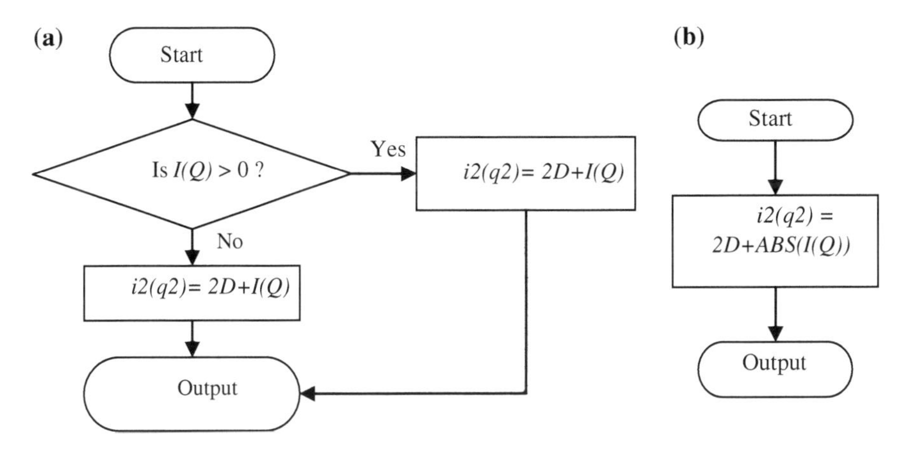

Fig. 2 De-mapper for i2 or q2 (**a**) the old scheme and (**b**) the new scheme

absolute words (PABSW) to replace the part of using the branch and also achieve the advantage of SIMD. Due to multiplication on the computer is more time-consuming, we use logical exclusive OR (PXOR) and packed add words (PADDW) to replace negate (NEG) operations, as shown in Fig. 3. The performance of de-mapper (C version), de-mapper_asm (old), and de-mapper_asm (new) are 277, 204, and 26.7 ms, respectively. The performance improvement is 6.64x

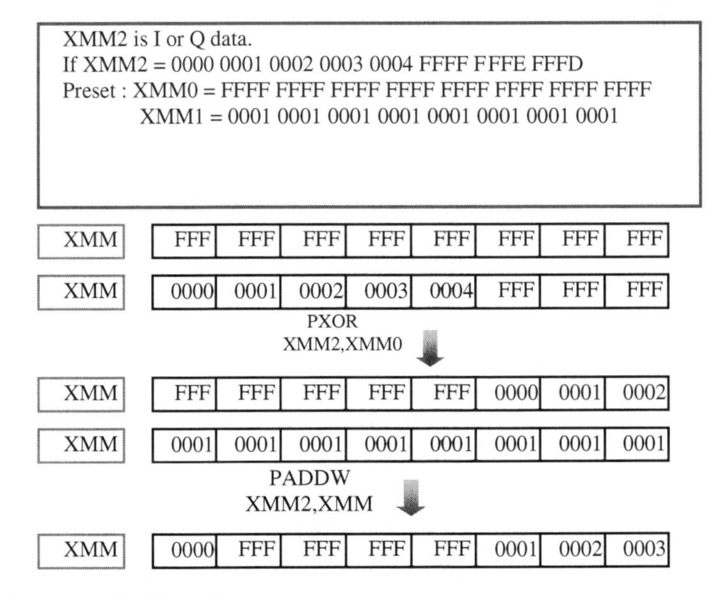

Fig. 3 De-mapper for i1 or q1

Optimization of Expanding_channel Coefficient Value

In Viterbi decoding of DVB-T, we use the maximum likelihood (ML) algorithm. The log likelihood function is given by:

$$p(r|x) = \log \frac{1}{\sqrt{2\pi\sigma^2}} e^{\frac{-|r-hx|^2}{2\sigma^2}} \tag{2}$$

We assume h is the channel coefficient, x is the transmitted data, and r is the received data.

Where $|r - hx|^2$ is the metric and can be expressed as follows:

$$|r - hx|^2 = |h|^2 \left|\frac{r}{h} - x\right|^2 = Q\left|\frac{r}{h} - x\right|^2 = metric \tag{3}$$

where $Q = |h|^2$.

The Q value must be processed in the de-mapper, In 16-QAM, non-hierarchical modulation requires de-mapper to 4 bits for Viterbi decoding. So we copy Q value fore times, as shown in Fig. 4. The new scheme uses unpack low packed data (PUNPCKLWD) SIMD instruction twice to speed up this operation, as shown in Fig. 5. The performance of expanding_Q_Value (C version), expanding_Q_Value_asm (old),and expanding_Q_Value_asm (new) are 119, 35.25, and 22.42 ms, resoectively. The performance is up to 0.57x.

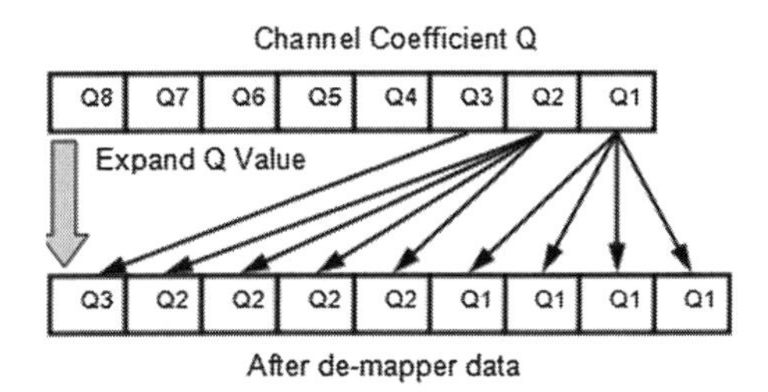

Fig. 4 The method of xpanding Q value (the old scheme)

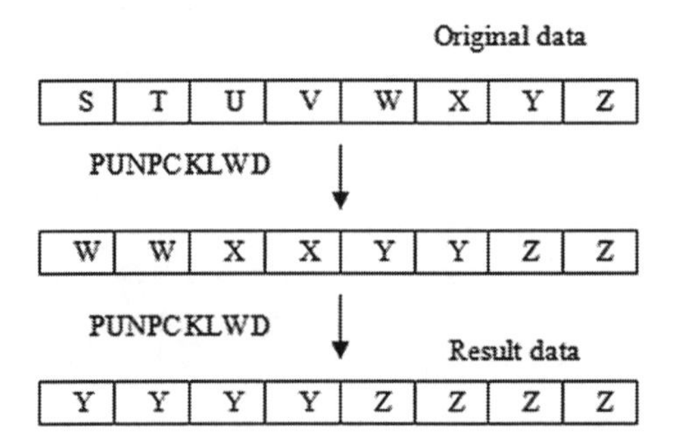

Fig. 5 The method of expanding Q value (the new scheme)

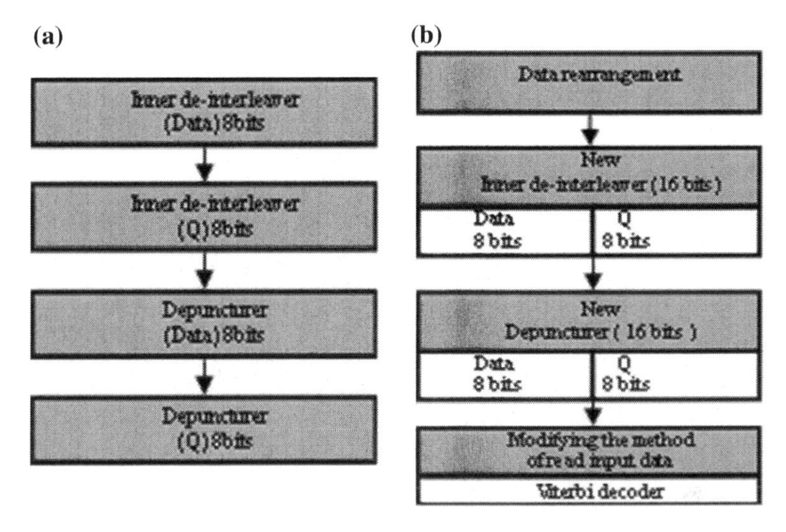

Fig. 6 **a** The old scheme of inner de-interleaver and depuncturer on software DVB-T receiver. **b** The new scheme of inner de-interleaver and depuncturer on software DVB-T receiver

Optimization of Inner De-interleaver and Depuncturer

In the old scheme, we execute inner de-interleaver and depuncturer twice, as shown in Fig. 6a. One for Q value, and the other for data (r/h). In Fig. 6, we show the new scheme. In Fig. 7, it shows that the arrangement of the old data format. In Fig. 8, it shows that we rearrange the data format from 8 to 16 bit. In the end, we modify the method of Viterbi to read input-data. Instead of processing date (8 bits) and Q value (8 bits) individually, we process them together (16 bits) in the inner

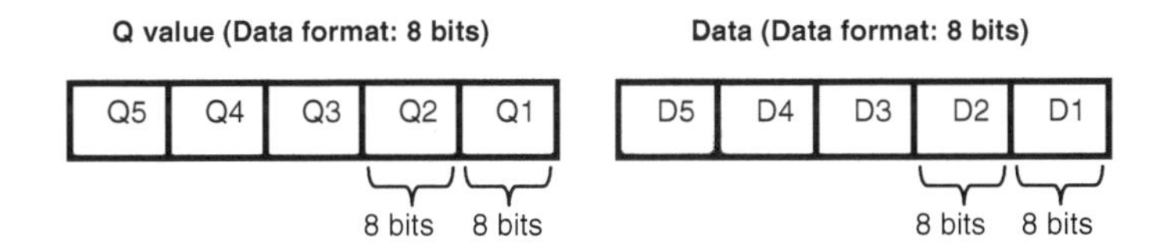

Fig. 7 Data arrange (the old scheme)

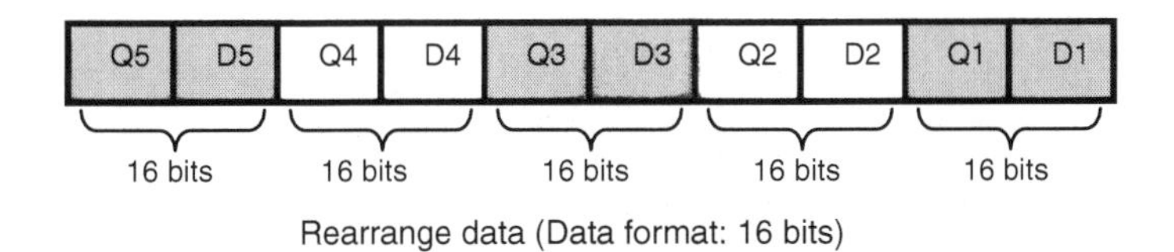

Fig. 8 The performance comparison between original method and reduction in operating time

DVB-T software radio receiver	Elapsed Time
Original	3662ms
Optimized implementation run on Intel® Core™ 2 Quad Q8200	2749ms
Optimized implementation run on Intel® Core™ i7-2600K (3.40GHz).	1819ms

Fig. 9 The elapsed time of the whole DVB-T software radio receiver (including all blocks including channel decoding) for 2,844 ms input data

de-interleaver and depuncturer operation. In Fig. 14, The performance of inner de-interleaver and depuncturer (C Version), inner de-interleaver and depuncturer asm (old), inner de-interleaver and depuncturer asm (new) are 767, 370, 197 ms, re-soectively. The performance is up to 0.88x.

Conclusion

We use SIMD instruction set, data rearrangement, etc. to speed up the three functional blocks of software DVB-T receiver. The 16QAM de-mapper, expanding channel coefficient value, and inner de-interleaver and depuncturer is 6.64, 0.57 and 0.88x faster. The quantity of input data is 28 frames which are equal to 2,844 ms. Our software DVB-T receiver takes 2,749 ms to process the input data. Thus, the overall optimization of baseband signal processing of PC-based software DVB-T receiver is now complete and achieve real-time reception. In addition, if we use the latest CPU and the performance can reach to 1,819 ms. The overall performance improvement including channel decoding is shown in Fig. 9.

Acknowledgments This work was supported in part by National Science Council, Taiwan, under grant NSC 96-2622-E-027-015-CC3.

References

1. Kubo N, Kondo S, Yasuda A (2005) Evaluation of code multipath mitigation using a software GPS receiver. IEICE Trans Commun E88-B(11): 4204–4211 (2005)
2. Li S, Zhai C, Zhan X, Wang B (2009) Implement of real-time GPS L1 software receiver. In: Proceedings of the 4th international conference on computer science & education (ICCSE '09), pp 1132–1137
3. Tseng SM, Hsu YT, Chang MC, Chan HL (206) A notebook PC based real-time software radio DAB receiver. IEICE Trans Commun E89B(12):3208–3214
4. Di N, Gao P, Wan G, Li J, Li J (2010) A common SDR platform for digital audio broadcasting system. In: Proceedings of the 3rd international congress on image and signal processing (CISP 2010), vol. 8, pp. 3708–3711

5. Tseng S-M, Kuo Y-C, Ku Y-C Hsu Y-T (2009) Software viterbi decoder with SSE4 parallel processing instructions for software DVB-T receiver. In: Proceedings of the 7th IEEE international symposium on parallel and distributed processing with applications (ISPA-09), Chengdu and Jiuzhai Valley, China, pp 102–105
6. Tseng S-M, Hsu Y-T, Shih J-Z (2012) Reed-solomon decoder optimization for PC-based DVB-T software radio receiver. Inf Int Inter J 15(8):3485–3498
7. Tseng S-M, Yu J-C, Hsu Y-T (2011) A real-time PC based software radio DVB-T receiver. In: Proceedings of the 3rd international conference on future computational technologies and applications (FUTURE COMPUTING 2011), Rome, Italy, pp 86–91
8. Tseng S-M, Hsu Y-T, Lin H-K (2013) Iterative channel decoding for PC-based software radio DVB-T receiver. Wirel Personal Commun 69(1):403–411
9. Asif M, Farooq M, Taj IA (2010) Optimized implementation of motion compensation for H.264 decoder. In: Proceedings of 5th international conference on computer sciences and convergence information technology (ICCIT), pp 216–221
10. Shahbahrami A, Juurlink B, Vassiliadis S (208) Implementing the 2-D wavelet transform on SIMD-enhanced general-purpose processors. IEEE Trans Multimedia 10(1):43–51
11. Digital Vide Broadcasting (DVB) (2009) Framing structure, channel coding and modulation for digital terrestrial television, ETSI EN 300 744 V1.6.1, pp 22

In-Time Transaction Accelerator Architecture for RDBMS

Su Jin Kim, Seong Mo Lee, Ji Hoon Jang, Yeong Seob Jeong, Sang Don Kim and Seung Eun Lee

Abstract In this paper, we propose a hardware architecture for in-time transaction accelerator that reduces the bottlenecks between the DB server and the DB storage in margin FX trading system's RDBMS (Relational Database Management System). In-time transaction accelerator located between the DB server and the DB storage analyzes and processes the queries used for margin FX trading system by co-processing of the CPU and the FPGA. The accelerator analyzes the patterns and the consistency of the queries to reduce the total database access in order to increase the RDBMS's throughput.

Keywords In-time transaction accelerator · Margin FX trading · RDBMS · FPGA

Introduction

The field of stock markets has heavy traffic volume. Especially, a foreign exchange margin trading (a.k.a. margin FX trading) market that is the buying and selling of currencies generates much heavier traffic volume than the general stock market does. Furthermore, the traffic volume is growing up. However, there are limitations to improve the performance of the existing database system. The reason is that the obstacle named power wall blocks to improve the performance of the CPU and the memory access time is still markedly slow. When financial traffic volume is increasing, bottlenecks can occur between the server and the storage. To resolve

---

S. J. Kim · S. M. Lee · J. H. Jang · Y. S. Jeong · S. D. Kim · S. E. Lee (✉)
Department of Electronic and Information Engineering, Seoul National University of Science and Technology, 172 Gongreung-2-dong, Nowon-gu, Seoul-si, Korea
e-mail: seung.lee@seoultech.ac.kr

Y.-M. Huang et al. (eds.), *Advanced Technologies, Embedded and Multimedia for Human-centric Computing*, Lecture Notes in Electrical Engineering 260,
DOI: 10.1007/978-94-007-7262-5_38,

the bottlenecks of DBMS, two groups of studies have been conducted. The first study is a new software architecture like NoSQL-based database such as the MongoDB [1] and database in memory such as the SAP HANA [2]. The second study is a new hardware architecture [3].

In this paper, we propose a hardware architecture for in-time transaction accelerator to reduce the bottlenecks that occur between the server and the storage. Our basic idea is using the Hardware parallelism by using the Field-Programmable Gate Array (FPGA). The FPGA is a possible parallel access and this feature facilitates the processing of streaming data [4].

In-time transaction accelerator architecture is located between the server and the storage. The accelerator is composed of CPU, FPGA, SSD and volatile memory. In order to have high-speed parallel access, we use the peripheral component interconnect (PCI) express interface between the CPU and the FPGA. Financial data (SQL query) is sequentially stored in the transaction accelerator. Then, the in-time transaction accelerator analyzes the *n*-query and optimizes the query to minimize the number of access to DB storage. Through this method, we can improve performance by extending the throughput between the DB server and DB storage.

We are working with the financial data from a financial solution company managing the margin fx trading system.

The rest of this paper is organized as follows. In section "Related Work", we review work related to improvement of the system by using FPGA. Section "RDBMS with In-Time Transaction Accelerator" describes the in-time transaction accelerator architecture. We conclude in section "Summary" by proposing the future works our architecture.

Related Work

The clock rate that is increasing at a high-speed is disturbed by the power wall. Hence, recently the researchers have been focusing on the architectural solutions to improve the CPU's performance. In [5], they proposed the co-processing of the CPU and the FPGA to improve the CPU's data processing performance. The performance is improved compared that of the original by using the FPGA's hardware parallelism. Furthermore, the system using a co-processing through the process where the system uses compressed the query along with the FPGA's decompression and process of the query increases the throughput by reducing the CPU usage about analyzing the query operation [6]. Our in-time transaction accelerator uses a co-processing of the CPU and the FPGA to process the data faster. Another approach uses a characteristic of the FPGA that is reconfigurable. When the queries are executed, the FPGA are reconfigured to the optimized design to process the data faster [7]. But the overhead of the configuration reduces the performance of the system. In the Netezza's FAST-engine, the FPGA fabrics are fixed and the internal registers of the FPGA are changed for the partial reconfiguration [8]. The Netezza's FAST-engine reduces the traffic volume between the memory and the CPU by using an engine's pipeline stage if the data pattern has a consistency. In our research, we hold a FPGA's design and use the [7]'s method to exclude a overhead when reconfiguration is needed.

RDBMS with In-Time Transaction Accelerator

Overview

The basic principle in order to improve the throughput of the RDBMS in in-time transaction accelerator architecture we proposed is to the commonalities between the requested queries and the pattern of the certain financial data. Therefore, we transfer the minimized queries to DB storage. As a result, we can reduce the DB access required to read and write. In-time transaction accelerator located between the DB server and the DB storage receives the queries and minimizes the query traffic volume. Minimizing the query traffic volume realizes CPU and co-processing by using the trait of parallelism of the FPGA. We explain detailed information in section "Query Analyzer". By reducing the number of the query to DB storage, we reduce the number of access to DB storage. As a result, we can improve the throughput of the RDBMS.

In-time transaction accelerator architecture is composed of the Query Analyzer (minimizing the query traffic volume), PCIe module (in order to parallel access of the FPGA) and Log buffer(saving the received queries). Figure 1 shows the RDBMS adding the in-time transaction accelerator.

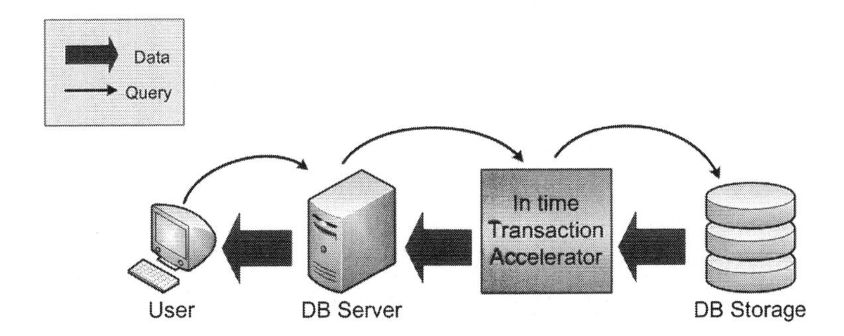

Fig. 1 RDBMS with in-time transaction accelerator

Query Analyzer

The Query Analyzer analyzes the pattern and consistency of the queries related the margin FX trading that is received sequentially from PCI Express and reduces the amount of query requesting the memory access to DB storage (Table 1).

For example, when above the query comes into Query Analyzer through the PCIe communication, the Query Analyzer analyzes the (1) and (2). The query (1) that requires the data can be realized through the query (2) that requires more data. Thus the query (2) is transferred to DB storage. In the case of the above example, the number of the query is reduced. In this way, the Query Analyzer reduces the number of access to DB storage. We reduce the number of the time-consuming memory access. We make the system to conduct to same work in lesser amount of time. In this way, we improve the entire throughput of the RDBMS.

PCIe Module

The PCI Express can realize the bandwidth from 200 MB/s to 6.4 GB/s as by serial bus system. The possible reason of the high bandwidth is that two PCI Express communication device can use a multi-lane link. A Lane an independent link, where can happen the single transmission at PCI Express, has an individual operating frequency. The multi-lane link is a large link comprised of the several independent Lane. Although the lane is serial bus, if multi-lane link consists the several Lane, it will transmit the data in the parallel form and be able to communicate with high bandwidth by operating frequency of an independent lane. The bidirectional data bandwidth is about 200 MB/s in a PCI Express 1× and the bidirectional data bandwidth is about 6.4 GB/s in a PCI Express 32x. A PCI module is the module to use the PCI Express which is based on high bandwidth. The data transfer system of the parallel form of PCI Express is made using the FPGA's hardware parallelism and supports the high speed transcipient from the CPU to the Query Analyzer.

Log Buffer

Log Buffer is responsible for storing the received queries to Solid State Disk (SSD). The Log buffer is designed to improve the safety of the financial system.

Table 1 Defined sample queries for explanation

select * from emp where ename like 'A %'; //Which starting 'A' -(1)
select * from emp where ename like ' %A %'; //Which includes 'A' -(2)

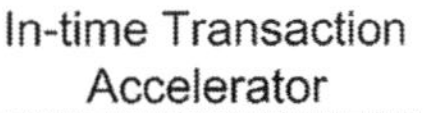

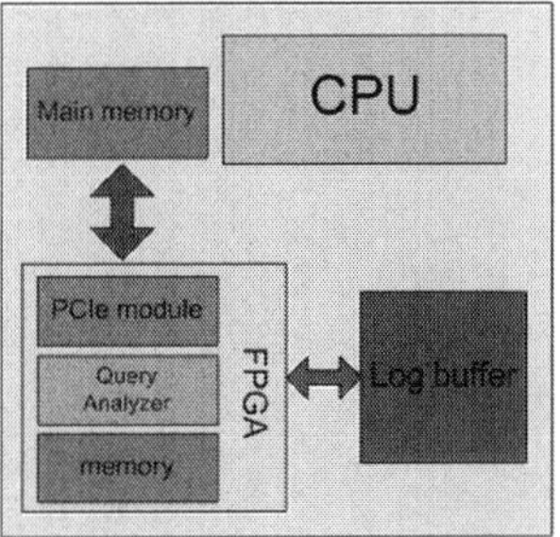

Fig. 2 In-time transaction accelerator

Uncompleted queries can disappear in unexpected situations (for example, system failure due to a disaster or outage and the system's internal computational error, etc.) during financial transaction. To prevent this, the queries coming into the FPGA are saved in SSD and are reloaded when we need it. Figure 2 shows architecture of the in-time transaction accelerator.

Summary

In this paper, we proposed the hardware architecture for in-time transaction accelerator that reduces the bottlenecks between the DB server and the DB storage in the RDBMS of margin FX trading system. PCIe interface is used for FPGA's hardware parallelism. The DB server transmits the queries to in-time transaction accelerator and then the CPU of the accelerator transmits the queries to the Query Analyzer of the FPGA through PCIe interface. The Query Analyzer analyzes and processes the queries received in a serial order. In consequence of the previous process, the primary queries are selected and are transmitted to the CPU with PCIe interface. And then the CPU transmits the primary queries received from the Query Analyzer to the DB storage. As a result, transmitting the primary queries can reduce the number of a memory access which takes a long time. This leads to improvement of RDBMS' throughput. Thus, performance of the RDBMS is enhanced. DB storage transmits the data demanded by the queries to the accelerator, and then process running on the CPU of accelerator transmits the data to all DB server that requests the data in the first step. Our In-time transaction accelerator architecture is appropriate for margin FX trading system's RDBMS as well as the big data systems that include a surge data stream. In the future, if the queries or the data are transmitted with InfiniBand which has wide Bandwidth from DB server to accelerator or from accelerator to DB storage on proposed our architecture, performance of in-time transaction accelerator architecture will be improved further.

Acknowledgments This study was supported in part by the IT R&D program of MKE/KEIT [10043896, Development of virtual memory system on multi-server and application software to provide real-time processing of exponential transaction and high availability service] and the Seoul National University of Science and Technology.

References

1. MongoDB, http://www.mongodb.org
2. SAP HANA, http://sap.com/HANA/
3. Nie C (2012) An FPGA-based smart database storage engine. Master's thesis, ETH zurich
4. Guha R, Al-Dabass D (2010) Performance prediction of parallel computation of streaming applications on FPGA platform. In: 12th international conference on computer modeling and simulation, UKSim, Cambridge pp 579–585
5. Mueller R, Teubner JM, Alonso G (2009) Data processing on FPGAs. J Proc VLDB Endowment, pp 910–921
6. Sukhwani B, Min H, Thoennes M, Dube P, Iyer B, Brezzo B, Dillenberger D, Asaad S (2012) Database analytics acceleration using FPGAs. In: Proceedings of the 21st international conference on parallel architectures and compilation techniques, ACM New York pp 411–420
7. Mueller R, Teubner J, Alonso G (2010) Glacier: a query-to-hardware compiler. In: Proceedings of the 2010 ACM SIGMOD international conference on management of data, ACM New York pp 1159–1162
8. Francisco P (2011) The Netezza data appliance architecture: a platform for high performance data warehousing and analytics. Technical Report, IBM

Intra-Body Communication for Personal Area Network

Sang Don Kim, Ju Seong Lee, Yeong Seob Jeong, Ji Hoon Jang and Seung Eun Lee

Abstract The intra-body communication uses the human body as a conducting wire, providing the simplicity and the security. Although the communication distance is limited within a body-area, it is useful on the construction of personal area network. In this paper, we introduce our prototype intra-body communication module using the FPGA. The proposed system has the FSK modulator and the demodulator. These modulation methods are chosen after body-channel analysis. The experimental results demonstrate the feasibility of our intra-body communication module for establishing the PAN (Personal Area Network).

Keywords Intra-body communication · PAN (personal area network) · FPGA

Introduction

The intra-body communication uses the human body as a conducting wire in order to conduct the signals. The intra-body communication has a higher security than conventional wireless communication like as a wire communication. Furthermore, the intra-body communication provides convenience thanks to its wireless connectivity. However, the communication distance is limited within body area.

Modern handheld devices have high performance and contain many features such as telephone conversation, multimedia contents player, and data communications. The handheld devices have the PAN composed of various wireless communications for efficiently usage of these features. The PAN using the radio frequency has a risk of wiretapping and has a possibility to collide with other PAN in crowded region. Therefore, the PAN with higher security is required in recent

S. D. Kim · J. S. Lee · Y. S. Jeong · J. H. Jang · S. E. Lee (✉)
Department of Information Engineering, Seoul National University of Science and Technology, 172 Gongreung-2-dong, Nowon-gu, Seoul-si, Korea
e-mail: seung.lee@seoultech.ac.kr

Y.-M. Huang et al. (eds.), *Advanced Technologies, Embedded and Multimedia for Human-centric Computing*, Lecture Notes in Electrical Engineering 260,
DOI: 10.1007/978-94-007-7262-5_39,

applications. Intra-body communication uses the human body instead of the radio frequency to transmit the electric signals thus it is free from the wiretapping and the radio interference. In addition, it has the simplicity because it does not require a conducting wire.

The intra-body communication was proposed in 1995 by Zimmerman [1]. He conducted various studies to analyze appropriate signals that can pass the human body. Recently, Schenk et al. have produced a prototyping that uses a human body as communication channel [2]. Joonsung Bae et al. have proposed about body channel communication from the surface of the human body by analyzing flow of signals in each of frequency band [3]. In this paper, we present the prototype of intra-body communication module to provide the PAN for handled devices.

The rest of this paper is organized as follows. We first briefly introduce the intra-body communication in section "Intra-Body Communication" and present the hardware features in section "Implementation". Section "Experimental Results" shows the experimental results on the data transmission. In section "Conclusion", we conclude this paper and introduce expected applications in the area of the intra-body communication module.

Intra-Body Communication

The intra-body communication is emerging as effective method which possesses the security and simplicity of the wireless communication. The human body has the conductivity because it includes water and some of electrolyte. The ECG is an example of the intra-body has the conductivity and the electric signals are conducted along the intra-body. Therefore, the external electric signal can be transmitted through the human body establishing communication channel. The weak biological signals are ignored or filtered. It is well known for the intra-body communication is not dangerous because the electric current passing through the human body is very low (about a milliampere).

Implementation

Human Body Channel Characterization

The signals to be transmitted should be modulated properly at a transmitter side for effective communication. A received signal from the human body is demodulated and data is reconstructed at the receiver module. In this case, the human body can be considered by one wire and provides the electric signal path from transmitter to receiver.

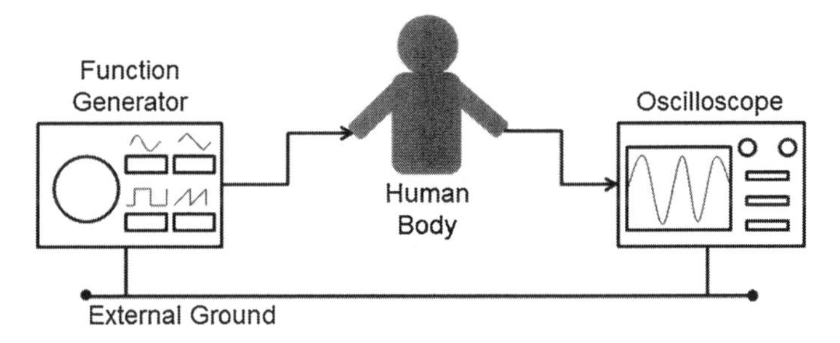

Fig. 1 Experimental environment of the signal transmission

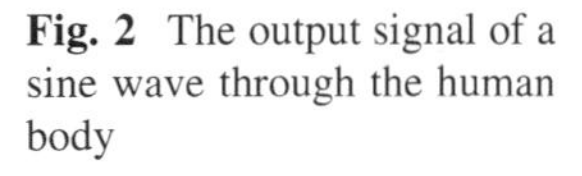

Fig. 2 The output signal of a sine wave through the human body

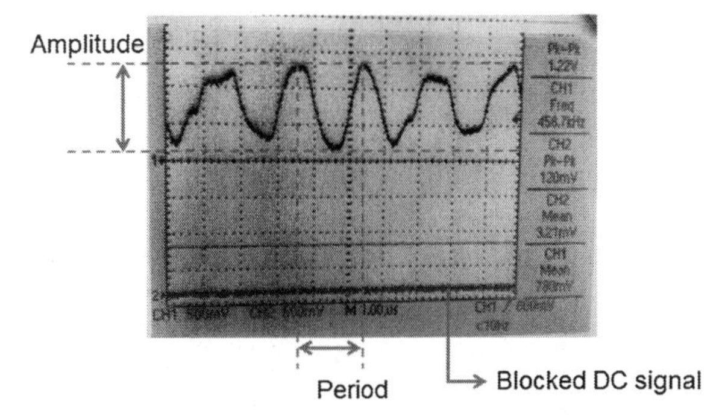

Figure 1 shows the experimental setup for human body channel characterization in order to decide the proper modulator scheme for the intra-body communication. Figure 2 illustrates the received waveform on oscilloscope. The DC electric signal is blocked and the AC electric signal is distorted when they pass through a human body. Therefore, the data have to be modulated into the AC signal. The modulator such as the ASK (Amplitude Shift Keying), the PSK (Phase Shift Keying) or the FSK (Frequency Shift Keying) can be applied to the data. The ASK is unsuitable for the intra-body communication because the distorted AC signal is not consistent at amplitude. The PSK has a more complex hardware than FSK and it may make an error when the transmitted signal is critically distorted. On the other hand, frequency at FSK is almost consistent although the signal was distorted thus we adopt FSK modulation in our intra-body connector.

The System Architecture

Our system consists of a transmitter and a receiver (See Fig. 3). The signal serializer in the transmitter block receives a parallel 8-bit data and serializes the data for serial communication. The modulator completes FSK modulation for intra-body communication. The modulated data is transmitted through the human body.

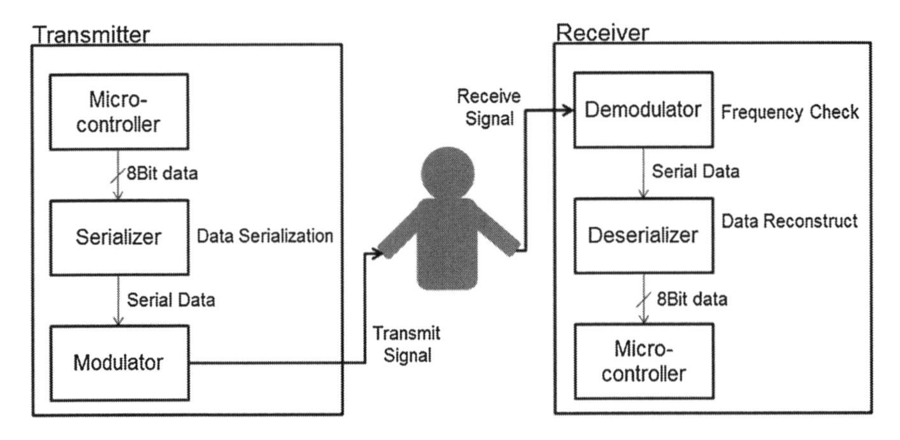

Fig. 3 The block diagram of a basic intra-body communication module

The demodulator in the receiver module checks the frequency of the received signal and reconstructs the data. A microcontroller at the transmitter side feeds the data stream should be transmitted through the intra-body communication and a microcontroller at the receiver side completes error rate.

Experimental Results

In order to verify the whole system, the intra-body communication module requires a communication protocol. We selected RS-232C protocol for serial communication. The data contains start bit and stop bit additionally. A byte data is transmitted through human body at a speed of 9,600 bps. We adopted Altera FPGA for realization of our proposed system. The electric signals which are modulated at transmitter are forwarded to the receiver through the human body. The restored data can be checked whether it is identical with transmitted data or not. An audio data was successfully transmitted to the receiver through the human body, demonstrating the feasibility of our proposal that the intra-body communication provides secure and simple wireless connection for the PAN (Fig. 4).

Conclusion

The intra-body communication has a simplicity and higher security compared with the conventional wire and wireless communication. In this paper, we presented the system of the intra-body communication by using the FSK modulation. The proper frequency band of conducting along the human body is selected by human body channel characterization. The audio data was successfully transmitted to the

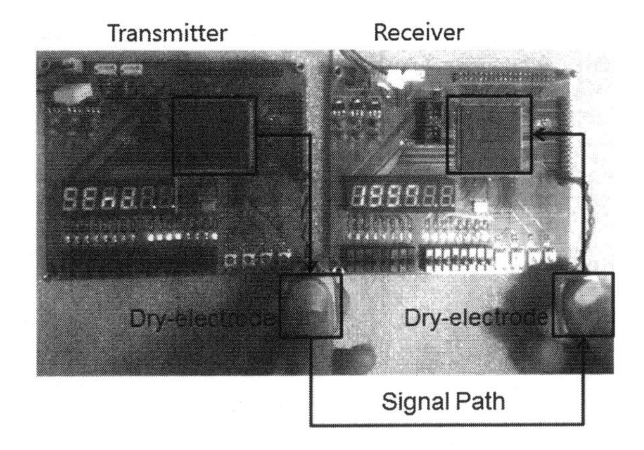

Fig. 4 The experimental environment of the data transfer

receiver through the human body. Experimental results demonstrated the feasibility of our proposal that the intra-body communication provides secure and simple wireless connection for the PAN.

In the applications such as financial services, where the higher security is required, the conventional wireless communications have the risk of the wiretapping. However, the intra-body communication can reduce the risk of the wiretapping in addition to can prevent other possible attacks. The safety can be improved by applying an encryption method. We expect the intra-body communication can be applied in various fields due to the simplicity and the safety.

Acknowledgments This work was supported by Seoul National University of Science and Technology, Korea.

References

1. Zimmerman TG (1995) Personal Area Networks (PAN): Near-field intra-body communication. Master's thesis, MIT, Cambridge, MA
2. Schenk TW, Mazloum NS, Tan L, Rutten P (2008) Experimental characterization of the body-coupled communications channel. Proc IEEE Int Symp Wearable Comput 234–239
3. Bae J, Cho H, Song K, Lee H, Yoo H-J (2012) The signal transmission mechanism on the surface of human body for body channel communication. IEEE Trans Microwave Theor Techn 60(3)

mrGlove: FPGA-Based Data Glove for Heterogeneous Devices

Seong Mo Lee, Ji Hoon Jang, Dae Young Park, Sang Don Kim, Ju Seong Lee, Seon Kyeong Kim and Seung Eun Lee

Abstract In this paper, we propose a glove based equipment (mrGlove) for a user interface that controls a device through hand motion recognition. The mrGlove provides more user experience compared to conventional devices such as a keyboard and a touchscreen. Our mrGlove is able to control a heterogeneous device (Windows or Android) that offers more convenience and user experience. Experimental results prove the feasibility of our proposal for enhancing convenience and user experience by obtaining control of an object through motion recognition in the PC and the smartphone.

Keywords Data glove · Heterogeneous device · Motion recognition · FPGA

Introduction

Over the past twenty years, many researches about a data glove have been done in entertainment field such as Mattel Power Glove or VPL Data Glove [1]. The glove-based controller controls a device through hand motion recognition. Control through motion recognition is intuitive and provides more user experience than a keyboard, mouse or touchscreen. Recently, controllers by using motion recognition such as Nintendo Wii [2] or Microsoft Kinect [3] have been widely used in entertainment field. Typically, motion recognition is classified into two groups. One is vision processing method and the other is the way to use mechanical equipment such as data glove. Data glove based motion recognition has the precise classification capability and fast response time compared to the vision processing method. Hence, it has been chosen for many systems [4].

S. M. Lee · J. H. Jang · D. Y. Park · S. D. Kim · J. S. Lee · S. K. Kim · S. E. Lee (✉)
Department of Electronic and Information Engineering, Seoul National University of Science and Technology, 172 Gongreung-2-dong, Nowon-gu, Seoul-si, Korea
e-mail: seung.lee@seoultech.ac.kr

Y.-M. Huang et al. (eds.), *Advanced Technologies, Embedded and Multimedia for Human-centric Computing*, Lecture Notes in Electrical Engineering 260,
DOI: 10.1007/978-94-007-7262-5_40,

In this paper, we propose FPGA-based data glove, motion recognition Glove (mrGlove), to enhance user experience and convenience compared to a keyboard and a touchscreen. The mrGlove is able to control a heterogeneous device. Experimental results prove the feasibility of our proposal for enhancing convenience and user experience by obtaining control of an object through motion recognition.

The rest of this paper is organized as following: Section mrGlove describes the mrGlove in detail. Section Experimental Results explains the experimental results. We conclude in section Conclusion by proposing the future works on the mrGlove.

mrGlove

The mrGlove captures the user's motion and transmits data to the PC or the smartphone through Bluetooth communication channel. Software in the PC or the smartphone realizes the motion and controls an application.

Our mrGlove uses flex sensors and acceleration sensors to recognize user's hand motion. Flex sensors and switches are located in each finger, and signals from flex sensors are transmitted to FPGA for signal processing. The switches located on each fingertip enable a flex sensor to transmit a data to FPGA when they are turned on. When flex sensor is bended, resistance is changed. Acceleration sensor is used to detect user's hand motion. Signals from the acceleration sensor are transmitted to FPGA. Signals from the flex sensors and acceleration sensor are aggregated and processed in FPGA. Terasic DE0-nano board [5] including cyclone IV FPGA, ADC and digital accelerometer is used for implementation of the mrGlove. Figure 1 shows a prototype of the mrGlove.

A module for motion recognition in FPGA is composed of a data capture, data processing and a data transmit units. The data capture unit captures the signal from sensors. The data processing unit is used for data processing and the transmit unit

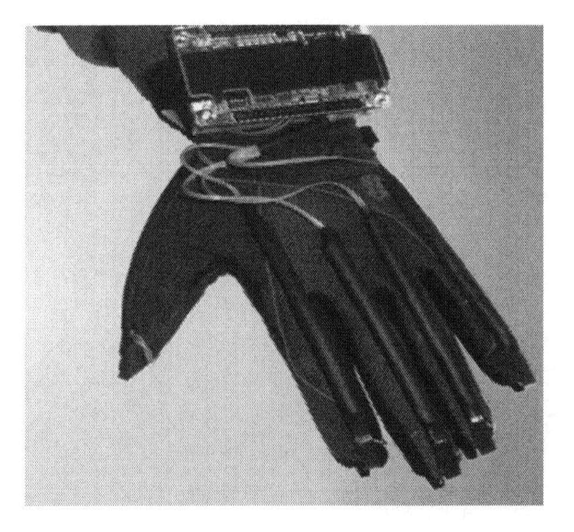

Fig. 1 Prototype of the mrGlove

forwards the data to Bluetooth module to communicate with a device. Figure 2 shows a system flow of the mrGlove.

Data Capture Unit. The data capture unit converts the analog flex signals into digital signals and brings the data from digital accelerometer by SPI (Serial to Peripheral Interface) communication. Calibration of resolution is possible in the analog-to-digital operation. The data capture unit includes the SPI controller to communicate with the digital accelerometer.

Data Processing Unit. The data processing unit sets a threshold of the data and assembles the data into a particular format for motion recognition. Figure 3 shows the data format for motion recognition. Basically, the data length is 8-bit for the RS232 communication standard. We determine five actions such as index finger, middle finger, balls of the feet, heel and tilt of the hand. We use only upper 5bits of the data according to five actions. When user generates each event, each bit corresponding to it is set high. For example, when an event corresponding to index finger is occur, the most significant bit of the data is set high.

Data Transmit Unit. The data transmit unit forwards the data from the data processing unit to the Bluetooth module. RS232 communication is used between the data transmit unit and the Bluetooth module. The data transmit unit includes the RS232 controller to communicate with the Bluetooth module. Calibration of a bps (bit per second) is possible.

Experimental Results

We demonstrate the motion recognition using the mrGlove on heterogeneous devices (PC and smartphone). The PC uses a Windows operating system and the smartphone uses an Android.

mrGlove with a PC

In order to demonstrate the mrGlove on the PC using Windows, we designed a middleware by using a LabVIEW [6]. The mrGlove communicates with the PC

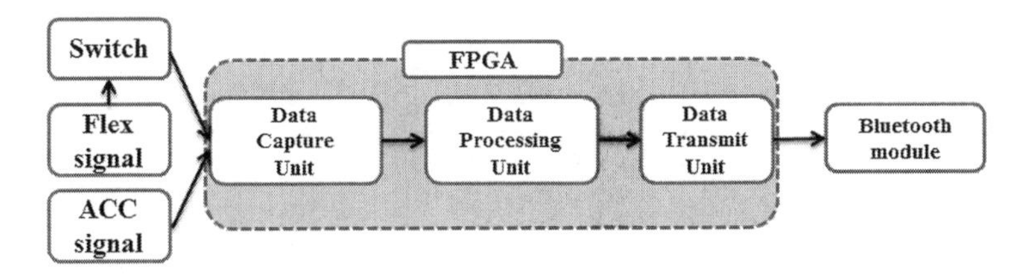

Fig. 2 System flow of the mrGlove

Bit	7	6	5	4	3	2	1	0
	Index finger	Middle finger	Balls of the feet	Heel	Tilt of the hand	-	-	-

Fig. 3 The data format for motion recognition

Fig. 4 A demonstration of motion recognition on the PC

Fig. 5 A demonstration of motion recognition on the smartphone

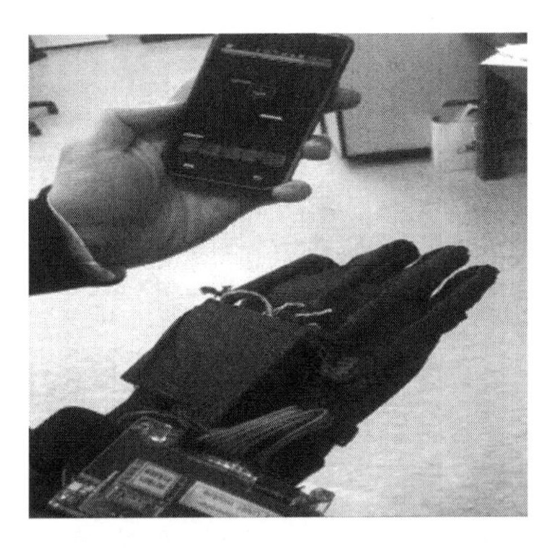

through a Bluetooth communication channel. The middleware processes the data and controls the PC as a keyboard does by using a Win32 library. Figure 4 shows the demonstration on the PC. The PC recognizes the hand motion through the mrGlove and controls the object in the racing game and the FPS (First-person shooter) game. Our mrGlove provides the way to control the most games with the same way.

mrGlove with a Smartphone

We designed an application (rhythm game) where five bar-shaped rectangles move down from top to bottom when music is selected. We tap the rectangles to the rhythm with the mrGlove. The application includes a class about a Bluetooth and a touch event. Figure 5 shows the touch event by using a switch of the mrGlove.

Conclusion

In this paper, we proposed a glove-based device for user interface that controls a heterogeneous device through hand motion recognition. We have succeeded in controlling the PC and a smartphone application. The mrGlove is able to control a smart device such as a tablet that has Bluetooth communication channel. We plan to add a feedback component in the mrGlove such as a vibration motor for a physical interaction game. The mrGlove is designed using a FPGA. Therefore, the units in the FPGA are easily converted to a single chip, minimizing the size of the mrGlove. We expect that the mrGlove brings a convenience and enhances a user experience in the entertainment field.

Acknowledgments This study was supported by Seoul National University of Science and Technology.

References

1. Dipietro L, Sabatini AM, Dario P (2008) A survey of glove-based systems and their applications. IEEE Trans Syst, Man, Cybernetics, Part C: Appli Rev 38(4)
2. Nintendo Wii http://www.nintendo.com/wii
3. Microsoft Kinect http://www.xbox.com/en-US/kinect
4. Jeong YM, Lim KT, Lee SE (2012) mGlove: Enhancing user experience through hand gesture recognition. Jin D, Lin S (eds) Advances in EECM. LNEE, vol 1 Springer, Heidelberg, pp 383–386
5. Terasic DE0-nano board, http://www.terasic.com.tw/en
6. LabVIEW, http://www.ni.com/labview

A Novel Wireless Context-Aware Network of Service Robot

Jianqi Liu, Qinruo Wang, Hehua Yan, Bi Zeng and Caifeng Zou

Abstract With the improvement of sensors, intelligence and wireless communication technologies, the service robot is faced with a new rapid development opportunity, which can utilize pervasive computing of wireless network to sensing everything happened in whole context. This paper presents a novel service robot global context-aware network, and puts emphasis on four key issues, such as ultra wideband radio, wireless positioning technology, wireless body area network and dynamic Bayesian network.

Keywords Service robot · Wireless sensor network · Wireless body area network · Context-aware · Dynamic bayesian network

Introduction

With the rapid development of context-aware technologies such as M2M communication technology [1], wireless positioning [2, 3] in recent years, many institutes and researchers have a strong interest in the service robot supported by

---

J. Liu · H. Yan (✉) · C. Zou
Guangdong Jidian Polytechnic, Guangzhou 510515, China
e-mail: hehua_yan@126.com

J. Liu
e-mail: liujianqi@ieee.org

C. Zou
e-mail: caifengzou@gmail.com

Q. Wang · B. Zeng
Guangdong University of Technology, Guangzhou 510515, China
e-mail: wangqr2006@gdut.edu.cn

B. Zeng
e-mail: z9215@163.com

Y.-M. Huang et al. (eds.), *Advanced Technologies, Embedded and Multimedia for Human-centric Computing*, Lecture Notes in Electrical Engineering 260,
DOI: 10.1007/978-94-007-7262-5_41,

context-aware technique. The independent context-aware service robot, that integrated calculation, inference, wireless communication and adaptive control, becomes one of leading-edge interdisciplinary research areas today [4]. People above the age of 60 years in China have now exceeded 10 % of the total population, and elderly population will be total 200 million by 2015, which is a new challenge. It is expected that in the near future, nursing care for the elderly will become an important burden, which requires a large number of service robots to complete nursing affairs such as delivery items, cleaning, rehabilitation services, to make up for the lack of nursing staff. The service robot can help to take care of the elderly daily life and improve their quality of life. Therefore, service Robotics market has great potential demand, and will produce good economic results.

In the past, the service robot primarily made use of sensors on the robot body to obtain cognition of the ambient environment, such as using laser range finder to achieve positioning, using acceleration meter to sense collision, and using gas sensor to detect dangerous gas leaks. The defect of these methods is that the environment information is partial around the robot. Fortunately, with the development of context-aware technology based on wireless sensor network (WSN), deploying multiple types of sensors to realize global contextual perception, the robot can understand the each of contextual changes and take appropriate action quickly. For example, once the fire disaster happened, the robot should give an alarm and open the fire hydrants. Similarly, if someone is ill, and is made a diagnosis by a doctor in hospital traditionally, but some disease such as stroke, should be dealed with quickly, which is a disaster for the elderly living alone; so the constant monitor is necessary. The service robot can be competent for this challenge supported by wireless body area network (WBAN). This paper presents a new architecture of context-aware service robot, which integrates sensing, positioning, and body monitor.

The remainder of this paper is organized as follows: We discuss the architecture of context-aware network in section Architecture of Context-Aware Network of Service Robot. In section Key Issues, we review key issues of building network such as UWB, wireless positioning technology, WBAN, and dynamic Bayesian network body status forecasting model. Section Conclusion concludes this paper and outlines some problems that need to be resolved in the future.

Architecture of Context-Aware Network of Service Robot

The scheme of service robot puts WSN, wireless positioning technology, WBAN, dynamic Bayesian network forecasting technology together to acquire the global context information. System deploys sensor to acquire real-time contextual data, and transmits the data to service robot via a wireless communication network. Then the service robot analyzes data and reasons the tasks, which need to cope with imminently. Finally, appropriate control action is taken to complete the task. For example, in kitchen collecting gas and smoke data to judge whether there is gas leak and fire disaster or not; in balcony deploying sensor to detect sun light strength and outdoor weather; in bedroom deploying sensor to collect temperatures humidity; in human body deploying WBAN, using medical sensor to collect people's physiological vital signs, monitor and judge people's body status by dynamic Bayesian network prediction model.

Wireless sensor nodes complete data collection and communication, at the same time, they should complete high accuracy positioning, which is primarily used for the service robot navigation and people's positioning and tracking. The architecture is shown as Fig. 1.

The whole network can be divided into WBAN and WSN according the radio transmit range. In WBAN, all nodes and hubs are organized into logical sets, and coordinated by their respective hubs for medium access and power management as illustrated in left of Fig. 1. There is one and only one hub in a WBAN, whereas the number of nodes in a WBAN is to range from zero to 2 m. In a one-hop star WBAN, frame exchanges are to occur directly between nodes and the hub of the WBAN. In a two-hop extended star WBAN, the hub and a node exchanges frames optionally via a relay-capable node.

A WSN consists of spatially distributed autonomous sensors to monitor physical or environmental conditions, such as temperature, sound, pressure, etc., and to cooperatively pass their data through the network to a main location. The modern networks are bi-directional, also enabling control of sensor activity. The development of wireless sensor networks was motivated by military applications such as battlefield surveillance; today such networks are used in many industrial and consumer applications, such as industrial process monitoring and control, machine health monitoring, and so on [5].

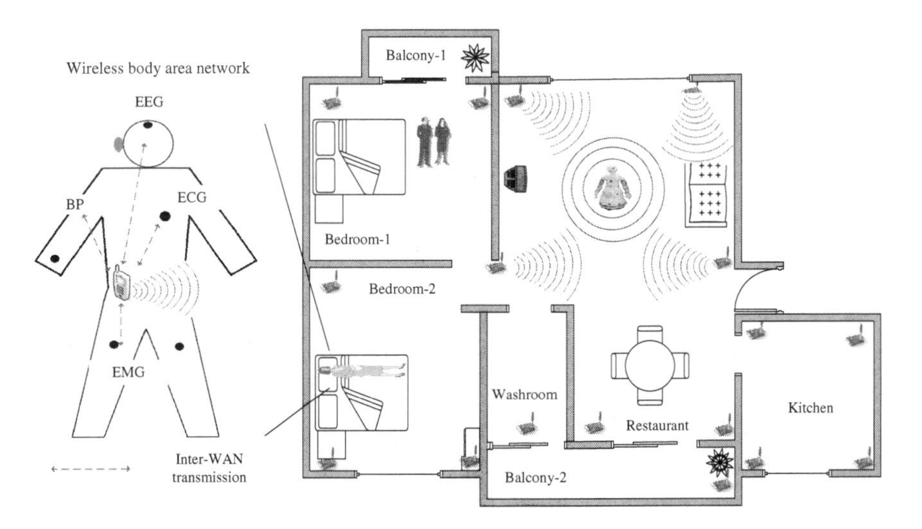

Fig. 1 Architecture of wireless context-aware network of service robot

Key Issues

The service robot wireless context-aware network mainly includes radio, wireless positioning technology, WBAN, and the physiological condition monitoring and prediction model. This paper will focus on four aspects of context-aware network.

UWB Radio

UWB was formerly known as "pulse radio". The FCC and the International Telecommunication Union Radio communication Sector currently define UWB in terms of a transmission from an antenna for which the emitted signal bandwidth exceeds the lesser of 500 MHz or 20 % of the center frequency. A significant difference between conventional radio transmissions and UWB is that conventional systems transmit information by varying the power level, frequency, and/or phase of a sinusoidal wave. UWB transmissions transmit information by generating radio energy at specific time intervals and occupying a large bandwidth, thus enabling pulse-position or time modulation. The information can also be modulated on UWB signals (pulses) by encoding the polarity of the pulse, its amplitude and/or by using orthogonal pulses. UWB pulses can be sent sporadically at relatively low pulse rates to support time or position modulation, but can also be sent at rates up to the inverse of the UWB pulse bandwidth. Pulse-UWB systems have been demonstrated at channel pulse rates in excess of 1.3 gig pulses per second using a continuous stream of UWB pulses (Continuous Pulse UWB or C-UWB), supporting forward error correction encoded data rates in excess of 675 Mbps [6].

A valuable aspect of UWB technology is the ability for a UWB radio system to determine the "time of flight" of the transmission at various frequencies. This helps overcome multipath propagation, as at least some of the frequencies have a line-of-sight trajectory. With a cooperative symmetric two-way metering technique, distances can be measured to high resolution and accuracy by compensating for local clock drift and stochastic inaccuracy.

Another feature of pulse-based UWB is that the pulses are very short (less than 60 cm for a 500 MHz-wide pulse, less than 23 cm for a 1.3 GHz-bandwidth pulse), so most signal reflections do not overlap the original pulse and the multipath fading of narrowband signals does not exist. However, there is still

Table 1 Comparison of UWB, CSS, and ZigBee

Radio	Accuracy of positioning	Bandwidth	Power dissipation	Anti-interference
UWB	10–30 cm	500 Mbps	Low	High
CSS	1–3 m	1 Mbps	Low	Medium
ZigBee	3–5 m	800 M: 20 kbps 2.4 G: 250 kbps	Medium	Low

multipath propagation and inter-pulse interference to fast-pulse systems which must be mitigated by coding techniques. Table 1 gives a comparison of UWB, CSS, and ZigBee.

Wireless Positioning Technology

The positioning plays an important role in system. We know that the positioning is difficult in non-line-of-sight (NLOS) environment. There are various obstacles, such as walls, which lead to multi-path effects [3]. Some interference and noise from other wireless networks such as WIFI, or electrical radiating equipment such as microwave ovens, degrade the accuracy of positioning. Irregular building geometry and the density of water vapor in the air leads to reflection, and extreme path loss. So indoor positioning is more complex.

The positioning algorithms can be classified into two categories: Rang-based and Rang-free. Range-Based algorithm positioning by measuring the distance between the anchor nodes and mobile node or angle information, use the trilateration, triangulation or Maximum Likelihood estimator (ML) positioning method for calculation of the node location; Range-Free location is not required to distance or angle information, thus only according to network connectivity information and so on can be realized. Range-based algorithms usually need some special hardware to obtain accurate absolute range measurements and can achieve higher positioning accuracy than range-free algorithms. Range-free algorithms, on the other hand, do not need special hardware and are at low cost [7]. Range-Based positioning algorithm includes Angle of Arrival (AOA) [8], Time of Arrival (TOA) [9], Time Difference of Arrival (TDOA) [10], Received Signal Strength Indication (RSSI) [11], Time of Flight (TOF) [12], Symmetrical Double Sided Two Way Ranging (SDS-TWR) [13], etc.

Ubisense 7,000 serial is an in-building UWB radio based tracking system developed by Ubisense company [14], which can determine the positions of people and objects to an accuracy of a few tens of centimeters, using small tags which are attached to objects and carried by personnel, and a network of receivers which are placed around buildings. UWB is well-suited to in-building of emergency field [15], because of its non-line-of-sight nature, 3D-position, modest infrastructure requirements and high tracking accuracy [16]. A properly-architected UWB tracking system is low-power, and the fundamental technology is simple and low-cost.

Wireless Body Area Network

A WBAN is a wireless network of wearable computing devices [17, 18]. WBAN is a basic technology, which can long-term monitor and record the signal of the human health. In particular, the network consists of several miniaturized body sensor units together with a single body central unit [19]. A typical WBAN requires vital sign monitoring sensors, motion detectors to help identify the location of the monitored individual and some form of communication, to transmit vital sign and motion readings to medical practitioners or care givers. A typical WBAN will consist of sensors, a processor, a transceiver and a battery. Physiological sensors, such as ECG and SpO2 sensors, have been developed. Other sensors such as a blood pressure sensor, EEG sensor and a PDA for BSN interface are under development [20].

Early application is used to continuously monitor and record of chronic diseases, such as diabetes, asthma and heart disease, health parameters in patients, to provide some way automatically therapy control. For example, once the insulin levels of the patients with diabetes drop down, his WBAN can immediately activate a pump automatically injecting insulin for patients, so that patients without a doctor can control insulin in the normal level. WBAN is the smallest coverage network, but it is benefiting a wide network. WBAN is applied into service robot, and robot use this advanced technology to monitor the physical condition of the elderly. Meanwhile, this technology may boost the real implementation of the telemedicine.

Dynamic Bayesian Network Monitor and Prediction Model

This scheme aims to discover the nonlinear association between the ill risk and body vital signs (such as temperature, blood sugar, blood pressure, heart rate and pulse, etc.). The dynamic Bayesian network model can be expressed for a probability dependent relationship between the physical status and body vital signs, which varies with time. Building health status monitoring and prediction model for the elderly consists of four main steps:

- Traditional static Bayesian network should be built in first;
- Collect the various physiological signs and analyze signs to find probability dependency relationship among them;
- Establish dynamic Bayesian network model, Probe the nonlinear association patterns between the ill risk and each of signs;
- Using system dynamics methods, computer simulation and clinical experience to validate and assess the dynamic nonlinear association model;
- Evaluate and verify the dynamic Bayesian network prediction model, attempting to make a reasonable medical explanation for the change of the human condition.

Conclusion

This paper discusses the service robot global context-aware network, gives an analysis of the limitations of the traditional robot, and proposes a framework for service robots perceive global environment. Four key issues, including UWB radio, wireless positioning technology, WBAN and dynamic Bayesian network model, are discussed. According to the progress of existing projects, if the four key technologies are grasped effectively, the establishment of the wireless context-aware network will become easy.

Acknowledgments The authors would like to thank the Natural Science Foundation of Guangdong Province, China (No.9151009001000021, S2011010001155), the Ministry of Education of Guangdong Province Special Fund Funded Projects through the Cooperative of China (No. 2009B090300 341), the National Natural Science Foundation of China (No. 61262013), the High-level Talent Project for Universities, Guangdong Province, China (No. 431, YueCaiJiao 2011), and the Chinese Society of Vocational and Technical Education 2012–2013 scientific research and planning projects (NO. 204921) for their support in this research.

References

1. Chen M et al (2012) Machine-to-machine communications: architectures, standards, and applications. KSII Trans Int Inf Syst 6:480–497
2. Liu J et al (2012) Towards real-time indoor localization in Wireless sensor networks, in Computer and Information Technology (CIT), 2012 IEEE 12th international conference on 2012, pp. 877–884
3. Suo H et al. (2012) Issues and challenges of wireless sensor networks localization in emerging applications, In: Proceedings of 2012 International Conference on Computer Science and Electronics Engineering, Hongzhou, pp 447–451
4. Wan J et al (2011) Advances in cyber-physical systems research. KSII Trans Int Inf Syst 5:1891–1908
5. Dargie W, Poellabauer C (2010) Fundamentals of wireless sensor networks: theory and practice: Wiley
6. Fernandes JR, Wentzloff D (2010) Recent advances in IR-UWB transceivers: an overview, in Circuits and Systems (ISCAS). In: Proceedings of IEEE International Symposium on 2010, pp 3284–3287
7. Zhang S et al (2010) Accurate and energy-efficient range-free localization for mobile sensor networks. IEEE Trans Mobile Comput 9:897–910
8. Priyantha NB et al. (2001) The cricket compass for context-aware mobile applications. In: Proceedings of the 7th annual international conference on mobile computing and networking, pp 1–14
9. Harter A et al (2002) The anatomy of a context-aware application. Wireless Netw 8:187–197
10. Girod L, Estrin D (2001) Robust range estimation using acoustic and multimodal sensing, in Intelligent robots and systems. In: Proceedings IEEE/RSJ International Conference on 2001, pp 1312–1320
11. Girod L et al (2002) Locating tiny sensors in time and space: A case study, In: Computer design: VLSI in computers and processors, 2002. Proceedings of IEEE International Conference on 2002, pp 214–219

12. Lanzisera S et al (2006) RF time of flight ranging for wireless sensor network localization in intelligent solutions in embedded systems, 2006 International Workshop on 2006, pp 1–12
13. *nanotron find*. Available: http://www.nanotron.com/EN/PR_find.php
14. *ubisense*. Available: http://www.ubisense.net/
15. Hill R et al (2004) A middleware architecture for securing ubiquitous computing cyber infrastructures. IEEE Distrib Syst Online 5:1
16. Mahfouz MR et al (2008) Investigation of high-accuracy indoor 3-D positioning using UWB technology. IEEE Trans Microw Theory Tech 56:1316–1330
17. Chen M et al (2011) Body area networks: a survey. Mobile Netw Appli 16:171–193
18. Chen M et al (2012) Machine-to-machine communications: architectures, Standards and applications
19. Ullah S et al (2012) A comprehensive survey of wireless body area networks, J Med Syst, vol 36, pp 1065–1094, 2012/06/01
20. O'Donovan T et al (2009) A context aware wireless body area network (BAN), In: Pervasive computing technologies for healthcare, 3rd International Conference on PervasiveHealth 2009, pp 1–8

Architecture of Desktop as a Service Supported by Cloud Computing

Jianqi Liu, Hehua Yan, Caifeng Zou and Hui Suo

Abstract Traditional desktop virtualization assumed the communication between client and cloud server happened on high-speed network, but as popularization of multimedia application, a large amount of data result in network congestion. With the development of cloud computing, desktop as a service is emerged as a new service for desktop delivery. This paper presents a new architecture of DaaS supported by client and protocol co-designed method, which deploys hardware microprocessor to encode and decode the channel data in client, and decreases the transmission data over the network by compression algorithm and protocol optimization.

Keywords Cloud computing · Virtual desktop infrastructure · Desktop as a service · Remote desktop protocol

Introduction

A new word, Desktop as a Service (DaaS) [1] is emerged, which is a natural evolution of desktop virtualization paradigm whereby desktops would be delivered as a service from a Desktop Cloud. It is similar to Software as a service (SaaS)

J. Liu · H. Yan (✉) · C. Zou · H. Suo
Guangdong Jidian Polytechnic, Guangzhou, China
e-mail: hehua_yan@126.com

J. Liu
e-mail: liujianqi@ieee.org

C. Zou
e-mail: caifengzou@gmail.com

H. Suo
e-mail: suohui79@163.com

Y.-M. Huang et al. (eds.), *Advanced Technologies, Embedded and Multimedia for Human-centric Computing*, Lecture Notes in Electrical Engineering 260,
DOI: 10.1007/978-94-007-7262-5_42,

[2, 3] model. DaaS provides benefits of virtual desktop infrastructure (VDI) [4, 5] without the extra costs or risks of owning and managing physical resources. Customers who need large number of desktops for their employees are no longer required to provision all of the required resources, such as servers and storage, but can outsource these tasks to desktop cloud and focus on business critical task instead.

The essence of DaaS is desktop virtualization [6, 7], which is the combination optimization of computing and Communication, and shows the feature of centralization computing mode. In brief, DaaS places display and control on the front-end client and the computation or storage on back-end "Cloud" (data or computing center), the communication between front-end client and backend cloud supported by high speed network. The application execution is migrated from local (to the user) client to a remote data center. Client becomes a lightweight computer that handles only keyboard, mouse and monitor, as well as locally attached devices such as scanners and printers. Communication between client and desktop cloud is handled by remote desktop protocols (RDP). The virtual desktop paradigm has several advantages over the typical "fat-desktop" approach. Administrative costs of the DaaS are significantly lower because operating system (OS) images, applications, and data are no longer installed on a large number of distributed systems but in a desktop could, which improves manageability of the system as well as data and application security. Moreover, since the local client device is stateless, and it is very easy to troubleshoot and replace, thus on-site labor is significantly reduced. In a word, DaaS is emerging as an alternative to traditional desktop delivery.

Currently, most of the proposed desktop virtualization systems [4, 8–11] are based on the technologies to provide the control desktop. The premise of traditional client is that commodity networks are fast enough to use a low-level protocol to remotely serve graphical displays of common, GUI-based applications without any noticeable performance degradation. This leads to take the notion of thin clients to the limit by removing all state and computation from the desktop and designing a low-level hardware- and software-independent protocol to connect all user-accessible devices to the system's computational resources over a low-cost commodity network, for example, Stateless, Low-level Interface Machine (SLIM) [12]. They pay more attention to supplying high quality display effects on clients regardless network bandwidth consumption. These protocols do not lay emphasis on optimizing multimedia data transmission, do not improve the interaction experiences between users and machines, do not reduce the delay and bandwidth consumption of network, and ignore co-design between protocol and client.

To address the above issues, we present a new architecture of DaaS supported by protocol and client collaborative design, which optimizes communication protocol by data compression, and utilizes the hardware unit to realize image and video decoding. The new architecture can improve the network performance drastically through increasing very few time delay generated by decompression and compression.

Overview

DaaS represents a kind of new computing service; users can enjoy high computing performance and huge storage capacity supported by cloud server only through a lightweight client. Desktop cloud is a set of physical resources (such as storage, servers, networking gear, etc.) together with virtualization, connection brokering, and management software allowing for remote access to large numbers of desktops (potentially tens or hundred of thousands). The overview of DaaS is presented in Fig. 1. Desktops execute in data centers and users access them via the Internet by RDP. The response speed of system depends on three portions: thin client, RDP, desktop cloud server. As the desktop cloud server has huge storage and super computing ability to process desktop delivery, data process ability of thin client and network capability are two key factors to the DaaS. One of the direct methods is increasing the bandwidth of network, but in WAN, bandwidth is limited by telecommunication operator. Moreover, using more bandwidth will cost more communication fee. The other alternative is decreasing communication data amount. Before data transfer, we deploy some encoding technology to compress data in cloud server. But this method will bring a new problem: software decoding process will result in time delay, as the process ability of embedded system on chip (SOC) is low in thin client. So we should take these issues into account, and design a new architecture.

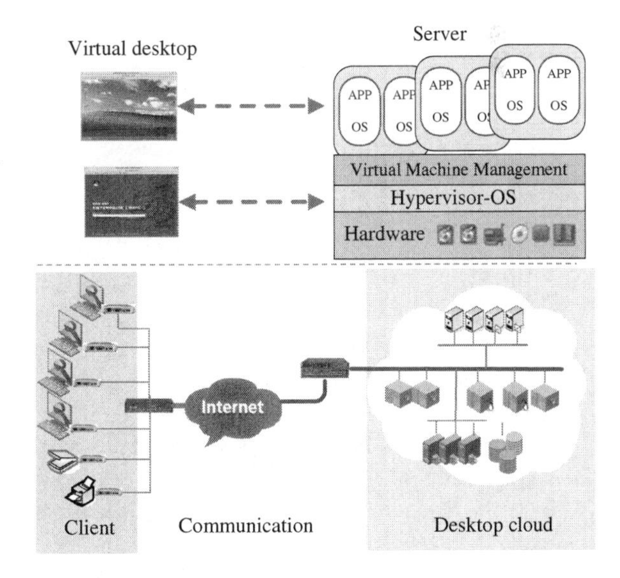

Fig. 1 Overview of DaaS. Users access their desktops and applications using thin client which forwards keyboard and mouse events and screen state updates over the local or wide area network. OS and applications run in virtual machine of the remote desktop cloud

Architecture

The desktop is delivered to the client from desktop cloud; the element of desktop (picture, text, mouse event, keyboard event etc.) from cloud should be processed by thin client quickly and transmit events and other data to desktop cloud promptly. The data is classified into image, video, sound, USB data, and keyboard event, mouse event. We deploy JPEG, MPEG4, MP3 encoding technology and lz4 to compress the data for transmission. In thin client, we deploy hardware decoding microchip to decode corresponding data. The details will be discussed in following three sections.

Thin Client

Formerly, we have become accustomed to the PC model, which allows every user to have their own CPU, hard disk, and memory to run their desktop application. However, in "Cloud" age, desktop virtualization is a modern way to delivery desktop, where multiple users share the processing power of cloud server by a simple client. The client usually deploys the system-on-chip (SoC) solution, which has several advantages over the traditional PC model, including lower costs, better energy efficiency, and simplified administration. Traditional independent communication protocol needs high-bandwidth network to transmit multimedia data. However, commodity wide-area networks are very slow, especially in some developing country, the independent protocol isolated by hardware result in bad user experience as the image and video time-consuming decoding. This paper deploys encoding and decoding chip to process channel data, such as image, sound and video.

In this architecture, S5PV210 chip of ARM CortexTM-A8 core is used as main control chip and also ARM V7 instruction set is adopted, relying on which, it can achieve 1 GHz of basic frequency. The hardware framework is shown as Fig. 2.

This thin client design has several features.

- High definition desktop display. Client is inbuilt with high-performance PowerVR SGX540 3Dgraphics engine and 2D graphics engine, supporting 2D/3D graphics acceleration. Its polygon formation rate is 28 million polygons per second, pixel fill rate can be up to 0.25 billion per second, and it supports PC level display technologies such as DX9, SM3.0, OpenGL2.0, etc. These can provide high performance desktop display update.
- Rapid encoding and decoding. Client's Multi Format Codec (MFC) supports the encoding and decoding of videos with MPEG4, H.263, H.264 and other formats, and supports simulated/digital TV output. With JPEG hardware encoding and decoding, the supported resolution ratio can be up to 8000×8000, and can play and record smoothly video documents of 1920×1080 pixel (1080p) at 30 frames per second. Audio encoding also supports MP3, WMA, EAAC + and

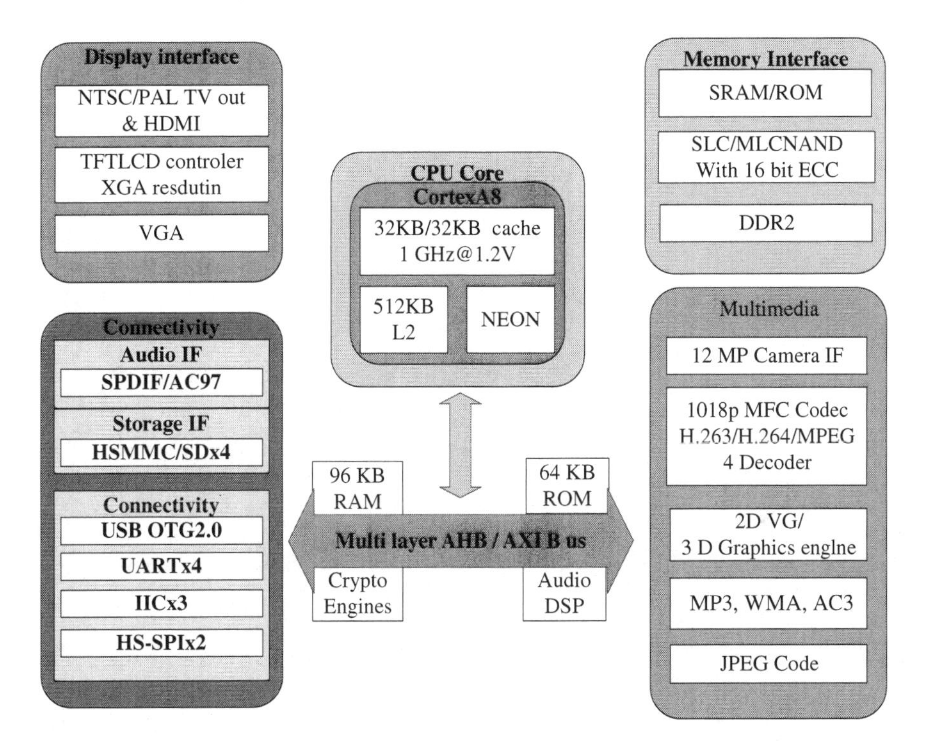

Fig. 2 Hardware framework

AC3. These hardware chips can complete encoding and decoding quickly, and people won't be affected by such a short time delay.

- Abundant interface. The client is inbuilt with HDMI interface, so that high-definition videos can be transmitted to external display. The client also provides three I2S, USB Host 2.0, and USB 2.0 OTG operating at high speed (480 Mbps), four SD Host and high-speed Multimedia Card interface.

Communication Protocol

Communication happened between front-end client and back-end cloud usually adopts the remote desktop protocols (RDP). The image, video, sound, keyboard event, mouse event and USB data are transmitted through the network. In order to allow some degree of flexibility in thin client and cloud server implementation, communication session is split into multiple communication channels (e.g., every channel is a remote device) to control communication and execution of messages according to the channel type, and to add and remove communication channels during run time. The following communication channels are defined: (a) the main channel services as the main session connection for sending control message; (b) events channel for sending mouse and keyboard events; (c) video channel for

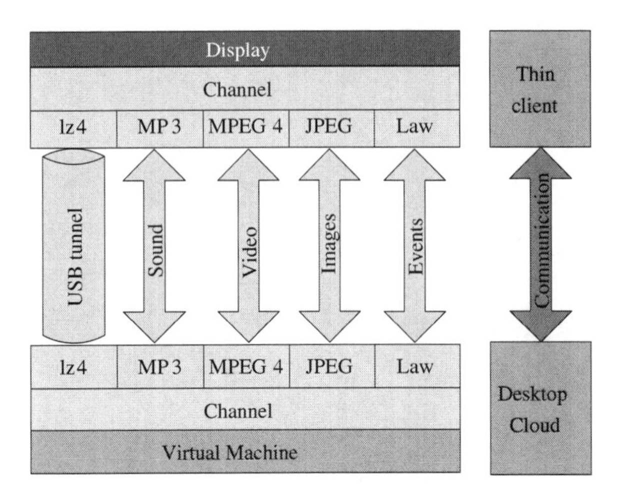

Fig. 3 Communication protocol of DaaS

receiving remote video updates; (d) sound channel for sending and receiving audio stream; (e) image channel for receiving display updates and (f) USB channel for sending and receiving USB data deploys as a tunnel. More channel types will be added as the protocol evolves expediently. In order to improve compatibility, communication protocol supports bidirectional exchange of channels compatibilities. In order to achieve higher performance, different data type deploy different compression algorithm. The Fig. 3 illustrates the communication protocol.

Desktop Cloud Server

The kernel of desktop cloud server is sharing hardware and software resource by virtual machine management module. Virtual machine can give better server utilization, better server management, and better power efficiency in desktop cloud. Now, there are some paradigms, such as VMware, Virtualbox, KVM, Xen and Hyper-V. In this paper, we utilize KVM as test platform. First, KVM is free software released under the general public license (GPL), and we can modify the remote desktop protocol (add compression algorithm in channel data process) to ease the pressure of communication network. Secondly, KVM enhances the performance by utilizing hardware virtual machine (HVM) technology, Intel VT and AMD-V, supported by CPU vendor. These instruction set extensions provide hardware assistance to virtual machine monitors. They enable running fully isolated virtual machines at native hardware speeds, for some workloads. Thirdly, KVM is part of Linux and uses the regular Linux scheduler and memory management. This means that KVM is much smaller and simpler to use.

Conclusion

Our main focus is to provide high-quality remote access from virtual machine of desktop cloud. We try to break down the barriers of traditional desktop virtualization. We present a new architecture of DaaS, which aims to decrease the transmission data over the network by compression algorithm and protocol optimization. In order to improve response speed, we deploy hardware encoding and decoding microchip in thin client to process data from or to communication channels. This architecture of DaaS supported by co-designed method overcomes the defects of traditional desktop virtualization, which has the following distinctive characteristics: (a) Low cost and energy-saving; (b) Multimedia supported; (c) Low bandwidth and low latency; (d) Good user experience. But some issues should be addressed in the near future, such as desktop migration, and dynamic resource allocation control.

Acknowledgements The authors would like to thank the Natural Science Foundation of Guangdong Province, China (No.9151009001000021, S2011010001155), the Ministry of Education of Guangdong Province Special Fund Funded Projects through the Cooperative of China (No. 2009B090300 341), the National Natural Science Foundation of China (No. 61262013), the High-level Talent Project for Universities, Guangdong Province, China (No. 431, YueCaiJiao 2011), and the Chinese Society of Vocational and Technical Education 2012–2013 scientific research and planning projects (NO. 204921) for their support in this research.

Reference

1. Cristofaro S et al. (2010) Virtual distro dispatcher: a light-weight desktop-as-a-service solution in cloud computing, Springer pp 247–260
2. Turner M et al (2003) Turning software into a service. Computer 36:38–44
3. Buxmann P et al (2008) Software as a service. Wirtschaftsinformatik 50:500–503
4. Velte A, Velte T (2009) Microsoft virtualization with Hyper-V: McGraw-Hill, Inc
5. Baratto RA et al. (2004) Mobidesk: mobile virtual desktop computing. In: Proceedings of the 10th annual international conference on mobile computing and networking, pp 1–15
6. Hazari S, Schnorr D (1999) Leveraging student feedback to improve teaching in web-based courses. The Journal 26:30–38
7. Watson J (2008) Virtualbox: bits and bytes masquerading as machines. Linux J 2008:1
8. Citrix Corporation,Citrix Application Delivery Infrastructure. Available: http://www.citrix.com/
9. Kivity et al. (2007) kvm: the Linux virtual machine monitor. In: Proceedings of the Linux Symposium, pp 225–230
10. Sugerman J et al. (2001) Virtualizing I/O devices on VMware workstation's hosted virtual machine monitor. In: Proceedings of the general track: 2002 USENIX annual technical conference, pp 1–14
11. Rosenblum M (1999) VMware's virtual platform^{TM}. In: Proceedings of hot chips, pp 185–196
12. Schmidt BK et al (1999) The interactive performance of SLIM: a stateless, thin-client architecture. ACM SIGOPS Oper Syst Rev 33:32–47

Lattice Boltzmann Method for the Velocity Analysis of Polymer Melt in Vane Extruder

Jianbo Li, Jinping Qu, Xiaoqiang Zhao and Guizhen Zhang

Abstract After some mixed methods in polymer processing are summarized, the new equipment is introduced which is called vane plasticizing extruder with elongational flow field. Then the unit of vane geometry is analyzed, and the analysis method of vane extruder conveying state is put forward by using the lattice Boltzmann method. On this base, physical model and mathematical model are established. Calculation of velocity distribution and quantitative description of the elongational deformation rate shows there has elongational rheological in the vane plasticizing extruder. Periodic extensional flow field has very good mixing effect in polymer processing.

Keywords Vane extruder · Lattice Boltzmann method · Positive displacement · Elongational rheology · Normal stress field

J. Li (✉) · J. Qu · X. Zhao · G. Zhang
The National Engineering Research Center of Novel Equipment for Polymer Processing, Guangzhou, China
e-mail: gdjdljb@126.com

J. Qu
e-mail: jpqu@scut.edu.cn

X. Zhao
e-mail: xqzhaocn@gmail.com

G. Zhang
e-mail: gaoyuanxuefen@126.com

J. Li · J. Qu · X. Zhao · G. Zhang
The Key Laboratory of Polymer Processing Engineering of Ministry of Education South China University of Technology, Guangzhou 510640, Guangdong, China

J. Li
Guangdong Jidian Polytechnic, Guangzhou 510515, China

Y.-M. Huang et al. (eds.), *Advanced Technologies, Embedded and Multimedia for Human-centric Computing*, Lecture Notes in Electrical Engineering 260,
DOI: 10.1007/978-94-007-7262-5_43,

Introduction

Mixing, which is an indispensable stage in polymer processing technology, to a great extent, determines the final performance of product. Components are scattered through physical movement in the process of polymer melt mixing [1]. In order to get better mixing effect, a lot of work has been done by predecessors.

Tadmor analyzes the solid-phase breakage mechanism, and compares the shear and tensile field dispersion ability. Dumbbell ball is studied under the action of external field separation. The adhesion strength of the two small balls is equivalent to the linkage between them. Particle separation occurs when the field force is greater than the binding force between spheres. From the point of view of the largest force of solid phase particles, elongational flow effect is double shear flow effect at the same deformation rate (tensile rate and shear rate). So it can be spread to get higher efficiency and more effective dispersion effect using stretch flow field [2].

Spherical agglomerates which have a certain intensity distribution is studied by Manas [3], under the low Reynolds number,in the shear flow field, elongation flow field, uniaxial tension, biaxial stretching field. Dispersing ability is compared in two-dimensional flow field and three-dimensional flow field with or without the curl. Manas thinks a plane of spherical agglomerates determines its strength, and aggregates disconnection along the plane when the flow force is greater than its strength [4]. His research shows that elongational dispersion efficiency is better than that of the simple shear field.

Chris Rauwendaal opens many tapered slot on the wedge screw arris. When through the screw edge and screw arris of the tapered slot, the melt is doubly stretching flow field effect, which can produce strong stretching flow [5].

Suzaka opens multiple convergent channels with stretch on the runner plate, and discusses the draw ratio of big diameter and path of convergent channel [6].

In order to get continuous stretch flow field, a novel polymer processing equipment is designed by Qu [7], a positive displacement conveying facility which totally alters the shear conveying mechanism. In this article, the working principle of this equipment will be described, with the melt flow speed in flow field as the key of the research.

Model of the Vane Extruder

Structure of the Vane Extruder

The structure schematic diagram of the vane extruder is shown in Fig. 1. The components of the vane extruder are a stator, a rotor, and four vanes. The four vanes are installed in pairs, each other contacting at the bottom surface. When the rotor is rotating, the volume is periodically changing which is constituted by the stator, the rotor, and the vanes.

The materials in the previous unit are fed into the current vane unit when the volume becomes large. The materials in the current unit are discharged to the next vane unit when the volume becomes small.

Geometric Structure of the Vane Unit

The geometric structure of the vane extruder unit is shown in Fig. 2.

The effective radius vector of the vane unit is defined as the clearance between the stator and the rotor radius. Assume that the radius of the stator is R, the radius of the rotor is r, the eccentricity between the stator and rotor is e, the angle between the effective radius vector and x axis is θ. So, the effective radius vector can be expressed in the following expression:

$$\rho(\theta) = \sqrt{R^2 - e^2 \sin^2 \theta} + e \cos \theta - r \tag{1}$$

The Lattice Boltzmann Method

It is more difficult to use the analytic solution to describe melt flow in the vane unit, while the Lattice Boltzmann Equation can effectively analyze the melt flow in the vane extruder.

Lattice Boltzmann Equation (LBM) originates from Lattice Gas Automata (LGA). LGA model, taking the fluid as a hypothetical particle on a regular lattice, asserts that particle collisions and migration occur according to certain rules in the grid, and macroscopic quantity is obtained by statistical, such as density velocity etc. [8].

The earliest LBM model was first proposed by McNamam and Zanetti in 1988, whose evolution process uses the Boolean variable statistical distribution function of particles in the LGA [9].

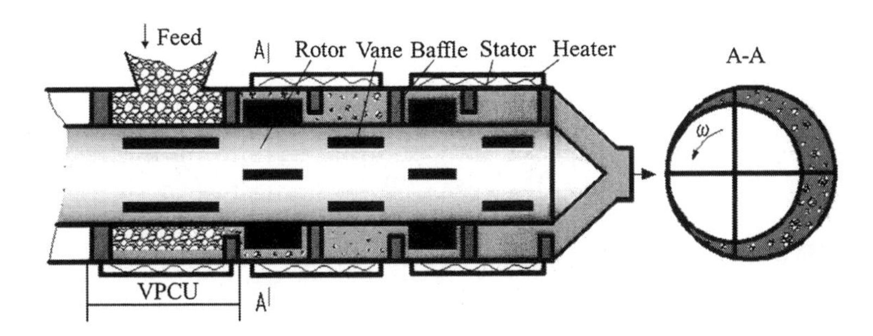

Fig. 1 The structure schematic diagram of the vane extruder

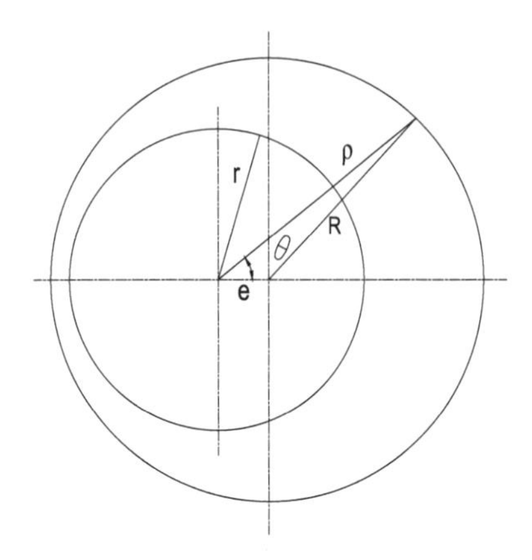

Fig. 2 Geometric structure of the vane unit

In 1991, single relaxation Bhatnagar-Gross-Krook model (BGK), a more convenient calculation, was proposed [10]. BGK model has a single relaxation time, and collision term convenient calculation, which make the LBGK model widely used.

DnQm model is the most representative model proposed by Qian [11], in which n is the dimension of the problem, m is the discrete velocity. In this paper, we use the D2Q9 model, two dimensions and nine discrete velocity, as shown in Fig. 3

The lattice Boltzmann equation of single relaxation can be approximated as [12]

$$f_k(x + \Delta x, t + \Delta t) = f_k(x, t) \cdot [1 - \omega] + \omega \cdot f_k^{eq}(x, t) \tag{2}$$

The equilibrium distribution function can be written as [12]

$$f_k^{eq}(x, t) = w_k \cdot \rho(x, t) \cdot \left[1 + \frac{\vec{c}_k \cdot \vec{u}}{c_s^2} + \frac{1}{2} \cdot \frac{(\vec{c}_k \cdot \vec{u})^2}{c_s^4} - \frac{1}{2} \cdot \frac{\vec{u}^2}{c_s^2}\right] \tag{3}$$

The macroscopic fluid density and the momentum can be represented as [12]

$$\rho = \sum_0^8 f_k \quad \vec{u} = \left(\sum_0^8 f_k \vec{c}_k\right) / \rho \tag{4}$$

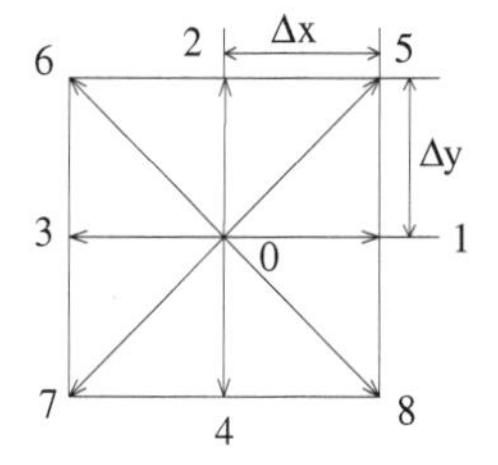

Fig. 3 The D2Q9 model

k	is the flow direction of the lattice, k = 0~8
w_k	is the weight factor
$\vec{c}_k$	is grid unit vector in the flow direction
c_s	is the speed of sound
ω	is a diffusion coefficient about LBM relaxation time
ρ	is the macro density in the grid at point x, at moment t
$f_k^{eq}(x,t)$	is the equilibrium distribution function in the grid at point x at moment t
$f_k(x,t)$	is the distribution function in the grid at point x, at moment t
$f_k(x+\Delta x, t+\Delta t)$.	is the distribution function in the grid at point x + Δx, at the moment t + Δt

Results and Discussion

The Velocity Distribution

The rotor radius of the vane extruders r is 15 mm, The stator radius R is 20 mm, the eccentricity e is 2 mm. The circumference of the rotor is 2*PI*r = 94 mm. For simple calculations, we divide 200 at the x direction, 100 at the y direction. Assume that the inlet velocity uMax is 0.02; the Reynolds number Re is 100; the kinematic viscosity nu is uMax*2*r/Re = 0.006; the relaxation parameter omega is 1.0/(3.0*nu + 0.5) = 1.9305.

After numerical computation, dimensionless inlet velocity is 1. The velocity distribution in vane unit is shown in Fig. 4 after numerical calculation.

The Numerical Analysis of Velocity

In order to get more detailed information of velocity in the flow field, the velocity is analyzed in the middle part of the effective radius vector. The graphics of x

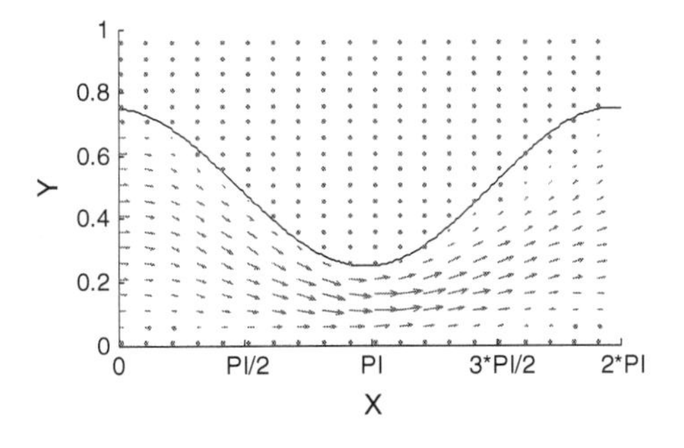

Fig. 4 The velocity distribution in vane unit

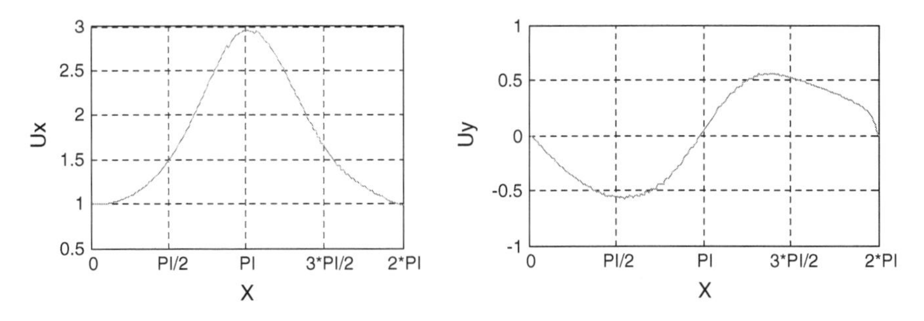

Fig. 5 The Ux and Uy in the middle of the effective radius vector

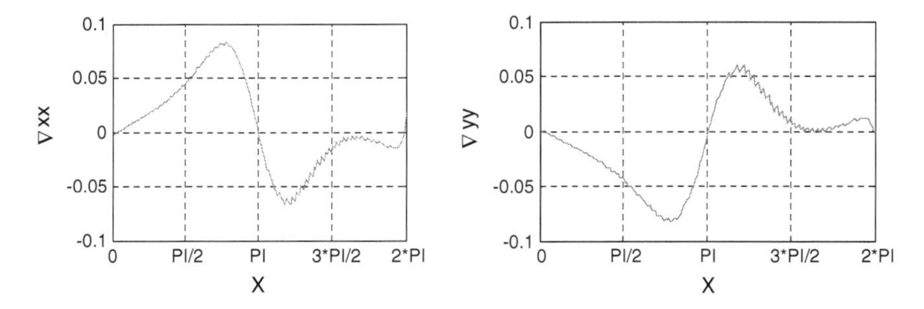

Fig. 6 The ∇xx and ∇yy in the middle of the effective radius vector

direction velocity and y in the middle part of the effective radius vector is shown in Fig. 5. The graphics of ∇xx and ∇yy is shown in Fig. 6.

Discussion

The Strain rate tensor component in the x direction ∇xx changes again in the vane unit. If $\nabla xx > 0$ in the lattice number $0 \sim$ PI of x direction, it has a maximum value 0.8 at about 130°; if $\nabla xx < 0$ in the number PI $\sim$ 2*PI, it has a minimum -0.8 at about $-130°$. Also in the y direction, if $Uy < 0$ in the lattice number $0 \sim$ PI of x direction, it has a minimum value -0.6 at about 130° lattice Point; if $Uy > 0$ in the number PI $\sim$ 2*PI, it has a maximum value 0.6 at about $-130°$.

In the lattice number 1–100 of x direction, the melt is stretched in the x axis, compressed in the y axis; or, the melt is compressed in the x axis, stretched in the y axis. This kind of phenomenon is repeated in the next unit, that is, the elongation deformation rate of the melt changes again and again in the vane extruder.

Conclusions

As a novel equipment based on elongation rheology, method and equipment of the vane extruder is a completely new theory and practice for polymer processing. The vane extruder totally alters the shear conveying mechanism of the screw extruder in polymer processing, presenting good effects of mixing, distribution and dispersion. Through some experimental investigations, high stability, wide adaptability, shorter polymer thermo-mechanical history, better plasticating and mixing effects are found in the vane extruder compared with the screw extruder.

Acknowledgments The authors wish to acknowledge the National Nature Science Foundation of China(Grant 10872071, 50973035, 50903033, 51073061, and 21174044), National Key Technology R&D Program of China (Grant 2009BAI84B05 and 2009BAI84B06), the Fundamental Research Funds for the Central Universities (NO. 2012ZM0047), Program for New Century Excellent Talents in University (No.NCET-11-0152), Pearl River Talent Fund for Young Sci-Tech Researchers of Guangzhou City (No.2011J2200058), 973 Program (No.2012CB025902) and National Natural Science Foundation of China-Guangdong Joint Foundation Project (U1201242) for their financial supports.

References

1. Tadmor Z, Gogos CG (2006) Principles of polymer processing, 2nd edn. Wiley, New Jersey, pp 473–475
2. Tadmor Z (1976) Forces in dispersive mixing. Ind Eng Chem Res Fundam 15(4):346–348
3. Zloczower IM, Tadmor Z (1994) Dispersive mixing of solid additives. Mixing Compd Polymer-Theory Pract 23:55–84
4. Manas-Zloczower Ica (1994) Studies of mixing efficiency in batch and continuous mixers. Rubber Chem Technol 67(3):504–528
5. Rauwendaal C, Osswald T, Gramann P (1999) Design of dispersive mixing devices. Int Polym Proc 14(1):28–34
6. Suzaka Y (1982) Mixing device. USP. 4334783
7. Jinping Q (2009) A method and a device for plasticating and transporting polymer material based on elongational flow. EP. 2113355
8. Wolfram S (1986) Cellular automaton fluids 1: basic theory. J Stat Mech 45:471–529
9. Mcnamara GR, Zanetti G (1988) Use of the Boltzmann equation to simulate lattice gas automata. Phys Rev Lett 61(20):2332–2335
10. Bhatnagar P, Gross EP, Krook MK (1954) a model for collision processes in gases. I. Small amplitude processes in charged and neutral one—component systems. Phys Rev 94(3):511–525
11. Qian Y, d'Humieres D, Lallemand P (1992) Lattice BGK models for Navier-Stokes equation. Europhysits Lett 17:479–484
12. Mohamad AA (2011) Lattice Boltzmann method fundamentals and engineering applications with computer codes. Springer, London, pp 69–72

Integrated Approach for Modeling Cyber Physical Systems

Shuguang Feng and Lichen Zhang

Abstract Cyber physical systems contain three parts: control part, communication part and physical part. In this paper, we propose an integrated approach for modeling cyber physical systems. Differential Dynamic Logic is used for modeling control part, Communicating Sequential Process (CSP) is applied to specify communication part, and Modelica is used for modeling physical part of cyber physical systems. The proposed approach is illustrated by a case study of the one-street style of vehicles on the street.

Introduction

The Differential Dynamic Logic, whose abbreviation is dL, is developed by Platzer [1]. Differential Dynamic Logic dL is a Dynamic logic for Hybrid Systems and it has some advantages over Dynamic Logic [2]. Differential Dynamic Logic dL has differential equations for Hybrid Systems which contains both discrete and continuous dynamics. It is suitable for Differential Dynamic Logic dL to model complex systems.

Communicating Sequential Processes (CSP) is a formal language and firstly described by Hoare [3]. CSP is a tool for specifying and verifying the concurrent processes of systems. CSP is famous for its mathematical theories of concurrency which is known as process algebras. In this paper, we use CSP to describe a signal system. The simplified syntax of CSP is as below, where P is a process and p is an event.

Modelica is an object-oriented language for system modeling and simulation [4]. Modelica is also non-causal modeling language, and there are tools that

S. Feng · L. Zhang (✉)
Shanghai Key Laboratory of Trustworthy Computing East China Normal University, Shanghai 200062, China
e-mail: zhanglichen1962@163.com

Y.-M. Huang et al. (eds.), *Advanced Technologies, Embedded and Multimedia for Human-centric Computing*, Lecture Notes in Electrical Engineering 260,
DOI: 10.1007/978-94-007-7262-5_44, © Springer Science+Business Media Dordrecht 2014

support the model generation of Modelica. Below is the simple example of Modelica.

In this paper, we propose an integrated approach for modeling cyber physical systems. We build a model about one-street driving using dL, CSP, and Modelica. With dL, we build the control component of the model, which supervises the

```
model simpleModel
equation
end simpleModel;
```

condition of the street and send signals to the cars on the street. With CSP, we build the signal transmission component of the model, which send data between control component and the cars on the street. The cars' data can be sent to control component via signal transmission component.

Model Description

The model is in Fig. 1. The street is unidirectional. Cars run on the street. No cars can run with another side by side but the moment of overtaking another car. Cars can accelerate and brake. And, one car can be driven to the street from outside or driven off the street. Control system calls the information of cars and sends the command to cars via signal system described by CSP. Signal system has four parts: CarReceive, CarSend, CtrlReceive and CtrlSend. THe CarSend part collects car's data. CtrlReceive part receives data from CarSend part. CtrlSend part sends

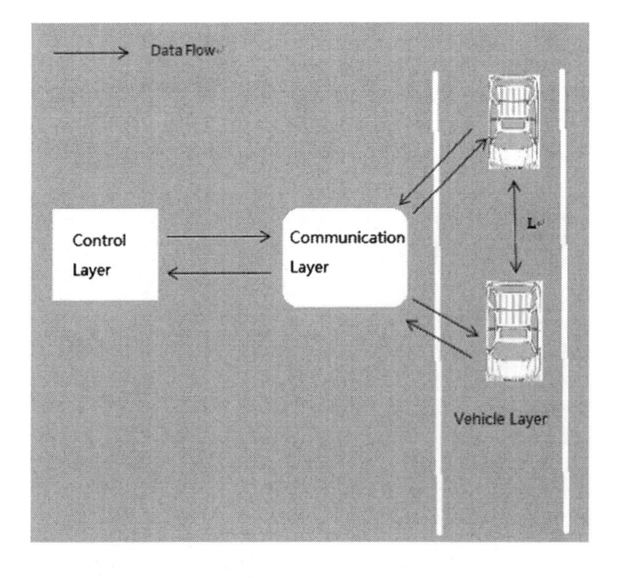

Fig. 1 Model of whole system

command from control system. CarReceive part receives command from CtrlSend part.

We introduce the variables needed: x is the position of the car, x is a scalar; v is the velocity of a car and a vector; a is the acceleration of a car and a vector; ID is the identification number of a car, which is assigned by the car's manufacturers and different from each other; state is the current state of a car.

Control System

Control system has five scenes to deal with. There is a standard time interval t and a length l. There is a length variable l. If two cars length is longer than l and the car can outstrip the one just in front of it within t, change the state of this car to ACCELERATION. If the distance between two cars is longer than I, keep the cars speed v to its current value and s to 0. If the distance between two cars is shorter than I and the car can't outstrip the one just in front of it within t, change the state of this car to DECELERATION.

The distance of a car runs in time t is:

$$s = \frac{1}{2}at^2 + vt$$

$$ControlSystem \equiv \begin{pmatrix} ?x_1 - x_2 < l \& s_1 - s_2 > x_1 - x_2; state = ACCELERATION; \\ a = b; \upsilon' = a; x' = \upsilon \end{pmatrix}$$

$$\cup$$

$$\begin{pmatrix} ?x_1 - x_2 < l \& s_1 - s_2 < x_1 - x_2; state = DECELERATION; \\ a = b_2; \upsilon' = a; x' = \upsilon; ?\upsilon > 0 \end{pmatrix}$$

$$(?x_1 - x_2 \geq l; state = CONSTANTSPEED; a = 0;)$$

$$(?state = SETOUT; a = b; \upsilon' = a; x' = \upsilon) \tag{1}$$

Signal System

The SignalSystem works in this way: process Car sends its data to process of CarSend; process CarSend sends this data to CtrlReceive; CtrlReceive sends this data to ControlSystem. Above is the way of transmitting car's data to the ControlSystem. After computing on this data, ControlSystem sends its command for the car, which is also data like: a, v, state and x. The sequence is: ControlSystem sends its command to process CtrlSend; CtrlSend sends its command to process CarReceive; CarReceive sends this data to Car.

Formula 2 models the processing of a car, which accepts a call and sends the information out. Formula 3 delivers the command that the length is 5 to the car. Formula 4 receives the command that the length is 5 from control system. Formula 5 models the control system. Formula 6 receives commands of control system. Formula 7 receives information of car.

The formulas are below:

$$\begin{aligned} Car = & left?call \rightarrow right!x \rightarrow right!a \rightarrow right!\upsilon \rightarrow right!ID \rightarrow \\ & right!state \rightarrow Car \end{aligned} \tag{2}$$

$$\begin{aligned} CarSend &= P\langle\rangle \\ & where \\ P_s &= right!s \rightarrow P\langle\rangle if \# s = 5 \\ P_s &= left?x \rightarrow P_{s \cap \langle x \rangle} \end{aligned} \tag{3}$$

And, x means the variable from Car, that is, x, a, v, ID and state.

$$\begin{aligned} Car\ Receive = & P\langle\rangle where \\ & P\langle\rangle = left!s \rightarrow P\{s\} \\ & P\langle x\rangle = right!x \rightarrow P\langle\rangle \\ & P\langle x\rangle^{smallfrown} s = right!x \rightarrow P_s \\ & left = \{s | s \in right * \# s = 5\} \end{aligned} \tag{4}$$

$$\begin{aligned} ControlSystem = & left?x \rightarrow left?a \rightarrow left?\upsilon \rightarrow left?ID \rightarrow left?state \rightarrow \\ & right!call \rightarrow right!s \rightarrow ControlSystem \\ & where \# s = 5 \end{aligned} \tag{5}$$

$$CtrlSend = left?call \rightarrow left?s \rightarrow right!call \rightarrow right!sCtrlSend \tag{6}$$

$$Ctrl\mathrm{Receive} = left?call \rightarrow left?s \rightarrow right!call \rightarrow CtrlReceive \tag{7}$$

Car

Car has its attributes: x for assigning it's current position, v for assigning its current velocity, a for assigning its acceleration value, state for assigning its state value and ID for assigning its identification value. Car send its information which is kept into the attributes x, v, a, state and ID, to the ControlSystem via SignalSystem. ControlSystem analyze these data using its algorithm and send the new data, which we can call it command, to the Car. Car set its attributes with the new data.

```
model Car
  input Real a_new;
  //Acceleration value from the Control System
  parameter Real x;
  //Current position of a car
  parameter Real v;
  //Current velocity of a car
  parameter Real a;
  //Current acceleration of a car
  parameter String state;
  //Current state of a car
  String ID;
  //The identification of a car
equation
        if a_new > a then
        state = Accelerating(a_new);
   elseif a_new < a then
        state = Deccelerating(a_new);
   end if;
   if a_new < 0 then
                state = Parking(a_new, state);
   else
             state = SETOUT(a_new, state);
   end if;
end Car;
```

Conclusion

In this paper, we use three modeling methods: Differential Dynamic Logic dL, CSP and Modelica to model a traffic system. The Differential Dynamic Logic dL reasons about possible behaviour of a complex system with temporal logic for reasoning about the temporal behaviour during their operation and first-order dynamic logic. The dL logic supports both discrete and continuous evolution. CSP is used to verify the concurrent processes of a system. And, Modelica can model the physical world properly. Modelica with its object-oriented, equation-based properties can be useful for the analyzing a system. We use dL to model the ControlSystem, CSP to model the Signal System and Modelica for the Car, which is a vehicle in our system.

Acknowledgments This work is supported by national high technology research and development program of China (No. 2011AA010101), national basic research program of China (No. 2011CB302904), the national science foundation of China under grant No. 61173046, No. 61021004, No. 61061130541), doctoral program foundation of institutions of higher education of China (No. 200802690018), national; science foundation of Guangdong province under grant No. S2011010004905.

References

1. Platzer A (2008) A differential dynamic logic for hybrid systems. J Autom Reas 41(2):143–189
2. Harel D, Kozen D, Tiuryn J (2000) Dynamic logic. MIT, Cambridge
3. Brookes SD, Hoare CAR, Roscoe AW (1984) A theory of communicating sequential processes. J ACM 31(3):560–599
4. Fritzson P, Engelson V (1998) Modelica—a unified object-oriented language for system modeling and simulation. ECOOP'98—Object-Oriented Programming, pp 67–90

Specification of Railway Cyber Physical Systems Using AADL

Lichen Zhang

Abstract Railway cyber physical systems involve interactions between software controllers, communication networks, and physical devices. These systems are among the most complex cyber physical systems being designed by humans, but the complexities of railway cyber physical systems make their development a significant technical challenge. Various development technologies are now indispensable for quickly developing safe and reliable transportation systems. In this paper, we apply AADL to specify railway cyber physical systems and give a detailed analysis and design of the CBTC system. The CBTC system is split into four subsystems and makes friendly communication between the other three subsystems connecting to the data communication subsystem. We apply AADL to model each subsystem and give a detailed analysis and modeling, and make an effective integration of all subsystems together to form a complete CBTC system finally.

Keywords Railway cyber physical systems · AADL · Specification · CBTC

Introduction

The problems that must be addressed in operating a railway are numerous in quantity, complex in nature, and highly inter-related [1–3]. For example, collision and derailment, rear-end, head-on and side-on collisions are very dangerous and may occur between trains. Trains may collide at level crossings. Derailment is caused by excess speed, a wrong switch position and so on. The purpose of train control is to carry the passengers and goods to their destination, while preventing

L. Zhang (✉)
Shanghai Key Laboratory of Trustworthy Computing, East China Normal University, Shanghai 200062, China
e-mail: zhanglichen1962@163.com

Y.-M. Huang et al. (eds.), *Advanced Technologies, Embedded and Multimedia for Human-centric Computing*, Lecture Notes in Electrical Engineering 260,
DOI: 10.1007/978-94-007-7262-5_45,

them from encountering these dangers. Because of the timeliness constraints, safety and availability of train systems, the design principles and implementation techniques adopted must ensure to a reasonable extent avoidance of design errors both in hardware and software. The train to train collision accident that happened on July 23, 2011 in one of the high speed lines gave a big hit to the high speed railway development in China. Besides a great surprise, everybody is eager to know what has happened, what went wrong, whose responsibility it is. The accident investigation report published in December 2011 described the events and the software and hardware failures of the train control system equipments. Thus, a specific methodology relevant, to design should be applied for train control systems development. The dependability of the railway cyber physical system should arouse more attention [4, 5].

The development of railway cyber physical systems is a challenging process. On the one hand, the railway domain experts have to make the requirement analysis for the railway cyber physical systems in such a way that they are implementable. On the other hand, the software engineer has to understand these domain-specific requirements to be able to implement them correctly. The high need for product quality is beyond dispute as human life may be endangered if a railway controller is malfunctioning. The struggle for high-quality software development methods is of highest importance in railway cyber physical area.

The SAE Architecture Analysis and Design Language (AADL) [6–8] defines a language for describing both the software architecture and the execution platform architectures of performance-critical, embedded, real-time systems. An AADL model describes a system as a hierarchy of components with their interfaces and their interconnections. Properties are associated to these constructions. AADL components fall into two major categories: those that represent the physical hardware and those representing the application software. The former is typified by processors, buses, memory, and devices, the latter by application software functions, data, threads, and processes. The model describes how these components interact and are integrated to form complete systems. It describes both functional interfaces and aspects critical for performance of individual components and assemblies of components. The changes to the runtime architecture are modeled as operational modes and mode transitions.

In this paper, we propose an approach to specify railway cyber physical systems based on AADL.

The Proposed Method for Specifying Railway Cyber Physical Systems Based on AADL

AADL [7–12] is an architecture description language developed to describe embedded systems is shown in Fig. 1. Architecture Analysis and Design Language (AADL), which is a modeling language that supports text and graphics, was

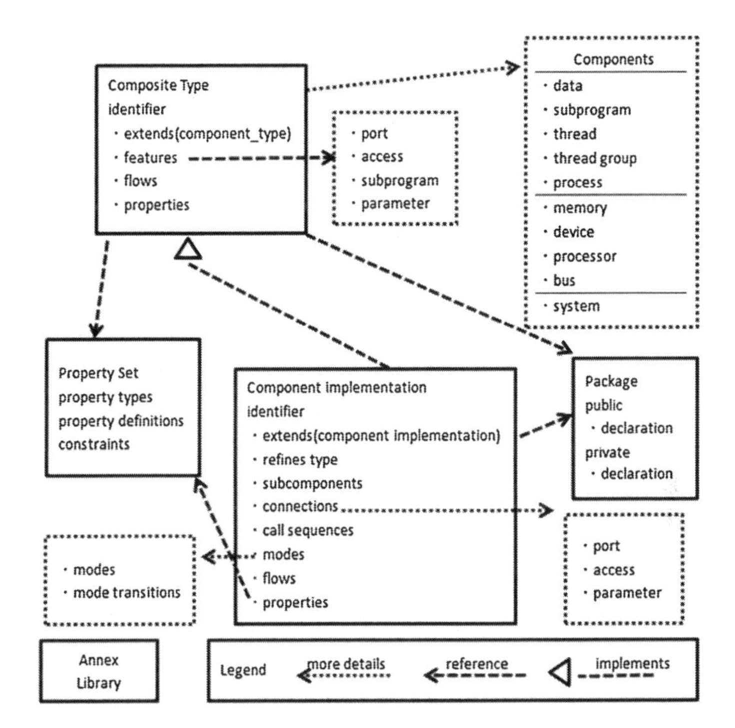

Fig. 1 AADL elements

approved as the industrial standard AS5506 in November 2004. Component is the most important concept in AADL. The main components in AADL are divided into three parts: software components, hardware components and composite components. Software components include data, thread, thread group, process and subprogram. Hardware components include processor, memory, bus and device. Composite components include system (Fig. 2).

In its conformity to the ADL definition, AADL provides support for various kinds of non-functional analyses along with conventional modeling [14].

Flow Latency Analysis: Understand the amount of time consumed for information flows within a system, particularly the end-to-end time consumed from a starting point to a destination.

Resource Consumption Analysis: Allows system architects to perform resource allocation for processors, memory, and network bandwidth and analyze the requirements against the available resources.

Real-Time Schedulability Analysis: AADL models bind software elements such as threads to hardware elements like processors. Schedulability analysis helps in examining such bindings and scheduling policies.

Safety Analysis: Checks the safety criticality level of system components and highlights potential safety hazards that may occur because of communication among components with different safety levels.

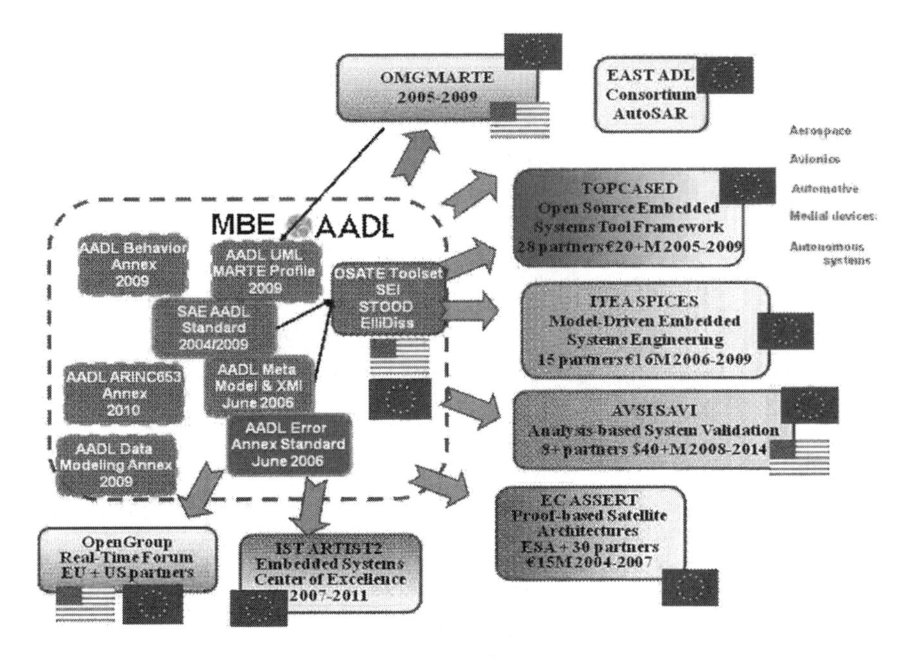

Fig. 2 Industry initiatives utilizing SAE AADL [13]

Security Analysis: Like safety levels, AADL components can be assigned various security levels. The analysis helps in identifying the security loopholes that may happen because of mismatches in security levels between a component and its subcomponents, and communication among components with different security levels.

AADL defines two main extension mechanisms: property sets and sublanguages (known as annexes). It is possible to extend the AADL concepts either by introducing new properties to the modeling elements, by addition of new modeling notations, or by developing a sublanguage as annex to the AADL standard Properties are label-value pairs used to annotate components. These properties can be grouped into named sets. These sets are then used in analysis tools that process AADL models to be able to verify characteristics of the modeled system. One example of a property set is the standard property set for RMA that contains period, execution time, and deadline among other properties. These properties are associated with threads to be able to derive a real-time task set amenable of timing analysis. Sublanguages, on the other hand, enable the encoding of complex statements about components for which syntactic verification makes sense. The syntax of the language is defined inside the annex that implements the language. Annexes and properties allow the addition of complex annotations to AADL models that accommodate the needs of multiple concerns. These annotations, along with their corresponding analysis plug-ins, provide a powerful combination for the architect to evaluate his/her design choices from different perspectives.

The extension mechanisms in AADL enable these perspectives to evolve in number and complexity as the knowledge on them also evolves [15, 16].

In this paper, we propose an approach to specify railway cyber physical systems based on AADL:

(1) Specify the physical world of railway cyber physical systems with Modelica [17–19].
(2) Combine SysML and AADL for the design, validation and implementation of railway cyber physical systems [20].
(3) Extend AADL to express the dynamic continous features by proposing new AADL annex.
(4) Specify spatial–temporal features by Cellular automata (CA) [21, 22].

Case Study: Specifying Communication Based Train Control Systems

It is known that CBTC system can be effectively split into four subsystems: automatic train supervision subsystem (ATS), zone control subsystem, vehicle on-board subsystem and data communication subsystem. The ATS subsystem, also named center control subsystem, includes a series of servers for different purposes, printers, displays, many workstations and so on. Via the data communication subsystem, the ATS subsystem gets respectively the information of databases, train position, wayside devices and movement authority from the database storage unit in data communication subsystem, vehicle on-board subsystem and zone control subsystem. After handling them, they are transferred to the correlated devices and subsystems. The vehicle on-board subsystem includes the vehicle on-board controller (VOBC), Driver Machine Interface (DMI) and so on. The sub-systems accepts many different types of data in zone control subsystem and ATS subsystem, then calculates the train movement curve, measures the train speed and movement distance associating with the guide way databases to protect the safety of train movement. The zone control subsystem contains zone controller, computer interlocking (CI) devices, axle counter, signals, platform doors, switches and other wayside devices. The subsystem gets and handles the useful data and statue information from other subsystems to generate the movement authority for the trains in control zones and update persistently as required. Then the subsystem transfers the movement authority to the vehicle on-board subsystem through the data communication subsystem in order to control the train movement. It also controls the switches, signals, and platform doors, and acknowledges the request of adjacent zone controller. The data communication subsystem mainly contains database storage unit, backbone fie-optical network, wayside access points, on-board wireless units and network switches. Data communication subsystem is the other subsystems communication bridge making normal communication between the subsystems and ensuring the safety of train operation. As shown in Fig. 3, the communications between the subsystems are via the data communication subsystem.

In OSATE environment, we will use the AADL to give the design and modeling of the CBTC system. The CBTC system's file structure is shown in Fig. 4.

Each subsystem corresponds to a file whose name ends with aadl. In view of the repeated use of some tools, the system provides a common tool file (Tools.aadl). Four subsystems are combined to form a complete CBTC system through a file (CBTCSystem.aadl). The four subsystems are achieved by using the system components in AADL, and the connections between the subsystems use the bus components. In AADL, components contain component type and component implementation. Component type specifies the external interfaces of component implementation, and component implementation describes the internal structure of the component. Each component implementation corresponds to a component type, a component type can have zero or more component implementations. The AADL code of CBTC system overall framework is as follows:

```
system CBTCSystem
end CBTCSystem;
system implementation CBTCSystem.Impl
  subcomponents
    ATSys: system ATSystem::ATSys.Impl;
    ZCSys: system ZCSystem::ZCSys.Impl;
    VOBSys: system VOBSystem::VOBSys.Impl;
    DCSys: system DCSystem::DCSys.Impl;
  connections
    conn1: bus access ATSys.toDCS -> DCSys.fromATS;
    conn2: bus access ZCSys.toDCS -> DCSys.fromZC;
    conn3: bus access VOBSys.ToDCS -> DCSys.fromVOBS;
    conn4: bus access DCSys.toATS -> ATSys.fromDCS;
    conn5: bus access DCSys.toZC -> ZCSys.fromDCS;
conn6: bus access DCSys.toVOBS -> VOBSys.FromDCS;
...
end CBTCSystem.Impl;
```

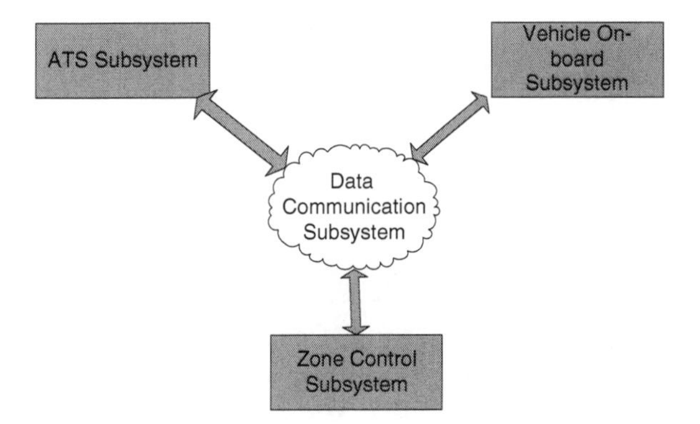

Fig. 3 The communication between the subsystems

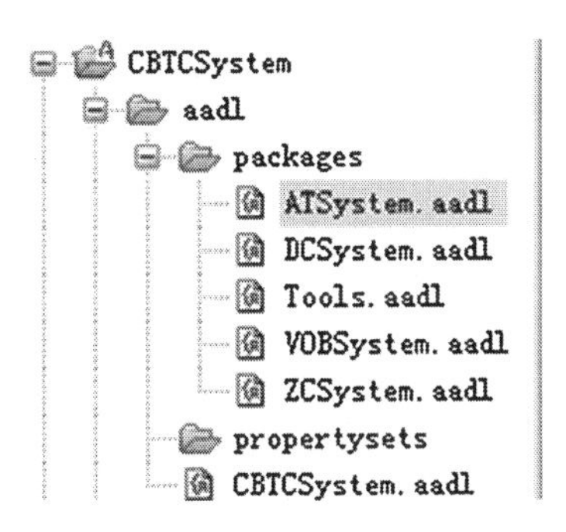

Fig. 4 CBTC system's file structure

As can be seen from the above code, the CBTC system contains four subsystems, data communication subsystem is a bridge of communication between the other three subsystems. OSATE will automatically generate an aaxl file for each aadl file, we can take advantage of aaxl file to new a corresponding graphics file (aaxldi file). The file lists the graphical representation of all components. And each implementation of the system components will generate a system instance diagram. The system instance diagram of the CBTC system implementation (CBTCSystem.Impl) is shown in Fig. 5.

The dynamic continuous features of the train is specified by AADL annexb as follows:

```
Device train
    Features
        In_data :in data port;
        Out_data: out data port;
    Properties
        Equation => {
```

$$F = 300 - 0.284v_t$$

$$W_0 = A + Bv_t + Cv_t^2;$$

$$W_i = 1000\tan\theta;$$

$$W_r = \frac{600}{R};$$

$$W_s = 0.00013 L_s;$$

$$W_f = W_i + W_r + W_s;$$

$$W1 = W_f * L_t;$$

$$B = M;$$

$$a = \frac{F - W_0 - W_1 - B}{(1+\gamma)M};$$

```
        };
        Const =>{M, γ};
        Const_value =>{M,γ};
        Var=>{vt,A,B,C,R,Ls,Lt, θ};
End Train;
```

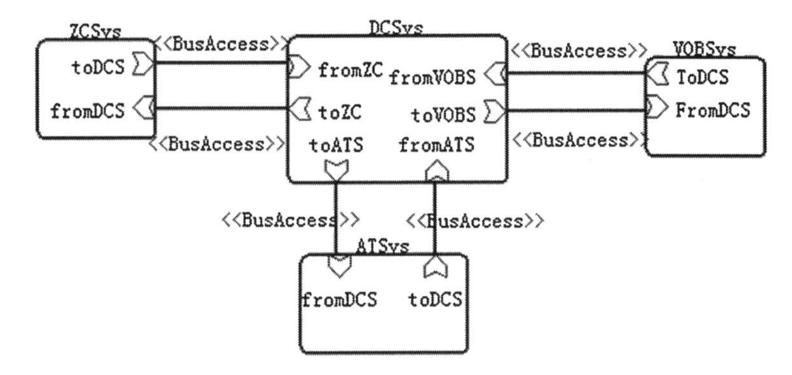

Fig. 5 System instance diagram of the of CBTC system implementation

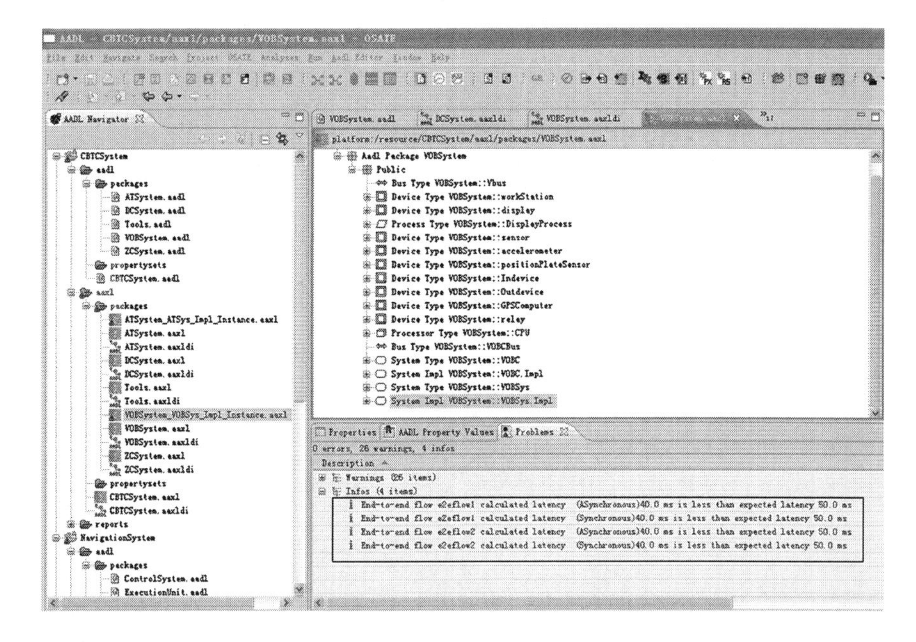

Fig. 6 End to end flow analysis of VOBS

The spatial requirements of railway cyber physical systems is specified as follows:

```
        dn: in parameter Types::float;
        dmin: in parameter Types::float;
        ltrain: in parameter Types::float;
        at: in parameter Types::float;
        bt: in parameter Types::float;
        vmax: in parameter Types::float;
        Xc: in parameter Types::float;
        Dgap: in parameter Types::float;
        vn: in parameter Types::float;
        xn: in parameter Types::float;
        vn: out parameter Types::float;
        xn: out parameter Types::float;
    end space;
    subprogram implementation space.default
        annex space_specification {**
    states
        s1: initial state;
        s2:return state;
    transitions
        normal: s1 -[ on 加速]-> s2 {
            if(isTrain){
                if(dn-ltrain>dmin){
                        vn=min(vn+at,vmax);
               }else if(dn-ltrain<dmin){
                  vn=max(vn-bt,0);
               }else{
                  vn=vn;
               }
               xn=xn+vn;
            }else{
                if( Dgap>Xc){
                        vn=min(vn+at,vmax);
               }else if( Dgap>Xc){
                  vn=max(vn-bt,0);
               } else{
                  vn=vn;
               }
               xn=xn+vn;
            }
        };
        normal: s1 -[on 减速 ]-> s2 {
            if(isTrain){
                    vn=min(vn,dn-ltrain-1);
                  xn=xn+vn;
              }else{
                    vn=min(vn,Dgap);
                  xn=xn+vn;
              }
          } ;
     **};
     end space.default;
```

The end to end flow analysis of VOBS is shown as Fig. 6.

Conclusion

In this paper, we proposed an approach to specify railway cyber physical systems based on AADL. We illustrated the proposed method by the specification of the communication based train control systems. The specification process of the communication based train control systems demonstrated the extension of AADL approach that can be used for modeling complex railway cyber physical system, effectively reduce the complexity of software development.

The further work is devoted to developing tools to support the automatic generation of model and code.

Acknowledgments This work is supported by Shanghai Knowledge Service Platform Project (No. ZF1213), national high technology research and development program of China (No. 2011AA010101), national basic research program of China (No. 2011CB302904), the national science foundation of China under grant (No. 61173046, No. 61021004, No. 61061130541, No. 91118008), doctoral program foundation of institutions of higher education of China (No. 20120076130003),national science foundation of Guangdong province under grant (No.S2011010004905).

References

1. IEC62278:2002 Railway applications: specification and demonstration of reliability, availability, maintainability and safety (RAMS)
2. IEC62279:2002 Railway applications: communications, signaling and processing systems– Software for railway control and protection systems
3. IEC62280:2002 Railway applications: communication, signaling and processing systems – Safety related electronic systems for signaling
4. Laprie C (1992) Dependability: basic concepts and terminology. Springer, Berlin
5. Svizienis A, Laprie JC, Randell B (2000) Dependability of computer systems: fundamental concepts, terminology, and examples. Technical report, LAAS-CNRS
6. Feiler PH, Gluch DP, Hudak JJ (2006) The architecture analysis and design language (AADL): an introduction. CARNEGIE-MELLON UNIV PITTSBURGH PA SOFTWARE ENGINEERING INST
7. Feiler PH, Lewis B, Vestal S et al (2005) An overview of the SAE architecture analysis and design language (AADL) standard: a basis for model-based architecture-driven embedded systems engineering. Springer, US, pp 3–15 (Architecture Description Languages)
8. Hudak JJ, Feiler PH (2007) Developing aadl models for control systems: a practitioner's guide
9. Feiler PH, Gluch DP (2012) Model-based engineering with AADL: an introduction to the SAE architecture analysis and design language. Addison-Wesley Professional
10. SAE AS-2C (2012) Architecture analysis and design language. SAE international document AS5506B(2012) Revision 2.1 of the SAE AADL standard
11. Delange J (2012) Towards a model-driven engineering software development framework. In: The third analytic virtual integration of cyber-physical systems workshop, 04 Dec 2012, Porto Rico
12. Feiler P, Hugues J, Sokolsky P (Eds) (2012) Oleg Architecture-driven semantic analysis of embedded systems. Dagstuhl Seminar 12272. Dagstuhl Report, 2(7):30–55. ISSN 2192-5283

13. The story of AADL (2012) AADL Wiki. Software Engineering Institute, 2010. Web. 06 Jan 2012
14. Muhammad N, Vandewoude Y, Berbers Y, van Loo S Modelling embedded systems with AADL: a practical study. www.intechopen.com/download/pdf/10732
15. de Niz D, Feiler PH Aspects in the industry standard AADL. In: AOM '07 Proceedings of the 10th international workshop on aspect-oriented modeling. pp 15–20
16. Michotte L, Vergnaud T, Feiler P, France R (2008) Aspect oriented modeling of component architectures using AADL. In: Proceedings of the second international conference on new technologies, mobility and security, 5–7 Nov 2008
17. Modelica Association (2002) Modelica—a unified object-oriented language for physical systems modelling. Language specification. Technical report
18. Modelica Association (2007) Modelica: A unified object- oriented language for physical systems modeling: language specification version 3.0. www.modelica.org
19. OMG OMG unified modeling language TM (OMG UML). Superstructure Version 2.2, February 20
20. De Saqui-Sannes P, Hugues J (2012) Combining SysML and AADL for the design, validation and implementation of critical systems. In: ERTSS 2012 (Embedded Real Time Software and Systems), Toulouse, France, 01–03 Feb 2012
21. Nagel K, Schreckenberg M (1992) A cellular automaton model for freeway traffic. Phys I France 2(12):2221–2229
22. Culik K, Hurd LP (1990) Formal languages and global cellular automaton behavior. Phys D 45(13):396–403

Formal Specification of Railway Control Systems

Bingqing Xu and Lichen Zhang

Abstract Train control systems must provide a high level of safety as they are a very important component and responsible for the safe operation of a train. To meet safety and reliability requirements, formal techniques must be used to specify train control systems. In this paper, we uses CSP, Object-Z and Clock to specify the Railway Control System concerning both the linear track and crossing area, especially the time delay between any two aspects of the railway system.

Keywords Railway control systems · Object-Z · CSP · Clock theory · Formal specification

Introduction

Train control systems must provide a high level of safety as they are a very important component and responsible for the safe operation of a train. To meet safety and reliability requirements, the relative international standards recommend the application of formal methods in specifying development specifications and design for train control systems [1, 2]. Complicated system such as railway control system is a system with many complex behavioral aspects. And the mechanism of communication between different aspects is hard to define. With help of formal methods, we now find a way to construct a detailed specification of each aspect and the link mechanism among various aspects. While a communication mechanism is not enough to describe the state change and data change in the system. Above all, the author tends to use Communicating Sequential Processes (CSP) [3] to specify the communication part of the Railway Control System. Concerning the

B. Xu · L. Zhang (✉)
Shanghai Key Laboratory of Trustworthy Computing, East China Normal University, Shanghai 200062, China
e-mail: zhanglichen1962@163.com

Y.-M. Huang et al. (eds.), *Advanced Technologies, Embedded and Multimedia for Human-centric Computing*, Lecture Notes in Electrical Engineering 260,
DOI: 10.1007/978-94-007-7262-5_46,

time characteristics in the system, Clock [4–7] specifies the system time requirements better. For the state and data changes, Object-Z is ideal for analysis in data change in a schema box form.

In this paper, we uses CSP, Object-Z and Clock to specify the Railway Control System concerning both the linear track and crossing area, especially the time delay between any two aspects of the railway system.

Relative Works

Hoenicke [8–12] uses a combination of three techniques for the specification of processes, data and time: CSP, Object-Z and Duration Calculus. The basic building block in our combined formalism CSP-OZ-DC is a class. First, the communication channels of the class are declared. Every channel has a type which restricts the values that it can communicate. There are also local channels that are visible only inside the class and that are used by the CSP, Z, and DC parts for interaction. Second, the CSP part follows; it is given by a system of (recursive) process equations. Third, the Z part is given which itself consists of the state space, the Init schema and communication schemas. For each communication event a corresponding communication schema specifies in which way the state should be changed when the event occurs. Finally, below a horizontal line the DC part is stated. The combination is used to specify parts of a novel case study on radio controlled railway crossings. Johannes Faber formally specifies a part of the European Train Control System (ETCS) with the specification language CSPOZ-DC treating the handling of emergency messages.

Gnesi et al. [13] described an important experiment in formal specification and validation, both performed in the context of an industrial project jointly performed by Ansaldobreda Segnalamento Ferroviario and CNR Institutes IEI and CNUCE of Pisa. Within this project they developed two formal models of a control system which is part of a wider safety–critical system for the management of medium-large railway networks. Each model describes different aspects of the system at a different level of abstraction. On these models they performed verification of both safety properties—in the hypothesis of Byzantine errors or in presence of some defined hardware faults—and liveness properties of a dependable communication protocols. The properties has been specified by means of assertions and temporal logical formulae. As a specification language we used Promela language while the verification was performed using the model checker Spin.

Hyun-Jeong Jo, Yong-Ki Yoon and Jong-Gyu Hwang proposed an eclectic approach to incorporate Z (Zed) formal language and 'Statemate MAGNUM', formal method tools using Statechart. Also they applied the proposed method to train control systems for the formal requirement specification and analyzed the specification results [1].

In Zafar paper [14], formal methods which are advanced software engineering techniques, in term of Z notation, are applied for the specification of critical components of automated train control system. At first graph theory is used for modeling of static components of the system and then integrated with Z notation to describe its entire state space. At first real topology is transferred to model topology in graph theory and then switches, crossings, and level crossing are formalized. At the end, these components are composed to define the entire interlocking system. Formal specification of the system is described in Z notation and the model is analyzed using Z/EVES tool.

Peleska et al. [15] motivate and illustrate an approach for domain-specific specification languages in the field of small local railway or tramway control systems. Adopting terms and concepts from the application domain, such languages are the ideal means of communication between users, system engineers and control computer specialists. Semantic rigour is provided by a transformation of the domain-specific representation into wide-spectrum formal specification languages. These transformations offer the possibilities of formal verification and testing against formal specifications, as well as automatic generation of executable programs from specifications. The construction of the transformation can be substantially simplified if the class of control systems is formalised in a generic way using design patterns and frameworks.

Haxthausen and Peleska [16] introduce the concept for a distributed railway control system and present the specification and verification of the main algorithm used for safe distributed control. Their design and verification approach is based on the RAISE method, starting with highly abstract algebraic specifications which are transformed into directly implementable distributed control processes by applying a series of refinement and verification steps. Concrete safety requirements are derived from an abstract version that can be easily validated with respect to soundness and completeness. Complexity is further reduced by separating the system model into a domain model and a controller model. The domain model describes the physical system in absence of control and the controller model introduces the safety-related control mechanisms as a separate entity monitoring observables of the physical system to decide whether it is safe for a train to move or for a point to be switched.

Guo Xie, Akira Asano, Sei Takahashi, Hideo Nakamura presents a formal specification of an Automatic Train Protection and Block (ATPB) model for local line railway system [17] and validates the model by internal consistency proving and systematic testing. The system consists of two parts, the on-board subsystem and ground subsystem. The former is to detect the basic state of train, such as position, speed and integrity, monitor the speed, communicate with ground equipment and record the relative events. And the latter is responsible for communicating with train, controlling the route and interlocking, and decision-making for train operation adjustment. The main purpose of this project is to improve the efficiency and guarantee that there is no collision, no derailment and no over speeding at the same. The formal language used in this project is VDM ++. And the state and specification of operation are all checked and validated using VDMTools. The results confirm the correctness of this system and the model throws new light on practical system design.

Bernardeschi et al. [18] outline an experience on formal specification and verification carried out in a pilot project aiming at the validation of a railway computer based interlocking system. Both the specification and the verification phases were carried out in the Just Another Concurrency Kit (JACK) integrated environment. The formal specification of the system was done by means of process algebra terms. The formal verification of the safety requirements was done first by giving a logical specification of such safety requirements, and then by means of model checking algorithms. Abstraction techniques were defined to make the problem of safety requirements validation tractable by the JACK environment.

The European Train Control System (ETCS) is a control system for the interoperability of the railways across Europe. A. Chiappini et al. report on the activities of the EuRailCheck project, promoted by the European Railway Agency [19], for the development of a methodology and tools for the formalization and validation of the ETCS specifications. Within the project, we achieved three main results. First, they developed a methodology for the formalization and validation of the ETCS specifications. The methodology is based on a three-phases approach that goes from the informal analysis of the requirements, to their formalization and validation. Second, they developed a set of support tools, covering the various phases of the methodology. Third, they formalized a realistic subset of the specification in an industrial setting. The results of the project were positively evaluated by domain experts from different manufacturing and railway companies.

Constance Heitmeyer and Nancy Lynch give a new solution to the Generalized Railroad Crossing problem [20], based on timed automata, invariants and simulation mappings, is presented and evaluated. The solution shows formally the correspondence between four system descriptions: an axiomatic specification, an operational specification, a discrete system implementation, and a system implementation that works with a continuous gate model.

Formal Specification of Train Control Systems

The Problem that must be addressed in operating a railway are numerous in quantity, complex in nature, and highly inter-related. For example, collision and derailment, rear-end, head-on and side-on collisions are very dangers and may occur between trains. Trains collide at level crossing. Derailment is caused by excess speed, wrong switch position and so on. The purpose of train control is to carry the passengers and goods to their destination, while preventing them from encountering these dangers. Because of the timeliness constraints, safety and availability of train systems, the design principles and implementation techniques adopted must ensure to a reasonable extent avoidance of design errors both in hardware and software. Thus, a formal technique relevant to design should be applied for train systems development. The purpose of our exercise is to apply aspect -oriented formal method s to develop a controller for train systems that tasks as input: a description of track configuration and a sequence of description of the moves of each of these trains.

The controller should take care of trains running over the track. It should control the safety of the configuration, i.e. No two trains may enter the critical section. When one critical section is occupied, some others, which share some part of section with this one, should be locked. The controlled can control the status, speed, position of trains.

In order to keep the description focused, we concentrate on some particular points in train control systems rather than the detailed descriptions of all development process. The specification is made by integrating Object-Z, CSP and Clock Theory.

Clock theory [4–7] puts forward the possibility to describe the event in physical world by using a clock, and can analyze, records the event by clock. To use clock to specify Cyber Physical Systems, the time description is clearer to every event and can link continuous world with discrete world better, the definition and linking mechanism of clock theory is provided as below.

Definition 1 *A clock c is an increasing sequence of real numbers. We define its low and high rates by*

$$\Delta(c) = df \quad \boldsymbol{inf}\{(c[i+1] - c[i])|i \in Nat\} \nabla(c) = df \quad \boldsymbol{sup}\{(c[i+1] - c[i])|i \in Nat\}$$

Here, c [1] and c' stand for the first element of c, and the resultant sequence after removal of c[1] from c respectively.

If c is a *healthy clock*, it does not speed up infinitely. Then

$$\Delta(c) > 0$$

If c runs faster than d if for all $i \in Nat$

$$c[i] \leq d[i]$$

then the relation of c and d can be denoted by c 主 d and this is a kind of partial order in clock.

Lemma

$$c \preccurlyeq c^{I}$$

Let c and d be clocks. We define the transition latency between the two clocks as

$$\rho(c,d) = df \quad \mathbf{sup}\{|c[i] - d[i]||i \in Nat\}$$

Lemma

$$\rho(c,d) \geq 0$$

Let e be an event. **clock**(e) denotes the clock that records the time instants when the event e occurs. And we use **clock(event**(c)) to denote the event that take place at every time instant c[i].

Definition 2 (Local Clock and Global Clock). *Let l be a label denoting a location, and c a clock. Then l:c denotes clock c that locates at l.*

$$l : c \preccurlyeq l : d = df \quad (c \preccurlyeq d)$$
$$\rho(l : c, l : d) = df \, \rho(c, d)$$

A global clock [l:c] is defined as an equivalent class of local clocks,

$$\rho([l1 : c1], [l2 : c2]) = df \quad \rho(l1 : c1, l2 : c2)$$

Definition 3 (Discrete Variable and Continuous Variable). *Some dynamic features of continuous variable can be described better.*

climb(u,r) is introduced to describe the time instants when the value of u rises up to r.
drop(u,r) is introduced to describe the times instants when the value of u falls below r.

Definition 4 (Linking Mechanism). *Here we assign continuous variables to discrete variables.*

$$x = u \; every \; c \; init \; x0$$

x is a discrete variable, u is a continuous variable, c is a clock with $c[1] \geq 0$, *and x0 is an initial value. In this equation, we assign the value of u to x at the every instant of clock c.*

A train controller limits the speed of the train, decides when it is time to switch points and secure crossings, and makes sure that the train does not enter them too early as shown in Fig. 1.

To extend state-based and behavioural techniques with real-time aspects, different approaches exist. One approach is unifying a state based language and Timed CSP, an extension of CSP with a time-out operator. In this paper, we take a different approach. We integrate CSP and Object-Z with clock theory to specify real-time aspects as shown in Figs. 2 and 9.

Rail crossing control is modeled as shown in Fig. 3.

Finally, woven model is shown as Fig. 4.

TrainController

position : [*p* ! : *Position*] **chan**
speed : [*s* ? : *Speed*] **chan**
t_now : [*t* ? : *Clock*] **local_chan**
t_out : [*t* ? : *Clock*] **chan**
t_wait : [*t* ? : *Clock*] **chan**
t_idle : [*t* ? : *Clock*] **chan**
record **:** [*r* ! : *record*]**chan**
state : [*st* ? : *State*] **chan**

Init

$t\_now = Clock_{now}$
$speed = 0$
$t\_out = 0$
$t\_wait = 0$
$t\_idle = 0$

sendPosition

pos ! **: *Position***

sendSpeed

speed ! **: *Speed***

updateClock

$\Delta(t\_out, t\_wait, t\_idle)$
t_out ? :***Clock***
t_wait? :***Clock***
t_idle? :***Clock***

$t\_out' = t\_out$
$t\_wait' = t\_wait$
$t\_idle' = t\_idle$

crossControl

$\Delta(speed, position)$
speed ?: *Speed*
position ?: *Position*

$speed' = speed \mid 0 < speed < max\_speed$
$position' = Position$

log

$\Delta(record)$

$record' = record$

Fig. 1 Model of train controler

ClockController[X]

$\uparrow$ ($Clock_{now}$, $Clock_{out}$,$Clock_{wait}$,$Clock_{idle}$, *Init*, *updateClock*)

$train_i$: ***Train***
$train_j$: ***Train***
t_pos_b : ***Track_Position***
t_pos_d : ***Track_Position***
t_now : ***Clock***
t_out : ***Clock***
t_wait : ***Clock***
t_idle : ***Clock***

Init

t_now = $Clock_{now}$
t_out = $Clock_{out}$
t_wait = $Clock_{wait}$
t_idle =$Clock_{idle}$

updateClock

Δ(*t_out*, *t_wait* ,*t_idle*)
$train_i$ **? :** ***Train***
$train_j$ **? :** ***Train***
t_pos_b **? :** ***Track_Position***
t_pos_d **? :** ***Track_Position***

t_out' : =$Clock_{now}$ + $Clock_{now}$- $Clock_{common}$
|$U\_C^{ij}_{fun\_pursued}$ Ú $U\_C^{ij}_{fun\_meet}$
t_idle' : = $Clock_{common}$ -$Clock_{now}$|$U\_C^{ij}_{fun\_pursued}$ Ú $U\_C^{ij}_{fun\_meet}$
t_wait' : = *Fun_Track*(*i*,*Track_Position*).$Clock_{wait}$-$Clock_{out}$
+$Clock_{idle}$ |$U\_C^{ij}_{fun\_pursued}$ Ú $U\_C^{ij}_{fun\_meet}$

Fig. 2 Model of clock

```
CrossController[X]
↑(count, Init, Leave, Enter)

  train : Train
  track_number : Track_number
  count : N
  switch : Switch

Init
  count = 0
  switch = down

Leaving
  Δ(count , switch )
  train ? : Train
  ---
  $count' : {count >0 Ù switch = up} = count -1

Leave
  Δ(switch )
  train ? : Train
  ---
  switch = up

Entering
  Δ(count , switch )
  train ? : Train
  ---
  $count' : {count < track_number Ù switch = up} = count +1

Enter
  Δ( switch )
  train ? : Train
  ---
  switch = up

Main ≙ Leave || Enter || Leaving || Entering
```

Fig. 3 Model of railway crossing control

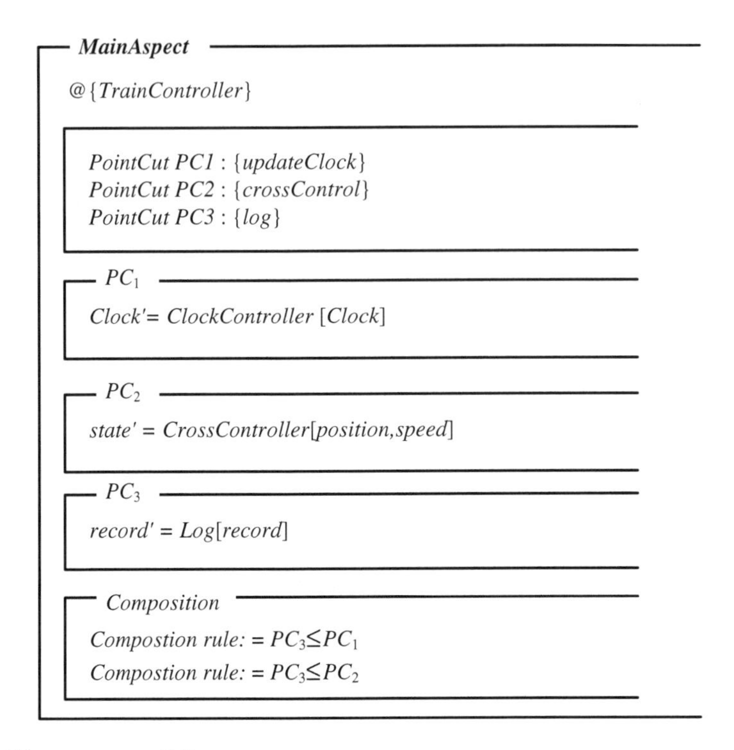

Fig. 4 Woven aspects of diagram

Conclusion

Train control systems must provide a high level of safety as they are a very important component and responsible for the safe operation of a train. To meet safety and reliability requirements, formal techniques must be used to specify train control systems. In this paper, we uses CSP, Object-Z and Clock to specify the Railway Control System concerning both the linear track and crossing area, especially the time delay between any two aspects of the railway system.

Future work focuses on the verification tool development of our proposed method.

Acknowledgments This work is supported by Shanghai Knowledge Service Platform Project (No.ZF1213), national high technology research and development program of China (No.2011AA010101), national basic research program of China (No.2011CB302904), the national science foundation of China under grant (No.61173046, No.61021004, No.61061130541, No.91118008), doctoral program foundation of institutions of higher education of China (No.20120076130003),national science foundation of Guangdong province under grant (No.S2011010004905).

References

1. Jo H-J, Yoon Y-K, Hwang J-G (2009) Analysis of the formal specification application for train control systems. J Electr Eng Technol 4(1):87–92
2. IEC62278:2002 Railway applications: Specification and demonstration of reliability, availability, maintainability and safety (RAMS)
3. Reed GM, Roseoe AW (1986) A timed model for communicating sequential processes. Pro ICALP'86. Lecture notes in computer science. Springer, Berlin
4. He J (2013) A clock-based framework for constructions of hybrid systems. Key talk. In the Proceedings of ICTAC'2013
5. Xu B, He J, Zhang L (2013) Specification of cyber physical systems based on clock theory. Int J Hybrid Inf Technol 6(3):45–54
6. Xu B et al (2013) Specification of cyber physical systems by clock. AST2013. ASTL 20: 111–114, Yeosu, South Korea
7. He J (2012) Link continuous world with discrete world. Shanghai Key Laboratory of Trustworthy Computing East China Normal University, China
8. Hoenicke J Specification of Radio based railway crossings with the combination of CSP, OZ, and DC. http://citeseerx.ist.psu.edu/viewdoc/summary?doi=10.1.1.21.4394
9. Hoenicke J (2006) Combination of processes, data, and time. PhD thesis, University of Oldenburg
10. Hoenicke J, Maier P (2005) Model-checking of specifications integrating processes, data and time. In: Fitzgerald JS, Hayes IJ, Tarlecki A (eds) FM 2005, volume 3582 of LNCS, Springer, pp 465–480
11. Hoenicke J, Olderog E-R (2002) CSP-OZ-DC: a combination of specification techniques for processes, data and time. Nordic J Comput 9(4):301–334
12. Hoenicke J, E-R Olderog (2002) Combining specification techniques for processes data and time. In: Butler M, Petre L, Sere K (eds) Integrated formal methods, volume 2335 of lecture notes in computer science, Springer, pp 245–266
13. Gnesi S, Latella D, Lenzini G, Amendola A, Abbaneo C, Marmo P (2000) A formal specification and validation of a safety critical railway control system. In: Fifth international workshop on formal methods for industrial critical systems, FMICS 2000, Berlin, Germany, April 3–4 2000
14. Zafar NA (2006) Modeling and formal specification of automated train control system using Z notation. Multitopic conference. INMIC '06. IEEE, pp 438–443, 23–24 December 2006
15. Peleska J Baer A, Haxthausen AE Towards domain-specific formal specification languages for railway control systems.http://www.informatik.uni-bremen.de/agbs/jp/papers/trans2000.html
16. Haxthausen AE, Peleska J (2000) Formal development and verification of a distributed railway control system. IEEE Trans Software Eng 26(8):687–70
17. Xie G, Asano A, Sei Takahashi, Hideo Nakamura, (2011) Study on formal specification of automatic train protection and block system for local line. ssiri-c, Fifth international conference on secure software integration and reliability improvement—companion, pp 35–40
18. Bernardeschi C, Fantechi A, Gnesi S, Larosa S, Mongardi G, Romano D (1998) A formal verification environment for railway signaling system design. Formal Methods Syst Design 12:139–161
19. Chiappini A et al (2010) Formalization and validation of a subset of the European train control system. ICSE'10, Cape Town, South Africa, 2–8 May 2010
20. Heitmeyer C, Lynch N (1994) The generalized railroad crossing: a case study in formal verification of real-time systems. In: Proceedings of real-time systems symposium, pp 120–131

A Clock Based Approach to the Formal Specification of Cyber Physical Systems

Bingqing Xu and Lichen Zhang

Abstract Many cyber-physical systems are under real-time constraints, and thus a critical research challenge is how to ensure time predictability of cyber physical systems. The paper describes the case studies in applying clock theory to the industrial problems. The clock theory described is very simple, in that it models clocks as potentially infinite lists of reals. Xeno's paradox and similar problems are avoided by specifying limits on clock rates, which effectively means that the model sits somewhere between a discrete synchronous model and a fully dense continuous time model as assumed by some other formalisms. Case studies show that using clock theory to specify cyber physical systems can give a more detailed description of the every subsystem and give a much more considerate observation of the time line and sequence of every event.

Keywords Cyber physical systems · Continuous and discrete · Clock · Time analysis

Introduction

Cyber physical systems related research is based on two, originally different world views: on the one hand the dynamics and control (DC) world view, and on the other hand the computer science (CS) world view [1]. The DC world view is that of a predominantly continuous-time system, which is modeled by means of differential (algebraic) equations, or by means of a set of trajectories. The evolution of a hybrid system in the continuous-time domain is considered as a set of piecewise continuous functions of time. The CS world view is that of a predominantly discrete-event system. A well-known model is a (hybrid) automaton, but modeling of discrete-event systems is also based on. As new CPS applications start to interact with the

B. Xu · L. Zhang (✉)
Shanghai Key Laboratory of Trustworthy Computing, East China Normal University, Shanghai 200062, China
e-mail: zhanglichen1962@163.com

Y.-M. Huang et al. (eds.), *Advanced Technologies, Embedded and Multimedia for Human-centric Computing*, Lecture Notes in Electrical Engineering 260,
DOI: 10.1007/978-94-007-7262-5_47,

physical world using sensors and actuators, there is a great need for ensuring that the actions initiated by the CPS is timely. This will require new time analysis functionality and mechanisms for CPS. Cyber physical systems are characterized by their stringent requirements for time constraints such as predictable end-to-end latencies, timeliness. Ensuring the time constraint needs of cyber physical systems entails the need to specify and analyze timing requirements correctly.

Since Cyber Physical Systems are dynamic systems that exhibit both continuous and discrete dynamic behavior, and the continuously bilateral interaction between discrete events and continuous time flow makes it hard to know the dynamic feature of the system. So specifying the timing issues is a really vital work in the early stage. This paper describes the case studies in applying clock theory [2, 3] to the industrial problems. The clock theory described is very simple, in that it models clocks as potentially infinite lists of reals. Xeno's paradox and similar problems are avoided by specifying limits on clock rates, which effectively means that the model sits somewhere between a discrete synchronous model and a fully dense continuous time model as assumed by some other formalisms. Case studies show that using clock theory to specify cyber physical systems can give a more detailed description of the every subsystem and give a much more considerate observation of the time line and sequence of every event.

Related Works

The UML profile for Modeling and Analysis of Real-Time and Embedded (Marte) [4] systems has been adopted by the OMG earlier this year. Marte supersedes the UML Profile for Schedulability, Performance and Time (SPT) [5] and extends the mainly untimed uml with several new constructs. Amongst several other things, Marte proposes a new resource model, an extensible way to express non functional properties and a time model. The time model adapts SPT timing mechanisms to the Unified Modeling Language (uml2) simple time and offers completely new features, like the support of logical and multiform time. The Clock Constraint Specification Language (CCSL) [6] defines a set of time patterns between clocks that apply to infinitely many instant relations. A CCSL specification consists of clock declarations and conjunctions of clock relations between clock expressions. A clock expression defines a set of new clocks from existing ones. Most expressions deterministically define one singleclock. In the paper of Lamport [7], concept of "happening before" defines an invariant partial ordering of the events in a distributed multiprocess system. He described an algorithm for extending that partial ordering to a somewhat arbitrary total ordering, and showed how this total ordering can be used to solve a simple synchronization problem. The representation of a closed finitary real-time system as a graph annotated with clock constraints is called a timed automaton [8], since clocks range over the nonnegative reals, every nontrivial timed automaton has infinitely many states. If the clocks of a finitary real-time system are permitted to drift with constant, rational drift bounds, one obtains a finitary drifting-clock system. The representation of a closed finitary drifting-clock system as a graph annotated with constraints on drifting clocks is called an initialized rectangular automaton [9]. Two popular specification languages for the algorithmic verification of untimed systems are finite automata and propositional temporal logics. In order to specify timing constraints, these languages can be extended by adding clock variables. If we judiciously add clocks to finite automata, One obtains the timed automata (TA); from propositional linear temporal logic, one obtains the real-time logic TPTL [10]; from the propositional, branching-time logic CTL, one obtains the real-time logic TCTL [11].

Silva and Krogh [12] introduced sampled data hybrid automata (SDHA) as a formal model of hybrid systems that result from clock-driven computer control of continuous dynamic systems. In contrast to standard hybrid automata, the discrete state transitions in the SDHA can occur only at valid sampling times when the guard conditions are evaluated. Sequences of valid sampling times are defined by a clock structure that specifies bounds on the possible initial phases, period variations and jitter. Approximate quotient transition systems are then defined for SDHA as a theoretical framework for performing formal verification.

Bujorianu et al. [13] presented a multiclock model for real time abstractions of hybrid systems. They call Hybrid Time systems the resulting model, which is constructed using category theory. Such systems are characterized by heterogeneous timing, some components having discrete time and others continuous time. They define a timed (or clock) system as a functor from a category of states to a category of time values. They further define concurrent composition operators and bisimulation.

Clock Theory

Clock theory [2, 14, 15] puts forward the possibility to describe the event in physical world by using a clock, and can analyze, records the event by clock. To use clock to specify Cyber Physical Systems, the time description is clearer to every event and can link continuous world with discrete world better, the definition and linking mechanism of clock theory is provided as below.

Definition 1 *A clock c is an increasing sequence of real numbers. We define its low and high rates by*

$$\Delta(c) =df\, \boldsymbol{inf}\{(c[i+1]-c[i])|i \in Nat\}\nabla(c)$$
$$=df\, \boldsymbol{sup}\{(c[i+1]-c[i])|i \in Nat\}$$

Here, c[1] and c' stand for the first element of c, and the resultant sequence after removal of c[1] from c respectively.

If c is a *healthy clock*, it does not speed up infinitely. Then

$$\Delta(c) > 0$$

If c runs faster than d if for all $i \in Nat$

$$c[i] \leq d[i]$$

then the relation of c and d can be denoted by c 王 d and this is a kind of partial order in clock.

Lemma

$$c \preccurlyeq c^{I}$$

Let c and d be clocks. We define the transition latency between the two clocks as

Lemma

$$\rho(c, d) = df\ \mathbf{sup}\{|c[i] - d[i]||i \in Nat\}$$

$$\rho(c, d) \geq 0$$

Let e be an event **clock**(e) denotes the clock that records the time instants when the event e occurs. And we use **clock**(**event**(c)) to denote the event that take place at every time instant c[i].

Definition 2 (Local Clock and Global Clock). *Let l be a label denoting a location, and c a clock. Then l:c denotes clock c that locates at l.*

$$\begin{aligned} l : c \preccurlyeq l : d =&df \quad (c \preccurlyeq d)\rho(l : c, l : d) \\ =&df \quad \rho(c, d) \end{aligned}$$

A global clock [l:c] is defined as an equivalent class of local clocks,

$$\rho([l1 : c1], [l2 : c2]) = df \quad \rho(l1 : c1, l2 : c2)$$

Definition 3 (Discrete Variable and Continuous Variable). *Some dynamic features of continuous variable can be described better.*

climb(u,r) is introduced to describe the time instants when the value of u rises up to r. drop(u,r) is introduced to describe the times instants when the value of u falls below r.

Definition 4 (Linking Mechanism). *Here we assign continuous variables to discrete variables.*

$$x = u\ every\ c\ init$$

$$x0$$

x is a discrete variable, u is a continuous variable, c is a clock with c[1] $\geq$ 0, *and* x0 *is an initial value. In this equation, we assign the value of u to x at the every instant of clock c.*

In differential equation, u is a continuous variable, and f is an expression.

$$\dot{u} = f \ init \ u0$$

sets up the relation between u and f

$$(\dot{u} = f) \wedge (u(0) = u0)$$

Implementation of Clock Function by Modelica

We use Modelica to implement the clock functions: climb(u,r), drop(u,r), cross(u, r). climb(u,r) and drop(u,r) is coded as follows:

```
( climb function )
function climb
  input Time stime;
  input Real r;
  input Time etime;
  input Time s;
  output Boolean b;
algorithm
  if u_time(stime) > r then
    b := false;
    return;
  end if;
  if _ut(true, s, etime) == r and u__t(true, s, stime) == r then
    b := true;
    return;
  end if;
end climb;

( drop function )
function drop
  input Time stime;
  input Real r;
  input Time etime;
  input Time s;
  output Boolean b;
algorithm
  if u_time(stime) < r then
    b := false;
    return;
  end if;
  if _ut(false, s, etime) == r and u__t(false, s, stime) == r then
    b := true;
    return;
  end if;
end drop;
```

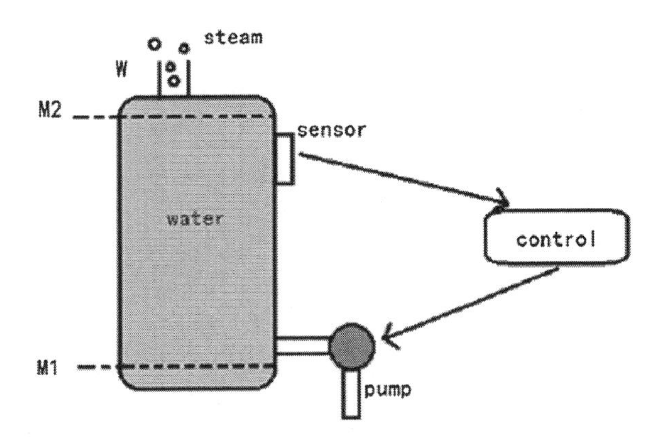

Fig. 1 Steam boiler control system

Case Study: Specification of the Steam Boiler Control System Based on Clock Theory

As Fig. 1 shows, the steam boiler control system consists of two parts. One is the steam boiler which acts continuously, and the other is the discrete control part which gives the command when it receives messages from sensors. To steam boiler, it has got limits of water level such as minimal water level and maximal water level, and the water level influences the steam rate directly since little water cannot produce steam and too much water is likely to break the limit of steam rate. Sensor collects the information of water level and steam rate and transfers the message to the controller. To the controller, it gives discrete commands to change the quantity of pumps so as to control the water volume and steam rate, and it has got some commands to reply the sensor fault and instrument fault. In addition, this paper only focuses on the basic steam boiler control system, but cannot tolerate the container volume expansion which is caused by heat etc. All the parameters are listed in Table 1.

Table 1 Parameter list for steam boiler control system

Parameter	Value
υ_p	Current speed of water
υ_s	Current speed of steam
V_w	Current volume of water
e	Volume of incoming water
z	Volume change of water
M_1	Minimal volume of water
M_2	Maximal volume of water
C	Capacity of steam boiler
W	Limit of steam rate

To guarantee the safety of the control system, the specification should obey the next two rules. The current steam rate is always less than the highest rate W and the maximal volume of water cannot exceed the capacity of the steam boiler.

$$0 < \upsilon_s < W$$
$$0 < M_1 < M_2 < C$$

First of all, we consider the water level; e}is the continuous variable which denotes the volume of incoming water; minus_pump} and add_pumpare events which control the quantity of pumps in order to adjust the steam rate. As the water volume is approaching the maximum, the controller is ready to make some pumps to stop, and the two events have time difference and sequence.

$$climb(e, M_2 - z) \preccurlyeq clock(minus\_pump)$$
$$\rho(climb(e, M_2 - z),\ clock(minus\_pump)) \le z/\upsilon_p$$
$$drop(e, M_1 + z) \preccurlyeq clock(add\_pump)$$
$$\rho(drop(e, M_1 + z),\ clock(add\_pump)) \le z/\upsilon_p$$
$$z \le min(M_2\ -\ V_w,\ V_w\ -\ M_1)/2$$

Since the water level can influence the steam rate, the following relationship is similar.

$$clock(minus\_pump) \preccurlyeq clock(low\_steam)$$
$$\rho(clock(minus\_pump),\ clock(low\_steam)) \le z/\upsilon_p$$
$$clock(add\_pump) \preccurlyeq clock(high\_steam)$$
$$\rho(clock(add\_pump),\ clock(high\_steam)) \le z/\upsilon_p$$

And it is vital that some couples of events in the equations above have noninterference.

$$clock(add\_pump)[1] > 0$$
$$clock(add\_pump),\ \preccurlyeq clock(minus\_pump) \preccurlyeq lock(add\_pump)'$$
$$clock(add\_pump) \wedge clock(minus\_pump) = \phi$$
$$clock(high\_steam) \wedge clock(low\_steam) = \phi$$

There are also some timing issues in the system, since sensor and controller all need time to transfer the message and command. sensor_delay and control_delay denote the time consumed to transfer information.

$$(clock(high_{water}) \preccurlyeq clock(sensor\_delay))$$
$$\le clock(minus\_pump) \preccurlyeq clock(control\_delay) \preccurlyeq clock(low\_steam)$$
$$clock(low_{water}) \preccurlyeq clock(sensor\_delay)$$
$$\le clock(add\_pump) \preccurlyeq clock(control\_delay) \preccurlyeq clock(high\_steam)$$

And no matter which part meets problem, the controller is fault-tolerant.

$$(clock(sensor\_error) \vee clock(pump\_error)) \\ \preccurlyeq clock(control\_stop) \\ \preccurlyeq clock(SYSTEM\_STOP)$$

In the equations, Ådenotes continuous speed change, and vp denotes discrete speed at each clock unit. The continuous variable and discrete variable can be linked as below:

$$\dot{e} = \upsilon_p \; init \; \upsilon_{po}$$
$$\upsilon_p = e \; every \; c \; init \; \upsilon_{po}$$

Conclusion

The paper presented the case study in applying clock theory to the industrial problems. Case study shows that using clock theory to specify cyber physical systems can give a more detailed description of the every subsystem and give a much more considerate observation of the time line and sequence of every event.

It is brilliant to connect the event with clock, while it is so difficult to handle so many local clocks with the global clock. In my point of view, it is always hard work to make local clocks keeping consistent with the global clock, and the verification of the security and accuracy of the synchronization is very complicated, and we need more ideas to do this work.

Acknowledgments This work is supported by Shanghai Knowledge Service Platform Project (No. ZF1213), national high technology research and development program of China (No. 2011AA010101), national basic research program of China (No. 2011CB302904), the national science foundation of China under grant (No. 61173046, No. 61021004, No. 61061130541, No. 91118008), doctoral program foundation of institutions of higher education of China (No. 20120076130003), national science foundation of Guangdong province under grant (No. S2011010004905).

References

1. Man KL, Schiffelers RRH (2006) Formal specification and analysis of hybrid systems. Universiteitsdrukkerij Technische Universiteit Eindhoven, ISBN-10: 90-386-2997-4
2. He J (2013) A clock-based framework for constructions of hybrid systems. Key talk. In the proceedings of ICTAC'2013
3. He J (2012) Link continuous world with discrete world. Shanghai Key Laboratory of Trustworthy Computing East China Normal University, China
4. Object Management Group (2009) UML profile for MARTE, v1.0.formal/2009-11-02
5. Object Management Group (2005) UML profile for schedulability, performance, and time specification. OMG document:formal/05-01-02 (v1.1)

6. Mallet F, DeAntoni J, AndrÅ C, de Simone R (2010) The clock constraint specification language for building timed causality models. Innovations Syst Softw Eng 6:99–106
7. Lamport L (1978) Time, clocks, and the ordering of events in a distributed system. Commun ACM 21(7):558–565
8. Alur R, Dill DL (1994) A theory of timed automata. Theor Comput Sci 126:183–235
9. Henzinger TA, Kopke PW, Puri A, Varaiya P (1995) What's decidable about hybrid automata? In: Proceedings of the 27th annual symposium on theory of computing, ACM Press, pp 373–382
10. Alur R, Henzinger TA (1994) A really temporal logic. J ACM 41(1):181–204
11. Alur R, Courcoubetis C, Dill DL (1993) Model checking in dense real time. Inf Comput 104 (1):2–34
12. Silva BI, Krogh BH Modeling and verification of sampled-data hybrid systems. www.ece.cmu.edu/~krogh/checkmate/.../adpm_sampled_data.p
13. Bujorianu MC, Bujorianu LM, Langerak R (2008) An interpretation of concurrent hybrid time systems over multi-clock systems. In: Proceedings of the 17th IFAC world congress, Seoul, Korea, 6–11 July 2008
14. Xu B, He J, Zhang L (2013) Specification of cyber physical systems based on clock theory. Int J Hybrid Inf Technol 6(3):45–54
15. Xu B et al (2013) Specification of cyber physical systems by clock. AST2013.Yeosu, South Korea, ASTL, vol 20, pp 111–114
16. Leeb G, Lynch N (1996) Proving safety properties of the steam boiler controller, lecture notes in computer science. 1165:318–338
17. Abrial J, Bger E, Langmaack H (eds) (1996) Formal methods for industrial applications—specifying and programming the steam boiler control, lecture notes in computer science, vol 1165, Springer, Berlin
18. Duval G, Cattel T (1996) Specifying and verifying the steam boiler problem with SPIN. In: Abrial J-R, Bger E, Langmaack H (eds) Formal methods for industrial applications—specifying and programming the steam boiler control, lecture notes in computer science, vol 1165. Springer, Berlin, pp 203–217
19. Willig A, Schieferdecker I (1996) Specifying and verifying the steam boiler control system with time extended LOTOS. In: Abrial J-R, Bger E, Langmaack H (eds) Formal methods for industrial applications—specifying and programming the steam boiler control, lecture notes in computer science, vol 1165. Springer, Berlin, pp 473–492
20. Carreira PJF, Costa MEF (2003) Automatically verifying an object-oriented specification of the steam-boiler system. Sci Comput Program 46:197–217

A Genetic-Based Load Balancing Algorithm in OpenFlow Network

Li-Der Chou, Yao-Tsung Yang, Yuan-Mao Hong, Jhih-Kai Hu and Bill Jean

Abstract Load balancing service is essential for distributing workload across server farms or data centers and mainly provided by dedicated hardware. In recent years, the concept of Software-Defined Networking (SDN) has been applied successfully in the real network environment, especially by OpenFlow designs. This paper presents an OpenFlow-based load balancing system with the genetic algorithm. This system can distribute large data from clients to different servers more efficiently according to load balancing policies. Furthermore, with the pre-configured flow table entries, each flow can be directed in advance. Once the traffic burst or server loading increased suddenly, the proposed genetic algorithm can help balance workload of server farms. The experiments demonstrate the better performance of the proposed method compared to other approaches.

Keywords Load balancing · Genetic algorithm · OpenFlow · Software-defined networking

Introduction

As growth of the Internet, server overloads always happen due to network traffic of excessive requests or malicious attacks increasing unexpectedly. Load balancing service is a necessary option to ease such burden or sufferings. It can distribute the incoming workload over server farms of duplicate servers to work for more clients.

L.-D. Chou (✉) · Y.-T. Yang · Y.-M. Hong
Department of Computer Science and Information Engineering,
National Central University, Taoyuan, Taiwan, Republic of China
e-mail: cld@csie.ncu.edu.tw

J.-K. Hu · B. Jean
Xinguard Company, Limited, Taipei, Taiwan, Republic of China
e-mail: openflow@xinguard.com

Y.-M. Huang et al. (eds.), *Advanced Technologies, Embedded and Multimedia for Human-centric Computing*, Lecture Notes in Electrical Engineering 260,
DOI: 10.1007/978-94-007-7262-5_48, © Springer Science+Business Media Dordrecht 2014

Such operations can be done through the process of rewriting the destination of header, and applied by several methods, such as random, round robin, or load-based, to keep the accesses with lower latency and higher throughput, even if more service requests submitted. However, load balancing function is always provided by dedicated software and hardware, which makes it less flexible.

In recent years, the idea of Software-Defined Networking (SDN) has prevailed and practiced successfully in several network environments [1]. OpenFlow is an open source software to implement the SDN architecture and a potential approach to provide abstract elements to rebuild or separate the network topology. The major purpose of using OpenFlow is to increase the flexibility of network management and traffic routing. Moreover, its programmable control interface can help quickly and easily deploy different policies around network devices on the fly instead of redundant manual configuration.

Therefore, we hope through using OpenFlow can manage and deploy the applications which satisfied with our demand to improve the performance of network transmission, raise the flexibility of packet transmission, simplify the network complexity and increase the network management capability even to define the self-network by ourselves.

In this paper, we use Pica8 Open Switch [2] to be the device which support OpenFlow and Open vSwitch (OVS) integration. And we deploy application control programs that using policy-based load balancing methods to achieve OpenFlow-based load balancing and raising the flexibility for network data transmission. We hope to achieve better load balancing performance through the OpenFlow capability and toward to the goal of Software Defined Networking finally.

This paper describes related work in Section Related Work, system design in Section System Design, and experimental results In Section Experimental Results. Finally in Section Conclusion and Future Work, the conclusion and future work are remarked.

Related Work

In this section, load balancing is first introduced; then the Open vSwitch and OpenFlow Controller are described in the final.

SDN is the innovative network architecture that provides a quick deployment instead of manually configuring policies. And OpenFlow is the best solution for achieving the objective of SDN now.

Load Balancing

Cloud computing becomes more and more popular in recent years, the reasons include not only the convenience for end users but also the virtualization technology and distributed computation. The number of computing nodes will also increase quickly because the need for support big data distributed processing. Therefore, it's much important for load balancing in the cloud network that distribute the traffic to separate servers. In cloud network architecture, the network traffic increase because of the big data from clients. So the single server can't afford the capability to process the large traffic. And in order to avoid wasting the resources with servers, load balancing technology was provided [3].

For network applications, it's necessary for load balancing and when the traffic are too large to handle, and we can also cut down the traffic load because of it. In [4], we can use the central load balancing decision module. It's not only increasing the decision speed but also allow single point of failure for the virtual machine. In [5], it present Plug-n-Server system which control the traffic load by OpenFlow-based routing methods in order to minimize the response time from client to server. But this system didn't consider the advanced load balancing algorithms to increase the network performance. On the other hand, someone also present the novel ideal for improving the bad performance for OpenFlow controller because the huge of flows caused by separating flows for each client connection [6]. The advantage for this paper is avoiding the cost for processing the huge of flows. On the contrary, there are scant physical OpenFlow switches support the module features presented by this paper as its drawbacks. And in OpenFlow-based load balancing technology, some researches present the comparison between simple load balancing algorithms such as random choice, time slice based choice and weighted balancing [7]. And from the research [8], it designs the load balancing architecture which apply in web services applications and it also presents three load-balancing policies which compared with our Genetic algorithm-based load balancing algorithm in this paper. The first is random-based load balancing policy that selects the registered servers randomly. Secondly, it uses the round robin load balancing policy to rotates the registered servers to serve the requests. Last, by using load-based load balancing algorithm which chooses the server with the current lowest load to serve the requests. In [9], it uses multiple OpenFlow controllers to distribute the different services that integrating the network and the load balancing functionality to reduce the maintenance effort. For example, one controller handles the load for e-mail servers and the other one processes the load for network traffic. It increases the efficiency by separating the load for multiple controllers which implemented by the open source project named FlowVisor [10].

Open vSwitch and OpenFlow Controller

OpenFlow is an open standard originated from the Clean slate project in Stanford University. And Open vSwitch (OVS) is a multilayer virtual switch licensed under the open source Apache 2.0 license which is also an Ethernet switch that has flow table inside to insert and delete flow entries [11]. It can operate both as a soft switch running within the hypervisor, and as the control stack for switching silicon.

Currently, it has many OpenFlow controllers which released with open source project, such as Beacon [12], NOX [13], POX [14], Helios, BigSwitch and so on. OpenFlow Controllers communicate to the OpenFlow switches by the secure channel which defined by OpenFlow protocol.

System Design

The proposed OpenFlow-based load balancing system is designed for redirecting the flows in order to balance the loads for the servers in load balancing pool and mirror the packets which have the abnormal traffic to backend detection server.

Figure 1 shows the OpenFlow-based load balancing system architecture, It separates two basic components: OpenFlow switch and OpenFlow controller. Two of components have multiple modules inside which perform its own functions and communicate each other.

In this paper, we propose an intelligent load balancing algorithm using Genetic Algorithm (GA). We assume each client sends multiple requests and the servers also have different workload on it. So the flow redirection problem become NP-Complete problem that we hope to find the best order for redirecting to servers to modify the flow rules. Suppose that we have *N* flows to do pre-configured, and

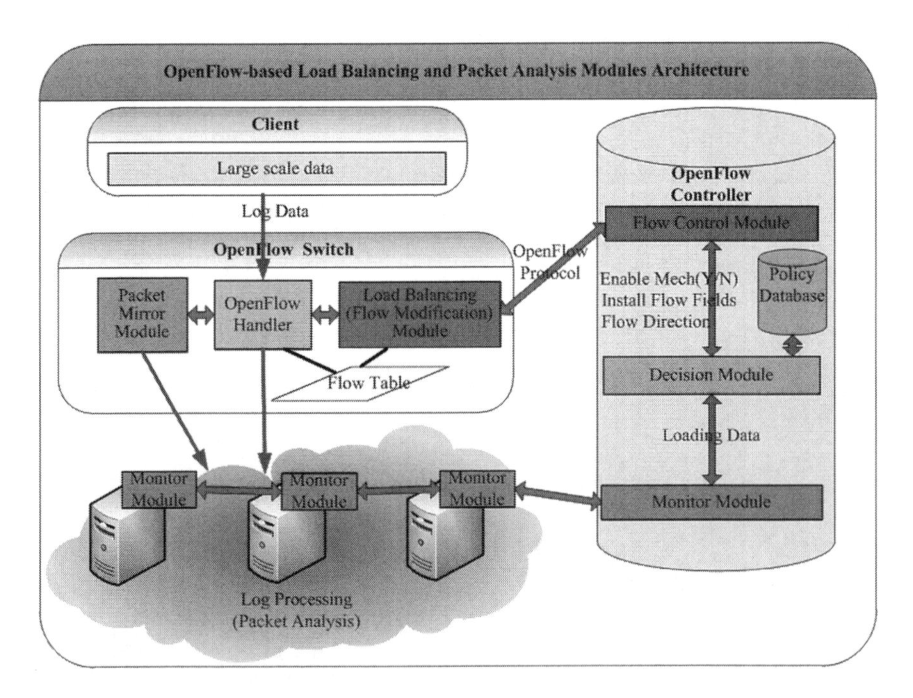

Fig. 1 System architecture

each flow have different load which noted the set of Ω. On the other hand, we have K servers to receive the data and each server has different workload which noted the set of X. Moreover, the flow redirection problem is considered with the objective of minimizing the server's coefficient of variation for traffic in OpenFlow environment. For proposed genetic-based load balancing algorithm in this paper, we define the fitness function by the coefficient of variation for server's traffic as the following formula:

$$\text{Min} \frac{\sqrt{\left(\left(\sum_{j=1}^{K} X[j]^2\right) - \left(\left(\sum_{j=1}^{K} X[j]\right)/K\right)^2\right)/K}}{\left(\sum_{j=1}^{K} X[j]\right)/K} \tag{1}$$

For the evolution processes in proposed algorithm, we choose the roulette wheel selection for reproduction, single point crossover and single point mutation.

Experimental Results

In order to compare the performance for proposed load balancing algorithm, we use the other three algorithms to be compared together. In order to achieve load balancing, we use the arithmetic average for coefficient of variation (CV) to be the indicator for the performance evaluation. The more lower of CV, it represents the extent of variability are lower. Therefore, if the CV calculated for load balancing algorithm is lower, it shows the algorithm get more better load balancing result. And we assume we can get the load for the flows at each clients, the simulation parameters in Table 1:

The following diagram, shown in Fig. 2, is the simulation results for the four load balancing algorithms. We can see the genetic-based load balancing algorithm has the best performance because it has the lowest percentage of arithmetic average for CV. In the other words, genetic-based load balancing algorithm can achieve the best load balancing result which compared with the other three algorithms.

Table 1 Simulation parameters

Experiment times Variables/parameters	20 times Input
K (number of servers)	4
N (number of flows)	4
Population size	5
Crossover rate	0.9
Mutation rate	0.2

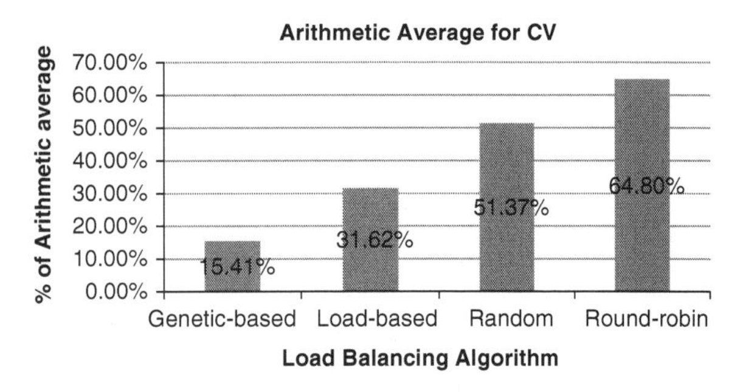

Fig. 2 Simulation results

Conclusion and Future Work

In this paper, we present flexible load balancing algorithms by our OpenFlow-based load balancing system. Through the proposed genetic-based load balancing algorithm and self-definition flow entries by this paper, the traffic follow the pre-configured flows and redirect to the nodes which in the load balancing pool. It can save the cost and avoid the bottleneck because of the throughput for the controller channel. In the simulation results, we can see the significant performance on our proposed genetic-based load balancing algorithm which evaluated by arithmetic average for coefficient of variation. In order to enhance the scalability and stability for our system, we will focus on scaling the network topology to enterprise class and providing the high availability properties for our system in the future.

Acknowledgments This work was supported by the National Science Council of the Republic of China under grant NSC 102-2622-E-008-002-CC2. And the author also thanks Xinguard for their support to Pica8 Open Switch.

References

1. Haleplidis E, Denazis S, Koufopavlou O, Halpern J, Salim JH (2012) Software-defined networking: experimenting with the control to forwarding plane interface. In: Proceedings of european workshop software defined networking (EWSDN), pp 91–96, Oct 2012
2. Pica8 Open switch http://www.pica8.com/open-switching/1-gbe-10gbe-open-switches.php
3. Maguluri ST, Srikant R, Lei Y (2012) Heavy traffic optimal resource allocation algorithms for cloud computing clusters. In: Proceedings of the 24th international teletraffic congress, pp 1–8, Sep 2012
4. Radojevic B, Zagar M (2011) Analysis of issues with load balancing algorithms in hosted (cloud) environments. In: Proceedings of the 34th international convention MIPRO, pp 416–420, 23–27 May 2011
5. Handigol N, Seetharaman S, Flajslik M, McKeown N, Johari R (2009) Plug-n-serve: load-balancing web traffic using openflow. In: Proceedings of demo at ACM SIGCOMM, Aug 2009

6. Wang R, Butnariu D, Rexford J (2011) OpenFlow-based server load balancing gone wild. In: Proceedings of the 11th USENIX conference on hot topics in management of internet, cloud, and enterprise networks and services, pp 12–12, Mar 2011
7. Marcon DS, Bays LR (2011) Flow based load balancing: optimizing web servers resource utilization. J Appl Comput Res Dec 2011
8. Uppal H, Brandon D (2010) OpenFlow based load balancing. In: Proceedings of CSE561: networking. project report. University of Washington, Spring 2010
9. Koerner M, Kao O (2012) Multiple service load-balancing with OpenFlow. In: Proceedings of the 13th international conference on high performance switching and routing, pp 210–214, 24–27 Jun 2012
10. Eokhong M, Eungju K, Yong L, Ngchul K, Taek H, Guk K (2012) Implementation of an openflow network virtualization for multi-controller environment. In: Proceedings of the 14th international conference on advanced communication technology, pp 589–592, Feb 2012
11. Open vSwitch. http://openvswitch.org/
12. Beacon: a java-based openflow control platform, Oct 2011. http://www.beaconcontroller.net/
13. Gude N, Koponen T, Pettit J, Pfaff B, Casado M, McKeown N, Shenker S (2008) Nox: towards an operating system for networks. ACM SIGCOMM Comput Commun Rev Jul 2008
14. POX www.noxrepo.org/pox/about-pox/
15. Jarschel M, Oechsner S, Schlosser D, Pries R, Goll S, Tran-Gia P (2011) Modeling and performance evaluation of an OpenFlow architecture. In: Proceedings of the 23rd international teletraffic congress, pp 1–7, Sep 2011

QoS Modeling of Cyber Physical Systems by the Integration of AADL and Aspect-Oriented Methods

Lichen Zhang

Abstract This paper proposes an aspect-oriented QoS modeling method based on AADL. Aspect-Oriented development method can decrease the complexity of models by separating its different concerns. In model-based development of cyber physical systems this separation of concerns is more important given the QoS concerns addressed by Cyber physical Systems. These concerns can include timeliness, fault-tolerance, and security Architecture Analysis and Design Language (AADL) is a standard architecture description language to design and evaluate software architectures for embedded systems already in use by a number of organizations around the world. In this paper, we present our current effort to extend AADL to include new features for separation of concerns., we make a in-depth study of AADL extension for QoS. Finally, we illustrate QoS aspect-oriented modeling via an example of transportation cyber physical system.

Keywords QoS · Cyber physical systems · Aspect-Oriented · AADL

Introduction

The dependability of the software [1] has become an international issue of universal concern, the impact of the recent software fault and failure is growing, such as the paralysis of the Beijing Olympics ticketing system and the recent plane crash of the President of Poland. Therefore, the importance and urgency of the digital computing system's dependability began arousing more and more attention. A digital computing system's dependability refers to the integrative competence of the system that can provide the comprehensive capacity services, mainly related to the reliability, availability, testability, maintainability and safety. With the

L. Zhang (✉)
Shanghai Key Laboratory of Trustworthy Computing, East China Normal University, Shanghai 200062, China
e-mail: zhanglichen1962@163.com

Y.-M. Huang et al. (eds.), *Advanced Technologies, Embedded and Multimedia for Human-centric Computing*, Lecture Notes in Electrical Engineering 260,
DOI: 10.1007/978-94-007-7262-5_49,

increasing of the importance and urgency of the software in any domain, the dependability of the distributed real-time system should arouse more attention.

Aspect-oriented programming (AOP) [2] is a new software development technique, which is based on the separation of concerns. Systems could be separated into different crosscutting concerns and designed independently by using AOP techniques. Every concern is called an "aspect". Before AOP, as applications became more sophisticated, important program design decisions were difficult to capture in actual code. The implementation of the design decisions were scattered throughout, resulting in tangled code that was hard to develop and maintain. But AOP techniques can solve the problem above well, and increase comprehensibility, adaptability, and reusability of the system. AOSD model separates systems into tow parts: the core component and aspects.

With the deepening of the dependable computing research, the system's dependability has becoming a important direction of cyber physical systems, the modeling and design of cyber physical systems has become a new field. The dependable cyber physical system has a high requirement of reliability, safety and timing, these non-functional properties dispersed in the various functional components of system, so the Object-Oriented design has lost its superiority very obviously. The QoS of dependable real-time system [3] is very complex, currently the QoS research still hasn't a completely and technical system, and there isn't any solution meeting all the QoS requirements. We design the QoS of dependable real-time system as a separate Aspect using AOP, and proposed the classification of complex QoS, divided into the timing, reliability and safety and other sub-aspects. These sub-aspects inherit t members and operations from the abstract QoS aspect. We design each sub-aspects through aspect-Oriented modeling, to ensure the Quality of dependable real-time system meeting the requirements of the dependability.

This paper proposes an aspect-oriented QoS modeling method based on AADL [4]. Aspect-Oriented development method can decrease the complexity of models by separating its different concerns. In model-based development of cyber physical systems this separation of concerns is more important given the QoS concerns addressed by Cyber physical Systems. These concerns can include timeliness, fault-tolerance, and security Architecture Analysis and Design Language (AADL) is a standard architecture description language to design and evaluate software architectures for embedded systems already in use by a number of organizations around the world. In this paper we present our current effort to extend AADL to include new features for separation of concerns., we make a in-depth study of AADL extension for QoS. Finally, we illustrate QoS aspect-oriented modeling via an example of transportation cyber physical system.

Aspect-Oriented Specification of QoS Based on AADL

AOP is a new modularity technique that aims to cleanly separate the implementation of crosscutting concerns. It builds on Object-Orientation, and addresses some of the points that are not addressed by OO. AOP provides mechanisms for decomposing a problem into functional components and aspectual components called aspects [5]. An aspect is a modular unit of crosscutting the functional components, which is designed to encapsulate state and behavior that affect multiple classes into reusable modules. Distribution, logging, fault tolerance, real-time and synchronization are examples of aspects. The AOP approach proposes a solution to the crosscutting concerns problem by encapsulating these into an aspect, and uses the weaving mechanism to combine them with the main components of the software system and produces the final system. We think that the phenomenon of handling multiple orthogonal design requirements is in the category of crosscutting concerns, which are well addressed by aspect oriented techniques. Hence, we believe that system architecture is one of the ideal places where we can apply aspect oriented programming (AOP) methods to obtain a modularity level that is unattainable via traditional programming techniques. To follow that theoretical conjecture, it is necessary to identify and to analyze these crosscutting phenomena in existing system implementations. Furthermore, by using aspect oriented languages, we should be able to resolve the concern crosscutting and to yield a system architecture that is more logically coherent. It is then possible to quantify and to closely approximate the benefit of applying AOP to the system architecture [6].

AADL [4] is an architecture description language developed to describe embedded systems is shown in Fig. 1. AADL (Architecture Analysis and Design Language), which is a modeling language that supports text and graphics, was approved as the industrial standard AS5506 in November 2004. Component is the most important concept in AADL. The main components in AADL are divided into three parts: software components, hardware components and composite components. Software components include data, thread, thread group, process and subprogram. Hardware components include processor, memory, bus and device. Composite components include system [7–9].

In its conformity to the ADL definition, AADL provides support for various kinds of non-functional analyses along with conventional modeling as shown in Fig. 2 [7–9]:

Flow Latency Analysis: Understand the amount of time consumed for information flows within a system, particularly the end-to-end time consumed from a starting point to a destination.

Resource Consumption Analysis: Allows system architects to perform resource allocation for processors, memory, and network bandwidth and analyze the requirements against the available resources.

Real-Time Schedulability Analysis: AADL models bind software elements such as threads to hardware elements like processors. Schedulability analysis helps in examining such bindings and scheduling policies.

Safety Analysis: Checks the safety criticality level of system components and highlights potential safety hazards that may occur because of communication among components with different safety levels.

Security Analysis: Like safety levels, AADL components can be assigned various security levels. The analysis helps in identifying the security loopholes that

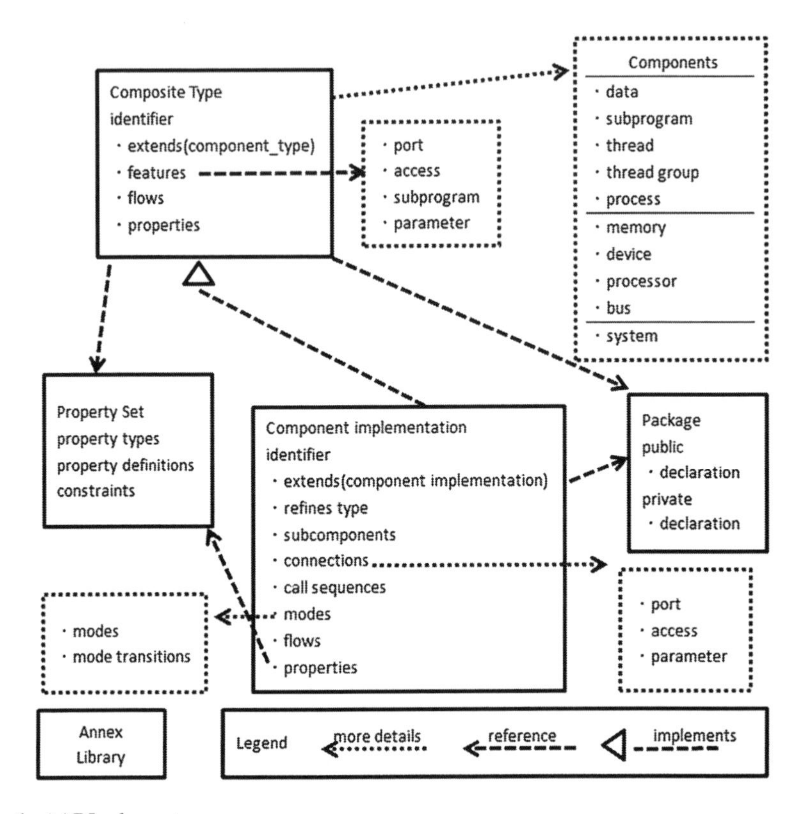

Fig. 1 AADL elements

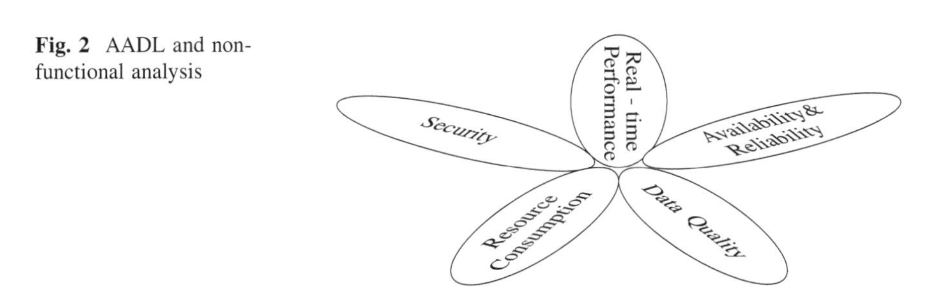

Fig. 2 AADL and non-functional analysis

may happen because of mismatches in security levels between a component and its subcomponents, and communication among components with different security levels.

AADL defines two main extension mechanisms: property sets as shown in Fig. 3 and sublanguages (known as annexes). Annexes and properties allow the addition of complex annotations to AADL models that accommodate the needs of multiple concerns. These annotations, along with their corresponding analysis plug-ins, provide a powerful combination for the architect to evaluate his/her

```
property set Clemson is
MbitPerSec : type units (MPS, GPS => MPS*1000);
Band_width: type aadlinteger units Clemson::MbitPerSec;
Radio_band_width: Clemson::Band_width applies to (all);
Band_width_802_11g: constant Clemson::Band_width => 54 MPS;
Band_width_802_11n: constant Clemson::Band_width => 300 MPS;
Band_width_fast_ethernet: constant Clemson::Band_width => 100 MPS;
end Clemson;
```

Fig. 3 Property sets of AADL

design choices from different perspectives. The extension mechanisms in AADL enable these perspectives to evolve in number and complexity as the knowledge on them also evolves [10, 11].

AO4AADL is an aspect oriented extension for AADL as shown in Fig. 4 [11–13]. This language considers aspects as an extension concept of AADL components called aspect annex. Instead of defining a new aspect oriented ADL, we extend AADL, a well-known ADL, with an aspect annex. So we consider, in this work, that aspects can be specified in a language other than AADL, and then integrated in AADL models as annexes. Based on the annex extension mechanism,

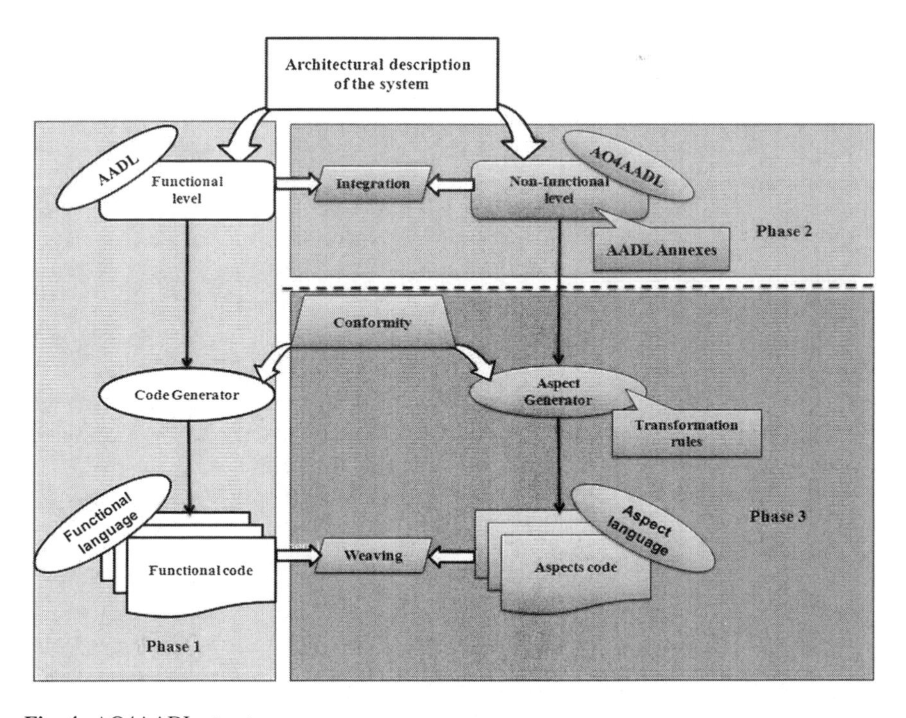

Fig. 4 AO4AADL structue

we propose to enrich AADL specifications with aspect concepts. AO4AADL Consists mainly in three phases:

Implementing Functional code: In this phase, the designer should focus on the main functionalities of his application without considering any non-functional properties.

Designing non-functional properties: At this phase, the designer defines the non functional safety properties and states the conditions under which his application operates correctly, such as security, availability, etc.

Generating non-functional properties code: AO4AADL aspects can be translated in different aspect languages such as AspectJ or JAC for Java language, AspectAda for Ada language, AspectC for C language, etc.

Ana-Elena Rugina, Karama Kanoun and Mohamed Kaâniche proposed a four step modeling dependability methods based on AADL [14]: The *first step* is devoted to the modeling of the application architecture in AADL. The *second step* concerns the specification of the application behavior in the presence of faults through AADL error models associated with components of the architecture model. The *third step* aims at building an analytical dependability evaluation model, from the AADL dependability model, based on model transformation rules. The *fourth step* is devoted to the dependability evaluation model processing that aims at evaluating quantitative measures characterizing dependability attributes.

In this paper, we extend AADL by aspect-oriented method in following aspect:

Physical world aspect: Cyber physical systems are often complex and span multiple physical domains, whereas mostly these systems are computer controlled.

Dynamic Continuous dynamics Aspect: Cyber physical systems are mixtures of continuous dynamic and discrete events. These continuous and discrete dynamics not only coexist, but interact and changes occur both in response to discrete, instantaneous, events and in response to dynamics as described differential or difference equations in time.

Formal Specification Aspect of Data: A formal specification aspect of data captures the static relation between the object and data. Formal data aspect emphasizes the static structure of the system using objects, attributes, operations and relationships based on formal techniques.

Formal Specification aspect of Information flow and control flow: Formal Specification aspect of Information flow and control flow aims at facilitating the description and evaluation of various flow properties measures. It provides an ideal semantic basis to characterize the behavior of the component communications in the system.

Spatial Aspect: The analysis and understanding of railway cyber physical systems spatial behavior—such as guiding, approaching, departing, or coordinating movements is very important.

Case Study: Aspect-Oriented Specification of QoS of VANET

Presently study in VANET is still at the preliminary stage [15–18]. We model the most promising system of VANET (Vehicular Ad-hoc NETwork) as shown in Fig. 5.

The AADL Model of Vehicle Station is shown in Fig. 6.

Delay Analysis of Data Flow is shown in Fig. 7.

```
flows

  ETE_F1: end to end flow BRAKE.Flow1
  -> C22 -> CC.brake_flow_1
  -> C29 -> TA.Flow1
  {
  Latency => 200 Ms;
  };
flows

  brake_flow_1: flow path brake_status
```

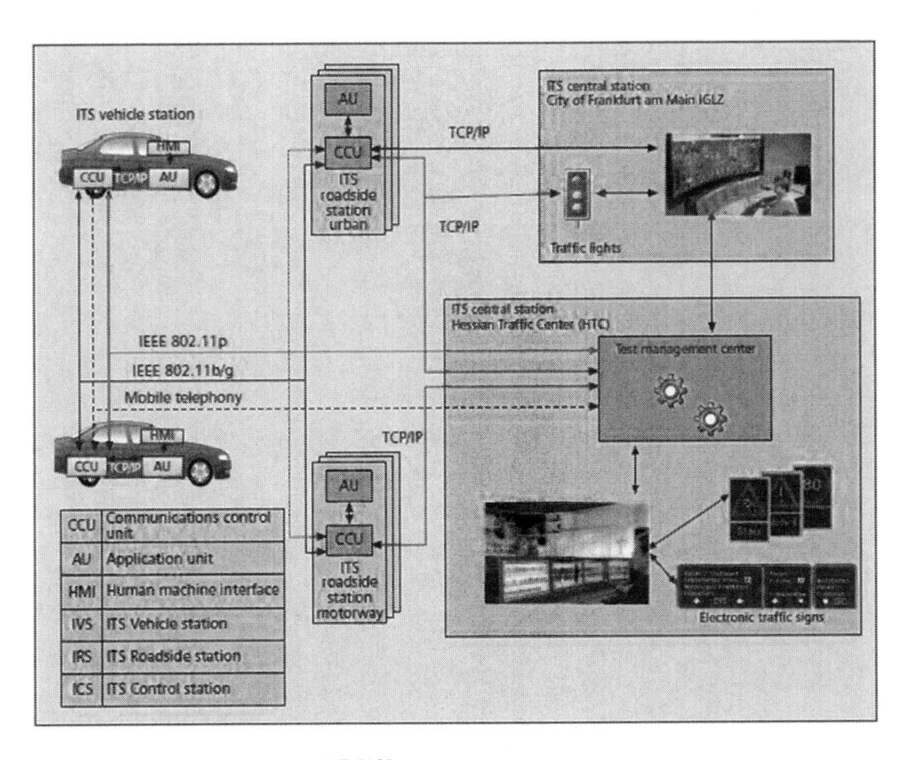

Fig. 5 The architecture of VANET [19]

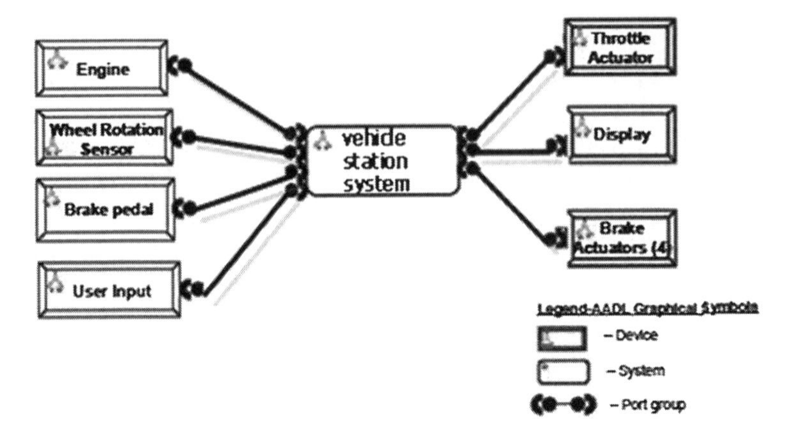

Fig. 6 AADL model of vehicle station

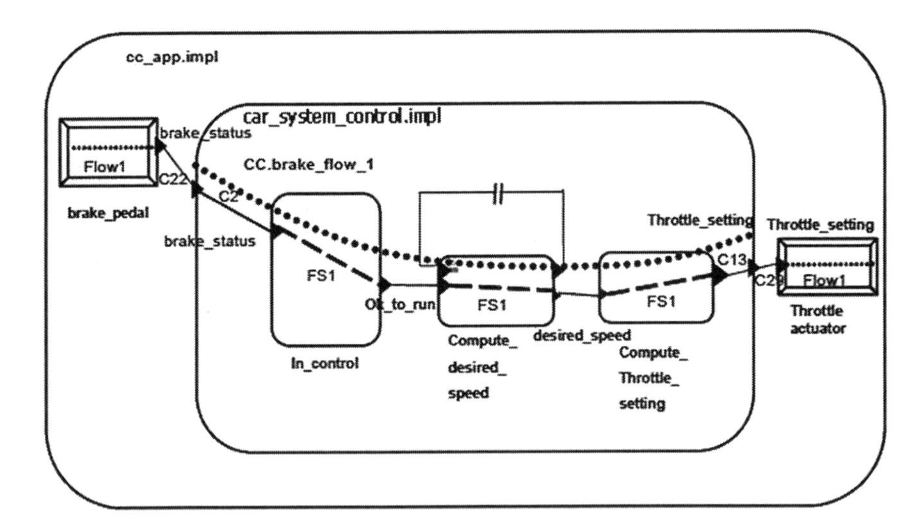

Fig. 7 Delay analysis of data flow

```
-> C2 -> I_C.FS1
-> C9 -> C_D_S.FS1
-> C12 -> C_T_S.FS1
-> C13 -> throttle_setting{
Latency => 130 Ms;
};
```

Conclusion

This paper proposed an aspect-oriented QoS modeling method based on AADL. Aspect-Oriented development method can decrease the complexity of models by separating its different concerns. In model-based development of cyber physical systems this separation of concerns is more important given the QoS concerns addressed by Cyber physical Systems. These concerns can include timeliness, fault-tolerance, and security. Architecture Analysis and Design Language (AADL) is a standard architecture description language to design and evaluate software architectures for embedded systems already in use by a number of organizations around the world. In this paper we present our current effort to extended AADL to include new features for separation of concerns., we made a in-depth study of AADL extension for QoS. Finally, we illustrated QoS aspect-oriented modeling via an example of the specification of VANET based on AADL.

The future work focuses on the integration AADL and formal techniques to specify and verification of QoS of cyber physical systems.

Acknowledgments This work is supported by Shanghai Knowledge Service Platform Project (No. ZF1213), national high technology research and development program of China (No. 2011AA010101), national basic research program of China (No. 2011CB302904), the national science foundation of China under grant (No. 61173046, No. 61021004, No. 61061130541, No. 91118008), doctoral program foundation of institutions of higher education of China (No. 20120076130003),national science foundation of Guangdong province under grant (No. S2011010004905).

References

1. Laprie JC (ed) (1992) Dependability: basic concepts and terminology. Springer, Berlin
2. Kiczales G et al (1997) Aspect-oriented programming. In: Proceedings of the 11th european conference on object-oriented programming, June 1997
3. Frolund Svend, Koistinen Jari (1998) Quality of service specification in distributed object systems. IEE/BCS Distrib Syst Eng J 5:179–202
4. AE Aerospace (2009) SAE AS5506A[S]: architecture analysis and design language V2.0
5. Aldawud O, Elrad T, Bader A (2001) A UML profile for aspect oriented modeling. In: Proceedings of workshop on AOP
6. Wehrmeister MA, Freitas EP, Pereira CE et al (2007) An aspect-oriented approach for dealing with non-functional requirements in a model-driven development of distributed embedded real-time systems. In: Proceedings of 10th IEEE international symposium on object and component-oriented real-time distributed computing, IEEE Computer Society. Santorini Island, Greece, 7–9 May 2007 pp 428–432
7. Feiler P, Hugues J, Sokolsky O (eds) (2012) Architecture-driven semantic analysis of embedded systems. Dagstuhl seminar 12272, Dagstuhl Report, vol 2 (7). pp 30–55. ISSN 2192-5283
8. The story of AADL (2010) AADL wiki. software engineering institute, Web. 06 Jan 2012
9. Muhammad N, Vandewoude Y, Berbers Y, van Loo S (2010) Modelling embedded systems with AADL: a practical study. www.intechopen.com/download/pdf/10732

10. de Niz D, Feiler PH (2007) Aspects in the industry standard AADL. In: Proceedings of AOM '07 Proceedings of the 10th international workshop on aspect-oriented modeling. pp 15–20
11. Michotte L, Vergnaud T, Feiler P, France R (2008) Aspect oriented modeling of component architectures using AADL. In: Proceedings of the 2nd international conference on new technologies, mobility and security, 5–7 Nov 2008
12. Loukil S, Kallel S, Zalila B, Jmaiel M (2010) Poster -AO4AADL: aspect oriented ADL for embedded systems. In: Proceedings of international conference on new technologies of distributed systems (NOTERE)
13. Loukil S, Kallel S, Zalila B, Jmaiel M (2010) AO4AADL: an aspect oriented ADL for embedded systems. In: Proceedings of the 4th european conference on software architecture (ECSA 2010), LNCS. Springer, Copenhagen
14. Rugina AE, Kanoun K, Kaaniche M (2006) An architecture-based dependability modeling framework using AADL. In: Proceedings of 10th IASTED international conference on software engineering and applications (SEA'2006), Dallas (USA), 13–15 Nov 2006 (13/11/2006), pp 222–227
15. Festag A, F̈ussler H, Hartenstein H, Sarma A, Schmitz R (2004) Fleet net: bringing car-to-car communication into the real world. In: Proceedings of the 11th ITS world congress and exhibtion [C], Nagoya, Japan, pp 1–8
16. CVIS. Cooperative vehicle-infrastructure systems [EB/OL]. http://www.cvisproject.org. Accessed on 19 Feb 2013
17. SAFESPOT. Cooperative vehicles and road infrastructure for road safety[EB/OL]. http://www.safespot-eu.org. Accessed on 19 Mar 2013
18. Kargl F, Papadimitratos P, Buttyan L (2008) Secure vehicular communication systems: implementation, performance, and research challenges. IEEE Commun Mag 46(11):110–118
19. Stubing Hagen (2010) Adam opel gmbh marc bechler.simTD: a car-to-x system architecture for field operational tests. IEEE Commun Mag 48(5):148–154

Assessment of Performance in Data Center Network Based on Maximum Flow

Kai Peng, Rongheng Lin, Binbin Huang, Hua Zou and Fangchun Yang

Abstract Recently, data center networks (DCN) have received significant attention from the academic and industry. However, researches of DCN are mainly concentrated on the improvement of network architectures and the design of routing protocols, or the performance evaluation from the perspective of node importance. In contrast to existing solutions, in this paper, we propose using maximum-flow theory to assess the network performance. Firstly, we abstract two kinds of typical DCN architectures and then formulate and convert the performance analysis of those architectures into a maximum-flow problem including a supersource and a supersink. Secondly, we get the value of maximum-flow by using Edmonds and Goldberg algorithm. Last but not the least, based on the theory of maximum-flow and Minimal cut sets, we get the critical edges for each architecture. Extended experiments and analysis show that our method is effective and indeed introduce low overhead on computation. In addition, the method and issues observed in this paper is generic and can be widely used in newly proposed DCN architectures.

Keywords DCN · Topologies · Maximum-flow · Assessment

K. Peng (✉) · R. Lin · B. Huang · H. Zou · F. Yang
State Key Laboratory of Networking and Switching Technology, Beijing University of Posts and Telecommunications, Beijing, China
e-mail: pkbupt@gmail.com

R. Lin
e-mail: rhlin@bupt.edu.cn

B. Huang
e-mail: binbinHUang@bupt.edu.cn

H. Zou
e-mail: zouhua@bupt.edu.cn

F. Yang
e-mail: fcyang@bupt.edu.cn

Y.-M. Huang et al. (eds.), *Advanced Technologies, Embedded and Multimedia for Human-centric Computing*, Lecture Notes in Electrical Engineering 260,
DOI: 10.1007/978-94-007-7262-5_50,

Introduction

Driven by the recent proliferation of cloud services [1, 2] and trend of IT systems, data center networks have gained a wide attention from both industry and research community. Several network architectures [3–8] have been proposed for these extensive data centers.

In general, all existing DCN architectures can be divided into two classes, hierarchical architecture and flat architecture. The hierarchical architecture is represented by the traditional Multi-rooted Tree Architecture and employs a layered structure which is constituted recursively by lower-level components. The other one (for example, FatTree [6] and VL2 [7] architectures) which organizes all servers into the same level by using some sort topologies. Actually, the main research of these new architectures is limited to load balancing or improvement of architectures as the increasing new services [9]. Although some of researchers are engaged in investigating how these architectures are affected in visualized environments [10], while a few of existing research concerned about the security of the data center network, especially for the vulnerability evaluation of these new architectures. In our previous research, we mainly focus on the assessment of node importance for those newly proposed ones. In contrast to previous solution, in this paper, we fill this void by conducting an experimental evaluation of the-state-of-the-art architectures from the perspective of network flow. The main idea is as follows. When the network traffic reaches the maximum, the paths or the edges which are in the saturated status are the critical ones. In addition, critical paths or edges are treated as the focus of attack and defense, which need to be given more attention. For one thing, NSP (Network service provider) should increase the traffic capabilities of the critical links in order to increase the capabilities of entire network. For another, for the attackers, once they get the information of critical paths, try to destroy the critical paths may greatly reduce the cost of attack so as to maximize their benefits. Unfortunately, edges importance of these new DCN architectures has never been investigated before.

In this paper, we try to conduct experiments on Multi-rooted Tree architecture and VL2 architecture, each respectively as a representative of hierarchical and flat architectures. Taking any of them for example, we first use a directed and weight graph to describe its topology. As we know, there are multi-source and multi-sink in this architecture, therefore, we make this be converted into an ordinary flow network with only one single source and one single sink by adding a supersource and a supersink. Secondly, we get the maximum-flow and corresponding paths by using Edmonds and Goldberg algorithm. Finally, based on maximum-flow and Minimal cut sets theory, we get the critical edges.

Our contributions are summarized as two aspects.

1. For one thing, we formulate and transform the DCN architecture as a maximum-flow problem from the perspective of network flow, and then show it feasibility.

2. We formulate and conduct an experimental evaluation of two typical DCN architectures. The results show that our proposed method indeed finds the maximum-flow and critical edges for each architecture. Furthermore, the method can be generalized and applied as guidelines for the newly proposed DCN architectures.

The remainders of this paper are as follows. In section Background, we give a brief introduction of current DCN architectures. Problem formulation and math mode are described in section Problem Formulation and Math Mode. Section Model Solution and Algorithm presents model solution and the algorithm process, followed by section Experiment Evaluation and Discussion; we show the experiment evaluation and the discussion. Finally, we conclude the paper in section Conclusion.

Background

In this section, we give a brief introduction of current data center architectures. Based on how the system is constructed, we classify existing data center network (DCN) architectures into two types, hierarchical one and flat one.

In a hierarchical architecture, servers are arranged in different levels. A higher-level network consists of multiple components of lower-levels. Actually, the most widely used hierarchical DCN architecture is traditional Multi-rooted Tree architecture, known as the typical three-tier architecture. As is shown in Fig. 1, the tree architecture generally consists of switches of three layers. We can see that such topology have 16 servers (has been marked as Tree architecture). At the bottom level, known as the edge layer, where each server connects to one or two edge switches. Each edge switches connect to one or two switches at the layer of aggregation, where all the hosts are placed on the edge layer, forming the leaf nodes within the whole networks, which are responsible for storage and computation. Above all, each aggregation switch connects with multiple switches at the core layer.

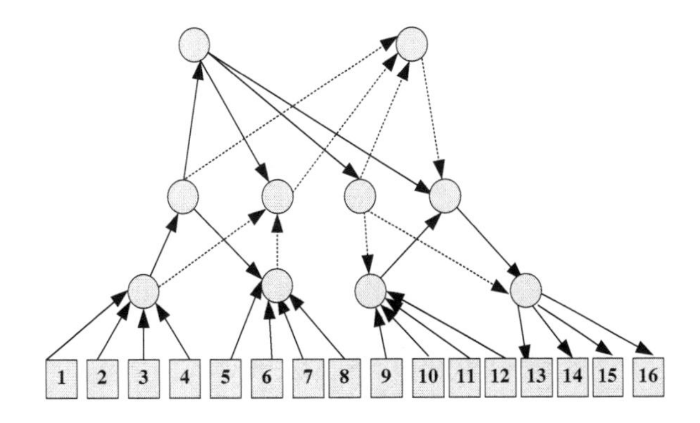

Fig. 1 Topology of tree architecture

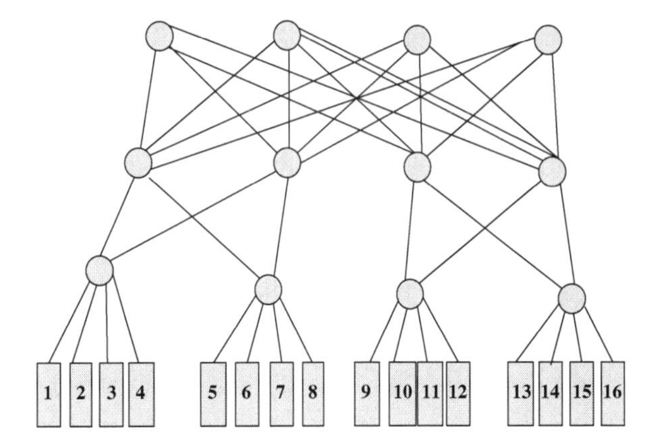

Fig. 2 Topology of VL2 architecture

In contrast to hierarchical architecture, the other type of DCN is flat one. It utilizes a flat organization of servers, which are placed into a single layer and are interconnected by switches. One typical representative of this class is the VL2 Architecture [7] which was introduced by Greenberg et al. In VL2 architecture, a complete bipartite graph was formed by the switches from the layers of aggregation and core. More specifically, interconnection topology of switches follows a folded Clos network where switches of intermediate and aggregation are respectively organized into each side of the graph. Figure 2 shows one example with 16 servers (marked as VL2).

These architectures have been designed independently for different goals in practice. In this paper, we aim to evaluate the performance of these architectures from the perspective of maximum-flow.

Problem Formulation and Math Mode

In this section, we formally propose and define the maximum flow problem and then mainly show the mode in details.

Problem Formulation

Based on the previous topologies in Section Background, we get the main structure of the data center network. Firstly, we choose one of the architecture for example. We use a directed graph G to describe the tree topology. As shown in Fig. 1, all of the nodes have been marked. We assume that the network flow goes through from left to right. As reflected in Fig. 1, The nodes of $\{v_1, v_2, v_3, v_4\}$ are all can be seen as the sources and the network flow eventually flows to the nodes of

$\{v_{13}, v_{14}, v_{15}, v_{16}\}$ which are the sinks in this topology. As maximum-flow theory only supports the graph which exists only one source and one sink, and thus when directly applied in tree architecture, they may suffer from drawback. Therefore, we will show how to convert this problem into a traditional maximum-flow one in the coming section.

The Mode of Maximum-Flow Problem

Transformation of Maximum-Flow Problem

Figure 3 shows how the network from Fig. 1 can be converted into an ordinary flow network with only a single source and a single sink. We first add a super-source s and add a directed edge (s, s_i) with capacity $c(s, s_i) = \infty$ for each $i = 1, 2, \ldots, m$. And then we create a new supersink t as well as add a directed edge (t_i, t) with capacity $(t_i, t) = \infty$ for each $i = 1, 2, \ldots, m$.

Intuitively, any flow in the network in Fig. 1 corresponds to a flow in this network in Fig. 3, and vice versa. The single source s provides the same flow as desired for the multiple sources s_i, and the single sink t as well as costs the same flow as desired for the multiple sinks t_i.

The Math Mode of Maximum Flow

We describe the establishment of the max-flow mode. Let $G = (V, E)$ be a flow network with a capacity function f, where V is the set of nodes and E is the set of edge.

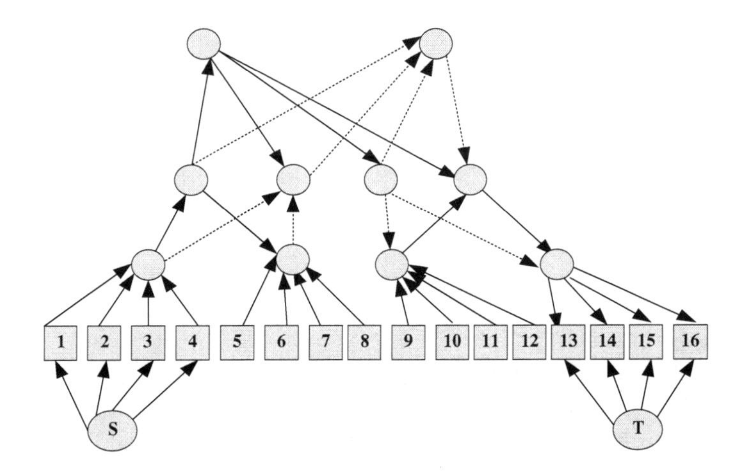

Fig. 3 Maximum-flow for tree

A flow in G is a function of real-valued.

We require that $f(a) : V \times V \rightarrow \mathbf{R}$.

Thus, $\mathrm{f(a)} = \mathrm{C(i,\ j)}$, where $a \in A$.

$\mathrm{C\ (i,\ j)}$ represents the capacity of network.

$\mathrm{f\ (i,\ v)}$ represents the total traffic that outflows from node v_i, and $\mathrm{f}\ (v,\ \mathrm{i})$ means the total flow which flows into the node v_i. In the maximum-flow problem, we are given a flow of network G, where s is source node and t is sink node, and our object is to find maximum value of the flow.

According to the conservation principle theory, when the node of v_i is the intermediate node, we can obtained that $f(\mathrm{i}, v) - \mathrm{f}(v, \mathrm{i}) = 0$.

Especially, for the source s and sink t, there will be the formulation,

$$\sum_{j \in N} f(s,j) = \sum_{j \in N} f(j,t)$$

For any network in a $\mathrm{(i,\ j)}$, when the rate of flow increases to the capacity of $\mathrm{C(i,\ j)}$. This path can be seen as saturated path or as unsaturated section. More specifically, if all paths between two nodes contain at least one saturated path, the traffic between two nodes is maximum flow. According to the above analysis, we can establish the maximum flow model.

$$\begin{cases} \max f(s,t) = \sum_{j \in N} f(s,j) \\ f(s,V) - f(V,s) = f(s,t) \\ f(t,V) - f(V,t) = -f(s,t) \\ f(i,V) - f(V,i) = 0, & where\ i \neq s,\ t,\ i \in V \\ \sum_{j \in N} f(s,j) = \sum_{j \in N} f(j,t) & where\ i = sort \\ 0 \leq f(i,j) \leq C(i,j) \end{cases}$$

When the network comes to the maximum-flow, according to maximum flow minimum cut theorem, we can get the critical edges for this flow network.

Model Solution and Algorithm

Model Solution

Based on the formulation which has been proposed in [11], we establish the entire cost matrix for each node in the topology. We show the new expressions of C for the two architectures in our paper.

Firstly, we return the cost function of tree architecture. For the Tree architecture, the cost between two nodes can be expressed as the function of the fan-out of the edge switches (p_0) as well as the fan-out of the aggregation ones (p_1),in addition, when v_i is the source there the weight between v_i and v_j will be ∞. Similarly, when the j is the sink, there the weight between v_i and v_j will be ∞ too. The function of Tree is named as C_{ij}^{Tree}.

$$C_{ij}^{Tree}\begin{cases} \infty & if\, i = s or j = t \\ 0 & if\, i = j \\ 1 & if \left\lfloor \frac{i}{p_0} \right\rfloor = \left\lfloor \frac{j}{p_0} \right\rfloor \\ 3 & if \left\lfloor \frac{i}{p_0} \right\rfloor \neq \left\lfloor \frac{j}{p_0} \right\rfloor \, and \left\lfloor \frac{i}{p_0 p_1} \right\rfloor = \left\lfloor \frac{j}{p_0 p_1} \right\rfloor \\ 5 & if \left\lfloor \frac{i}{p_0 p_1} \right\rfloor \neq \left\lfloor \frac{j}{p_0 p_1} \right\rfloor \end{cases} \qquad C_{ij}^{VL2} = \begin{cases} \infty & if\, i = s \; or \; j = t \\ 0 & if\, i = j \\ 1 & if \left\lfloor \frac{i}{p_0} \right\rfloor = \left\lfloor \frac{j}{p_0} \right\rfloor \\ 5 & if \left\lfloor \frac{i}{p_0} \right\rfloor \neq \left\lfloor \frac{j}{p_0} \right\rfloor \end{cases}$$

Next, we return the function of VL2 architecture. In the VL2, the cost is a function only about the fan-out of the edge switches (p_0), given the traffic that departs the edge switches always passes the core switches. The function is named as C_{ij}^{VL2}.

Maximum-Flow Solution

In this paper, we use Ford-Fulkerson algorithm for the calculation of maximum-flow, which repeatedly increase the flow through the augmenting path until find the maximum flow.

The Ford-Fulkerson method is based on that a flow is a maximum flow if and only if its residual network does not contain the augmenting path.

It is an iterative method. Firstly, it initializes all the flow f to zero. Secondly, we find an augmenting path to increase traffic at every iteration. Repeat this while there exists an augmenting path p unit all the paths have been found. Finally, we will find the value of maximum flow. Specially, use BFS to find each augmenting path at every iteration will greatly improve the efficiency.

The details of the algorithm are described in pseudo-code. It has two major components. (1) Graphmaxflow. (2) Edmonds-Karp Algorithm.

In Alogrithm1, we input the matrix s, t and e, where s and t each represents the source and sink while exists an edge between S and E. In addition, e is the cost matrix for all the nodes in s and t. After the sparse operation, we use: Graphmaxflow (Algorithm 2) for the calculation of maximum-flow.

Algorithm1: Maximum-Flow Algorithm
Input: *s, t, e, n*
Output: {Maximum-flow f and Critical Paths p}
matr = sparse (s, t, e, n, n)
[f, p] = Graphmaxflow (matr, s, t)
return {f, p}

Algorithm2: Graphmaxflow
Input: *(G, s, t)*
Step1: for each edge(u, v) ∈ E(G)
Step2: do f [u, v] ← 0
Step3: f [v, u] ← 0
Step4: Breadth-first search
while there exists a path p from s to t in the residual network of G_f.
Step5: do augment flow f along p
return f[s, t]

Experiment Evaluation and Discussion

In this section, we demonstrate a thorough experimental evaluation of the proposed technique on Multi-rooted Tree and VL2 Architectures. The whole experiment project is implemented by Matlab 7.0 on a Windows 7 Operating System with Intel Core i3 Processor 2.10 GHz. Based on the classification of architecture which described in Section Background, we divided into two groups, the former one is Tree architecture, and the other one is VL2 architecture.

Experimental Evaluation

Tree Experiment Design

As shown in Fig. 1, since the nodes within a cluster got the same nature, in order to simplify the complexity of our maximum-flow problem, we choose two nodes in each cluster for our experiment. There are four groups of $\{v_1, v_2, v_3, v_4\}$ $\{v_5, v_6, v_7, v_8\}$ $\{v_{13}, v_{14}, v_{15}, v_{16}\}$ $\{v_9, v_{10}, v_{11}, v_{12}\}$ in the topology. When designing experiments, we only need to choose three groups as group three and group four got the same cost (six nodes for the test, where two nodes in each cluster). Based on the formula cost matrix in Section Model Solution and the Fig. 3, we can get the adjacency matrix for those six nodes of tree topologies. Figure 4 show the initial capacity of Tree architecture. According to the algorithm 1 and algorithm 2, we get the capacity of maximum-flow which is 32, and the corresponding critical edges of $\{e_{24}, e_{25}, e_{26}, e_{27}, e_{34}, e_{35}, e_{36}, e_{37}\}$.

VL2 Experiment Result

Figure 5 shows the capacity figure of the VL2 architecture. The capacity of maximum-flow is 36, and the critical edges are $\{e_{26}, e_{27}, e_{36}, e_{37}, e_{46}, e_{47}, e_{56}, e_{57}\}$

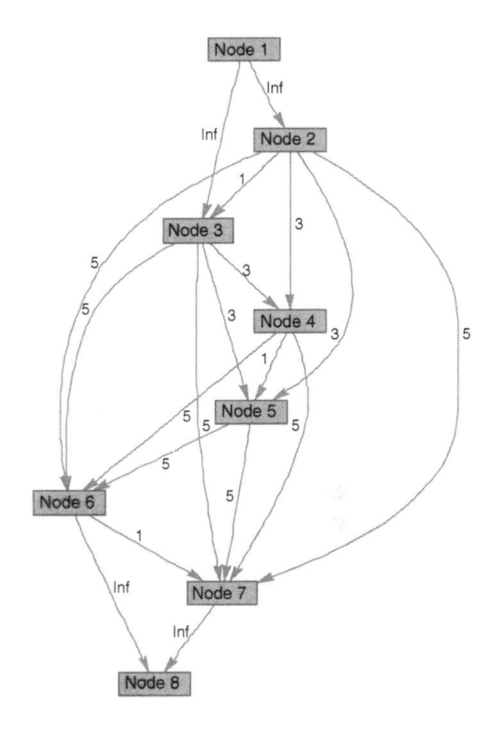

Fig. 4 The capacity of tree architecture

Complexity Analysis and Discussion

Complexity Analysis

The computational complexity of this Algorithm is determined by Edmonds and Karp algorithm. Implementation is based on a variation called the "labeling algorithm". Time complexity is O (n*e^2), where n is the number of the nodes and e is the corresponding edges in the given architecture.

Discussion

As shown in the above experiments (see Section Experimental Evaluation), we can see that our method can find the maximum flow and the corresponding critical edges. The result of complexity analysis also shows that our approach is effective and efficient for the evaluation of the maximum flow. From the perspective of network attack and defense, NSP (Network service provider) should increase the traffic capabilities of the critical links in order to increase the capabilities of whole network.

In this paper, although we choose two kinds of data center network (DCN) topology for our experiments, our method should be generic and can be properly used in any new architecture. Furthermore, we use the cost matrix which has been

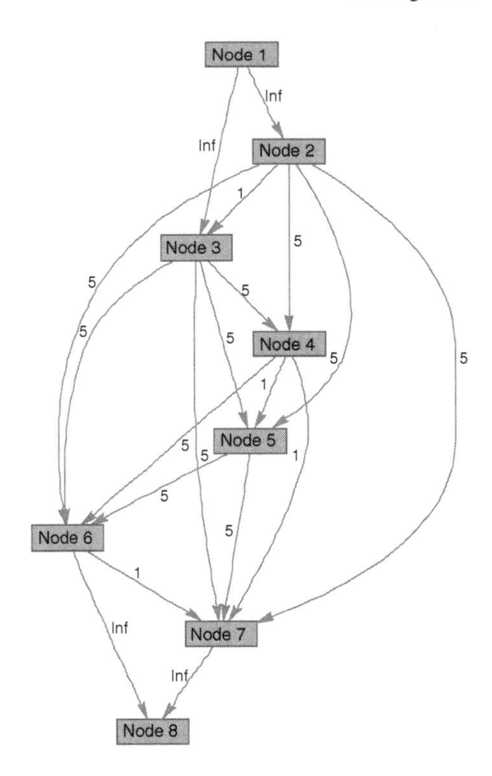

Fig. 5 The capacity of VL2 architecture

proposed in the literature [11]. Considering the changes of cost matrix in different scenario, we only need to change the input of the algorithm. Thus, our method can be widely used for the evaluation of maximum-flow for new DCN architectures.

Conclusion

In this paper, we investigate the problem of topology vulnerability in Data Center Network (DCN) Architectures. We abstract two kinds of typical DCN architectures and formulate the performance analysis of architecture as a maximum-flow problem by adding a supersource and a supersink, and then we propose a method to efficiently solve this problem. Furthermore, the experimental results show that our algorithm is effective. In our paper, although we only choose two kinds of architectures while the method and observed issues can be generic and widely used in the newly proposed ones.

Acknowledgments This work is supported by the National Natural Science Fund China under Grant No. 2009CB320406, the National 863 High-tech Project of China under Grant No. 2011AA01A102, Funds for Creative Research Groups of China (60821001) and State Key Lab of Networking and Switching Technology. Ph.D. Programs Foundation of Ministry of Education (20110005130001).

References

1. Vaquero LM, Rodero-Merino L, Caceres J, Lindner M (2008) A break in the clouds: towards a cloud definition. ACM SIGCOMM Comput Commun Rev 39:50–55
2. Peng K, Zou H, Lin R, Yang F (2012) Small business-oriented index construction of cloud data. In: Proceedings in 12th international conference on algorithms and architectures for parallel processing, pp 156–165
3. Guo C, Wu H, Tan K, Shi L, Zhang Y, Lu S (2008) Dcell: a scalable and fault-tolerant network structure for data centers. ACM SIGCOMM Comput Commun Rev 75–86
4. Li D, Guo C, Wu H, Tan K, Zhang Y, Lu S (2009) FiConn: using backup port for server interconnection in data centers. In: Proceedings in 28th conference on computer communications, pp 2276–2285
5. Guo C, Lu G, Li D, Wu H, Zhang X, Shi Y, Tian C, Zhang Y, Lu S (2009) BCube: a high performance, server-centric network architecture for modular data centers. ACM SIGCOMM Comput Commun Rev 39:63–74
6. Leiserson CE (1985) Fat-trees: universal networks for hardware-efficient supercomputing. Comput IEEE Trans 100(10):892–901
7. Greenberg A, Hamilton JR, Jain N, Kandula S, Kim C, Lahiri P, Maltz DA, Patel P, Sengupta S (2009) VL2: a scalable and flexible data center network. In: Proceedings of ACM SIGCOMM computer communication review, pp 51–62
8. Liao Y, Yin D, Gao L (2010) Dpillar: scalable dual-port server interconnection for data center networks. In: Proceedings in 19th international conference on computer communications and networks, pp 1–6
9. Shangguang W, Zheng Z, Qibo S, Hua Z, Fangchun Y (2011) Cloud model for service selection. In: Proceedings in 30th IEEE conference on computer communications workshops on cloud computing computer communications workshops, pp 666–671
10. Zhang Y, Su AJ, Jiang G (2011) Understanding data center network architectures in virtualized environments: a view from multi-tier applications. Comput Netw 55:2196–2208
11. Meng X, Pappas V, Zhang L (2010) Improving the scalability of data center networks with traffic-aware virtual machine placement. In: Proceedings in 29th IEEE conference on computer communications, pp 1–9

A Situation-Oriented IoT Middleware for Resolution of Conflict Contexts Based on Combination of Priorities

Z. Cheng, J. Wang, T. Huang, P. Li, N. Yen, J. Tsai, Y. Zhou and L. Jing

Abstract Situation-aware service is recognized as an emerging research issue in ubiquitous computing. It becomes more important and significant with the recent progress in IoT (Internet of Things), since the situations considered in IoT are more complex, become global, and cause more conflict. In this paper, a middleware for management conflict situations was designed, to prompt the development of context-aware services. It is characterized by its ability of situation-oriented, paying attention to relations among users (and situations as well) and smart objects around. Eventually, following issues were solved: (a) a method for detecting (i.e., being aware of) a specific situation, and triggering corresponding service; and (b) an algorithm for conflict situations/contexts management. A diagram of situation state transition (DSST) was proposed to specify states of a situation. A set of situation-oriented ECA rules are presented to reason the situations' states based on sensed data. Policies based on DSST for resolving conflicts were also given. The experiment results demonstrate the feasibility of proposed method, and the performance of proposed situation-oriented policies.

Keywords Internet of things · Context-aware service · Conflict situations · Diagram of situation state transition · Reasonably fair policy

Z. Cheng (✉) · J. Wang · T. Huang · P. Li · N. Yen · J. Tsai · Y. Zhou · L. Jing
School of Computer Science and Engineering, The University of Aizu, Aizu-Wakamatsu City, Fukushima-ken 965-8580, Japan
e-mail: z-cheng@u-aizu.ac.jp

J. Wang
e-mail: j-wang@u-aizu.ac.jp

N. Yen
e-mail: nyyen@u-aizu.ac.jp

J. Tsai
e-mail: jctsai@u-aizu.ac.jp

Y.-M. Huang et al. (eds.), *Advanced Technologies, Embedded and Multimedia for Human-centric Computing*, Lecture Notes in Electrical Engineering 260,
DOI: 10.1007/978-94-007-7262-5_51,

Introduction

Situation-aware service is a hot research topic in ubiquitous systems. For examples, a travel information service is presented in [1], which can recommend attractions to users, based on their context information, such as current location, time, and weather etc. It becomes more important and significant with the recent progress in IoT (Internet of Things), since the situations considered in IoT are more complex, become global, and cause more conflict, e.g. many elderly need help in the same time, but the helpers are not enough, especially when a disaster happens in a city.

In this paper, our goal is to design a middleware of management for complex and conflict situations. Especially, we focus on resolution of conflicts among different users' situations, to enhance the development of context-aware applications.

The requests for services by different users may be related or even conflict in the same location and time, in order to provide situation-aware services to each of the user. The conflict often happens, when situation-aware services to different users need common resources. Conflict should be resolved and coordination or scheduling of those services are necessary.

To this end, the following problems have to be solved. (a) a method for detecting (being aware of) a situation, and triggering corresponding services, (b) a resolution method for conflict situations, and (c) a coordination mechanism between related situations. In this paper, we mainly focus on the solutions of the first two problems.

Many approaches have been studied for situation-aware applications [1–6], which can be divided into the following two approaches. The first is applications without using a platform/middleware [1, 2]. Model-based methods are employed to well design applications. However, common functions for handling context have to be realized in each application. The second develops a middleware for dealing with the functions and builds the applications above the platform, to simplify the development of situation-aware services [3–6].

The platform approaches can be further classified into two, depending on whether solutions for the conflicts are provided. Though researches [3, 4] can provide a platform, they cannot be used in conflict situations. On the other hand, [5, 6] can resolve the conflict situations to provide context-aware services. In [5], managing resources is discussed for resolution of conflict between electric devices, e.g., lamp, display and so on. However, the resolution policies are not clear. In real application, the more urgent situation should have more chance to get the services. In [6], Shin and Woo proposed a method based on weights to the historical conflicts. However, it is hard to deal with interruption to a current service, due to lack of sit state manage.

We first present a diagram of situation state transition (DSST) to specify the state of a situation. The diagram shows the phases and states of a situation and the change of them. Executable functions and degree of urgency are bound with each state of each phase. Therefore, the coordination and resolution of conflicts can be simplified by using the DSST. We give policies based on DSST for resolving conflicts, since different states reflect different levels of request urgency. We design a mechanism for reasoning phases and states of the situation based on sensed data. The change of states of a situation is triggered by situation-oriented ECA (Event-Condition-Action) rules, which are an extension of ECA rules for context-aware applications.

In this paper, we first present the Diagram of Situation State Transition (DSST), to describe the change the situation status, and give Situation-oriented ECA rules in order to trigger the creation and management of DSST. Based on DSST, reasonably fair policies are discussed and employed for resolution of the conflict situations of different users. Our advantages of are conflict-free, situation oriented solution, which can deal with interruption of services. In addition, the rate of successful request is reasonably fair, from the viewpoint of urgency and importance, and times of waiting.

The Requirement Model for the Management of Situations

Figure 1 shows the requirement for the management of situations. In Fig. 1, there are elderly or handicapped persons living in their homes. There are also helper(s) and care-robot(s) for support of the persons. Around a person, there are appliances, furniture, etc. It is supposed that some sensors are embedded into appliances, furniture, and rooms. For examples, under the floor, there is RFID sensors, such as U-Tiles [7], which can detect the position and person's ID for collecting the data of situations. A person may wear some wireless sensor device, e.g. a Magic Ring [8], which can be worn on a figure to detect the person's intension for control appliances or a care-robot, and call for the helpers. Some sensors, such as temperature sensor is employed in the room.

By using those sensors, data can be collected, and situation can be reasoned based on the data. For examples, a simple situation could be the handicapped person fell the temperature of the room is too low, the system is aware of his situation, and automatically turn on the heater in the room.

Generally, a situation of a person reflects the user's intension and requirement. The intention or requirement may be directly shown by some gestures, e.g. a person calls the help by shaking his hand, which can be detected by Magic Ring put on her figure. The requirement or needs of the person may not be shown clearly but hidden in his/her activities of life. In other words, they are shown unconsciously.

Another situation might be the person is approaching the table and want to drink his favorite tea; the care-robot will be aware of the requirement and serve the tea for the person. However, there may be a conflict that the user on the bed in the same home may feel bad and ask the care-robot to pick up a medicine, trigged by the signal from her Magic Ring on her finger. Therefore, sometime situations

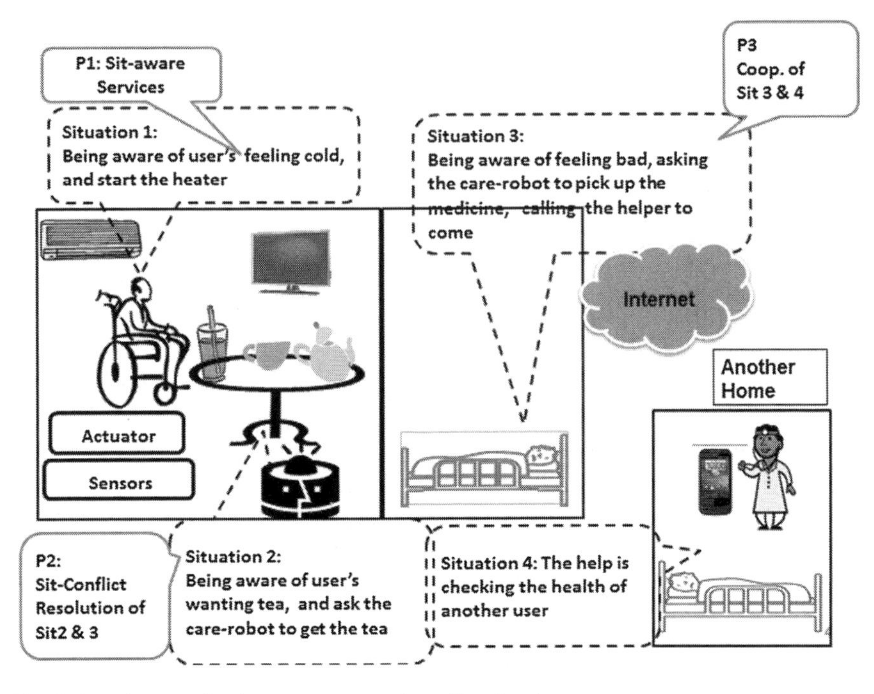

Fig. 1 Requirement for the management of conflict situations

happened around the two persons may compete for some resources, e.g. the care robot for provision of the situation-aware services. In addition, she may also call the helper for direct service by the helper. However, the helper is taking care of another person in another home. That is to say the situations happened in different homes are related and need coordination.

The Middleware and Conflict Resolution

The Outline of the Middleware

Figure 2 shows the architecture and the position of the middleware in the architecture. There are three layers in the architecture. The lowest one is the sensors and actuators; the middle is the middleware; the highest one is users' applications. The application is situation-awareness, which means they can automatically provide services according to the situations around the users, such as sending a warning signal when a danger may happen, a reminder just before a pre-defined situation happens, or a message to hospital/care center when an elderly person need help. To this end, the sensors are employed to get the data for reasoning the situations, and actuators are employed to perform the services.

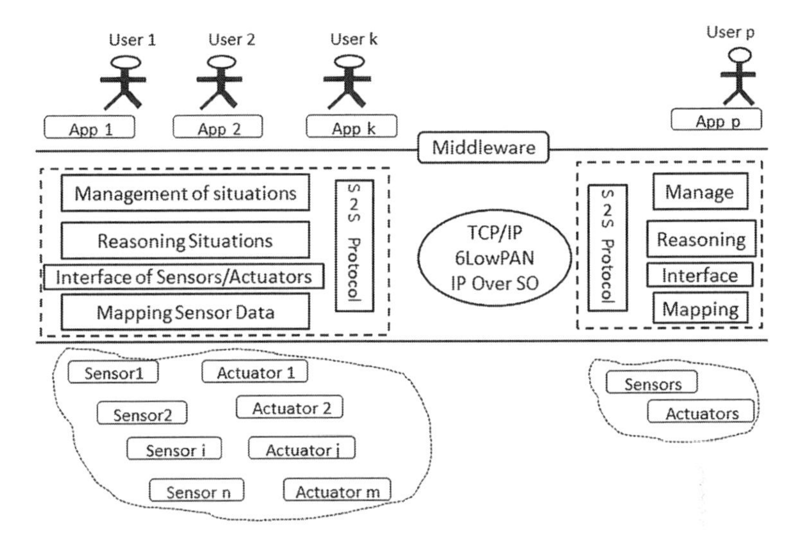

Fig. 2 The architecture and outline of the middleware

The middleware is to help the applications. Many common functions for dealing with the situations can be encapsulated in the middleware, such as creating situations, maintaining a situation, resolving the conflict situations, and coordination of situations. So the developer can concentrate on the design of services and conditions for triggering the services considering real situations.

The middleware is built above the internet currently. It will be built above some core technologies of IoT such as 6LowPAN and IP over smart object in the future. The main functions of the middleware are,

- For each kind of sensors, we assume driver software of the sensors is available to collect the data and transform them into the pre-defined regular data form.
- Reasoning the situations based on those data, which means to explain the meaning the data and relation of the data, by some formal way, such as SoECA.
- Each situation has a life time from start to the end. During life time, there are a series of the states, in each of which different functions or services can be triggered to provide services to user, resolve conflict, etc.
- A protocol called *situation to situation* communication (S2S) is necessary, for dealing with the coordination between related situations happening in different locations. (The protocol will be developed in the future)

Diagram of Situation State Transition

Each situation will be managed by a situation manager, which is called *situator*. There are more than one *situators* corresponding to multiple situations in the middleware. There is a special *situator* called *arbiter*, which is for solving the

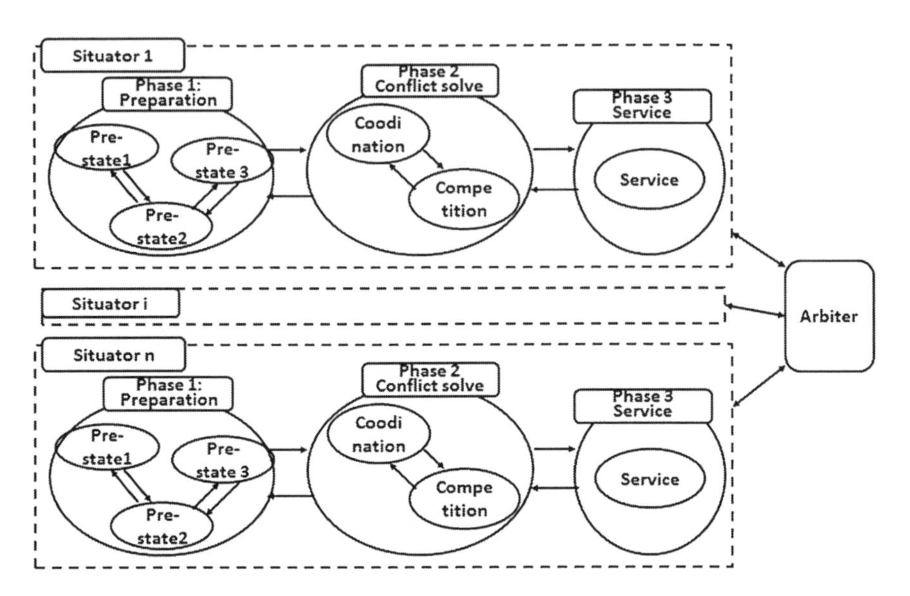

Fig. 3 Diagram of situation state transition

conflicts among the situations. Each *situator* communicates with the *arbiter*, as shown in Fig. 3.

A situator for a situation is created, when at least one condition defined for the situation becomes true, which can be detected by sensors. Each situator has three phases in its life time, i.e. preparation, coordination, and service. The first phase preparation starts, when some data related with a situation is detected by some sensors. Though the conditions for triggering the awareness service defined on the situation are partially satisfied, preparation for provision the pre-defined service, e.g. to find and reserve the necessary device and resource will be performed at this phase.

The second phase is mainly for resolving the conflict with other situations, e.g. completing for a device to provide the services for different users' situations. At this phase, each *situator* sends messages to the arbiter about their requests with the importance and urgency of the requests. The arbiter will make the decision to solve the conflict, based on priority scheme/policies, described later.

The 3rd phase is to provide services to the user, e.g. sending a warning message to avoid danger, etc.

The transition between phases is bi-directional, e.g. when the situator is providing a service, some new requests from other users with higher priority happen, the phase 3 will return to phase 2 for resolving the conflicts. Moreover, if the conditions for triggering a service are changed based on newly detected data, the situator will change its phase to preparation.

A situator will be terminated, when the all services for such situation have been finished and/or all the conditions for the situation become false.

There may be one or more states in each phase, to represent the stage in each phase, such as the level of preparation, degree of urgency for the services, etc. The transition between the states is also bi-directional, and the situator changes its state depending on the changes of situation conditions.

Policies for Resolving Conflicts

The *arbiter* will resolve the conflict based on some policy. Generally speaking, *fairness* is the first policy for resolving the conflict. That is, every user's situation should be allocated the resources in a fair way. For example, a *turn* based resolution is to give the high priority to each user in turn. However, if urgency of the requests from different users is not the same, the *turn* based policy is not really fair and reasonable, in the sense that some users should be paid more attention, since his/her situation is more urgent and crucial. Therefore, we use *degree of urgency*, as a criterion for the resolution. The urgency of the request is different in various situations, when a user need help in his/her everyday life, in an accident, or in a disaster. The user's situation for the services can be detected and reasoned by the system, or directly expressed by the user by some way, such as gestures detected by wearable sensor device. The types of different gestures can express the degree of the urgency, e.g. asking the system to help, call for helper, or SOS. If the degree of urgency of two requests is the same, degree of the importance of the users will be also considered, e.g. the worse the user's health condition is, the higher priority will be given, or the ages of the users (say 90', 80', 70', 60') are used.

The policies are managed based on the situation diagrams. Different states reflect level of satisfaction of conditions for providing services, and degree of urgency for the service. In the same level of state, the more important request will be served with higher priority. Preemption is also employed, which means a request with higher priority will interrupt the service of a lower priority being provided. And the interrupted and pending service will be resumed when the higher priority service is finished. First-come-first-served policy will be used, if the levels of the importance and urgency of two situations are the same.

The arbiter runs the following algorithms.

//* Dealing with the request and put it into a queue

When the arbiter receives a request (user's phase and state, etc.) from a situator

Check User Context Check carried by the parameters of the request.

Put the Request into the priority Queue based on the following combined policies

//* Check the requests in the queue, and compare them with the new request

Prior-based Policy

step 1. Check degree of urgency (depending on the phases/states reflecting urgency)

step 2. Check degree of importance (users' profile information such as ages)

step 3. Update request queue, by inserting the request into the proper position.

Fairness-Tuning Rule
step 1. Check if taking over request exist (times of waiting over default 5)
step 2. Update the priority of request being taken over.
//* Services Provision
Assign devices to provide services based on the request queue.
Preemption-based Policy is used to interrupt the current service if necessary.
Except a user's services is preempted too many times (default at 5)

Reasoning of Situations

As shown in Fig. 4, there are four main parts for reasoning situations. They can be divided into *specification* and *detection*.

Specification

The description is based on the concepts of situation theory [9], which specifies a situation using 4 elements, i.e. individual, properties, relations, and spatial-time location. The specification is presented in a formal way, such as first-order predicates.

An example of situation could be a handicapped person feels thirsty and wants to drink a cup of the tea, since it is a hot day and high humidity. In this example, the individual is the person as well as the smart objects, such as the cup embedded with sensors to know if it is empty or not. The properties are the person is

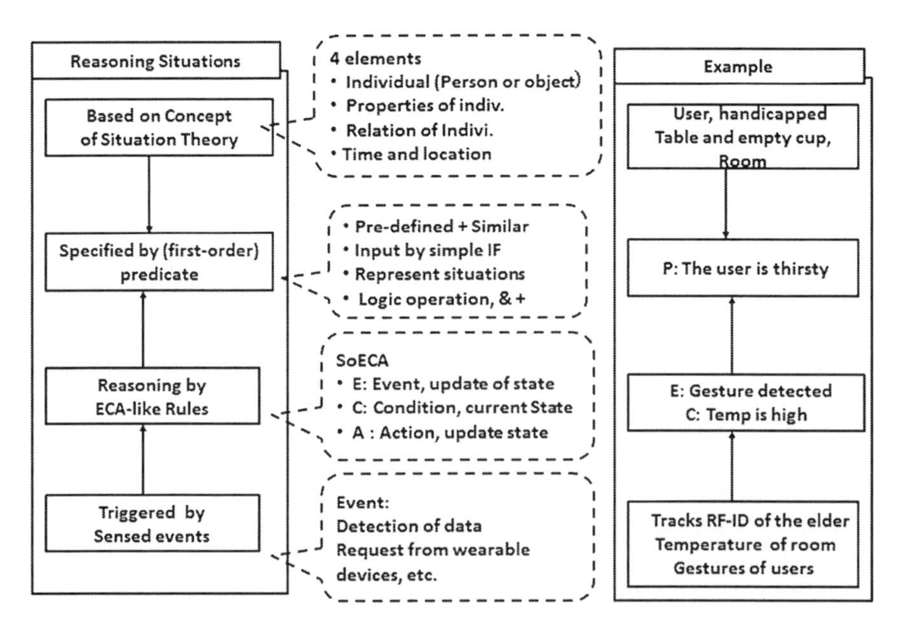

Fig. 4 Reasoning situations

handicapped so that it is hard to get the tea by himself, and the tea cup is empty. The spatial-time location is current room and time. Thus, the situation could be represented as follows

P The user feels thirsty

Q The user is hard to pick up the cup of the tea.

Services actions are defined on a situation, e.g.

P and Q => SA

where, SA (service action) is to ask the robot to get the tea.

Detection (Situation-oriented ECA, simply called SoECA)

Sensors are used to detect the raw data for reasoning a situation, and ECA rules are employed to reason the situation based on the data. The ECA is situation-oriented, which means *Event*, *Conditions,* and *Actions* are related with the situation states. *Event* could be detection of a new data or an update of situation states. *Condition* could be a condition for the service on the situation and current state. *Action* could be a trigger for service, or update of the state of the situation for giving the service later.

SoECA rule:

E an event or an update of situation state

C a condition for a service and current state of the situation

A a service action and/or updating state of the situation.

Some examples of ECA are as follows.

SoECA rule 1:

E Temperature is high than a threshold & humidity high than threshold

C *situator* has not been created

A Create *situator* and set its state to 1st state (S1) of 1st phase with the lowest priority.

SoECA rule 2:

E User1 is approving the table (detected by RFIDs and its position)

C The cup is empty & the current state is S1 of 1st phase

A Ask the robot to get the tea & the state to 2nd phase.

SoECA rule 3:

E User2 did gesture *G*1 for help with picking up medicine

C Situator has not been created

A Creation of the *situator* and set the state to 2nd phase.

SoECA rule 4:

E User2 did a gesture *G*2 for help with getting a drink

C Situator has not been created)

A Creation of the *situator* and set the state to 2nd phase.

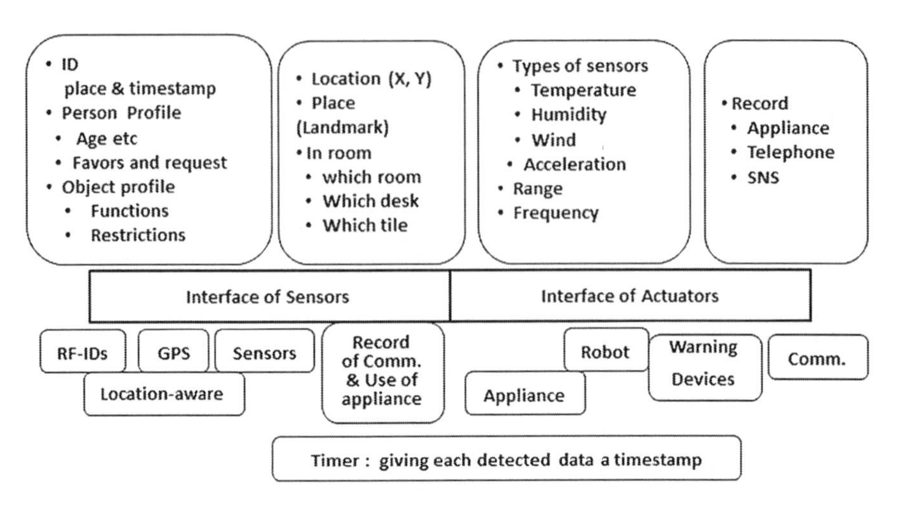

Fig. 5 Interface of sensing/actuation

Generally, a collection of SoECA rules will be designed by developer of the application and/or the services providers who has the experts in the domain, based on domain knowledge and engineering design knowledge.

Ideal Interface of Sensors and Actuators

As shown in Fig. 5, many sensors are available in current technologies, e.g. RFIDs, GPS, various environment sensors, such as temperature, humidity, wind, as well as acceleration, etc. The mobile or wearable devices can detect demands of a user, activities of the user, and vitality data of a user. In addition, the usage of appliance, furniture, care-robots, and other actuators can be recorded and used as historical data. The sensed data can be transformed into a regular form. The time and places when the data is detected is also useful for situation reasoning.

Feasibility Study

Example of Conflict Situations

In this paper, we show how the conflict resolution is realized using an example to investigate the feasibility of the implementation. As shown in Fig. 6, we assume that User1 is a handicapped person in the left room and User2 is an elderly person

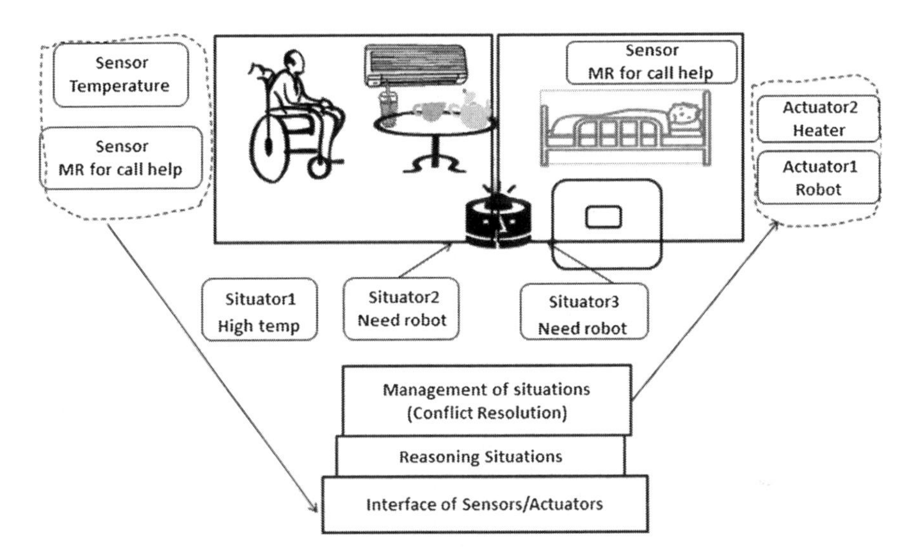

Fig. 6 Implementation example

lying on the bed in the right room, and care-robot is employed in the home for services.

We assume that User1's situation is he wants to drink the tea, since the weather is hot and he has not drunk for a while. The temperature can be detected by the sensor, the User1 can be detected by RFID on his body, and the time duration from last drinking can be detected, by a timer and smart cups (with sensor). User2's situation is she wants help by the care-robot to pick up drink or medicine. The request can be sent by using gestures, which can be detected by the figure-worn device Magic Ring [8]. In this example, two gestures are used. Gesture 1 and 2 mean requesting a drink only or medicine, respectively.

Implementation of the Example

We have implemented the example using C++ on Window 7. *Situators* of User1, User2, and the care-robot are implemented as classes. The *arbiter* is implemented as another class, which receives the requests from classes of User1 and User2, and controls the care-robot for the service.

We first show the software for conflict situations. In Fig. 7, three windows show the running information of objects of User1, User2, and the robot classes. Initially, the robot character is on the middle of the window, which means there is no request for it. The *progress bar* in the left window is used to show the thirsty degree. After a drink, the progress bar is set to *full* (*blue*), to mean the user is not thirsty. With time elapses, the blue part is reducing, which means the degree of

Fig. 7 User1, user2 and robot

thirsty increases. The speed of the reducing is depending on the temperature. User2 can push the button to ask the robot to get help directly, which can be realized by a gesture detected by Magic Ring.

In Fig. 8, when user1 feels thirsty (the progress bar becomes lower than 20 % of the full), the robot will walk to him to provide a service. The progress bar in the middle window show the service by the robot is going on.

In Fig. 9, if User2 uses the button to call the robot, the robot will break the service to user1 and walks to User2 to help her, even when the robot is providing service to User1, since the priority of User2 is higher than User1's. In Fig. 10, the User2 can really use a wireless sensor to make request. The conflict is resolved, same as the above discussed.

To evaluate our method is workable and solve the conflicts reasonably. We use rate of satisfied requests denoted with R_i to represent the percentage of getting the resource when conflict. That is, $R_i = N_i/N$, i belongs to $\{1, 2\}$. The number N_i is the times $user_i$ won the competing resource. N is the number of total competing times.

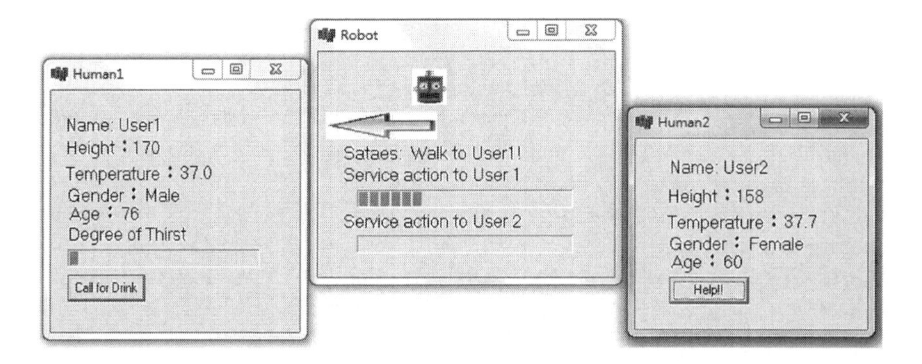

Fig. 8 User1 becomes thirsty and robot comes to user1

Fig. 9 User2 asks robot for service

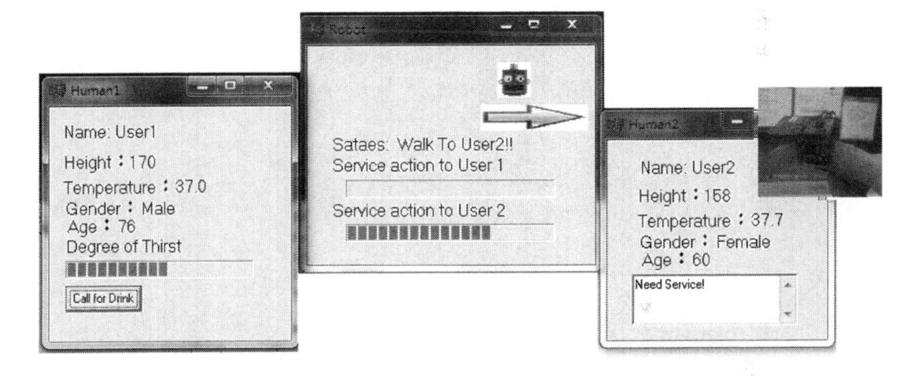

Fig. 10 User2 uses gestures detected by MR to make request

We compare User1 and User2 in various states. For each case, we run the example mentioned above 1,000 times, record the result, and compute R_i, as shown in Table 1.

The degree of User1's urgency, is in such as a way that no-need < thirty < request < drinking. The urgency of User2's gesture is "for a drink < medicine. User1's request for a tea is almost equal to User2's request for a drink. A pure turn based police means to give the higher priority to each user alternately. Generally, a turn based policy lead to Ri = 50 %. Our method also combines an extension of turn based method. Times of waiting for recourse are used, when the waiting times increase the priority will be increased.

Table 1 Simulation results of conflict resolution

User1's state	Thirsty (%)	Request (%)	Drinking (%)	No-need (%)
User2 requests for a drink	R1: 29.9 R2: 71.1	R1: 54.7 R2: 46.3	R1: 94.8 R2: 5.2	R1: 0 R2: 100
User2 requests for medicine	R1: 16.1 R2: 83.9	R1: 20.7 R2: 79.3	R1: 20 R2: 80	R1: 0 R2: 100

Conclusion

The contributions of this paper are as follows. At first, we proposed architecture of IoT middleware, which described composition of the middleware and relations with situation-aware applications. The main contribution is that we presented the *diagram of situation state transition* (DSST) and conflict resolution policies based DSST. We also presented SoECA rules for reasoning states of situations. Simulation shows our proposal is feasible and effective for solving conflicts.

Out proposal is significant, in the sense that it can well represent changes of user's situation using state transition and degree of urgency in each state, and resolve the conflict between situations, based on the urgency represented by phases and states. The middleware can be a platform for developing situation-aware service application.

In the future, we will complete the middleware and evaluate of the proposal extensively. We will improve DSST in order to manage the situations elegantly. Furthermore, we will give a protocol of situation to situation (S2S) to enhance the coordination of different situations happening in different locations.

References

1. Kapitsaki GM, Prezerakos GN, Tselikas ND, Venieris IS (2009) Context-aware service engineering: a survey. J Syst Softw 82(8):1285–1297
2. Prezerakos GN, Tselikas N, Cortese G (2007) Model-driven composition of context-ware web services using ContextUML and aspects. In: Proceedings of the IEEE international conference on web services 2007 (ICWS'07), Salt Lake City, USA, pp 320–329, 9–13 Jul 2007
3. Zhu J, Oliya M, Pung HK, Wai CW (2010) LASPD: a framework for location-aware service provision and discovery in mobile environments. In: Proceedings of the 5th IEEE Asia-Pacific services computing conference (APSCC 2010), Hangzhou, China, 6–10 Dec 2010
4. Pung HK, Gu T, Xue W, Palmes PP, Zhu J, Ng WL, Tang CW, Chung NH (2009) Context-aware middleware for pervasive elderly homecare. IEEE J Sel Areas Commun 27(4):510–524
5. Hanssens N, Kulkarni A, Tuchida R, Horton T (2002) Building agent-based intelligent workspaces. In: Proceedings of ABA conference, pp 675–681, June 2002
6. Shin C, Woo W (2005) Conflict resolution method utilizing context history for context-aware applications. In: Proceedings of pervasive 2005 workshop, Munich, Germany, pp 105–110
7. Wang J, Cheng Z, Jing L, Ota K, Kansen M (2010) A two-stage composition method for danger-aware services based on context similarity. IEICE Trans Inf Syst E93-D(6):1521–1539, Jun 2010
8. Jing L, Zhou Y, Cheng Z, Wang J (2011) A recognition method for one-stroke finger gestures using a MEMS 3D accelerometer. IEICE Trans Inf Syst E94-D(5):1062–1072
9. Barwise J, Perry J (1983) Situation and attitudes. Center for the Study of Language and Information Stanford, CSLI Publication, California

Data Transmission Mechanism in Cluster-Based Wireless Sensor Networks with Mobile Sink

Ying-Hong Wang, Yu-Jie Lin and Shao-Wei Tsao

Abstract In traditional wireless sensor networks, the sensed data will be forwarded to sink by multi-hop, so that the hot spot problem will happen on the nodes near the sink. The power consumption of these nodes is higher than others, these nodes will be dead sooner. It leads to the network lifetime decreasing. We proposed a data transmission mechanism using mobile sink to solve this problem. We use two threshold values to prevent data overflow. If the data buffer of cluster head exceed to 1st threshold, the node will call sink to come and transmit data. If the sink didn't arrive the transmission range of the node, and the data buffer exceed to 2nd threshold. Instead of forwarding data to sink directly by multi-hop, we use forwarding data to cluster head of neighboring cluster to help buffer the exceeded data. It can decrease the number of node which has to help to help to forward data, so that it can decrease the power consumption of network, and prolong the network lifetime.

Keywords Wireless sensor networks · Multi-hop · Sink · Hot spot · Overflow · Cluster head · Network lifetime

Y.-H. Wang (✉) · Y.-J. Lin · S.-W. Tsao
Department of Computer Science and Information Engineering, Tamkang University, New Taipei, Taiwan, Republic of China
e-mail: inhon@mail.tku.edu.tw

Y.-J. Lin
e-mail: yeanling319@gmail.com

S.-W. Tsao
e-mail: how4205@yahoo.com.tw

Y.-M. Huang et al. (eds.), *Advanced Technologies, Embedded and Multimedia for Human-centric Computing*, Lecture Notes in Electrical Engineering 260,
DOI: 10.1007/978-94-007-7262-5_52,

Introduction

In the operation of the wireless sensor network, how to response the sensed data to the sink is one of the most important issues [1–4]. In the network environment of traditional wireless sensor network, sensors near the sink have to forward the sensed data of other sensors frequently, so the power consumption of these nodes is very seriously. It may lead to sensor dead sooner. This phenomenon is called hot spot problem. To solve the hot spot problem, [5] it has proposed a lot of paper which use mobile sink to solve this problem recently [6, 7]. With the movement of the sink, the neighbor node of the mobile sink will be different, so it can balance the power consumption. In cluster based wireless sensor network system, the sensed data is forwarded to cluster head, and transmit to sink when the sink is in the transmission range of the cluster head. Because the number of sensors in each clusters are different, so the amount of transmitted data in each cluster are different too. However, before mobile sink moves to the destination node, it will lead to data overflow because of the data transmission too much, that will make data loss [6]. To avoid data loss which is because of data overflow, the exceeded data has to transmit to sink before data overflow, but it has to consume additional power to transmit data. So we propose a mechanism to control the data amount of the cluster head. We use two thresholds to achieve the proposed mechanism. First threshold use to call sink to come, and the second threshold use to find the neighboring cluster head to help buffer the exceeded data.

Related Works

Mobile wireless sensor network can be divided into three category, the first one is mobile sink with fixed sensor nodes, the second one is fixed sink with mobile sensor nodes, the last one is both sink and sensor are mobility. Our paper is fall in first category. The existence type of mobile sink with fixed sensor nodes wireless sensor network routing protocols can be divided into two types [7]. The first type is the sensed data through multi-hop transmission to the mobile sink. Another type is the sensed data store in memory of sensor, and waiting for the mobile sink moves to the communication range of the sensor node and then transmit data to the mobile sink directly. By multi-hop, it can response the data immediately, but we can't measure the moving direction of the mobile sink, so we usually use broadcast to forward data, but it will consume lots of energy because of the unnecessary transmission [7]. By one hop, the advantage of this type is the data through other sensor nodes do not have to jump through a multi-point to generation therefore can reduce the wear and tear of the overall network power, but it can't provide real-time transmission, and the sensed data has to store in the memory and waiting for the mobile sink moves to its transmission range, it may lead to data overflow [6].

The Proposed Algorithm

In this paper, we proposed a Non-Real Time Data Transmission Mechanism with Mobile Sink in Cluster-based Wireless Sensor Networks (NTDT). In the proposed mechanism, we use two thresholds to control the data buffer. When the data amount exceed to first threshold, it will broadcast a query to call sink. When the data amount exceed to second threshold, it will broadcast a query to ask neighboring cluster head to help buffer the exceeded data, it can decrease the data loss rate caused by data overflow. The Fig. 1 is the system architecture.

Network Architecture and Assumptions

We deploy hundreds to thousands of sensor randomly in our interested environment and the sensor has fixed on the location. We assume all the sensors are homogeneous, which has the same energy, memory size, computing ability, and transmission range. And every sensor has the following node information table (NIT), as Table 1.

And there is one mobile sink, it has no resource constrain. When the mobile sink moving, it broadcast a time stamp to its one hop neighboring sensor every 2R

Fig. 1 The system architecture

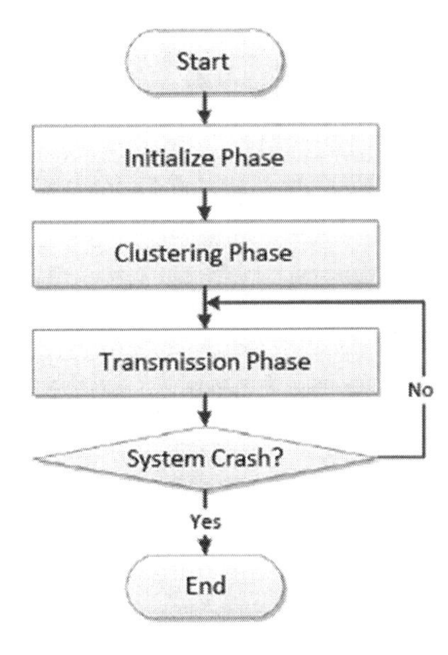

Table 1 Node information table (*NIT*)

Node ID	Cluster ID	N mode	TS

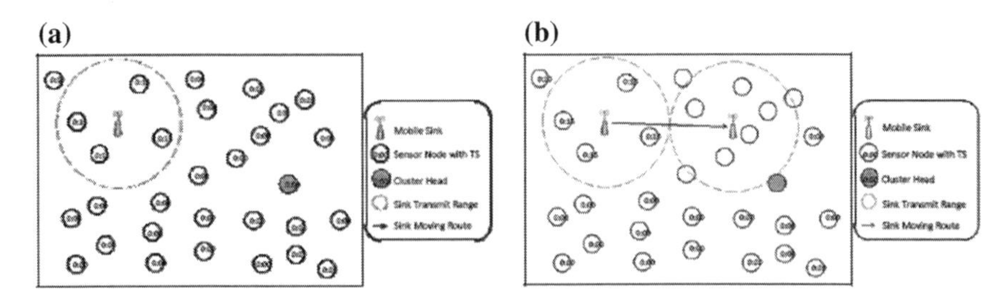

Fig. 2 **a**, **b** Sink broadcasts time stamp

distance (R is the transmission range of the sink). When the sensors receive a time stamp, they will update their TS in NIT. Figure 2a, b show the action of sink broadcasts the time stamp and sensors receive the time stamp, respectively.

The TS value updates by sink broadcasts time stamp, so that the greater TS value indicates that the node is the closer the data collection point.

The Definitions of Message Packets and Record Tables

Except the sensed data, we use some message and record table to execute our mechanism. The first one is Query Message, and there are two query types, the first one is 1st threshold query (Q_1) which is uses to call sink, the second one is 2nd threshold query (Q_2) which is used to ask neighboring cluster head to help buffer exceeded data. The two types of query use the same format packet as Table 2.

When a sensor receives a Q_1 or Q_2 query, if the query hasn't been received, it will be store in Query Record Table (QRT) which is used to avoid the duplicated query. It's shown in Table 3.

After a cluster head receive a Q_2 query which is sent from other cluster, it will reply a Residual Memory Reply (RMR) to the Q_2 query generator. RMR message format is shown in Table 4.

Table 2 Query message

Cluster ID	Q ID	Query type	Q TH

Table 3 Query record table (*QRT*)

Cluster ID	Q ID	Last hop	Query type

Table 4 Residual memory reply (*RMR*)

Cluster ID	RMR ID	TS	Residual memory

Table 5 Ack record table (*ART*)

Next hop	Q ID	Cluster ID	Ack type

Table 6 Residual memory table (*RMT*)

TS	Cluster ID	Residual memory	Last hop

Table 7 Exceeded data packet

Destination cluster	Data payload

Table 8 Query table (*QT*)

Cluster ID	Q ID

When a sensor node receives a message from the other sensor node, it will reply an Ack to the last hop of the message that the sensor nodes have received this message receives the Ack will be recorded in Ack Record Table (ART), the tabular format as shown in Table 5.

When the generator of Q_2 query receives a RMR from other cluster, the RMR will record in Residual Memory Table (RMT), the tabular format as shown in Table 6.

After the generator of Q_2 query receives the RMR, it will forward the exceeded data to the chosen neighboring cluster head according to the TS value in the RMT. And the data packet format shows in Table 7.

When sink receive a Q_1 query, it will be record in Query Table (QT) which is a table built in sink. The format is shown in Table 8.

Data Transmission Phase

After initial phase, it will begin the data transmission phase, we will divide this phase to cluster head, normal sensor, and sink to introduce.

Cluster Head Receive Message Processing Strategy

After cluster head receive a message, first it will determine what kind of the message is, if it is a query, it will execute query dissemination scheme in CH (QDSCH), if the received message is a data, it will execute Data Packet Dissemination Scheme in CH (DPSCH), if the received message is an Ack, it will record in ART, if the received message is a RMR, it will execute RMR Dissemination Scheme in CH (RDSCH), if the received message is a time stamp from sink, it will update the TS value in NIT, and transmit data to sink.

In QDSCH, it will determine the query has received or not by QRT. If not, it will determine the query type, if it is Q_2, it will determine where the query comes from, if it is from other cluster, it will reply a RMR. If it is Q_1, if the Q_{TH} is equal or less than TS in NIT, it will be forwarded continue.

In DPHSCH, after receiving data, it will be stored in memory of sensor, if the data amount exceed to 1st threshold, it will execute Data Buffer Control (DBC), when data exceeded to first threshold, it will broadcast a Q_1 query to sink to ask sink some to transmit data. Before sink arrive, if the data amount exceed to 2nd threshold, it will broadcast a Q_2 query to ask neighboring cluster head to help buffer data.

In RDSCH, it will determine the RM value of the RMR, if the value less than a threshold RM_{TH}, it won't be record in RMT.

Normal Sensor Receive Message Processing Strategy

After normal sensor receive a message, as cluster head, it will determine what kind the message is, if it is a query, it will execute Query Dissemination Scheme in Sensor (QDSS), if it is a data, it will execute Data Packet Dissemination Scheme in Sensor (DPDSS), if it is an Ack, it will be record in ART, if it is a RMR, it will execute RMR Dissemination Scheme in Sensor (RDSS), if it is a time stamp from sink, it will update the TS value in NIT.

In QDSS, it will determine that the query has received or not, if it hasn't received, it will determine what kind the query type is, and determine whether it will be forwarded or not, as the judgment of cluster head. In DPDSS, it will determine where the data come from, if it comes from the same cluster, it will be forwarded to cluster head, if it comes from other cluster, and it will determine that the data packet is on the right route, if it is not on a right route, it will be discarded. In RDSS, it will determine that the RMR is on a right route, if it is, it will be forwarded.

Sink Moving Scheme

Moving scheme of mobile sin in our mechanism, it uses random way point, after receive a Q_1 query it will change its way toward to the Q_1 query generator. When sink is moving into a new cluster, it will find the cluster head of current cluster to transmit data, after receiving data it will delete the cluster information from QT until the QT is empty.

Simulation an Analysis

We use the network simulator tool NS2 2.29 to show our performance. We compare the two papers, "An energy and delay aware data collection for mobile sink Wireless Sensor Networks based on clusters" (HDCA) and "An Efficient Data-Driven Routing Protocol for Wireless Sensor Networks with Mobile Sinks" (DDRP).

We will introduce the assumptions and the parameter setting in Section Assumptions and Parameter Setting. And we will show the comparison with different data generation ratio and different moving speed of the mobile sink.

Assumptions and Parameter Setting

The environment settings and the parameters used in our simulation process is as follows.

- Network size is 150*150 m
- Deploy 100 sensors randomly
- Number of mobile sink is 1
- Transmission range of sensor is 15 m
- Initial power of sensor is 2 J
- Memory size of sensor is 50 K Bytes
- Data packet size is 100 Bytes
- Simulation time is 3,600 s.

To find the most appropriate number of cluster, we first compare the data transmit successful ratio between different thresholds and different number of clusters. The moving speed of sink is fixed to 5 m/s, and we set the first threshold as 50 and 60 % will the second threshold different to 20 and 30 %, and the number of cluster set to 3, 4, 5 to do the comparison, the result shows in Fig. 3.

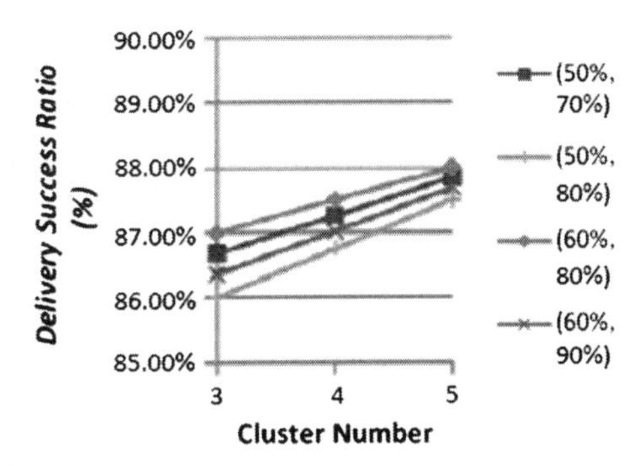

Fig. 3 Data delivery success ratio with different threshold and number of cluster

Because the data in 1st threshold is transmitting by 1-hop, the higher value of the first threshold is, the more complete data is. The data between the 2nd and 1st threshold is transmitted to neighboring cluster head by multi-hop, so that it will lead to data loss caused by the national loss. In different number of the cluster, the less number of clusters is, the greater cluster size is, so that the data will be forwarded by much more node, the data loss ratio will increase.

Simulation and Analysis

In this section, we will compare the performance between, HDCA, DDRP, and our mechanism.

First, we compare the difference of the packet generation interval, and the cluster number fixed to 5, speed of mobile sink fixed to 5 m/s, the difference is shown on Fig. 4. By simulation, we can see the packet loss ratio in DDRP is the highest, 10 % more on average than our proposed NTDT, because DDRP uses multi-hop to transmit data, the major factor is the route time expired. However, if the data generate too fast, it will lead to data overflow, 3 % on average than our proposed NTDT.

Next, we compare different moving speed of mobile sink with packet generation ratio fixed to 3 s., the performance is shown on Fig. 5. By simulation, we can see the packet loss ratio of DDRP is still the highest one. In HDCA, although the packet generation interval is slow, it will happened data overflow caused by moving speed of mobile sink, and the loss ratio is 5 % more than our proposed NTDT.

The last one we compare the system residual energy, moving speed of mobile sink is different, and the packet generation ratio is fixed to 3 s. The performance is shown on Fig. 6.

The major factor of power consumption is data transmission. By simulation, we can see DDRP is the most serious one, because it uses multi-hop to transmit data, and the energy consumption is 15 % more than our proposed NTDT. HDCA use the same type of mobile WSN, as ours, it buffers data in cluster head, and wait sink

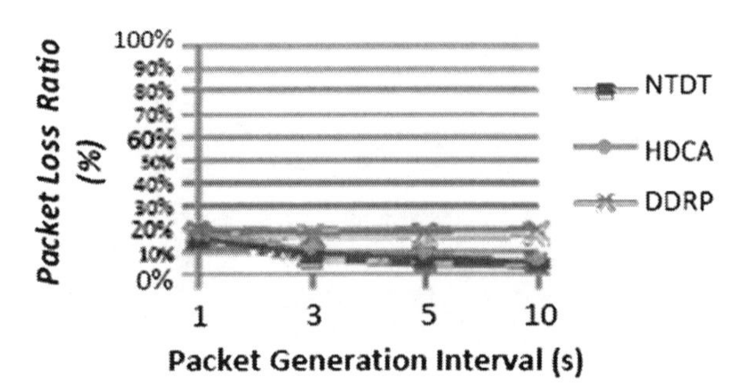

Fig. 4 Packet loss ratio with different packet generation interval

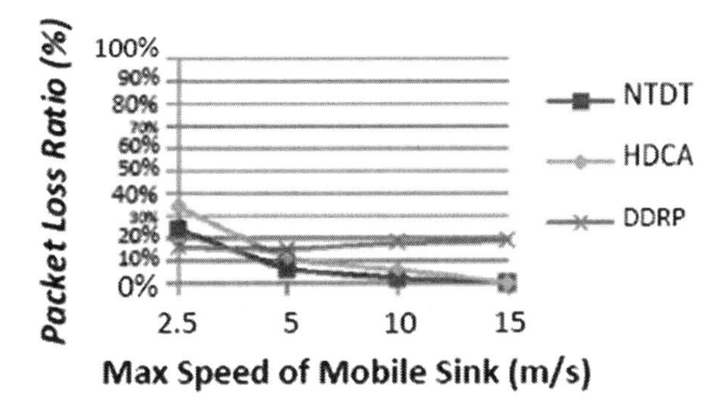

Fig. 5 Packet loss ratio with different moving speed of mobile sink

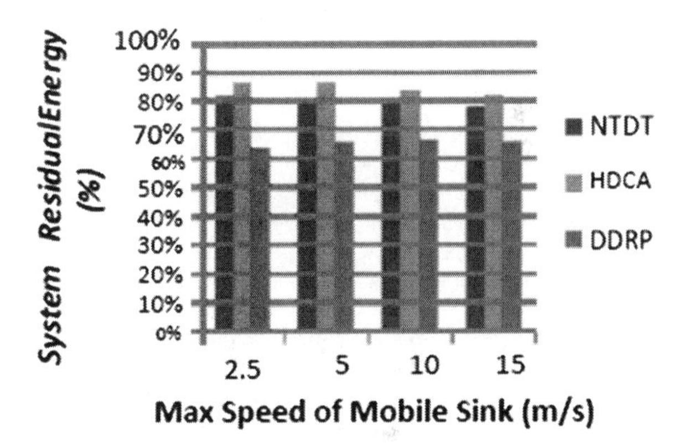

Fig. 6 System residual energy with different moving speed of mobile sink

move to its 1-hop range to transmit data. To avoid data overflow, we forward the exceeded data, so the power consumption of our proposed mechanism is more than HDCA about 5 %.

Conclusion

The most important problem in wireless sensor networks is the energy and the integrity of the data collection; however, the two types of the mobile wireless sensor networks have their own advantages and disadvantages. With non-real time data transmission, the efficacy of the type of data buffered in memory and use one-hop to transmit data to sink is better than the type of data transmitted directly by multi-hop. In this paper, we propose a mechanism, which can avoid data loss caused by data overflow. And we consider the power consumption problem, instead of transmitting the sensed data by multi-hop, we use the query whose packet is smaller than the data packet, to call sink, to decrease the power consumption of the data transmission. By simulation, our power consumption is slightly higher than the other papers, but the difference is not much. We provide the better data protection avoid data loss caused by data overflow.

References

1. Lin CJ, Chou PL, Chou CF (2006) HCDD:hierarchical cluster-based data dissemination in wireless sensor networks with mobile sink. In: Proceedings of the international conference on wireless communications and mobile computing, pp 1189–1194, Jul 2006
2. Lee D, Park S, Lee E, Choi Y, Kim SH (2007) Continuous data dissemination protocol supporting mobile sinks with a sink location manager. In: Proceedings of the Asia-Pacific conference on communications, pp 299–302, Oct 2007
3. Soyturk M, Altilar T (2007) A routing algorithm for mobile multiple sinks in large-scale wireless sensor networks. In: Proceedings of 2nd international symposium on wireless pervasive computing, Feb 2007
4. Heinzelman WR, Chandrakasan A, Balakrishnan H (2002) An application-specific protocol architecture for wireless microsensor networks. IEEE Trans Wireless Commun 1(4):660–670
5. Ye M, Chen G, Wu J (2005) An energy-efficient unequal clustering mechanism for wireless sensor networks. In: Proceedings of the IEEE international conference on mobile adhoc and sensor systems conference, pp 597–604, Nov 2005
6. Yu H, Kuo M (2010) An energy and delay aware data collection for mobile sink wireless sensor networks based on clusters. In: Proceedings of the international conference on computer application and system modeling (ICCASM), pp 536–540, Oct 2010
7. Shi L, Zhang B, Huang K, Ma J (2011) An efficient data-driven routing protocol for wireless sensor networks with mobile sinks. In: Proceedings of the IEEE international conference on communications (ICC), pp 1–5, Jun 2011
8. Chang Sh, Merabti M, Mokhtar HM (2007) Coordinate magnetic outing for mobile sinks wireless sensor networks. In: Proceedings of the international conference on advanced information networking and applications workshops, vol 1. pp. 846–851, May 2007
9. Zhou Z, Xang X, Wang X, Pan J (2006) An energy-efficient data-dissemination protocol in wireless sensor networks. In: Proceedings of the international symposium on a world of wireless mobile and multimedia networks, pp 10–22, Jun 2006
10. Luo H, Ye F, Cheng J, Lu S, Zhang L (2002) TTDD: two-tier data dissemination in large-scale wireless sensor networks. In: Proceedings of the 8th international ACM conference on mobile computing and networking, pp 148–159, Sep 2002

Border Detection of Skin Lesions on a Single System on Chip

Peyman Sabouri, Hamid GholamHosseini and John Collins

Abstract High speed image processing is becoming increasingly important in medical imaging. Using the state-of-the-art ZYNQ-7000 system on chip (SoC) has made it possible to design powerful vision systems running software on an ARM processor and accelerating it from hardware resources on a single chip. In this paper, we take the advantage of accelerating an embedded system design on a single SoC, which offers the required features for real-time processing of skin cancer images. Different edge detection approaches such as Sobel, Kirsch, Canny and LoG have been implemented on ZYNQ-7000 for border detection of skin lesions, which can be used in early diagnosis of melanoma. The results show that the extended 5×5 canny edge detection implemented on the proposed embedded platform has better performance in compare with other reported methods. The performance evaluation of this approach has shown good processing time of 60 fps for real time applications.

Keywords Border detection · Edge detection · ZYNQ-7000 · Medical imaging

Introduction

To date, skin cancers have been one of the most common form of cancers particularly in New Zealand [1, 2]. Skin cancers are divided into two main categories: melanoma and non-melanoma. Early diagnosis of malignant melanoma can

P. Sabouri (✉) · H. GholamHosseini · J. Collins
Department of Electrical and Electronics Engineering, Auckland University of Technology, Private Bag 92006, Auckland 1142, New Zealand
e-mail: psabouri@aut.ac.nz

H. GholamHosseini
e-mail: hgholamh@aut.ac.nz

J. Collins
e-mail: jcollins@aut.ac.nz

Y.-M. Huang et al. (eds.), *Advanced Technologies, Embedded and Multimedia for Human-centric Computing*, Lecture Notes in Electrical Engineering 260,
DOI: 10.1007/978-94-007-7262-5_53,

significantly decrease the morbidity, death and cost of the treatments [3]. Dermoscopy is a non-invasive method for diagnosis of melanoma and pigmented skin lesions. Although this device illustrates features of pigmented lesions, it is a challenging task for dermatologist to diagnose melanoma from other skin lesions [4]. Image processing techniques can be applied to skin images for better diagnosis of melanoma. For example, image features such as Asymmetry, Border irregularity, Color variation and regions with Diameter greater than 6 mm (ABCD rule) can be extracted using high performance image processing techniques [5].

In recent years, real-time vision systems have been used in a wide range of applications such as sophisticated medical imaging. Most computer-based vision applications have been developed based on a Graphics Processing Unit (GPU). However, recent developments in the field of powerful, low cost and energy-efficient embedded systems have led to the implementation of image/video applications into Digital Signal Processors (DSPs) and Field Programmable Gate Arrays (FPGAs). Nonetheless, each technique has its advantages and tradeoffs based on the nature of the algorithm, performance requirement, power consumption, cost, productivity, flexibility and design cycle time. In addition, the complexities of these techniques can create bottlenecks for developers. Moreover, working with both hardware and software design components as one system may lead to additional challenges. One of the recent products from Xilinx, ZYNQ-7000, can be considered as a promising solution to overcome these challenges. The ZYNQ-7000 consists of a dual-core ARM processor which is surrounded by a new 7-series Xilinx FPGA-based on 28 nm technology. The close integration between processing system (PS) and programmable logic provides the flexibility of ASIC technology and performance of FPGA technology on a single System on Chip (SoC). The ZYNQ-7000 SoC may be suitable for designing a handheld medical imaging system for skin cancer detection.

In this paper, different edge detection methods have been implemented in the programmable logic of ZYNQ using the VIVADO HLS tool (ver. 12.3). Border detection is widely used dermoscopic image analysis with the aim of detecting the boundaries between melanoma regions and the background.

ZYNG-7000 and System Implementation

The ZYNQ-7000 SoC consists of a dual-core ARM Processor and a Xilinx FPGA based on 28 nm technology [6]. The proposed system consists of the Zynq-7000 AP SoC ZC702 evaluation kit (ZC702). A CCD camera is connected to the board by a high-performance video I/O FPGA mezzanine card (FMC module) with HDMI input and output. Several video soft IP cores, such as defective pixel removal, de-mosaic, color correction matrices and video input/output, are used in order to design an image processing pipeline and border detection. In addition, in the processing system (PS) DDR3 memory is used to create input/output images via the ARM processor using AXI Video Direct Memory Access (VDMA).

Moreover, the VIVADO HLS tool (ver. 12.3) is used to implement the real-time border detection accelerator on the programmable logic (PL) of the ZYNQ AP SoC.

Unlike the traditional FPGA design flows, High Level Synthesis tools (HLS) can improve design quality, accelerate the design and verification tasks by applying optimal synthesis directives to transform C/C++/SystemC code specifications to register transfer level (RTL) implementation [7, 8]. Although C code is used for implementation of the algorithm, several directives such as DataFlow, memory/interface, and for loop optimisations [9] are applied to obtain the required performance and area utilisation in PL. In this article functional verification of the border detection algorithms was performed using C++ testbenches.

Border Detection

Edge detection is one the most important algorithms which is widely used in image analysis with the aim of detection of boundaries between objects and the background in gray level images. Although there are many approaches for border detection and object recognition in medical imaging, the calculation of the gradient level value for each pixel using neighboring masks and comparing with a given threshold, is a common approach. If the calculated value is greater than the threshold, the pixel is considered as an edge. The most important categories of edge detection algorithms are [10]:

- Gradient edge detectors (first derivative such as Sobel, Prewitt and Kirsch)
- Zero crossing (second derivative)
- Laplacian of Gaussian (LoG)
- Gaussian edge detectors (such as Canny)
- Colored edge detectors

In this paper, Sobel, Kirsch, LoG and Canny edge detectors were chosen for border detection of the skin cancer images. It was found that 3×3 kernels, which are normally used in edge detection methods, are highly localized. To achieve the best method for border detection, the extended edge operators (5×5 kernels) were used to include more neighborhood pixels [11]. These extended operators are described in more details in the following sections.

Sobel Operator

The Sobel operator consists of two vertical and horizontal kernel components, which can be obtained by 3×3 vertical and horizontal kernels (Eq. 1). The kernels are applied to each pixel in the image, and then a threshold is applied to create the final output pixel (Fig. 1b). To achieve better results, the Gx and Gy

Kernels are extended to 5 × 5 masks (Eq. 2) and Fig. 1c shows the results of applying this filter to the input image in Fig. 1a.

$$G_x = \begin{bmatrix} -1 & 0 & +1 \\ -2 & 0 & +2 \\ -1 & 0 & +1 \end{bmatrix}, G_y = \begin{bmatrix} +1 & +2 & +1 \\ 0 & 0 & 0 \\ -1 & -2 & -1 \end{bmatrix} \tag{1}$$

$$G_x = \begin{bmatrix} 2 & 2 & 4 & 2 & 2 \\ 1 & 1 & 2 & 1 & 1 \\ 0 & 0 & 0 & 0 & 0 \\ -1 & -1 & -1 & -1 & -1 \\ -2 & -2 & -4 & -2 & -2 \end{bmatrix}, \mathrm{G}_y = \begin{bmatrix} 2 & 1 & 0 & -1 & -2 \\ 2 & 1 & 0 & -1 & -2 \\ 4 & 2 & 0 & -2 & -4 \\ 2 & 1 & 0 & -1 & -2 \\ -2 & -2 & -4 & -2 & -2 \end{bmatrix} \tag{2}$$

Kirsch Operator

The Kirsch operator consists of eight basic convolution kernels components [8]. The vertical and horizontal edges can be obtained using the 5 × 5 Gx and Gy kernels in Eq. (3) and Fig. 1d shows the results of applying this filter to the input image.

$$G_x = \begin{bmatrix} 9 & 9 & 9 & 9 & 9 \\ 9 & 5 & 5 & 5 & 9 \\ -7 & -3 & 0 & -3 & -7 \\ -7 & -3 & -3 & -3 & -7 \\ -7 & -7 & -7 & -7 & -7 \end{bmatrix}, G_y = \begin{bmatrix} 9 & 9 & -7 & -7 & -7 \\ 9 & 5 & -3 & -3 & -7 \\ 9 & 5 & 0 & -3 & -7 \\ 9 & 5 & -3 & -3 & -7 \\ 9 & 9 & -7 & -7 & -7 \end{bmatrix} \tag{3}$$

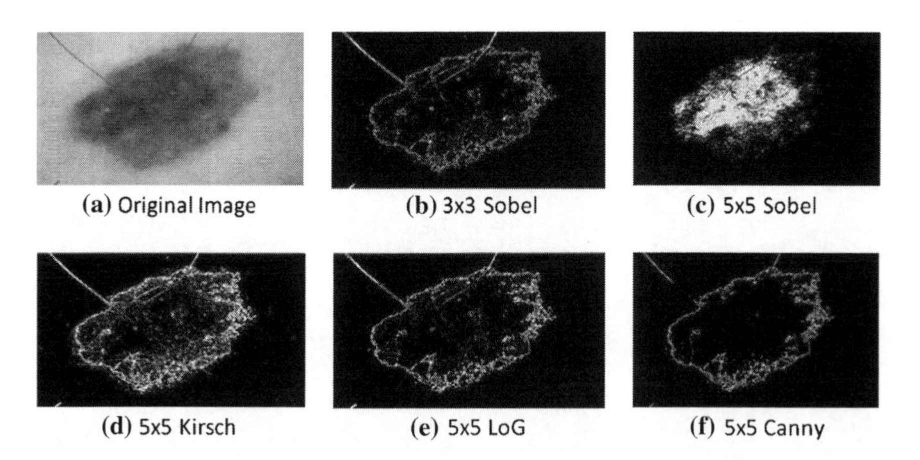

Fig. 1 The resulting images from the border detection IP cores: **a** original image, **b** 3 × 3 Sobel operator, **c** 5 × 5 Sobel operator, **d** 5 × 5 Kirsch operator, **e** 5 × 5 LoG operator and **f** 5 × 5 Canny operator

LoG Operator

The second-order gradient LoG edge detector can be used by convoluting the LoG filter to the image [12]. The mask in Eq. (4) implements the LoG filter and Fig. 1e shows the results of applying this filter to the input image.

$$G_x = \begin{bmatrix} 0 & 0 & -1 & 0 & 0 \\ 0 & -1 & -2 & -1 & 0 \\ 0 & -2 & 16 & 0 & -1 \\ 0 & -1 & -2 & -1 & 0 \\ 0 & 0 & -1 & 0 & 0 \end{bmatrix} \tag{4}$$

Canny Operator

The canny edge detector is an optimal approach to find the edges in images using a Gaussian filter to reduce noise when the raw image is convolved with this filter [13]. The masks in Eq. (5) implement 5×5 canny edge detection and Fig. 1f shows the results of applying this filter to the input image.

$$G_x = \begin{bmatrix} 15 & 69 & 114 & 69 & 15 \\ 35 & 155 & 255 & 155 & 35 \\ 0 & 0 & 0 & 0 & 0 \\ -35 & -155 & -255 & -155 & -35 \\ 15 & 69 & -144 & -69 & 15 \end{bmatrix},$$

$$G_y = \begin{bmatrix} 15 & 35 & 0 & -35 & -15 \\ 69 & 155 & 0 & 155 & 69 \\ 114 & 255 & 0 & -255 & -114 \\ 69 & 155 & 0 & -155 & -69 \\ -15 & -35 & 0 & -35 & -15 \end{bmatrix}$$

Experimental Results

The original skin cancer image used for testing of border detection algorithms is shown in Fig. 1a. In addition, the results for Sobel, Kirsch, LoG and Canny edge detection implementations are illustrated in Figs. 1b–f. The results show that the extended 5×5 canny edge detection has better performance than other methods. After function verification of the algorithms, the IP core is generated using C-to-FPGA (VIVADO HLS) to be implemented into the programmable logic of ZYNQ-7000. Using this feature the development time was significantly reduced during the functional verification stage. The generated core segments border of skin lesions for skin cancer images. Therefore, it can be used for feature extraction step such as asymmetrical shape analysis. Furthermore, the created IP performed better border-detection in real-time (60 fps) using VIVADO HLS in compare with the traditional FPGA design flow.

Conclusions

The results show that the extended 5 × 5 canny edge detection implemented on the proposed embedded platform performs better than other reported methods. Consequently, it can segment border of the lesion before applying image analysis and feature extraction for melanoma detection. The generated IP for lesion border detection presented good processing time of 60 fps for real time applications using VIVADO HLS. Therefore, the proposed ZYNQ-7000 EPP can be considered not only as a high-performance and cost-effective solution for hardware/software implementation of image processing algorithm, but also a promising alternative for computer aided diagnosis (CAD) systems as a single SoC for portable solutions.

References

1. Diepgen T, Mahler V (2002) The epidemiology of skin cancer. Br J Dermatol 146:1–6
2. Kopf AW, Salopek TG, Slade J, Marghoob AA, Bart RS (1995) Techniques of cutaneous examination for the detection of skin cancer. Cancer 75:684–690
3. Goldsmith LA, Askin FB, Chang AE, Cohen C, Dutcher JP, Gilgor RS, Green S, Harris EL, Havas S, Robinson JK (1992) Diagnosis and treatment of early melanoma. JAMA: J Am Med Assoc 268:1314–1319
4. Argenziano G, Soyer HP, Chimenti S, Talamini R, Corona R, Sera F, Binder M, Cerroni L, De Rosa G, Ferrara G (2003) Dermoscopy of pigmented skin lesions: results of a consensus meeting via the internet. J Am Acad Dermatol 48:679
5. Xu L, Jackowski M, Goshtasby A, Roseman D, Bines S, Yu C, Dhawan A, Huntley A (1999) Segmentation of skin cancer images. Image Vis Comput 17:65–74
6. Zynq-7000 Extensible Processing Platform. http://www.xilinx.com/products/silicon-devices/soc/zynq-7000/index.htm
7. Chen J, Cong J, Yan M, Zou Y (2011) FPGA-accelerated 3D reconstruction using compressive sensing. In: Proceedings of the ACM/SIGDA international symposium on field programmable gate arrays, pp 163–166
8. Noguera J, Neuendorffer S, Van Haastregt S, Barba J, Vissers K, Dick C (2011) Implementation of sphere decoder for MIMO-OFDM on FPGAs using high-level synthesis tools. Analog Integr Circ Sig Process 69:119–129
9. Cong J, Zhang P, Zou Y (2011) Combined loop transformation and hierarchy allocation for data reuse optimization. In: IEEE/ACM international conference on computer-aided design (ICCAD), pp 185–192
10. Setayesh M, Mengjie Z, Johnston M (2012) Effects of static and dynamic topologies in particle swarm optimisation for edge detection in noisy images. In: IEEE congress on evolutionary computation (CEC), pp 1–8

11. Kekre DHB, Gharge MSM (2010) Image segmentation using extended edge operator for mammographic images. Int J Comput Sci Eng 2:1086–1091
12. Dhawan AP (2011) Medical image analysis. Wiley-IEEE Press
13. Canny J (1986) A computational approach to edge detection. IEEE transactions on PAMI-8 pattern analysis and machine intelligence, pp 679–698

A New Computer Based Differential Relay Framework for Power Transformer

Rachid Bouderbala and Hamid Bentarzi

Abstract A differential relay that is very sensitive relay operating even at its limits may be used for protecting a power transformer. However, this characteristic may lead to unnecessary tripping due to transient currents. In order to avoid this unnecessary tripping, estimated harmonics of these currents may be required which need great computation efforts. In this paper, a new frame work is proposed using PC interfaced with a data acquisition card AD622, which acquires real-time signals of the currents, process them numerically in the computer and outputs tripping signal to the circuit breaker. All algorithms of differential protection function and blocking techniques have been implemented using the Simulink/ Matlab. To validate the present work, the performance of developed relay is tested by signals generated by Simulink/MATLAB simulator under different conditions. The test results show that this proposed scheme provides good discrimination between the transient currents and the internal fault currents.

Keywords Differential protection · Fault · Inrush current · Data acquisition card · Real time toolbox

Introduction

Differential protection is one of the most reliable and popular techniques in power equipment protection such as transformer and generator. It is so important for clearing rapidly all types of fault that occur in the power equipment. Differential

R. Bouderbala
Department of Instrumentation, Algerian Institute of Petroleum (IAP), Boumerdès, Algeria
e-mail: sisylab@yahoo.com

H. Bentarzi (✉)
Signals and Systemes Laboratory Lab, IGEE, UMBB University, Boumerdes, Algeria
e-mail: bentarzi_hamid@yahoo. com

Y.-M. Huang et al. (eds.), *Advanced Technologies, Embedded and Multimedia for Human-centric Computing*, Lecture Notes in Electrical Engineering 260,
DOI: 10.1007/978-94-007-7262-5_54,

protection operates and isolates a faulted part from the power system during an internal fault. However, the differential relay is blocked during the transient conditions due to the magnetizing inrush phenomenon which produces current only in the energized winding side otherwise the differential protection operates unnecessary. Hence, it is unable to distinguish between the transient current and the current that is due to an internal fault. The bias setting is one technique that has been used for overcoming this problem which is not effective. An increase in the protection setting to a value that would avoid operation may lead to make the protection insensitive. However, techniques of delaying, restraining or blocking of the differential element may be used to prevent mal-operation of the protection [1, 2].

In this work, a differential relay has been implemented using Simulink/MATLAB, which ensures security for external fault, inrush, and over-excitation conditions and provides dependability for internal faults, for protecting a three phase power transformer. This work combines harmonic restraint technique with a percentage blocking technique. The harmonic based dual slope characteristic differential relay is modeled using also Simulink. A power transformer Simulink model is used for evaluating the proposed differential relay performance under different operation conditions. The interfacing of this model with the differential relay based on PC is accomplished via acquisition card AD622 associated with real-time toolbox.

Differential Relay Principle Operation

Differential protection compares the currents that enter with the currents that leave a zone or element to be protected. If the net sum of the currents is zero, then the protected equipment is under normal condition. However, if the net sum is different from zero, the differential relay operates due to a fault existing within the equipment and isolates it from the power system.

Even differential protection is relatively simple to be implemented, but it has drawbacks. One of these drawbacks is its unnecessary tripping due to transformer magnetizing current. An inrush current is the surge of transient current that appears in a transformer due to inrush and over excitation conditions. The exciting voltage applied to the primary of the transformer forces the flux to build up to a maximum theoretical value of double the steady state flux plus reminisce as follows,

$$\phi_{MAX} = 2\phi_M + \phi_R \tag{1}$$

Therefore, the transformer is greatly saturated and draws more current which can be in excess of the full load rating of the transformer windings.

The slope, magnitude and duration of inrush current depend on several factors [3], such as size of a transformer, magnetic properties of the core material, and way a transformer is switched on (inner, outer winding, type of switchgear).

The most recent technique used for preventing false tripping is the use of second and fifth harmonic restraint. If these harmonic contents of the differential

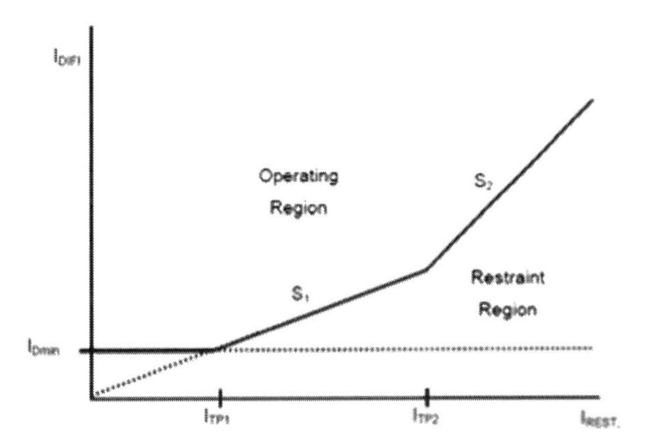

Fig. 1 Typical restraint and operating characteristic of a differential relay

current exceed pre-defined percentages of the fundamental, inrush current is detected and the relay is blocked from tripping [1].

One study reported the minimum possible level of second harmonic in inrush current is about 17 % [4], a 15 % threshold is a good choice.

The developments in digital technology lead to the incorporation of microprocessors in the construction of relays and to the investigation of protection systems with the capacity to record signals during faults, monitor them and communicate with their peers.

Differential Relay Settings

Low impedance differential protection systems typically have 3–5 settings required to properly define the restraint characteristic of the relay (see Fig. 1). Where, I_{Dmin} = minimum differential current required to operate the relay,

I_{TP1} = turning point 1, I_{TP2} = turning point 2, S_1 = Slope 1 setting,

S_2 = Slope 2 setting, I_{RST} = Total current through the differential system,

I_{DIF} = For a given I_{TOT}, the Mini. Diff. current required to operate the relay.

The settings to be considered are I_{Dmin}, I_{TP1}, I_{TP2}, S_1, and S_2. Besides, second harmonic (H2) and fifth harmonic (H5) may be used.

Differential Relay Implementation

Software Structure

Differential protection algorithm, which has been implemented using Simulink/Matlab, its flow chart is shown in Fig. 2. Besides, Graphical User Interface (GUI)

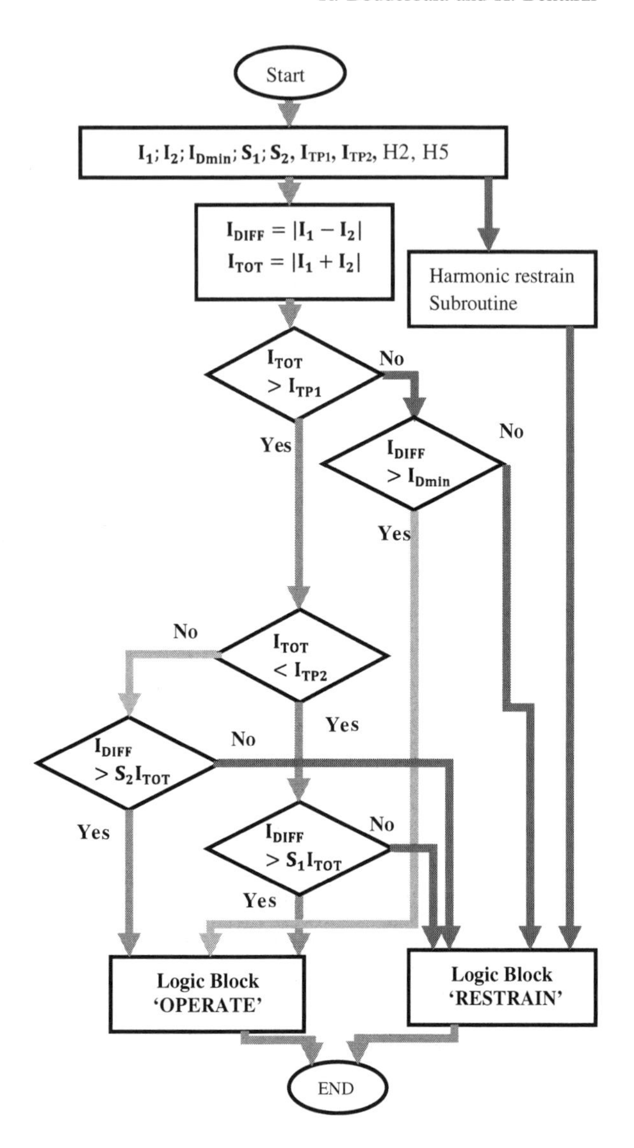

Fig. 2 Differential protection algorithm flowchart

has been developed using the same software tool, the user can select and set the desired parameters, and makes a test by running program and displaying the tripping signal (see Fig. 3).

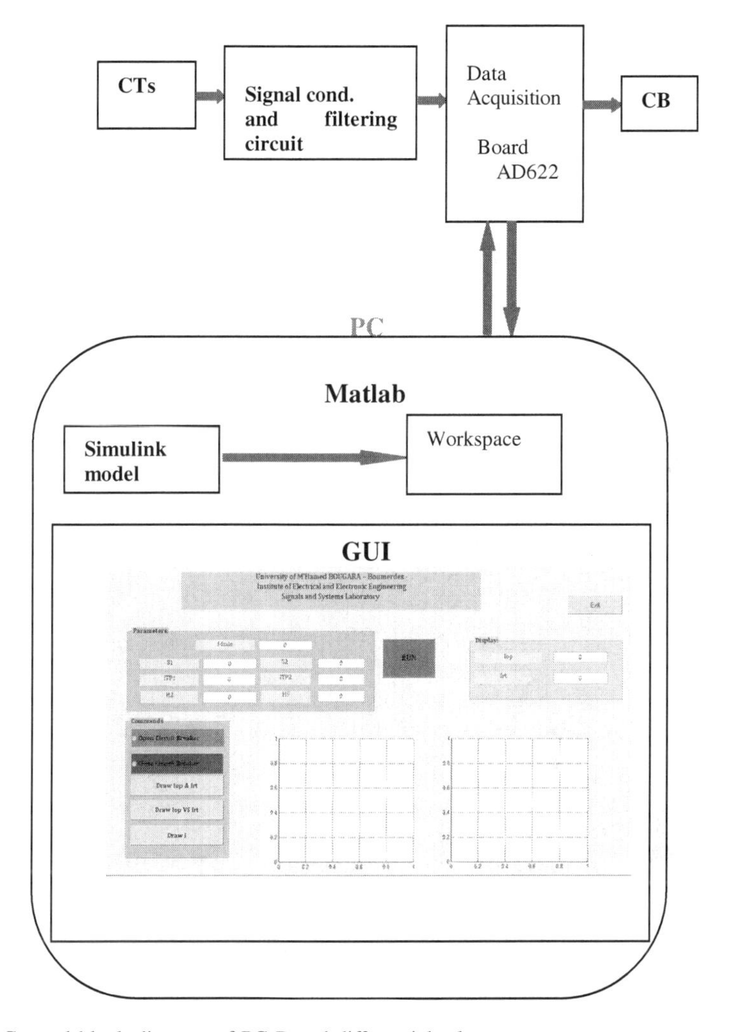

Fig. 3 General block diagram of PC Based differential relay

Hardware Architecture

In protection field, current transformers are used to measure the current and provide the measured quantity as analog voltage signal to the input of protective relay. Circuit breaker is used as actuator.

Thus, the differential protection hardware whose block diagram shown in Fig. 3 consists of:

- Signal transformation: current transformer (CT) transforms currents of the power system into voltage with low safe magnitudes.

- Signal Conditioning and filtering Circuit: the measured values of the power system fed from CTs in analog forms are passed through an anti-aliasing filter (low pass filter).
- Data acquisition boards: Sample and hold circuits and analog multiplexed are used to sample the three different signals of the three phase lines supplied by instrument transformers at the same time. The sampled signals are converted into digital form using ADC.
- PC: the digital signals are fed from data acquisition board AD 622 to the PC where they will be numerically processed as shown in Fig. 3 [5, 6].

Test Results and Discussion

The experimental setup system that may be used for testing the developed differential relay is composed of a 25 kV three-phase voltage source in series with a three-phase power transformer feeding a RL load. One circuit breaker is connected to the primary side of the transformer and takes its tripping signal from the differential relay so that it may open the circuit during faulty condition. The inputs to the relay are the primary side current of the transformer and the current of the secondary side. A simple system is used to check the validity of the proposed algorithm, and it is mainly composed of three phase star/star transformer with saturable core and initial fluxes. The main parameters of the used power transformer are given in Table 1. Two block sets of three phase faults may be applied, one is considered as internal fault and the other is the external fault.

Figure 5 shows the Simulink model of experimental test setup that may be used for generating the current signals under different conditions that may be injected via acquisition card to test the PC based differential relay (Fig. 4) [7].

In order to avoid the source transients, the transformer is connected at time $t = 1$ s. After running the program, the magnetizing current appears only in the primary side as shown in Fig. 6. This may produce a great difference in the input currents, but, the relay is blocked to operate as shown in Fig. 7.

Table 1 Simulated power transformer main parameters

Parameters	Values
Rated power	250 MVA
Voltage ratio	25,000/250 V
Rated frequency	60 Hz
Primary impedance/phase	0.002 + j0.08 pu
Secondary impedance/phase	0.002 + j0.08 pu
Magnetization resistance	500.2 pu
Connection	*Y/Y*

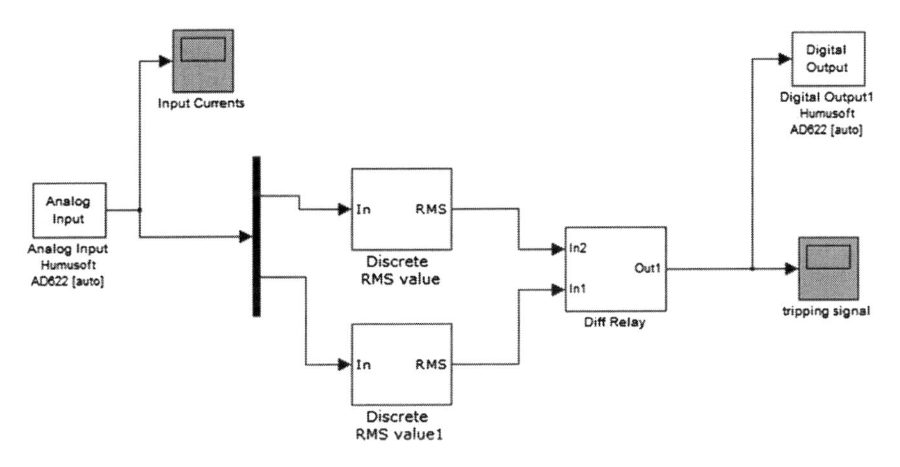

Fig. 4 Differential relay real time Simulink model

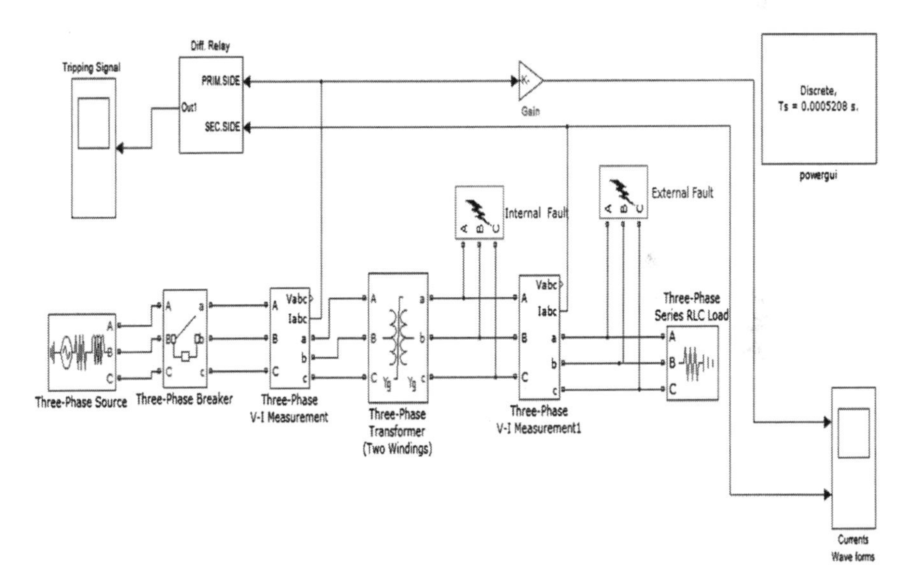

Fig. 5 Simulink model of the three phase transformer for experimental test

However, by applying an internal fault during an interval time $t_1 = 4\,s$ to $t_2 = 5\,s\,5\,s$, a large difference in the current amplitudes of both sides of the power transformer may be produced as shown in Fig. 6 and hence the differential relay operates in this situation as shown in Fig. 7. In the other hand, when an external fault is applied at the time $t_1 = 6\,s\,6\,s$, an increase of the current amplitudes on both sides of the power transformer can be remarked as shown in Fig. 6, in this case the differential relay is blocked to operate.

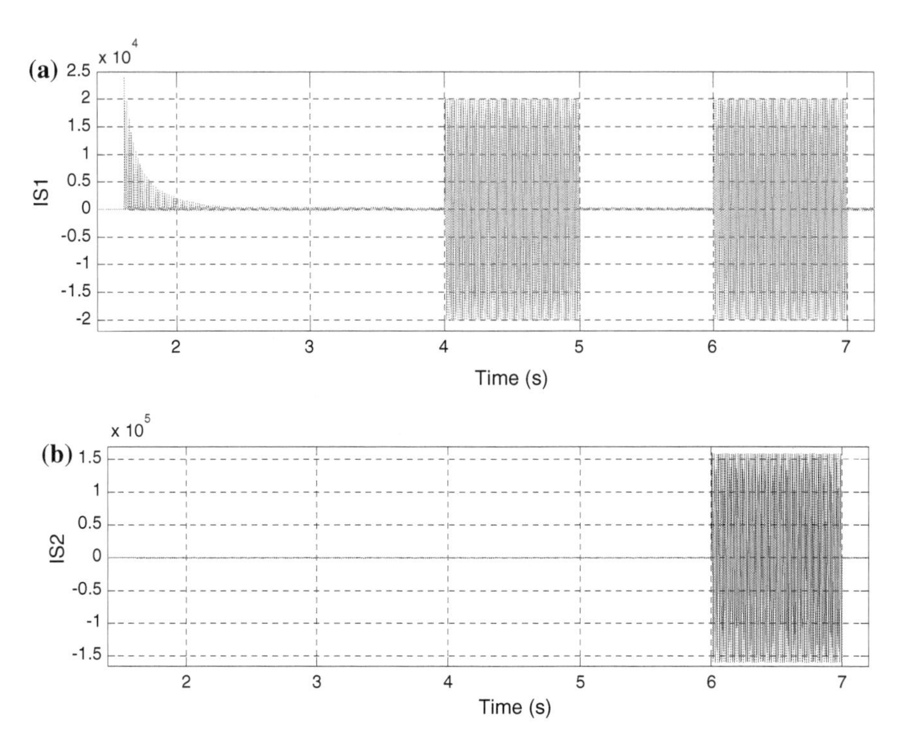

Fig. 6 The currents of (**a**) the primary side and (**b**) the secondary side by applying internal and external faults

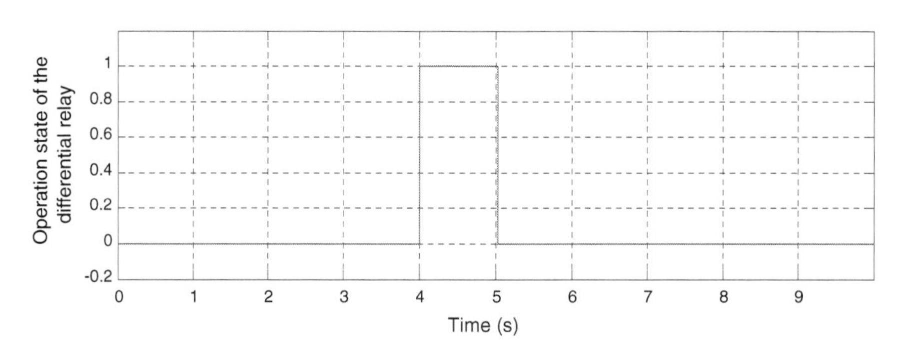

Fig. 7 The differential relay tripping signal

Conclusion

The PC-based differential relay prototype has been realized in this work. Where the algorithm has been implemented using the Simulink/Matlab and interfaced with the real world via the data acquisition card AD622.

After the test, the obtained results satisfy the principle operation of numerical differential relay and its characteristics using this new frame work.

Moreover, it can be concluded that this framework has the advantage; it can easily be implemented and tested using any protection function associated with recent the blocking function.

References

1. Bouderbala R, Bentarzi H, Ouadi A (2011) Digital differential relay reliability enhancement of power transformer. Int J Circ, Syst Sig Process, ISSN: 1998-446, vol 5(1), pp 263–270
2. Bentarzi H Some new aspects of protective relaying in modern electric power system (unpublished)
3. Karsai K, Kerenyi D, Kiss L (1987) Large power transformers. Elsevier, New York
4. Sonnemann WK, Wagner CL, Rockefeller GD (1958) Magnetizing Inrush Phenomena in transformer banks. AIEE Trans 77:884–892
5. IEEE power system relaying committee: understanding microprocessor-based technology applied relaying (2009) WG I-01 report, IEEE Organization
6. AD622 data acquisition card user's manual
7. Chafai M, Bentarzi H, Ouadi A, Zitouni A, Recioui A (2013) PC based testing system for protective relay. Accepted to be presented in 4th IEEE international conference on power engineering, May 13–17, Istanbul, Turkey
8. Blackburn J (1998) Lewis, protective relaying: principles and applications, 2nd edn. Marcel Dekker, New York, pp 275–280

A New Computer Based Quadrilateral Distance Relay Framework for Power Transmission Lines

Abderrahmane Ouadi and Hamid Bentarzi

Abstract In conventional transmission line protection, a three–zone stepped directional distance scheme is used to provide the primary as well as remote backup protection. The voltage and current phasors are needed by the distance relay for determining the impedance. In this paper, a new frame work is proposed using Data Acquisition Card, which acquires real-time signals of the voltages and currents, processes numerically them in PC and outputs tripping signal to the circuit breaker. Algorithm of quadrilateral distance protection has been implemented using the Simulink/Matlab. To validate the present work, the performance of developed relay is tested by signals generated by Simulink/MATLAB simulator under different conditions. The obtained simulation results are satisfactory.

Keywords Quadrilateral distance protection · Numerical relay · Data acquisition card · Power swing · Real time toolbox

Introduction

Power grid protection is the process of making the production, transmission, and distribution of electrical energy as safe as possible from the effects of equipment failures and events that place the power system at risk. When the faults occur in such power grid, protection systems are designed to isolate faulted part of the power grid, and leave the healthy parts of the system connected in order to ensure the continuity of the power supply. The operational security of the power system depends upon the successful performance of the thousands of relays that protect equipments and hence protect the whole system from cascading failures. Thus, the

A. Ouadi · H. Bentarzi (✉)
Signals and Systemes Laboratory Lab, IGEE, UMBB University, Boumerdes, Algeria
e-mail: sisylab@yahoo.com

Y.-M. Huang et al. (eds.), *Advanced Technologies, Embedded and Multimedia for Human-centric Computing*, Lecture Notes in Electrical Engineering 260,
DOI: 10.1007/978-94-007-7262-5_55,

failure of a relay to operate as intended may jeopardize the stability of the entire power system and its equipment.

Disturbances may affect on transmission line relays such as overcurrent, directional overcurrent and distance relays which may respond to the variations of voltage and currents and their phase angle relationship. Under this condition, these relays operate even for stable power swings for which the system can recover [1]. In order to overcome the drawbacks of these traditional relaying systems, many techniques have been developed such as a blinder that blocks the relay to operate during the power swing [2] or by using a quadrilateral distance relay.

The mal-operation of this relay which is generally due to unnecessary tripping during power swing reduces the security of protection system and hence its reliability. The ideal characteristic for preventing this mal-operation is a quadrilateral characteristic but it can be implemented with difficulty.

The basic principle behind distance relay is calculating the line impedance seen by the relay at location using the fundamental frequency components of voltage and current signals. The Fourier transform based method can be used to estimate the fundamental components of the voltage and current. In this paper the basic theory behind the impedance measurement and the DFT algorithms are discussed. Besides, the simulations of the proposed scheme and its implementation using Acquisition card are presented.

Distance Relay Principle Operation

Protections based on distance relaying have been used in the power grid generally and in transmission lines particularly in order to detect the fault rapidly and disconnect the faulted part only. This maintains a reliable operation of the power grid by ensuring continuity of power supply [3–5].

The basic principle governing the operation of a distance relay is the ratio between the voltage V and the current I at the relaying point as shown in Fig. 1. The ratio (V/I) represents the measured impedance Z of the faulty line between the relay location and the point of fault occurrence. Then, the measured impedance is

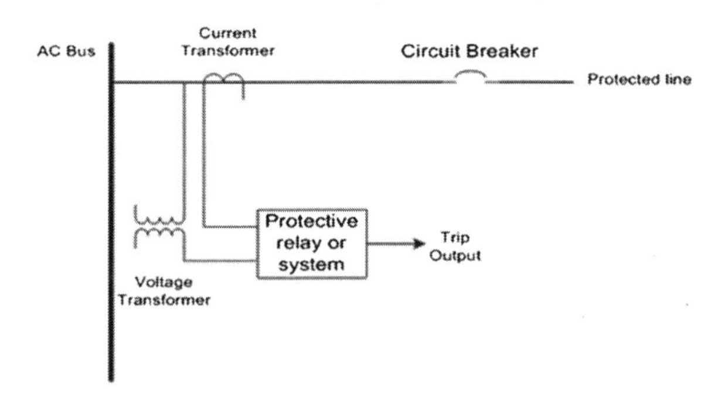

Fig. 1 A typical single line AC connection of a protective distance relay

compared to the set impedance, and if this Z is within the reach of the relay then the fault will be cleared. However, zone-three relay may be affected by the power swing because its impedance enters the characteristic of the relay and its time delay is longest.

A distance relay can be set to protect different zones of a transmission line either in forward direction (mho characteristic) or in both forward and backward direction (Offset mho characteristic). Figure 2a shows relay characteristics of an offset Mho and Fig. 2b illustrates Mho characteristics.

Faults are permanent compared to power swings that are more gradual events and disappear after a short period. This fact is used in the distance relay to distinguish between short circuit faults and stable or unstable power swings [6].

Traditional impedance-based characteristics for detecting power swings on a transmission system are shown in Fig. 3. Almost of these methods involve measuring apparent impedance and introducing a time delay between two measuring elements.

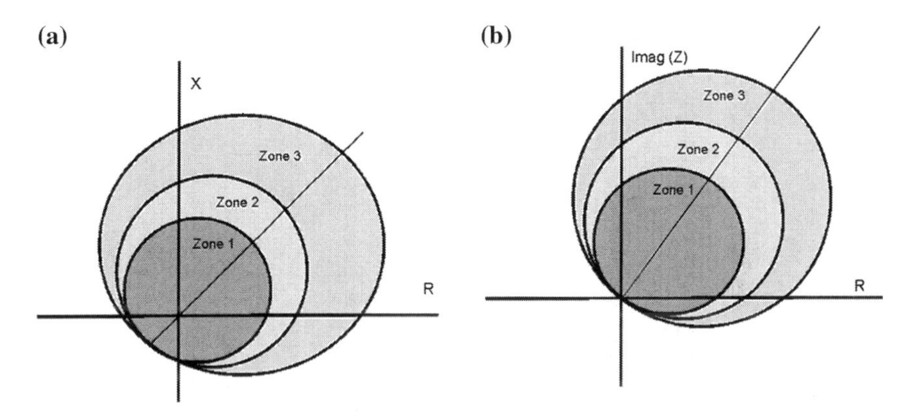

Fig. 2 **a** Offset Mho, and **b** Mho relays characteristics

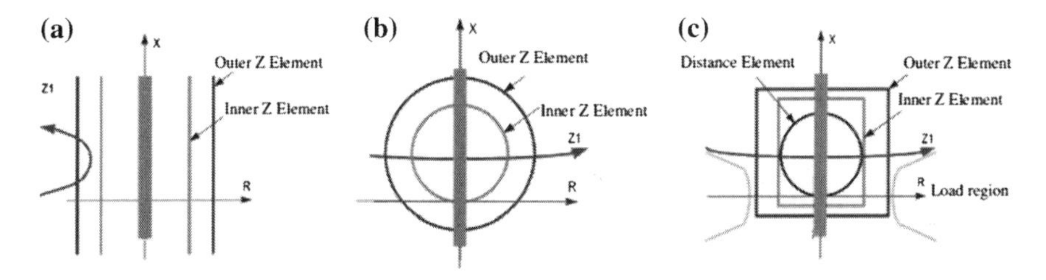

Fig. 3 Different power swing protection schemes. **a** Double blinder, power swing. **b** Offset mho, power swing characteristic. **c** Quadrilateral, power swing characteristic

However, when a quadrilateral characteristic is used, a blinder and a time delay are not required for avoiding the power swing as shown in Fig. 3c.

Impedance Measurement Algorithm

Generally, Fourier analysis can be used to calculate the fundamental frequency components of the voltage and current which in turn they will be used for estimating the impedance. In digital algorithms, sampled values are used instead of continuous variables. By taking N samples within time interval T which immediately precedes the sampling instant in question. The digital domain is given by [7],

$$x(k) = A_0 + \sum_{n=0}^{(N-1)/2} \mathrm{A_n} \cos\left(\frac{2\pi}{\mathrm{N}} nk\right) + \sum_{n=0}^{(N-1)/2} \mathrm{B_n} \sin\left(\frac{2\pi}{\mathrm{N}} nk\right) \quad (1)$$

where, n = 0,1,2,.......(N-1)/2 and k = 1,2,3,........(N-1). The discrete from of the integrals are given by:

$$\begin{aligned} \mathrm{A_n} &= \frac{2}{\mathrm{N}} \sum_{k=0}^{(N-1)} \mathrm{x(k)} \cos\left(\frac{2\pi}{\mathrm{N}} nk\right) \\ \mathrm{B_n} Z &= \frac{2}{\mathrm{N}} \sum_{k=0}^{(N-1)} \mathrm{x(k)} \sin\left(\frac{2\pi}{\mathrm{N}} nk\right) \\ \mathrm{A_0} &= \frac{1}{\mathrm{N}} \sum_{k=0}^{(N-1)} \mathrm{x(k)} \end{aligned} \quad (2)$$

where x(k) represents individual samples within the sampling window using the discrete Fourier series. The amplitude and the phase of the nth harmonic are given by,

$$|\mathrm{C_n}| = \sqrt{(A_n^2 + B_n^2)}, \quad (3)$$

$$\theta_n = \tan^{-1}\left(\frac{B_n}{A_n}\right) \quad (4)$$

The fundamental frequency components are given by $\mathrm{A_1}$ and $\mathrm{B_1}$. The discrete form of Eq. (1) is given by

$$\mathrm{X}(n) = \frac{1}{\mathrm{N}} \sum_{k=0}^{(N-1)} \mathrm{x(k)} e^{-j\left(\frac{2\pi}{\mathrm{N}} nk\right)} \quad (5)$$

where k, n = 1,2, ...N. Using Eq. (5), the fundamental components of the voltage and current can be obtained.

$$|\mathrm{C_1}| = \sqrt{(A_1^2 + B_1^2)}, \quad (6)$$

$$\theta_1 = \tan^{-1}\left(\frac{B_1}{A_1}\right) \tag{7}$$

Using Eqs. (6) and (7), the voltage C_V, θ_V and current C_I, θ_I fundamental phasor components can be obtained. The impedance and phase angle can be derived by:

$$|\bar{Z}| = \frac{c_V}{c_1} \tag{8}$$

$$\theta_z = \theta_V - \theta_1 \tag{9}$$

The above computation is carried out at each sampling interval. This will give an impedance value based on the voltage and current samples of the previous cycle.

Implementation is made through the use of Quadrilateral technique considered as the best suitable one that can be implemented only in the Numerical relays.

Simulink/MATLAB is used to model the distance relay components such as ADC and digital filters. Figure 4 shows the block diagram of the developed distance relay. The voltage and current data are derived using the power system simulator (power system block set for SIMULINK/MATLAB) as illustrated in Fig. 5. This system may include three phase source, transmission line (represented using π model), current transformers, voltage transformers and voltage and current measurement units. The voltage and current input signals are inserted in the distance relay block.

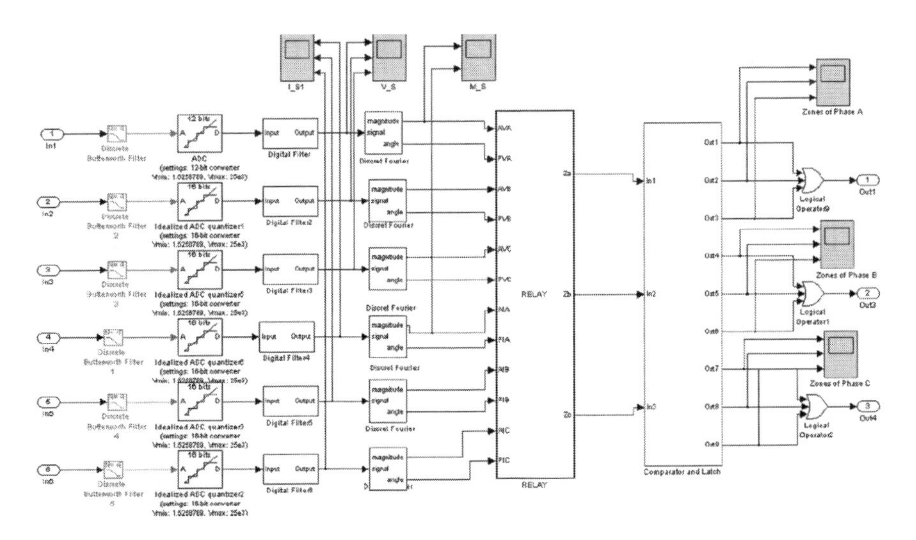

Fig. 4 General bock diagram of the designed quadrilateral relay

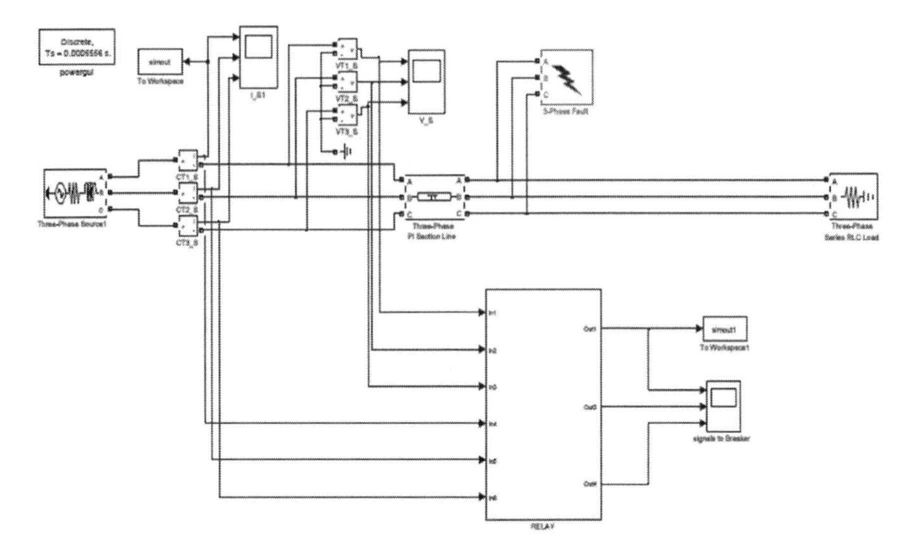

Fig. 5 Simulink model of the experimental test setup

Distance Relay Implementation

Software Structure

A quadrilateral characteristic is best suited for the numerical protection of HV transmission lines as it possesses an ideal distance relay characteristic. It excludes all the conditions for which the tripping is undesirable such as power swings, fault resistance and overloads. In the present work, is designed and implemented.

The quadrilateral distance relay algorithm that is implemented, its flowchart is illustrated in Fig. 6 [8, 9].

Hardware Architecture

In protection field, current transformers and potential transformer are used to measure the current and the voltage and provide the measured quantity as analog voltage signal to the input of protective relay. Circuit breaker is used as actuator.

Thus, the distance protection hardware whose block diagram shown in Fig. 9 consists of:

- Signal transformation: current transformer (CT) transforms currents of the power system into voltage with low safe magnitudes.
- Signal Conditioning and filtering Circuit: the measured values of the power system fed from CTs in analog forms are passed through an anti-aliasing filter (low pass filter).

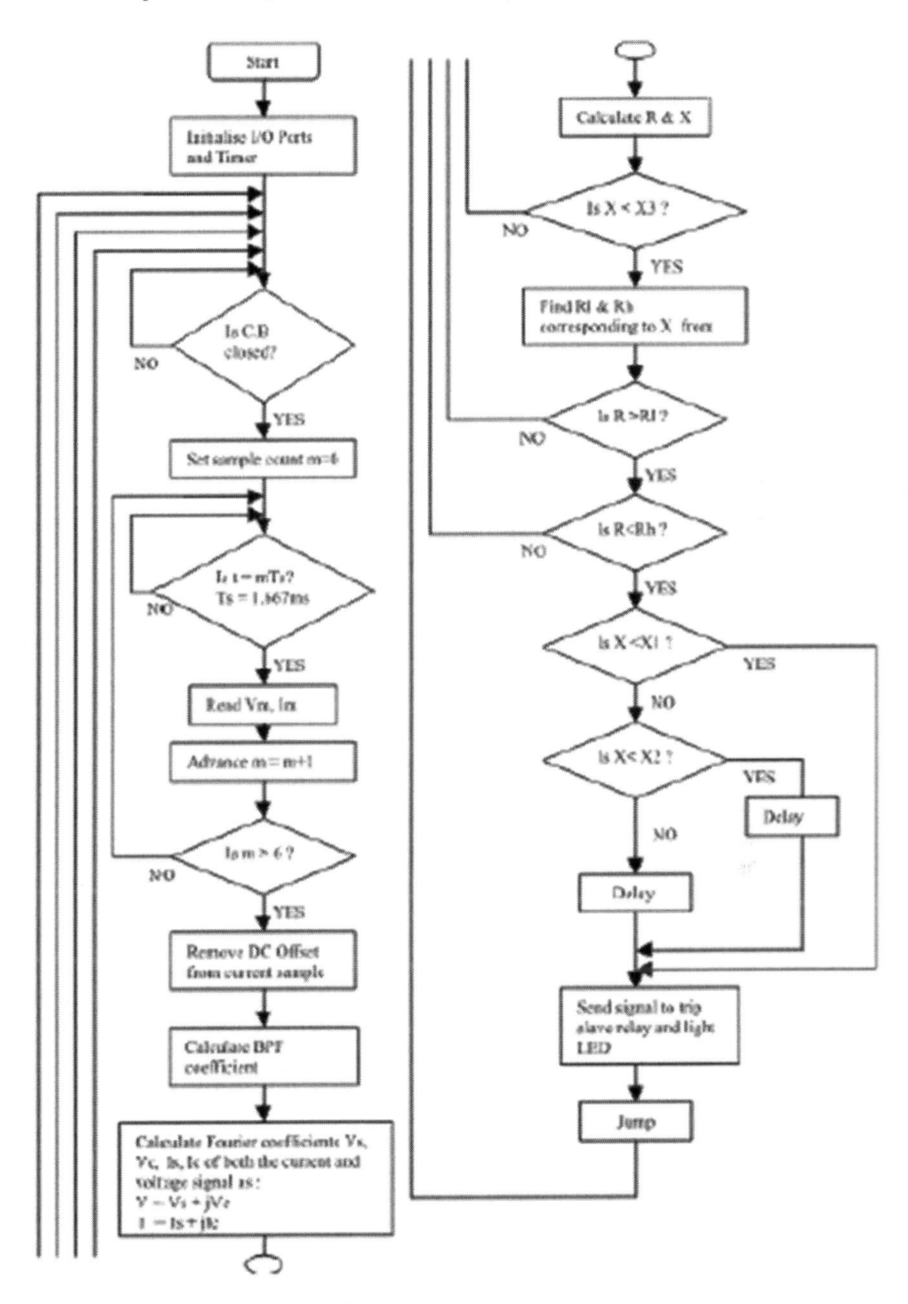

Fig. 6 Flowchart of quadrilateral distance relay algorithm

- Data acquisition boards: Sample and hold circuits and analog multiplexed are used to sample the three different signals of the three phase lines supplied by instrument transformers at the same time. The sampled signals are converted into digital form using ADC.
- PC: the digital signals are fed from data acquisition board to the PC where they will be numerically processed.

The developed distance protection relay has been implemented in PC associated with acquisition card AD 622 using Real time toolbox of Simulink as shown in Fig. 4 [6, 10].

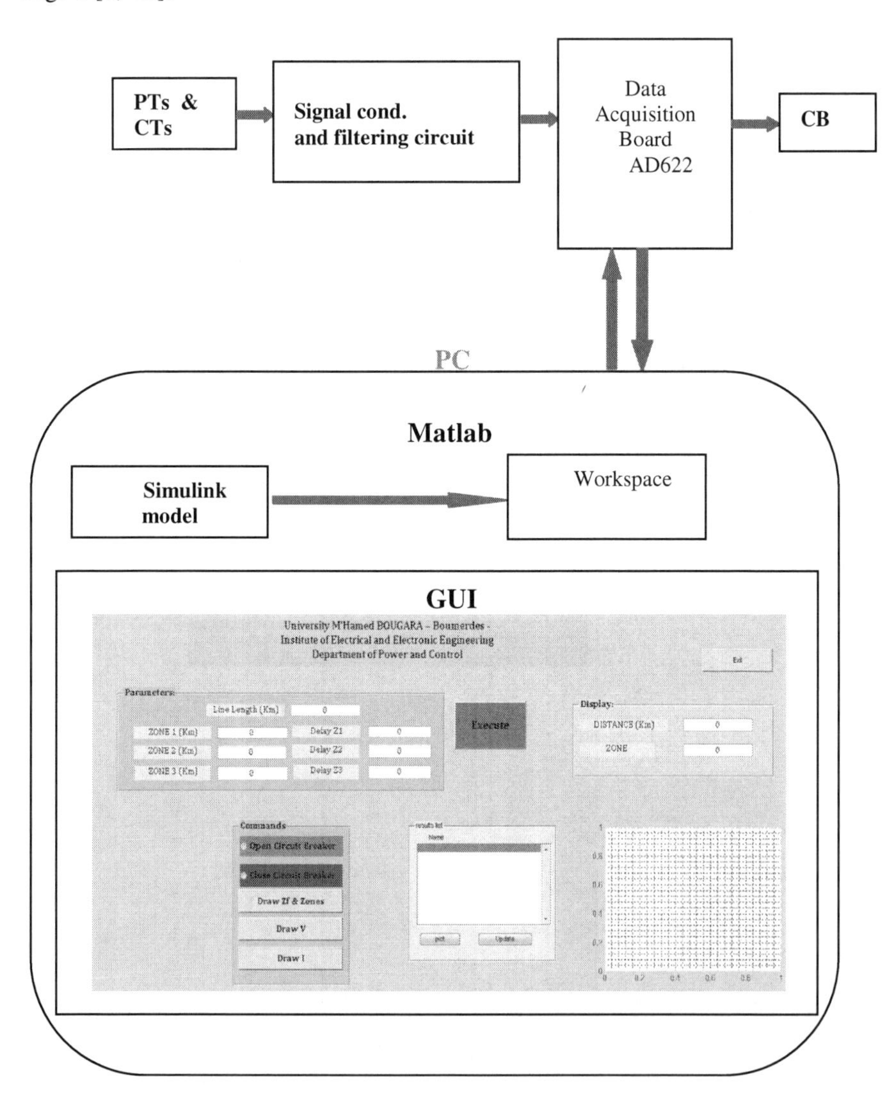

Fig. 7 General block diagram of PC based differential relay

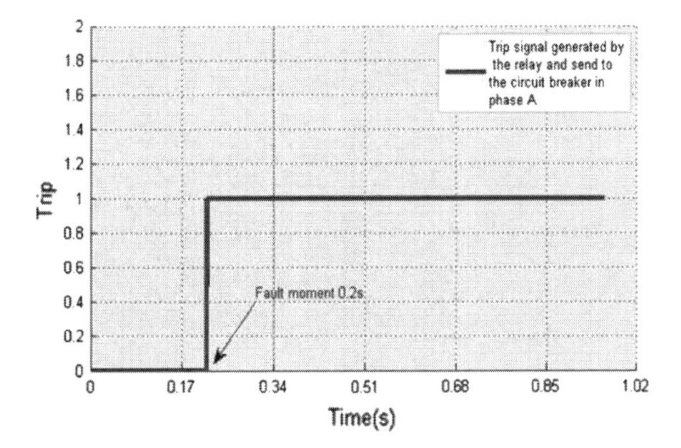

Fig. 8 Trip signal generated by the distance relay to the circuit breaker during the fault

Testing Results

The experimental setup system that may be used for testing the developed distance relay is shown in Figs. 5 and 7.

Figure 8 shows the trip signal generated by the relay when the fault occurs in phase A.

Conclusion

The PC-based distance relay prototype has been realized in this work. Where the quadrilateral characteristic algorithm has been implemented using the Simulink/ Matlab and interfaced with the real world via the data acquisition card AD622.

After the test, it can be noticed that the obtained results satisfy the principle operation of numerical quadrilateral distance relay and its characteristics using this new frame work.

Moreover, it can be concluded that this proposed framework has the advantage; it can easily be used implemented and tested even when complex protection algorithms such as quadrilateral characteristic are used.

References

1. Power system relaying committee of the IEEE power engineering society: power swing and out-of-step consideration on transmission lines. IEEE PSRC WG D6 (2005)
2. Bentarzi H, Ouadi A, Chafai M, Zitouni A (2011) Distance protective system performance enhancement using optimized digital filter. In: Proceedings of CSECS '11 the 10th WSEAS international conference on circuits, systems, electronics, control and signal processing, Montreux University, Switzerland, 29–31 Dec

3. Sachdev MS, Baribeau MA (1979) A new algorithm for digital impedance relay. IEEE Trans Power Apparatus Syst 98:2232–2240
4. Isaksson A (1988) Digital protective relaying through recursive least-squares identification. In: IEE proceedings, Part C: generation, transmission, and distribution, vol 135, pp 441–449
5. Phadke AG, Hlibka T, Ibrahim M (1976) A digital computer system for EHV substation: analysis and field tests. IEEE Trans Power Apparatus Syst, vol PAS-95, pp 291–301
6. Internal PNR report (2012)
7. Ouadi A, Bentarzi H, Maun JC (2011) Phasor measurement unit reliability enhancement using real-time digital filter. Int J Circ, Syst, Sig Process 5(1):1–8
8. Shrivastava K, Vishwakarma DN (2007) Microcontroller-based numerical quadrilateral relay for the transmission line protection, electric power components and systems, pp 1301–1315
9. Ziegler G (2006) Numerical distance protection, 2nd edn. Public is Corporate Publishing, Siemens, Erlangen
10. Ouadi A, Bentarzi H, Maun JC (2009) A new computer based phasor measurement unit framework (published conference proceedings style, IEEE explorer). In: Proceedings 6th international conference SSD'09, Djerba, Tunisia, pp 1–6
11. Gadgil K (2010) A numerical protection relay solution, texas instruments application report, SLAA466—Sept

A Study for a Low-Power Way Predictor for Embedded Data Caches

Yul Chu

Abstract This paper introduces an enhanced predictor to reduce power consumption for a way-prediction cache used for embedded systems. The proposed predictor shows better prediction accuracy and lower power consumption compared to any conventional data caches. In addition, two representative cache replacement policies, *LRU* (Least recently Used) and *random*, are examined for low-power data caches; simulation results show that *random* reduce power consumption more than *LRU* for highly-associative way-prediction caches. SimpleScalar and Cacti simulators are used for these simulations with SPEC benchmark programs.

Introduction

During the last decade, low-power consumption has been a critical issue for embedded systems, especially for hand-held devices. For most embedded systems, microprocessor and cache memory might consume most of the power in the system. According to [1–3], modern cache memories occupy more than 60 % of the microprocessors' die area and cause more than 40 % of total power dissipation. To reduce power dissipation, it is necessary to design low-power cache memory by lessening access times to charge bit-lines of cache memory [3, 4]. Typically, on-chip caches in mobile devices are highly-associative, which is greater than 16-way. Therefore, a cache miss results in a costly access to the memory. Highly-associative caches in [5–7] are specifically designed for low-power embedded systems to provide better performance by reducing the conflict misses, due to imperfect allocation of entries in a cache. However, they significantly increase power

Y. Chu (✉)
Department of Electrical Engineering, University of Texas Pan American, 1201 W. University Dr, Edinburg, TX 78539, USA
e-mail: chuy@utpa.edu

Y.-M. Huang et al. (eds.), *Advanced Technologies, Embedded and Multimedia for Human-centric Computing*, Lecture Notes in Electrical Engineering 260,
DOI: 10.1007/978-94-007-7262-5_56,

consumption because of simultaneous accesses to all the banks in parallel, i.e., the n-way cache has n banks to access simultaneously. This paper aims to reduce the power consumption by accurately predicting only one bank out of n banks (i.e., n-way) and by restricting access to the predicted bank only. This not only saves the power consumption but also reduces the latency since the cache is accessed as a direct-mapped cache.

The rest of the paper is organized as follows: section Related Works explains some related works done in this area; section Dual-Access Way-Prediction Cache presents the proposed way predictor; section Simulations and Performance Metrics deals with simulations and performance metrics; section Experimental Results provides the experimental results and discussion; and finally the conclusions are provided in section Conclusions.

Related Works

Much research has been proposed how to reduce the power consumption for highly associated cache memories. One popular approach is to use a phased cache, which divided into two caches, tags and data caches [8]; in the phased cache, the tag bits for all tag banks are powered, active to be accessed, and compared with a referenced memory address; on a hit from a bank, the bank is powered and accessed during the next cycle. Even though the phased cache can reduce power consumption, it has a fixed latency to all cache accesses. Another approach is to use a way-prediction cache [9]; the way predictor logic is added on top of a conventional cache scheme, such as a 2-way, 4-way, etc. This logic predicts a way, which should be accessed; on a miss, all the rest of the banks are accessed. *Therefore, the way-prediction can reduce power consumption more effectively than the phased cache* [9]. However, the prediction logic might increase power consumption slightly compared to a conventional cache memory. The way predictor logic implemented in [9] is a simple 'most recently accessed logic', having the same number of index table entries as the banks in a cache. This results in better power savings than a conventional cache. The power savings from such a design is entirely dependent on a prediction-hit rate of a way predictor. Such a scheme also has the advantage of reducing latency [10].

Dual-Access Way-Prediction Cache

This section proposes an enhanced way-prediction cache called Dual-Access Way-Prediction (DAWP) to select one of banks in a highly-associative data cache to reduce power consumption.

Figure 1 shows a conventional way-prediction (WP) cache: The way-prediction cache speculatively chooses one way from the index table, and then accesses the predicted bank as shown in Fig. 1. If the prediction is a hit in the bank, the cache access will be completed. Otherwise, the other remaining banks will be accessed using a normal cache access process, which is a case of prediction miss.

Figure 2 shows the proposed DAWP cache, which has two major parts. The first part is a direct-mapped index table, which stores previously accessed bank (way) address. The referenced output (as a predicted way) in the table goes to a multiplexer. The second part of this scheme consists of a global history register and a small fully-associative cache. The register stores a group of cache ways [e.g., 0001 (bank 1) and 0010 (bank 2)], which accessed recently. The small fully-associative cache has a fixed number of lines and each line holds a selected global history and a predicted way for the given global history. The output of this also goes to the multiplexer. The global history and fully-associative cache are rarely enabled, *valid* in Fig. 2; the *valid* will be '1' only when there are constant trashing of data in the index table; after then, if the global history is matched, the predicted way is accessed to see if the data is present in it. If the data is not present, the rest of the banks will be checked. On a cache miss, the data is retrieved from the memory and updated to the cache, and the way information is updated in the index table and/or the fully-associative cache. On a way-prediction miss but a cache hit, i.e., if the data is found in one of unpredicted banks, the correct bank is updated to the index table.

This scheme works based on the locality of data. A sequence of data, spatial locality, can be located and accessed in a cache line of a bank (way). As we discussed, the way information is updated via memory address (index) and global history. To reduce power consumption, it is required to reduce conflicts in the table and cache; highly biased accesses can be filtered out using a global history; in addition, the hit time can be reduced while accessing only one way instead of accessing n-way caches [9] for a prediction hit.

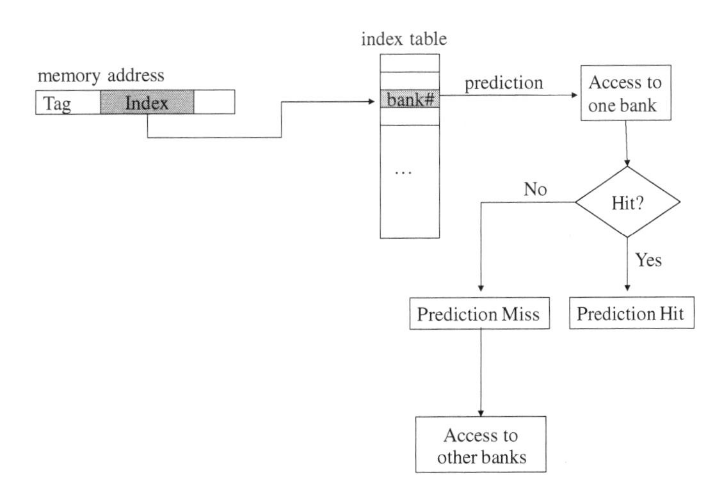

Fig. 1 Conventional way-prediction cache (WP)

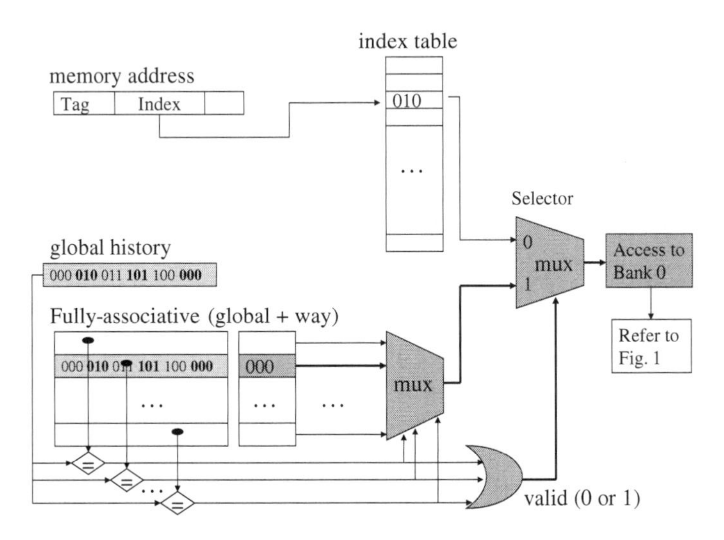

Fig. 2 Dual-access way-prediction (DAWP) cache

Simulations and Performance Metrics

The simulations are done on 16 KB, 32 KB and 64 KB cache sizes with configurations of direct mapped, 2-way, 4-way, 8-way, 16-way, and 32-way set-associative cache memories. SimpleScalar is used for these simulations with 9 SPEC2000 programs (art, ammp, equake, mesa, mcf, vpr, vortex, gcc, and gzip) [11].

For the simulation, the size of the index table is optimized as 512 bytes since we found that it is a good compromise between prediction-hit rate and power consumption through our simulation results.

To compare the miss rates, we use one more metrics, latency. In order to calculate this, the cache accesses are grouped into three categories: 1) the first one is the case of a correct prediction. For this case, the latency and power dissipation is based on accesses to the index table and the predicted bank like a direct mapped cache; 2) the second one is the case of a wrong prediction but a cache hit. In this case, the predictor predicts a wrong way but the data is found in a different bank (way). So after the data access, the correct way needs to be updated to the index table. The latency and power dissipation is due to an access to the table, an access to one way of the cache, accesses to other banks, and an access to update the index table; and 3) the last case is a cache miss, which results in the above accesses and a costly access to memory to fetch data.

Cache latencies and power dissipations are calculated using Cacti [10]. The memory latencies and power dissipations are estimated from [12]. All the hit rates, prediction misses and cache miss rates are obtained from Simplescalar simulator [11].

Experimental Results

The way-prediction (WP) cache described in [8] has an index table with the same number of entries as the number of sets in a cache. In Fig. 3, index size of 1X represents the WP cache and other sizes (2X to 16X) are based on the DAWP cache.

As the associativity of the cache increases such as 16-way or 32-way, the number of sets in a cache reduces drastically; then, the reduced entries of the index table degrade prediction rate because of conflicts in the table. For example, index table of 16-way is 1/16 of the direct-map (one-way) index table size. For this reason, the number of entries in the index table is scaled up by multiplication factors of 1X, 2X, 4X, 8X and 16X. Since each entry of the index table is ranged only from 1 bit to 5 bits, the scaling up of the index table does not affect total power consumption and latency significantly. Our experimental result shows prediction miss rates according to variable index sizes for a benchmark program (swim); the prediction miss rates for a 32 KB and 16-way cache are decreased as the index table is scaled up. However, the prediction rate is saturated to 5 % after 16X. Therefore, we select a scaling factor of 16X as an optimal value for DAWP cache to run all the benchmarks with the cache size of 512B.

The next optimizing parameter for our simulation is to determine the replacement policy for data caches, *LRU* (least recently used) and *random* replacement policies. *LRU* replacement policy conventionally provides low miss rates at the cost of having more complex hardware. For WP and DAWP caches, *random* replacement policy provides better performance than *LRU* according to our experimental results.

Figure 4 compares the total power dissipation using arbitrary unit (a relative unit of measurement) for 16- and 32-way cache memories between *LRU* and *random* with SPEC2000 benchmark programs. The figure shows *random* can reduce more power consumption than *LRU*; therefore, *random* will be used for the simulations for WP and DAWP caches in this paper.

According to [8] and Fig. 3, the DAWP cache gives better results than a conventional cache, which has the same number of banks (e.g., n banks for n-way

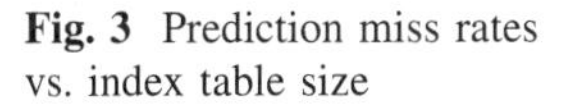

Fig. 3 Prediction miss rates vs. index table size

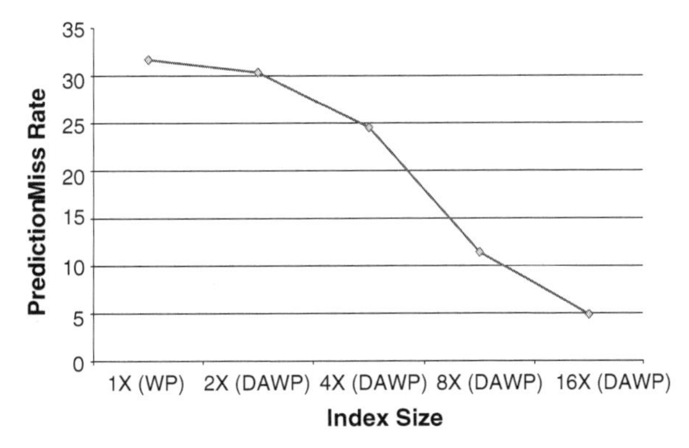

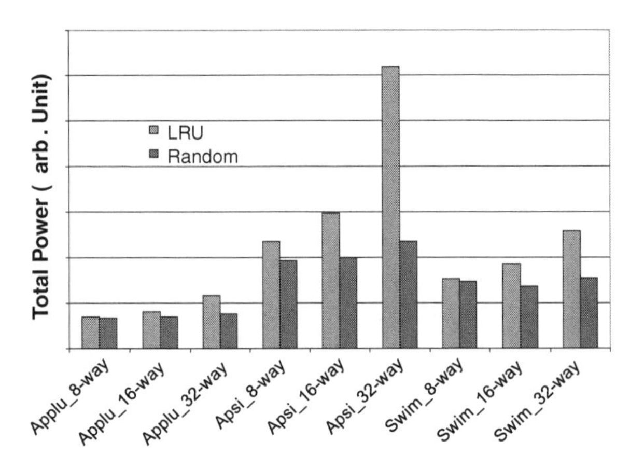

Fig. 4 LRU vs. random power dissipation

set-associative and n-way way-predictive cache) for SPEC benchmark programs. Therefore, we implement DAWP cache instead of WP cache in this paper.

In all the benchmarks, the DAWP cache has equal or lesser latency than conventional caches. In addition, power consumption savings for a highly-associative DAWP cache is much lower than any conventional caches, such as 16-way or 32-way.

Figure 5 shows total power consumption for SPEC2000 benchmark programs. The power consumptions are normalized to that of a direct-mapped cache in the figures. That means the total power dissipation of the nine SPEC2000 benchmark programs normalized to a direct mapped cache; power consumption for a direct-mapped cache is ‘1’. From our experimental results, we found that the DAWP caches offer lower power consumption than the highly-associative conventional caches because the DAWP cache is accessed as a direct-mapped cache for prediction hits; this leads to both low latency and power consumption especially when the prediction miss rate is low.

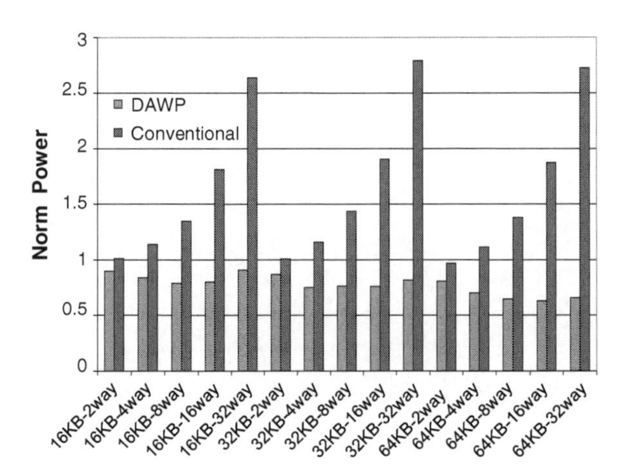

Fig. 5 Harmonic mean of power for SPEC2000

Conclusions

An enhanced way-prediction cache, DAWP, is implemented and simulated using SPEC 2000 benchmark programs. The WP and DAWP caches are optimized for the best prediction rates by scaling the index table and choosing the *random* replacement policy. The DAWP cache has equal or lesser latency than a conventional cache for all the SPEC benchmark programs. This cache also reduces power dissipation compared to any conventional caches, making it ideal for embedded systems. Our simulation results also show that a 16-way DAWP cache is found to have the best compromise for low-power consumption.

References

1. Montanaro J, Witek RT, Anne K, Black AJ, Cooper EM, Dobberpuhl DW, Donahue PM, Eno J, Hoeppner W, Kruckemyer D, Lee TH, Lin PCM, Madden L, Murray D, Pearce MH, Santhanam S, Snyder KJ, Stephany R, Thierauf SC (1996) A 160-MHz, 32-b, 0.5-W CMOS RISC microprocessor. IEEE J Solid-State Circ 31(11):1703–1714
2. Flynn MJ, Hung P (2005) Microprocessor design issues: thoughts on the road ahead. IEEE Micro 25(3):16–31
3. Zhang C (2006) A low power highly associative cache for embedded systems. IEEE international conference on computer design (ICCD), San Jose, CA, pp 31–36, Dec 2006
4. Powell M, Agarwal A, Vijaykumar TN, Falsafi B, Roy K (2001) Reducing set-associative cache energy via way-prediction and selective direct-mapping. IEEE 34th international symposium on microarchitecture (MICRO), Austin, TX, pp 54–65
5. Furber S et al (1989) ARM3—32b RISC processor with 4 kbyte on-chip cache. In: Musgrave G, Lauther U (eds) Proceedings IFIP TC 10/WG 10.5 international conference on VLSI, pp 35–44. Elsevier (North Holland)
6. Santhanam S et al (1998) A low-cost, 300-MHz, RISC CPU with attached media processor. IEEE JSSC 33(11):1829–1838
7. Intel XScale Technology. http://www.intel.com/design/intelxscale. Accessed on April 2013
8. Inoue K, Ishihara T, Murakami K (1999) Way-predicting set-associative cache for high performance and low energy consumption. In: Proceedings of international symposium on low power electronics and design, pp 273–275
9. Batson B, Vijaykumar TN (2001) Reactive-associative caches. Parallel architectures and compilation techniques. In: Proceedings of international conference on 8–12 Sept 2001 pp 49–60
10. Shivakumar P, Jouppi N (2001) CACTI 3.0: an integrated cache timing, power, and area model, Compaq, Palo Alto, CA, WRL Res. report 2001/2
11. Burger DC, Austin TM (1997) The SimpleScalar tool set, version 2.0, computer architecture news, vol 25(3), pp 13–25, June 1997
12. Contreras G, Martonosi M (2005) Power prediction for Intel Xscale processors using performance monitoring unit events, ISPLED '05, San Diego, CA, August 2005

A Density Control Scheme Based on Disjoint Wakeup Scheduling in Wireless Sensor Network

EunHwa Kim

Abstract Wireless sensor networks consist of many nodes with sensing, computation, and wireless communications capabilities. Due to difficulty of recharging battery, energy efficiency is essential problem in wireless sensor network. We propose a density control scheme based on disjoint wakeup scheduling which can provide a full connectivity to sink node with a minimum set of active nodes in a highly dense network to prolong the network lifetime.

Keywords Wireless sensor network · Scheduling · Connectivity · Energy efficiency · Density control

Introduction

Advanced sensor technologies and wireless communications have enabled the development of wireless sensor networks which can be used for various applications such as earth-quake, forest fire, the battlefield surveillance, machine failure diagnosis, biological detection, home security, smart spaces, inventory tracking, etc. [1, 2]. A wireless sensor network consists of tiny sensing device which has capability of detecting some phenomena.

In a large-scale sensor network, we need to deploy numerous sensor nodes densely in the sensing field and maintain them in proper way. One of the most important issues in such high-density sensor networks is density control [3, 4].

It can save energy by turning off redundant nodes and it can prolong the system lifetime by replacing the failed nodes with some sleeping nodes. Density control is

E. Kim (✉)
YongIn University, Yongin, South Korea
e-mail: ehkimanna@yongin.ac.kr

Y.-M. Huang et al. (eds.), *Advanced Technologies, Embedded and Multimedia for Human-centric Computing*, Lecture Notes in Electrical Engineering 260,
DOI: 10.1007/978-94-007-7262-5_57,

very important for energy efficiency in topology control with wireless sensor networks.

In GAF [5], total network is divided a smaller virtual grid cell enough to communicate with its neighbors and one sensor node is selected as active node in a virtual grid cell. OGDC [6] prove that if communication radius is greater than twice of sensing radius, the connectivity is guaranteed by full coverage problem. In ASCENT [7], sensor node participate the network topology with parameter as the number of neighbors and packet loss ratio. Joint Scheduling method [8] awakes sensor nodes with random time slot which provide statistical coverage ratio and it awakes the other extra time slot that belongs to its downstream neighbor for connectivity to the sink node.

In this paper, we propose a density control scheme based on disjoint wakeup scheduling to improve energy efficiency while satisfying the requirements for sensing coverage and network connectivity. Our scheme works in a distributed manner at each sensor node and does not require the location information.

The rest of this paper is organized as follows. In paper A Vocabulary Learning Game Using a Serious-Game Approach, we propose our scheme, and evaluate it through simulation in paper An Improved Method for Measurement of Gross National Happiness Using Social Network Services. Finally, we conclude the paper in Advanced Comb Filtering for Robust Speech Recognition.

Scheduling Algorithm

In this paper, we propose a density control scheme based on disjoint scheduling algorithm which can provide a full connectivity to sink node with a minimum set of active nodes to prolong the system lifetime. It is assumed that every sensor node initially knows the hop count to the sink node. The best way to deliver data to sink node is forwarding its data to one of its upstream neighbor nodes. Upstream neighbors of a node are defined as nodes with smaller hop count by one than itself within its transmission range. A sensor node has only to communicate with upstream neighbor nodes to deliver data to sink node or receive a query from to sink node. In our algorithm, only one of nodes with the same hop count within transmission range (hereinafter, we call them 'peer neighbors') is allowed to wake up. To elect such an active node, each sensor node initially turns on its radio and keeps listening, and performs a random back-off. Once the back-off process is finished, it sends a *GREETING_MSG* to declare that it wants to become an active node and thus prevents other peer neighbors from activating. When a node hears a *GREETING_MSG* from its peer neighbors, it immediately stops the back-off process and goes to sleep. In this way, we can minimize the number of active nodes in the network field. However, it cannot guarantee path connectivity from a node to the sink node if there is no active upstream neighbor node around it.

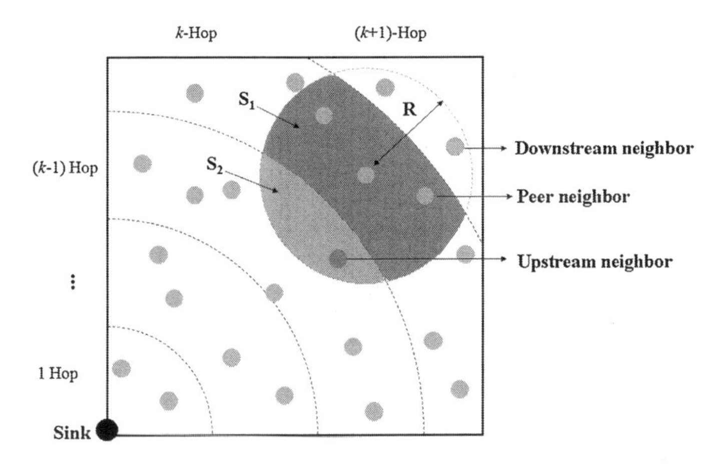

Fig. 1 Wireless sensor network

As shown in Fig. 1, the average number of active upstream neighbor nodes in our disjoint scheduling method is given by

$$N_{active\_up} = \frac{S_2}{S_1} \quad \text{where} \quad \begin{aligned} S_1 &= 2\int_{-\frac{1}{2}R}^{\frac{1}{2}R} (\sqrt{R^2 - y^2})\, dy * \frac{A(N_{adj})}{\pi R^2} \\ S_2 &= 2\int_{\frac{1}{2}R}^{R} (\sqrt{R^2 - y^2})\, dy \end{aligned}$$

where *A(k)* means area of a regular k-polygon, N_{adj} means the average number of adjacent active nodes, and *R* denotes communication radius. From [9], N_{adj} is given by

$$N_{adj} = \frac{\pi}{\theta} \approx 3.3 \quad where \quad \cos\theta = \frac{{}^{7}\!/_{6}R}{R} = \frac{7}{6}$$

So, we have $N_{active\_up} = 0.64$ approximately. This indicates that there is a possibility of 0.36 that an active node can not find another active neighbor in the upstream direction toward the sink. Actually, the number of upstream neighbors required to additionally become active is not so many that it is not much burden to the network. In our algorithm, if an active node does not hear a *GREETING_MSG* from upstream neighbor nodes, it selects an upstream neighbor node at random, and wakes up it by sending a *SELECT_MSG*.

Performance Evaluation

To evaluate our algorithm, we carried out experiments to measure coverage and connectivity of the network. To validate our scheme, we implemented a simulator in C/Java and compare it with GAF method and Joint Scheduling method [8].

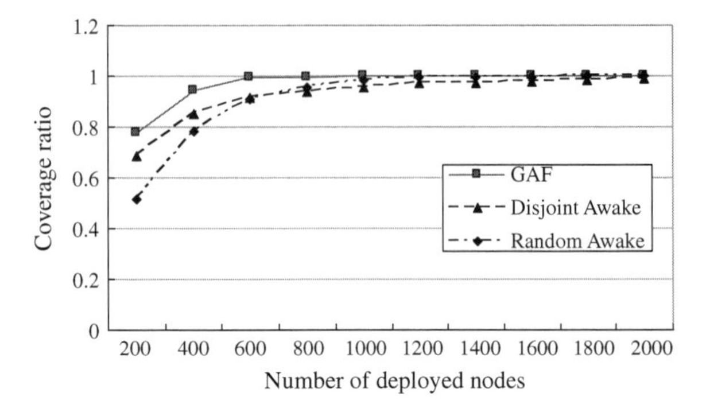

Fig. 2 Coverage ratio

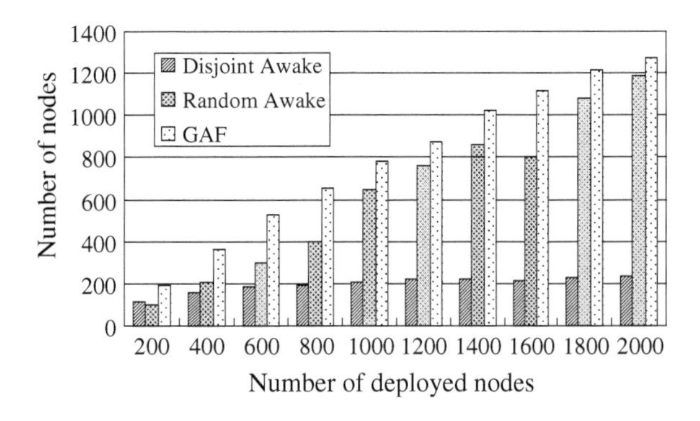

Fig. 3 Number of active nodes

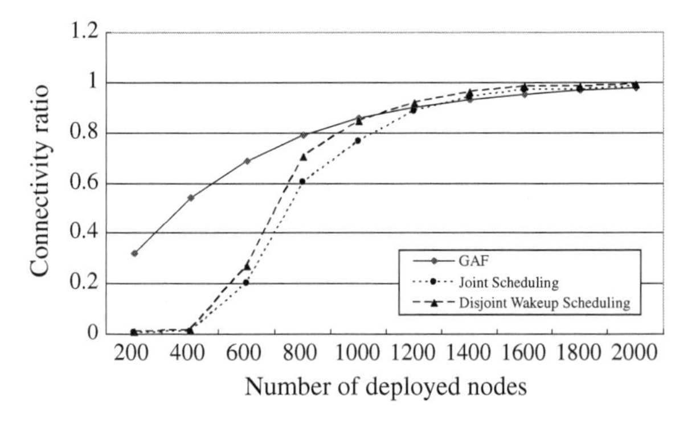

Fig. 4 Path connectivity ratio

Sensor nodes are deployed uniformly in a random fashion in a 200 * 200 square region with a communication radius of 10. The sink node is located at the corner of network.

Figure 2 shows that our method gives a satisfactory coverage performance. GAF and random awake method also provide full coverage but they require more active nodes as shown in Fig. 3. In a dense network, our method performs well with the coverage problem.

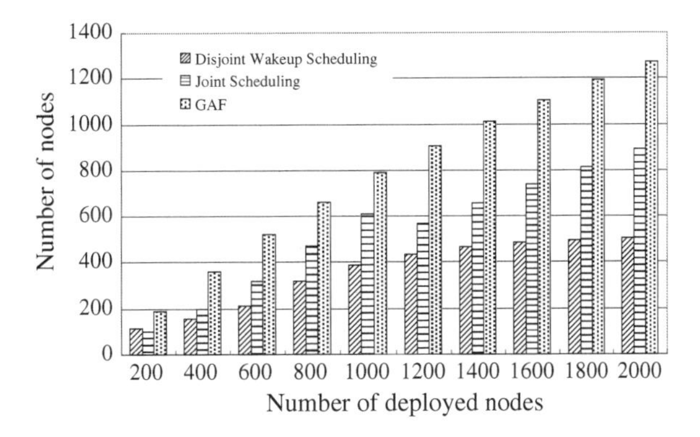

Fig. 5 Number of active nodes

Figure 4 shows that our method provides successful connectivity in a dense network. It also awakes smaller nodes than GAF and Joint Scheduling method as shown in Fig. 5. In a dense network, our method performs well more apparently in terms of coverage and connectivity performances.

Conclusion

In this paper, we propose a disjoint scheduling algorithm for density control in wireless sensor network. It selects a small set of working nodes to avoid wasting excessive energy by turning off too many redundant nodes. It selects a small set of active nodes to improve energy efficiency while providing path connectivity to sink node.

Simulations showed that our scheme achieves the desired robust coverage as well as satisfactory connectivity to the sink with a smaller number of working nodes in an energy efficient fashion than GAF and Joint Scheduling method.

References

1. Akyildiz IF, Su W, Sankarasubramaniam Y, Cayirci E (2002) Wireless sensor networks: a survey. In: Computer networks, vol 38, pp 393–422
2. Chong C, Kumar SP (2003) Sensor networks: evolution, opportunities, and challenges. In: Proceedings of the IEEE, vol 91, pp 1247–1256
3. Ye F, Zhong G, Lu S, Zhang L (2003) Peas: a robust energy conserving protocol for long-lived sensor networks. In: ICDCS
4. Zhang H, Hou JC (2003) Maintaining sensing coverage and connectivity in large sensor networks. Technical report UIUC, UIUCDCS-R-2003-2351
5. Xu Y, Bien S, Mori Y, Heidemann J (2003) Topology control protocols to conserve energy in wireless Ad Hoc networks. Technical report CENS 0006
6. Zhang H, Hou JC (2003) Maintaining sensing coverage and connectivity in large sensor networks. Technical report UIUC, UIUCDCS-R-2003-2351

7. Cerpa A, Estrin D (2004) ASCENT: adaptive self-configuring sensor networks topologies. J IEEE Trans Mob Comput 3:272–285
8. Liu C, Wu K, Xiao Y, Sun B (2006) Random coverage with guaranteed connectivity: joint scheduling for wireless sensor networks. J IEEE Trans Parallel Distrib Syst 17(6):562–575
9. Li H, Yu D (2002) A statistical study of neighbor node properties in ad hoc network. In: Proceedings of ICPPW, pp 103–108

Part VII
Multimedia Computing

A Human Voice Song Requesting System Based on Connected Vehicle in Cloud Computing

Ding Yi and Jian Zhang

Abstract Traditional on-board vehicle computer (OVC) systems do not support speech recognition and do not have massive song library due to their limited storage capacity and computing power. Furthermore, it is very inconvenient to update the song library to keep it up-to-date. The advent of cloud computing in conjunction with mobile computing has enabled human voice song requesting and mass song playing from on-board computer in a vehicle. A system of human voice song requesting from OVC connected to the music cloud has been proposed. Self-adaptive speech recognition and speech recognition technology based on keywords from song titles has been adopted. By recognizing the voice on a song title in a cloud environment, the digital content of the selected song stored in the cloud is streamed to and then played back in the OVC. As a result, the system improves the safety of driving and enables automobile entertainment more humane.

Keywords Networking of vehicles · On-board vehicle computer (OVC) · Speech recognition · Human voice song requesting (HVSR) · Cloud computing · Music cloud

Introduction

Vehicle drivers tend to listen to songs while driving. However, it is inconvenient for a driver to operate keyboard, even a remoter while steering the vehicle wheel. Thus, drivers cannot choose songs on demand while driving a traditional vehicle.

Traditional on-board vehicle computer (OVC) system has small word bank and small linguistic model. It does not support speech recognition and does not have

D. Yi (✉) · J. Zhang
School of Computer Engineering, Shenzhen Polytechnic, Shenzhen, Guangdong, China
e-mail: yiding300@gmail.com

Y.-M. Huang et al. (eds.), *Advanced Technologies, Embedded and Multimedia for Human-centric Computing*, Lecture Notes in Electrical Engineering 260,
DOI: 10.1007/978-94-007-7262-5_58,

massive song library due to their limited storage capacity and computing power. Furthermore, it is very inconvenient to update the song library to keep it up-to-date. The advent of cloud computing in conjunction with mobile computing has eliminated the above limitations, enabling human voice song requesting (HVSR) and massive song playing in OVC.

In this paper, a system of HVSR based on connected vehicle in cloud computing has been proposed and implemented. Self-adaptive speech recognition technology in combination with song title keyword speech recognition has been adopted. By recognizing the voice of a song title in the cloud environment, the digital content of the selected song stored in the cloud is streamed to and then played back in the OVC. Vehicle drivers can request songs by only speaking out its song title. As a result, the system improves the safety of driving and renders automobile entertainment more humane as well.

General Scheme

A HVSR system based on connected vehicle in cloud computing (as shown in Fig. 1) consists of OVC, HVSR management cloud server, music cloud servers and networking of vehicles. Among these components, the networking of vehicles and music cloud servers are built from existing online resources while the design and implementation of the OVC and the HVSR management server are our own innovations.

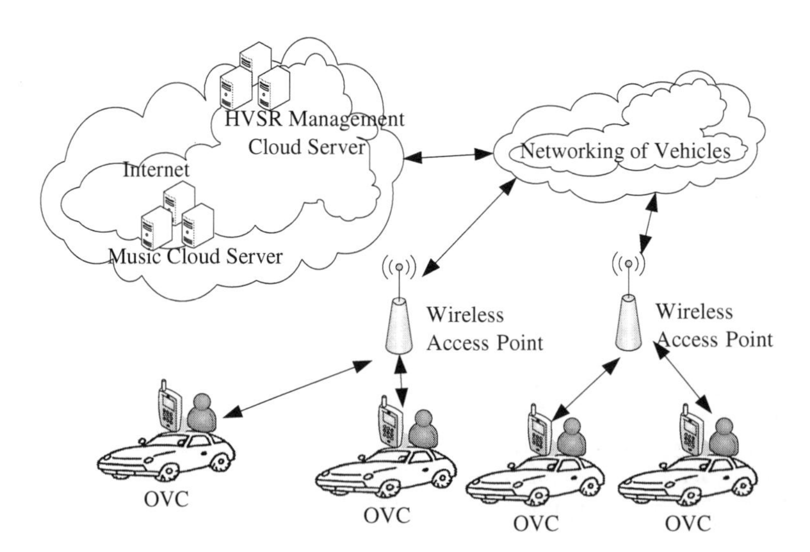

Fig. 1 General scheme

The networking of vehicles, provided by network service providers, is the access network by which an OVC is connected to the Internet.

The music cloud servers located in the Internet are the existing public resources. These servers host and control audio streaming data that is served as audio streaming source. In the Internet there have been many music servers that possess massive audio streaming files. An OVC needs to playback these audio streaming data from these public music cloud servers.

The HVSR management cloud server provides the service of speech recognition and song titles management. The speech recognition and song titles management is realized in cloud computing.

The client, OVC, installed in an automobile consists of hardware and software. It receives the service of speech recognition and song titles management from cloud services, and obtains audio streaming data from the cloud services to playback.

The Workflow and Dataflow

There is a one-push button in the OVC. When the button is pushed down, it starts to run, and when pushed again it stops working. The following is the workflow when the system starts:

- The OVC accesses the Internet via the networking of vehicles. It establishes connection to the HVSR management cloud server.
- The OVC receives song requesting voice spoken by a driver, and then sends the digital voice data to the HVSR management cloud server;
- The HVSR management cloud server receives the digital voice data of song requesting from the OVC and identifies the song's title through speech recognition. Then the music server that hosts the song audio streaming data is queried and the corresponding network address is sent back to the OVC.
- The OVC receives the above network address, establishes a connection to the music server of the given address. It then receives and plays the audio streaming from this music server [1].

Corresponding to the above workflow, the dataflow is shown in Fig. 2.

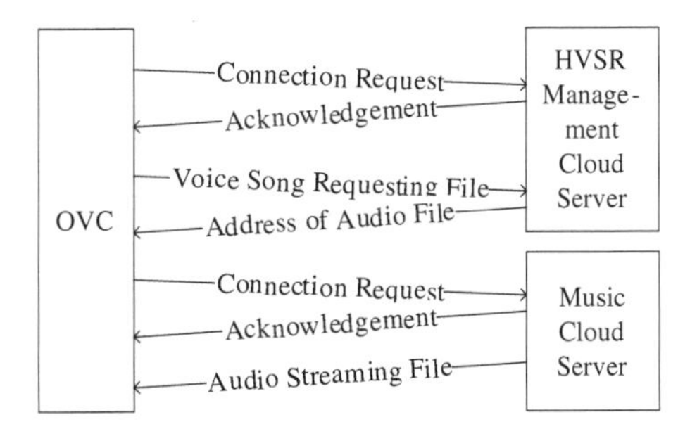

Fig. 2 Dataflow

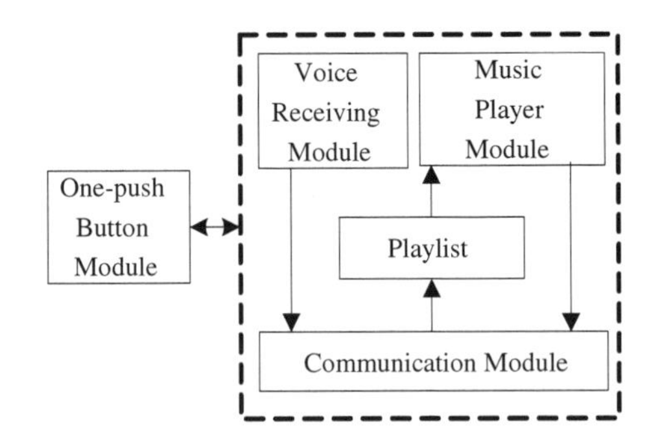

Fig. 3 OVC structure

The On-board Vehicle Computer

The client, OVC, installed in an automobile consists of both hardware and software [2]. It obtains the service of speech recognition and song titles management from cloud services, and receives audio streaming from the cloud services.

The OVC provides the following functions:

- Access to the Internet through the networking of vehicles,
- Receive users' voice song requesting,
- Communicate with HVSR management cloud server,
- Communicate with public online cloud music servers,
- Manage the playlist,
- Play the audio streaming data from cloud music servers.

The OVC consists of one-push button module, communication module, voice receiving module, playlist and music player module (as shown in Fig. 3).

The one-push button module in the OVC is used to start/stop the system.

The communication module in the OVC has a unique network address with which the vehicle is connected into the Internet. It is responsible for accessing to HVSR management cloud server and music cloud servers.

The voice receiving module in the OVC is responsible for detecting voice, and recording effective song requesting voice. The communication module sends this recording file to the HVSR management server.

The communication module in the OVC receives the song audio streaming file's network address from the HVSR management server, and then appends this address to the playlist.

The music player module in the OVC retrieves the server's network address attached to the current selected song from playlist. Then the communication module establishes connection to that music server, and receives the audio streaming data from the music server. Once the streaming data is received, the player module plays this audio streaming data.

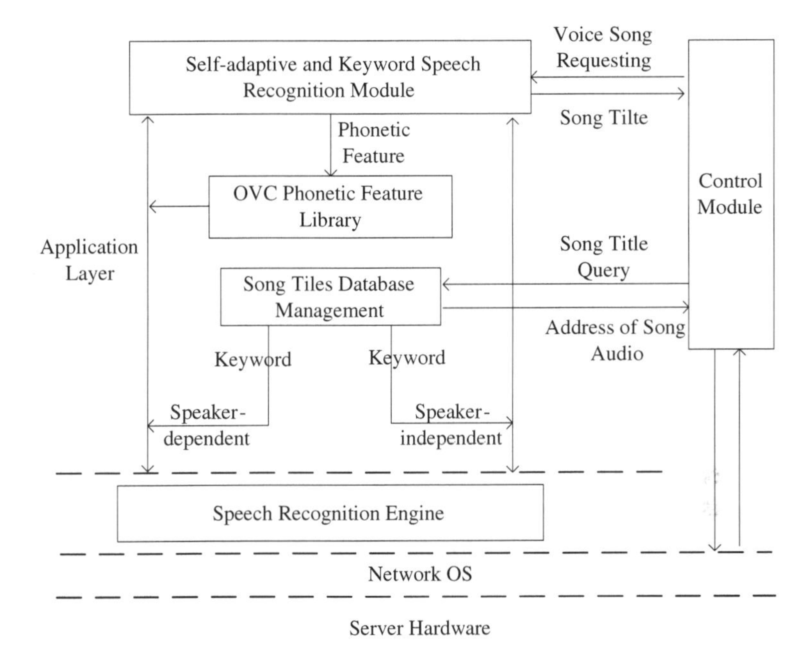

Fig. 4 HVSR management cloud server

The HVSR Management Cloud Server

The HVSR management cloud server (as shown in Fig. 4) provides the service of speech recognition and song titles management [3]. It adopts third party professional speech engine product that directly runs on top of the operating system [4, 5]. Speech engine is a middleware with the capabilities of both speech recognition and synthesis. Leveraging its services of speech recognition and synthesis in the application layer, the legacy keyboard operations such as text input, menu selection, etc. are now substituted by a user's voice requesting, which implements truly man–machine dialogue.

The service of speech recognition and song titles management that works in the application layer is mainly discussed in the paper.

The Service of Speech Recognition and Song Titles Management

The service of speech recognition and song titles management works in the application layer of network architecture (Fig. 4). It is realized by self-adaptive speech recognition and song titles keyword speech recognition technology module, song titles database management module, OVC phonetic feature library and control module.

The control module sends HVSR digital file from the OVC to the speech recognition module. The "song title" text is identified in the speech recognition module. Then the song audio streaming file's network address in the song titles database is queried using this text as keyword. Finally the control module sends the server address back to the OVC.

The Song Titles Database Management

Each record in song titles database includes at least three fields: the song title, the singer name, the network address of the song audio streaming file.

There are many music servers in the Internet that provide massive song audio streaming files. Therefore it is sufficient to store the network address of song audio streaming files instead of complete audio streaming files in song titles database.

The song titles management implements the real-time maintenance of song titles database, renews the fields of the database in time. This is to ensure that new songs are appended up-to-date and the song audio streaming file's network address is valid.

The 'song title' and the 'singer name + song title' in song titles database are searched as keywords in keyword speech recognition. So the song titles database has two fundamental functions: keyword speech recognition, and the query of the network address for a song audio streaming file.

The Song Title Keyword Speech Recognition

Applying keyword speech recognition technology, rather than applying continuous natural language speech recognition technology, is more effective in the HVSR application. For a continuous natural language, the same semantics can have different ways of expression. If continuous natural language speech recognition technology is adopted, the system becomes difficult to implement, and the rate of recognition accuracy will be low. Aimed at the special application environment of the HVSR, sufficient information is collected by using 'song tile' or 'singer name + song title' spoken by the user without the full natural language expression. Therefore, the technology of keyword speech recognition simplifies the problem and is easy to realize with higher accuracy rate of speech recognition.

The OVC Phonetic Feature Library

The OVC phonetic feature library includes OVC address field and its phonetic features field. It is used to support speaker-dependent speech recognition.

Every networking OVC has a unique address. In most cases, an automobile has one fixed driver or multiple drivers. Therefore, the phonetic characteristics from the same OVC are limited. These limited features can be stored in the phonetic features library. If an OVC accesses the system for the first time, the system creates its phonetic characteristics information record. Later new phonetic features from the same OVC are appended to the record.

Speaker-Independent/Speaker-Dependent Self-adaptive and Song Title Keyword Speech Recognition Technology

During speech recognition, the song title keyword in speaker-dependent speech recognition technology is used first [6]. If succeeded, it means that the phonetic feature of the OVC has already been stored in the library. If failed, it could be the case that this is a new access, thus there is not such phonetic feature recorded in the phonetic feature library. In such case, the system automatically switches to use speaker-independent speech recognition and song title keyword speech recognition technology to identify the selected song. After a song is identified, the new phonetic feature is appended into the record of the phonetic feature library.

Conclusions

While driving a vehicle driver needs to control the steering wheel. Thus it is inconvenient to operate buttons. In this paper, a system of HVSR based on OVC connecting to the music cloud has been proposed. A driver can request a song by only speaking song title. The song title is then recognized by speech recognition technology in a cloud environment. Then the digital audio content of the selected song stored in the cloud is streamed to and finally played back in the OVC.

The OVC installed in automobile is a combination of hardware and software. It obtains the services of speech recognition and song titles management from cloud services, and accesses audio streaming from cloud services as well.

Using speaker-independent/speaker-dependent self-adaptive in combination with song title keyword speech recognition technology in cloud computing is an important characteristic of this system.

The advent of cloud computing in conjunction with mobile networking of OVC has eliminated the traditional limitation, making HVSR and mass song playing from OVC into reality. The methodology and the implemented system in the paper allow a driver to request a song by only speaking the song title. We believe this renders automobile entertainment more humane while considerably improving the safety of driving as well.

References

1. Wang Z (2004) Research in the application of streaming media technology. Comput Eng 5:34–36
2. Chen X (2005) The design of embedded streaming media player. Zhejiang University, pp 17–19
3. Zhao B, Yan F, Zhang L, Wang J (2012) The construction of dependable cloud computing environment. China Comput Fed Commun 8(7):28
4. Windows Azure blog: http://blogs.msdn.com/b/azchina/
5. Microsoft Tellme Speech Innovation: http://www.microsoft.com/en-us/Tellme/default.aspx
6. Li N, Xu S, Ma Z, Shi L (2011) Simulation research on adaptive speech recognition algorithm. Comput Simul 8:181–185

Access Control System by Face Recognition Based on S3C2440

Yongling Liu, Yong Lu and Yue Song

Abstract Face recognition technology has become more and more important in the field of biometric identification because of its advantages. A face recognition system combining hardware and software based on S3C2440 is presented in this paper. We adopt S3C2440 embedded development board as the hardware platform. This paper creates an embedded Linux software platform and our system is divided into four sections by their function: graphical user interface, image capture, face detection and face recognition. The whole system is tested as an access control system in practical situation. Experimental results show that our system has a good performance in both accuracy and efficiency of face recognition in access control.

Keywords S3C2440 · Face recognition · Access control system · Linux

Introduction

Face recognition has been widely used in access control and authentication because of its benefits such as accuracy and fast identification, high usability and security. Previous face recognition system was designed under the platform of personal computer at the cost of portability. Therefore, with the rapid development

Y. Liu · Y. Lu · Y. Song (✉)
School of Electronic and Information Engineering, Beijing Jiaotong University, Beijing, China
e-mail: sycat7@gmail.com

Y. Liu
e-mail: 10120009@bjtu.edu.cn

Y. Lu
e-mail: ylu@bjtu.edu.cn

Y.-M. Huang et al. (eds.), *Advanced Technologies, Embedded and Multimedia for Human-centric Computing*, Lecture Notes in Electrical Engineering 260,
DOI: 10.1007/978-94-007-7262-5_59,

of embedded technology, an embedded face recognition system is more popular. An embedded face recognition system would entail many advantages including low cost, integration with other technologies and optimization for real-time operations.

Now, access control system has evolved as an intelligent management system of Entry/Exit Control instead of physical key management. With such a smart access control system, it can ensure more security on the condition that the people lose or forget the ID card. Therefore, it is meaningful and useful to carry out research on embedded face recognition system for future security and convenience.

System Design

Hardware framework of the access control system is shown in Fig. 1. The face images are extracted from the USB camera and then processed with S3C2440 embedded processor by means of face recognition algorithm. SDRAM and FLASH work together as memory devices which host temporary files data and permanent data. Besides, the human–computer interaction module consists of LCD and touch screen. The RS232 serial port or JTAG port connects the access control system and PC to communicate each other.

And our system is divided into four sections by their software function: graphical user interface, image capture, face detection and face recognition.

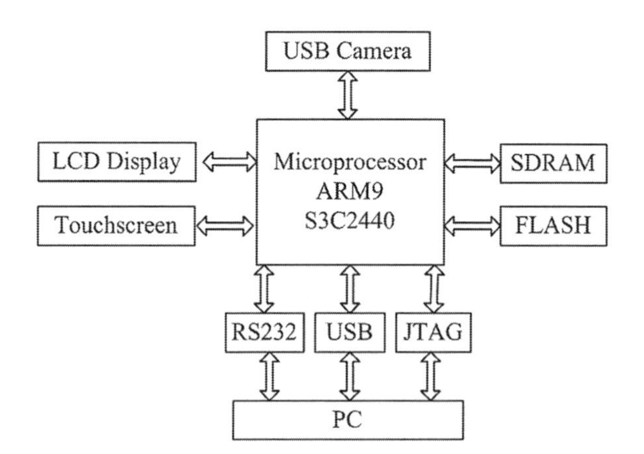

Fig. 1 Framework of the access control

Hardware Platform

Our system is based on FL2440 FORLINX embedded development board. The major descriptions of FL2440 are following:

- CPU: S3C2440A processor made by Samsung, main frequency 400 MHz.
- Memory: SDRAM, 4 Banks * 4 Mbits * 16 bits.
- Flash Memory: 256 M Nandflash, 4 M Norflash.
- System clock: 12 MHz system external clock source.
- LCD: 3.5 inch touchable screen supports TFT type.
- Interface: 2 pieces of RS232 serial ports and 4 pieces of USB host interfaces.

Besides, we add a USB camera into the system to capture face images. This camera (ZC301) is created by VIMICRO Corporation.

Software Platform

Our platform is based on Linux operating system. To complete the platform, the following work is been done:

1. Building up cross compiling environment

In this paper, the operating system for the prototype machine is Red Hat 9. We install cross compiler of version 3.4.1.

2. Portability of Boot loader and kernel

Our system adopts the U-Boot of version 1.3.3 as boot loader and Linux version 2.6.12 as our kernel. An executable file named U-Boot.bin has been generated finally. Besides,

3. Root file system

Our system chooses the document system with yaffs2 format based on NAND FLASH. We complete the file system by means of BusyBox to generate an executable file.

4. Device drivers support

There are two cases in Linux device drivers. One is the hardware supported by kernel. The other one is the hardware not supported by kernel such as the USB camera used in our system. The drivers can be available by the third party and dynamic load into the kernel.

Software Design for Face Recognition System

1. User interface Module

Our user interface is designed by QT. First, we download the source code of QT-x11 and QT-embedded. The QT-x11 will generate QT development tools and qvfb by which could enable embedded development environment simulation to develop QT projects under the situation of x86. Particularly, the QT-embedded is optimized for embedded development. Then, QT development environment will be established by compiling, installing and configuring.

The flow chart of user control interface is followed as Fig. 2.

The system interface is in two parts: window button and display area. When a user click the button, the system will call program such as pre-processing and face extracting to process face images and then output results in the display area.

2. Image capture Module

Our system will collect images by means of Video4Linux (hereafter referred to as V4L). The image capture module is designed based on API functions offered by V4L. The program of image capture is realized as Fig. 3 and the capture result in real solution is shown in Fig. 4.

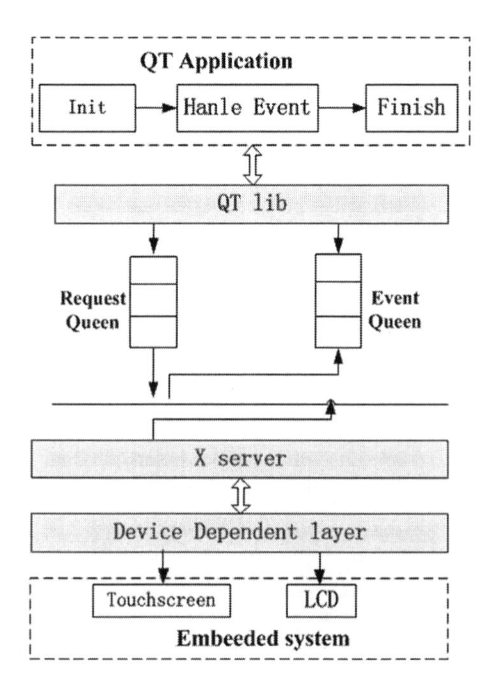

Fig. 2 The flow chart of user control interface

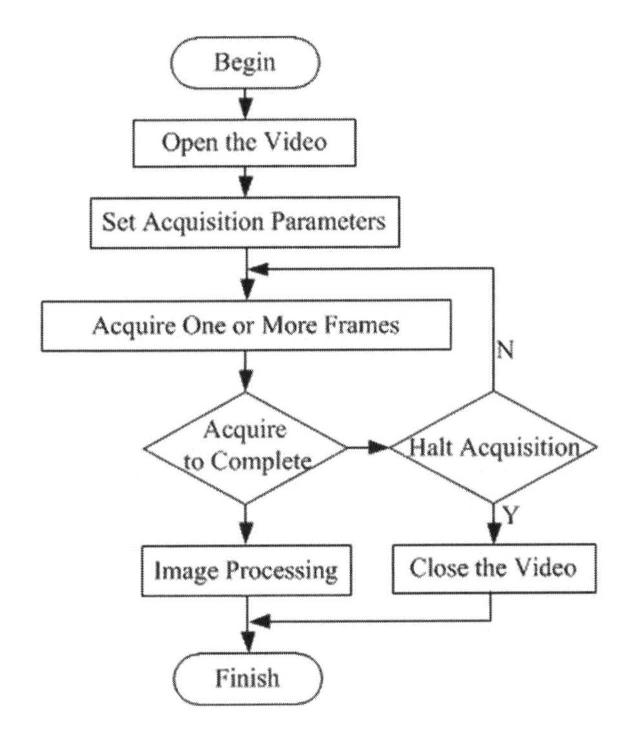

Fig. 3 The flow chart of image captures

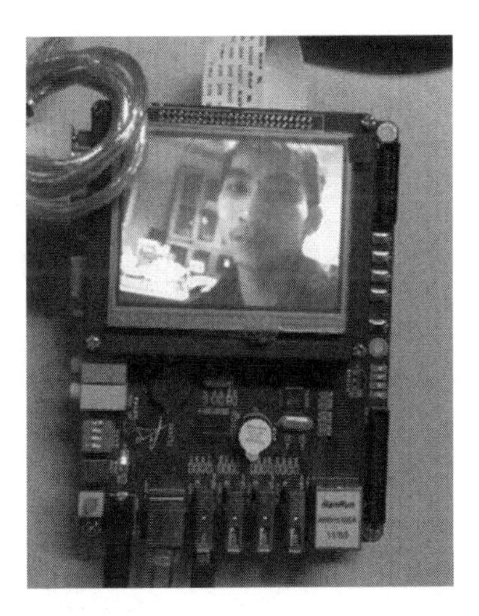

Fig. 4 Capture results

3. Face detection Module

The major function of face detection module is to locate the face in a captured image from the USB camera. The flow chart of face detection is followed as Fig. 5.

In this paper, we adopt method of haar+Adaboost as detection algorithm. Calculating the features of haar-like by such a method will be more efficient.

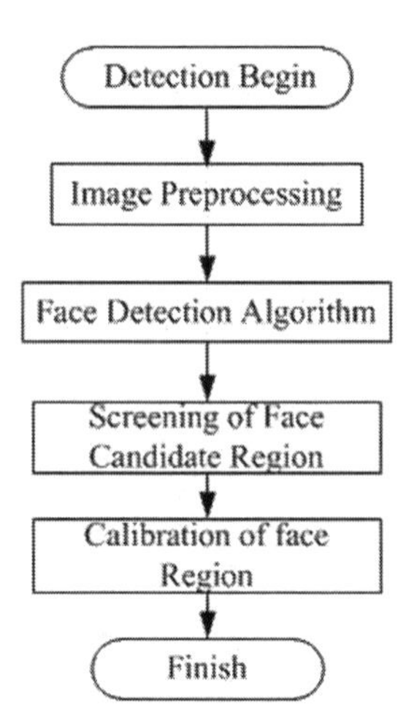

Fig. 5 The flow chart of face detection

Furthermore, Adaboost learning algorithm could combine weak classifiers into one strong classifier with feature selecting and classification training. We use the algorithm based on cascade classifiers instead of skin color because in comparison to skin-color-based algorithm, cascade-classifier-based method will offer a more accurate location in the situation that the background is similar to skin color or skin on other parts of body exposed besides human face. Major steps of detection are following:

- Select a proper classifier.
- Read images or videos.
- Detect human faces.
- Display the target and locate the human face.

From the Fig. 6, we can see the detection performance from a dynamic image captured by the USB camera.

4. Face recognition Module

Now we need to extract features describing the face for recognition by face recognition module. Principal Component Analysis (PCA) and 2 Dimension Principal Component Analysis have a wide range of applications for feature extraction algorithms. Compared to PCA, 2DPCA constructs covariance matrix by its image matrix directly. Therefore, the 2DPCA algorithm has a better performance in both accuracy and computation speed. However, the advantages of 2DPCA are obtained at the cost of a high memory usage during the period of image reconstruction. In this paper, we combine PCA and 2DPCA to face recognition. That is a two-stage strategy that we apply 2DPCA to reduce image's dimensionality at first stage and then extract features from prior outputs based on PCA algorithm. This strategy will help the system to save memory and reduce computational cost.

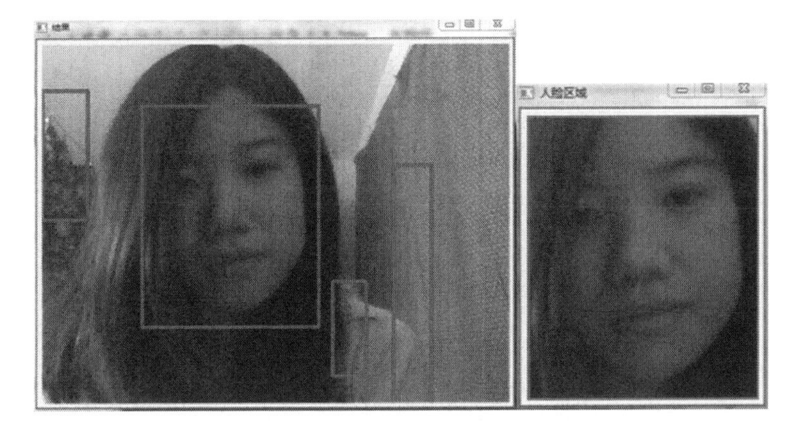

Fig. 6 Face detection performance

Besides, our face recognition algorithm is based on gray scale images while the captured images are color images of RGB. We must make pre-processing such as grey level transformation and normalization before training. In our system, the image resolution is 65 × 80 and the training images after pre-processing are shown in Fig. 7.

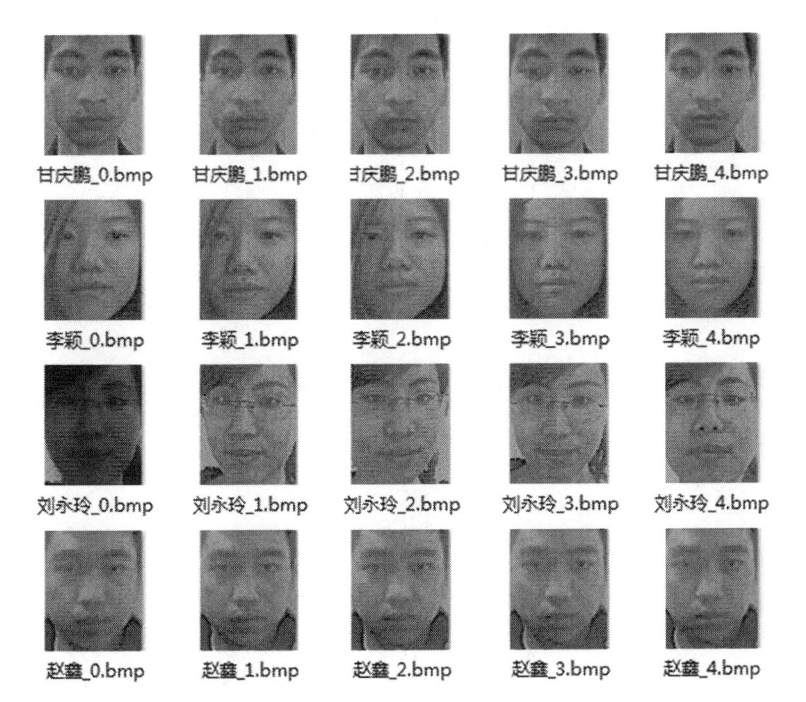

Fig. 7 Normalized face images

There are two parts in face recognition: the training process and the matching process.

Firstly, the training procedure works as following:

- Create feature subspaces from the training set by function prototype shown below.
  ```
  bool CreateEigenFaceSpaces(IMG *src, MAT *dest, MAT *EigenVect, int num);
  ```
- The eigen subspaces can be constructed by image projection. The function prototype is:
  ```
  bool ProjectSubEigenFace(IMG *src, MAT *dest, MAT *SubSpace, int num);
  ```

In addition, the recognition procedure is following:

- Project the test image into the eigen subspace by function:
  ```
  bool ProjectSubEigenFaceForSingle(IMG *src, MAT *dest, MAT *SubSpaceForSingle);
  ```
- Compare it with training images. Then output the matching result.
  ```
  int Compare_Ma(MAT *SubSpaceForSingle,MAT *SubSpace, MAT*EigenVect,int num);
  ```

From the Fig. 8, we can see the face recognition system performs well. In the window, the left one is detected human face while the right one is registered face image. If the two images are matched, the system will display the registered information of the person and the authorized person will be allowed to enter the secure area.

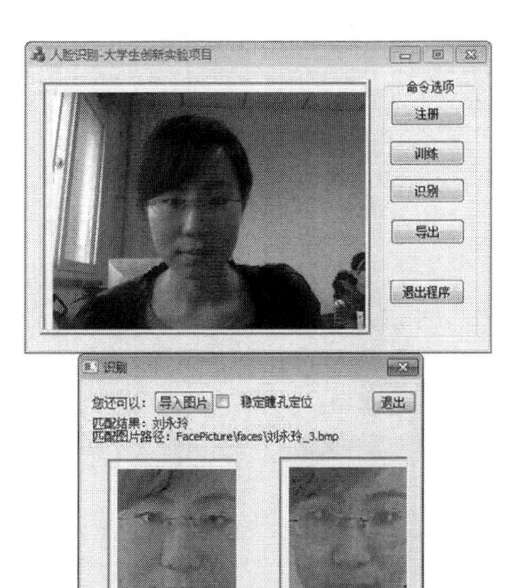

Fig. 8 Face recognition

Table 1 The results of face recognition

Training samples of per person	1	2	3	4	5
Recognition rate	0.80	0.83	0.85	0.87	0.91

System Test

We can set up a database by 100 photos to do a test with this system. These photos are taken of 20 people in the lab, each with 5. The dimensionality is 55. We need to observe the changes of recognition rate when the samples are changed from one to five. When we conduct the experiment, we need to ensure that light is much enough and that the faces of the tested person stay a little while in front of the camera. We can come to the conclusion in the following sheet1 (Table 1).

The more the samples are, the higher the recognition rate is. After a series of experiments based on the five samples, the recognition rate can be increased to 90 % or higher. But if the light is not enough, shooting angle change is big, or facial expression changes great, it is easy for us to make mistakes on recognition. Therefore, when doing a registration, we need to try to take different photos from different angles with different facial expressions, which benefits increasing recognition rate.

Conclusion

This thesis mainly covers a research on face recognition access control system based on S3C2440. During the process, I divided the whole system into four functional parts, and respectively design and fulfill each of the tasks, including developing graphic user interface, image collection, face detection, and the algorithm of face detection rate. The results manifest that the system needs to be improved in the following aspects:

1. Algorithm needs to be improved or new algorithm needs to be proposed, in order to get a better effect on face detection.
2. Fully take advantage of the resources of the system, for example, increasing network transmission patterns.

References

1. Zhang J, Yan Y, Lades M (1997) Face recognition: eigenface, elastic matching and neural nets. Proc IEEE 85(9):1422–1435
2. Meng L, Nguyen TQ, Castanon DA (2000) An image-based Bayesian framework for face detection. In: Proceedings of IEEE conference on computer vision and pattern recognition, Hilton Head Island, South Carolina, USA, pp 302–307

3. Lu H, Shi W (2005) Accurate ASM for human face image search. The 17th IEEE international conference on tools with artificial intelligence, pp 642–647
4. Xin X, Lansun S, Kongqiao W (2000) Automatic human face detection. ISO/IECJTCI/SC29/WGll MPEG99/M6144, Beijing, China, July 2000
5. Hsu RL, Abdel-mottaleb M, Jain AK (2002) Face detection in color images. IEEE Trans Pattern Anal Mach Intell 24(5):696–706
6. Zhang J, Yan Y, Lades M (1997) Face recognition: eigenface, elastic matching and neural nets. Proc IEEE 85(9):1422–1435
7. Hadid A, Pietikinen M, Ahonen T (2004) A discriminative feature space for detecting and recognizing faces. In: IEEE conference on computer vision and pattern recognition (CVPR)

A Secure Digital Watermark Skeleton Based on Cloud Computing Web Services

Jian Zhang and Ding Yi

Abstract More and more digital watermark systems provide the services via the internet. The watermark embedding and detection technology components are the parts of the web services of the sites. This paper designs a secure digital watermark skeleton based on cloud computing, pushing these web services to the cloud. The entire skeleton is based on cloud computing architecture, including service delivery, data storage and security, data lifetime, etc. The skeleton combines the digital watermark and cloud computing technologies with the certain reference value, it also provides efficient digital watermark within the large number of complicating requests to protect digital products in the marketing of multimedia network environment.

Keywords Digital watermark · Cloud computing · Web service

Introduction

Digital watermark is an important direction of the multimedia technology within the field of information security. It protects the copyright of the original data by embedding secret information, such as the watermark in the original data. This watermark can be embedded into the original data as a paragraph of text, logos, serial numbers, etc. The watermark is with raw data (images, audio, video, etc.) and hides them closely. It can be retrieved via the specific algorithm so that it can protect the copyright of the multimedia products.

With the internet technology and object-oriented component technology's rapid development, it promotes the internet within the distributed computing technology

J. Zhang (✉) · D. Yi
School of Computer Engineering, Shenzhen Polytechnic, Shenzhen, China
e-mail: jzha930@szpt.edu.cn

Y.-M. Huang et al. (eds.), *Advanced Technologies, Embedded and Multimedia for Human-centric Computing*, Lecture Notes in Electrical Engineering 260,
DOI: 10.1007/978-94-007-7262-5_60,

changes. In this environment, the web services-based digital watermarking technology, as a new distributed computing systems, has been proposed and so that enterprises in property rights protection, quickly deploy solutions and explore new opportunities. There for, many of the original system designs and developments are getting re-deployed through again as a web services system there. Some of digital watermarking systems are web-based. They can provide the services of watermark embedding and detecting via the internet [1, 2].

According to the characteristics of web services, this paper provides a cloud computing web services based secure digital watermark skeleton. It combines the digital watermark and cloud computing technologies, providing the watermark embedding and detecting services via the cloud. The author of the multimedia products just uploads the products to the web site, the cloud will call the relevant applications to embed the watermark or detect the watermark. In the course of using the services there are many aspects of security problems should be concerned: the safety of the web services environment and the security of digital watermarking itself. These are service delivery, data storage and security, data lifetime, etc. This paper will analyze these issues in detail and propose a workable solution.

The Background of the Project

There is a company which provides a network platform for picture trading. The system is embedding the lightweight watermark into the picture when authors upload their own products to ensure non-destructive and products of digital watermarking robustness, and the system can provide the detecting services [2]. The company is getting very successful with the picture trading. Now the company wants to provide the professional digital watermarking services to many commercial entities via the internet. According to their marketing records, they found it is difficult to provide the complicating requests services via the traditional web services. They want to deploy the web services and combine them as a cloud so that they can complete the mission.

Web Services System and Its Key Technology

Web services are self-contained self-describing modular components which can be publishing, locating and calling through the Web. Web Services system in a number of existing technologies (such as HTTP, XML) added some new standards. Here are the four key technologies [3].

- Simple Object Access Protocol, SOAP, is XML based for a decentralized, distributed environment, a lightweight protocol for exchanging information.

SOAP in between requestor and provider objects defines a communication protocol.

- Web Services Description Language, WSDL, provides a description of the service mode. Through a complete description of the service, the service requester can access the service methods and know the specific location; service developers also can use this interface is a standard compatible development services.
- Universal Description, Discovery, and Integration, UDDI, provides Web Services Publishing and discovery methods. Through open standards, Internet on the entity to find each other and call each other's Web Services.
- Web Services Security [4]. In the W3C XML Signature and XML Encryption based on the specification provides three mechanisms to extend the SOAP message: trustlike transmission, message integration and message confidentiality. WS Security and other Web services protocols can work together to meet a variety of types of application security needs.

In the environments of dynamic e-commerce, the original digital watermarking system migrates to web services system and releases it as a service and registration. The process will use SOAP, WSDL, UDDI and other key technologies. Because the digital watermark itself and security issues are closely related, so how to ensure that web services-based digital watermarking system security is a particularly important issue.

Cloud Computing Skeleton and Its Key Tehchnology

According to the characteristics and key technologies of web services, this paper proposes a digital watermark skeleton based on cloud computing web services. This skeleton combines the digital watermark and cloud computing technologies. It provides the watermark embedding and detecting services from the cloud. Innovations include:

- The watermark embedding and watermark detection technology as web services components in the form of a "cloud" released over the internet.
- The whole system is based on cloud computing architecture Web service platform, through the Simple Object Access Protocol (SOAP) exchange service requestor and service information between providers.
- Digital certificates, XML encryption and digital signatures, and other security technology to ensure the security of information exchange.
- Data storage, security and life time will be the key aspect of the skeleton.

The skeleton proposed to reduce the cost of the development and use of digital watermarking system, effectively protect the property rights of digital media products in an open network environment.

Paper [2] discuss the detail of the web services-based security model for digital watermarking. The following will discuss the key technologies of this skeleton involves service delivery, data storage and security, data lifetime, etc.

Service Delivery

The model is based on cloud computing and SaaS technology. The core of the model is service level. It will provide better availability management, capacity management, service level management, financial and risk management, continuity management in the field of SMS/MMS/CRBT/OSS/BSS. It delivers the five management processes and basic infrastructure [5].

Figure 1 shows the features of the A&S Cloud structure, they are: Business logic and calculation of storage are separated; Resources abstraction and sharing; Intelligent and automatic control. These features will be the direction of future research.

Data Storage and Security

Cloud computing usually consists of front-end users who possess clients devices and back-end cloud servers. This paradigm empowers users to pervasively access a large volume of storage resources with portable devices in a distributed and cooperative manner. During the period between uploading and downloading files

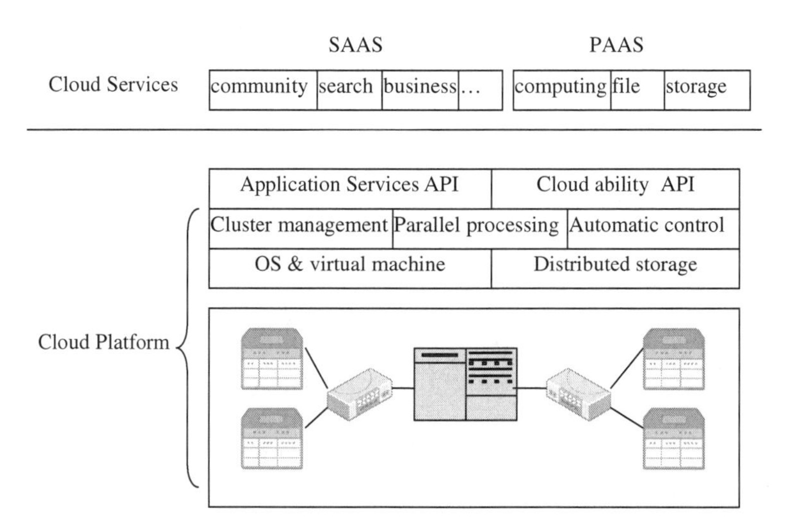

Fig. 1 A&S cloud structure (*Source* Huawei A&S cloud computing solutions report, p. 11)

(data), the privacy and integrity of files need to be guaranteed. To this end, paper [6] proposed a family of schemes for different situations. All schemes are lightweight in terms of computational overhead, resilient to storage compromise on clients devices, and do not assume that trusted cloud servers are present. Corresponding algorithms are proposed in detail for guiding off-the-shelf implementation. The evaluation of security and performance is also extensively analyzed, justifying the applicability of the proposed schemes.

As cloud providers have priority access to data, it is difficult to guarantee the confidentiality and integrity of users data. For this problem the paper [7] presents an architecture to protect user data security by encryption. Searching cipher text and integrity verification will be used to ensure the availability and integrity of users data. This mechanism provides reliable security support for the massive data in the cloud computing system.

Data Lifetime

Data privacy protection is one of the primary concerns and major challenges for online services, such as cloud computing and outsourced data center. The concern is getting serious with the computing practices shifting towards cloud computing. Once users data is uploaded, end users are hard to guarantee that the data is protected and can be completely destructed by any means. Users can only rely on blind trust on the online service vendors. However, the privacy of user data can be compromised in multiple ways including careless operations of cloud administrators, bugs and vulnerabilities inside cloud infrastructure and even malicious cloud vendors. Paper [8] seeks to provide users with a concrete way to protect or destroy uploaded data. It utilizes the technique of trusted computing as the trusted root in the hardware layer, and the hypervisor as the trusted agent in the software layer. The trusted hypervisor is responsible for protecting sensitive user data or destructing them at users command. Even administrators of the cloud cannot bypass the protection. This paper presents Dissolver, a novel system that keeps the data privacy in the whole life time and ensures the destruction at the users command. Performance evaluation shows that the prototype system imposes reasonably low runtime overhead.

Web Services Cloud Computing-Based Digital Watermarking Model and Security Analysis

The purpose of the skeleton is to migrate the web services environment to cloud level. Any service requester can publish service requests by searching through the UDDI information, the service interface description and SOAP protocol. From the service requester perspective, embedding a watermark to the image request to the server needs to provide three aspects:

- Original image data: it is the carrier of the watermark.
- Watermark: It is the identification of the copyright owner of images and image basis for legitimacy.
- Watermark extended information: In the detection of watermark, digital watermarking service using the extended information to identify the real watermark, the watermark and the extracted contrast, to determine whether the watermark has been tampered with.

The algorithm of the digital watermark has been achieved. The development of corresponding interfaces, the original components of the digital watermark embedding and watermark detection as a service released [1, 2]. According to paper [1] discussed security-based digital watermark over the internet, the scheme can provide the encryption and digital signature information in the secure transmission of services over the internet. Paper [6–8] proposed relevant algorithms to secure the data up in the cloud.

Conclusion

Digital watermark embedding and watermark detection over the cloud can reduce development and spending costs. The single web service of prototype digital watermarking systems using the Java language has been implemented on the Apache server, and is in trial operation [2]. The system uses digital watermarking for multimedia services, online services and e-work applications such as instances of rights management to provide a safe and feasible solution. The system will effectively protect digital property rights in an open network environment. Once the different web services from different company upload over the cloud, the followings will be the key issues to study: the security of the digital watermark, cloud computing security and compliance, web service dynamic balance.

References

1. Hu W, Tong C, Chen D, Li Z, Kou W (2004) Secure digital watermark scheme based on web services. J Zhejiang Univ (Eng Sci) 38(11):1441–1445. Zhejiang University, Hangzhou
2. Zhang J (2011) A web services-based security model for digital watermarking. In: 2011 international conference on multimedia technology (ICMT 2011), pp 4805–4808. IEEE press, New Jersey
3. Booth D, Haas H, McCabe F, Newcomer E, Champion M, Ferris C, Orchard D (2011) Web services architecture. http://www.w3.org/TR/ws-arch/, 6 Jun 2011
4. Atkinson B, Della G, Hada S (2002) Specification: web services security (WS-Security). http://www-106.ibm.com/developerworks/library/ws2secure/, 4 May 2002

5. Huawei Technologies Co., Ltd (2009) Huawei A&S cloud computing solutions. In: Huawei technical report
6. Wei R, Linchen Y, Ren G, Feng X (2011) Lightweight and compromise resilient storage outsourcing with distributed secure accessibility in mobile cloud computing. Tsinghua Sci Technol 16(5):520–528. Tsinghua University, Beijing
7. Dapeng Z, Ke C, Min Z, Zhen X (2011) The research of cloud computing data security support platform architecture. J Comput Res Dev 48(Suppl):261–267. Institute of Computing Technology, Beijing
8. Fengzhe Z, Jin C, Haibo C, Binyu Z (2011) Lifetime privacy and self-destruction of data in the cloud. J Comput Res Dev 48(7):1155—1167. Institute of Computing Technology, Beijing

Video Transmission Quality Improvement Under Multi-Hop Wireless Network Architecture

Chih-Ang Huang, Chih-Cheng Wei, Kawuu W. Lin and Chih-Heng Ke

Abstract As wireless local area networks have been widely deployed and smart devices are becoming common in human everyday life, the demanding for video applications, such as video conference, video on demand, and video games is still increasing. However, the bandwidth in wireless networks is limited and channel quality is varying. As a result, how to provide a better delivered video quality over wireless networks is challenging. Also, the current solutions to better video transmission are almost only considering single hop transmission. When those solutions are applied in multi-hop wireless networks, such as wireless mesh network or VANET, the delivered video quality may not be good as one hop transmission. Therefore, we propose a new mechanism to improve video transmission in multi-hop wireless networks. The new mechanism contains two methods. One is to privilege the packets that have already traversed more hops. The other is control the number of retransmission for video packets with different importance. Through NS2 simulations, the results show the effectiveness of proposed mechanism.

Keywords Multi-hop wireless networks · Video transmission · MPEG · TTL

Introduction

Advances in wireless network technology with increased bandwidth and the benefits of unwired connectivity have overcome numerous inconveniences commonly experienced in cabled networking. For example, through wireless

C.-A. Huang (✉) · C.-C. Wei · K. W. Lin
Department of Computer Science and Information Engineering, National Kaohsiung University of Applied Sciences, Kaohsiung, Taiwan, Republic of China
e-mail: smallko@gmail.com

C.-H. Ke
Department of Computer Science and Information Engineering, National Quemoy University, Kinmen, Taiwan, Republic of China

Y.-M. Huang et al. (eds.), *Advanced Technologies, Embedded and Multimedia for Human-centric Computing*, Lecture Notes in Electrical Engineering 260,
DOI: 10.1007/978-94-007-7262-5_61, © Springer Science+Business Media Dordrecht 2014

distribution systems (WDSs), a wireless network might continually extend its coverage by joining two or more WDS-capable nodes. The network architecture that enables data transmission by relaying packets from one node to the next between the sending and receiving end is termed a multi-hop wireless network. Most studies regarding multi-hop wireless networks have focused on increasing the volume or speed of data transmission [1–6], but few have examining improving the transmission quality of videos. Therefore, in this paper, the transmission quality of video streams through IEEE 802.11 [7, 8] and a multi-hop wireless network is examined.

Related Work

In wireless local area network communications, the IEEE 802.11 protocol is employed to access a shared microwave band of the transmission medium. Because of this characteristic of multi-hop wireless networks, preceding nodes and subsequent nodes compete for access to the channel, as shown in Fig. 1. Packet collisions resulting from this competition might cause latency and additional back-off time, thereby continually reducing the network bandwidth. Because of the competition between nodes, multi-hop wireless networks provide less bandwidth and longer latencies than single-hop wireless networks do. In this paper [9], a novel node-based priority calculation mechanism is presented to ease competition by granting higher priorities of channel access to packets that are sent earlier than later ones. When packets must be transmitted to a destination through a node, packets that are sent earlier compete with subsequent packets for channel access. However, if subsequent packets obtain the channel first, the utility is minimal because they cannot start transmission before the earlier packets finish transmission. Therefore, by granting channel access to earlier packets, the latency can be reduced and the bandwidth increased. Therefore, regarding access to the communication channel, the method presented here assigns a higher priority to earlier packets. In conventional techniques, all packets have equal priority for packet transmission, and early packets might compete with those sent later for channel access. The method proposed in this study increases the priority of the packets after passing through each node to reduce the competition between packets. In other words, the transmission priority is determined by the distance to the

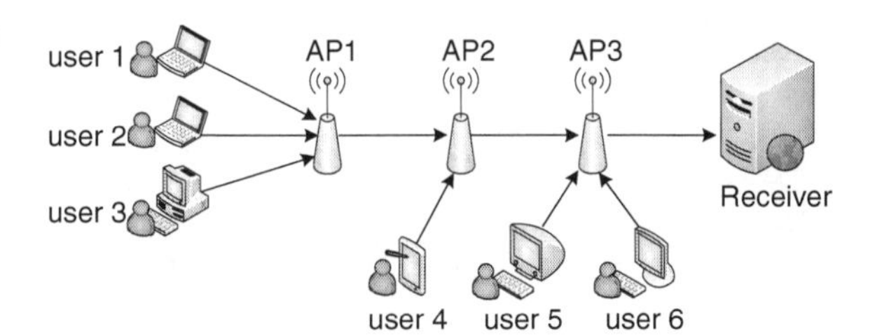

Fig. 1 Network environment of multi-hop wireless network

receiving end: Packets with shorter travel distances are assigned higher priority. By specifying a smaller contention window, packets closer to the receiving end gain access to the channel more easily. Similarly, by specifying a larger contention window, packets farther from the receiving end have a lower priority of access to the channel, and gaining access to the channel becomes more difficult. In this manner, higher transmission quality and operational efficiency in multi-hop wireless networks are achieved.

The Proposed Method

In addition to existing IEEE 802.11 specifications, this paper proposes adding two new mechanisms to improve the quality of video transmission in multi-hop wireless networks. The two mechanisms are described as follows (Tables 1 and 2):

Table 1 TTL and *retransmission* mechanism algorithm

TTL mechanism algorithm	Retransmission mechanism algorithm
If (qlen>threshold)	*I frame packet*
{	set RetryLimit = normal;
△TTL = default_TTL – current_TTL;	*P frame packet*
prob = ((qmax-qlen)/qmax)*(△TTL/maxhops);	if(qlen>threshold)
tmpx=random(0.0 ~ 1.0)	set RetryLimit = less;
if(prob≥tmpx)	otherwise
enque(packet)	set RetryLimit = normal;
else	*B frame packet*
drop(packet)	if(qlen>threshold)
}	set RetryLimit = min;
	otherwise
	set RetryLimit = normal;

Table 2 Description of mechanism parameters

Parameter	Description
qlen	Current total queue length
threshold	Threshold of network congestion
default_TTL	Default TTL value of packet
current_TTL	Current TTL value of the packet
prob	Probability that the packet enters the queue
qmax	Total queue length
maxhops	Total number of nodes
△TTL	Total number of nodes the packet passes through
Tmpx	Randomly generated probability value (0.0–1.0)
RetryLimit	Maximum number of packet retransmissions
normal	Normal number of retransmissions
less	Fewer numbers of retransmissions
min	Minimum number of retransmissions

- **Time to Live (TTL) Mechanism**

Time to live is employed to protect packets traveling through numerous hops. Having passed through multiple contentions and approaching the receiving end, such packets should be granted higher priority through protective measures to increase the opportunity to enter the queue and the success rate for transmission, otherwise the video transmission bandwidth is affected and transmission quality compromised.

- **Retransmission mechanism**

Vital packets are protected by altering the number of retransmissions. Because the space for node queuing is limited, if lower-order packets are retransmitted continually, the packet queue might overload and lose the higher-order packets because of its inability to accommodate them in the queue. The loss of higher-order packets might have severe consequences on video decoding and transmission quality.

TTL Mechanism

TTL refers to the longest time that a packet can live in a network. The TTL value of each packet varies in different operating systems. TTL primarily works by assigning a value to the packet transmitted by the sending end; the value decreases by 1 each time the packet passes through a node. When the TTL value drops to zero, the packet is discarded by the node that receives it. Specifying a TTL value prevents packets from random continuous travel among various network nodes, which is caused by unpredicted factors, before reaching the intended destination; this avoids affecting the overall network bandwidth and quality.

This mechanism first determines the number of nodes the packet has traveled through, and based on an algorithm, it assigns a higher priority to packets that have traveled through more nodes (i.e., by assigning a smaller TTL value) so that these packets have a greater opportunity to enter the queue for transmission. By assigning varying weights to packets through TTL values, packets that have traveled more nodes have a greater possibility of being transmitted to the receiving end. This mechanism is based on the concept that in a wireless network environment, each packet must pass through at least one node to obtain a transmission opportunity before it is transmitted again. In a multi-hop wireless network, all packets must pass through multiple nodes before reaching the receiving end; in other words, the packets must compete for priority in each node to be successfully transmitted to the receiving end. Conversely, when the wireless network environment is congested, the ability of the packets to travel through multiple nodes and successfully arrive at the receiving end decreases, thereby reducing the quality of the video transmission. Therefore, the use of the TTL mechanism could protect packets that have traveled through numerous hops by prioritizing them in the transmission queue, thus increasing the chance of successful transmission and greatly enhancing the bandwidth and video transmission quality.

According to the TTL mechanism algorithm, when the queue length surpasses a set threshold of network congestion, the number of nodes that a packet has passed through is equal to the default TTL value minus the current value. The probability of the packet entering the queue is determined by two factors: (1) the ratio of vacant space to the total length of the queue; a smaller remaining space represents a smaller probability of the packet being granted entry to the queue; and (2) the ratio of the number of nodes the packet has passed through to the total number of nodes (△TTL). The more nodes that a packet passes through increase the probability of the packet entering the queue; conversely, fewer nodes reduce the probability of a packet entering the queue. In addition, a probability value (tmpx) is randomly generated, with a value ranging between 0.0 and 1.0. When the probability of entering the queue (prob) is greater than or equal to the tmpx, the packet enters the queue for subsequent transmission; otherwise, the packet is discarded.

Retransmission Mechanism

Both IEEE 802.11 Distributed Coordination Function (DCF) and IEEE 802.11e [10] Enhanced Distributed Channel Access (EDCA) consist of a retransmission mechanism that might appear in two forms: the time limit and retry limit. This study is based on the retry limit mechanism because after compression by MPEG-4, video streams are encoded into I-frames, P-frames, and B-frames, which are subsequently encapsulated into frame packets. During transmission, these frame packets may experience unsuccessful transmission because of collisions or other factors. Under such circumstances, the working node determines the number of packet retransmissions based on the retransmission mechanism. For example, suppose an I-frame packet has a retry limit set to four. In the event that the frame packet transmission fails, the working node competes for its right to transfer again and retransmits the packet once retransmission permission is granted. If the frame packet cannot be transmitted to the next node after four attempts, the working node would discard the frame packet, thereby protecting the priority rights of other frame packets.

The retransmission mechanism is primarily based on adjusting the number of packet retransmissions. When a packet transmission fails, the so-called retransmission mechanism activates; when a packet fails to be correctly transferred to the receiving end, the node initializes packet retransmissions. In the current IEEE 802.11 configuration, the number of retransmissions is fixed and a priority is not assigned based on packet importance. In the present study, the varying effects of frames with differing importance levels on video transmission are considered; therefore, different probabilities for retransmission are assigned to frame packets based on their priority to enhance the quality of video transmission. When the network is busy, frame packets with higher priority are assigned more retransmissions, and lower priority frame packets have fewer opportunities for retransmission. In this manner, the probability of losing frame packets with higher priority is significantly reduced, and the likelihood of successfully transmitting frame packets with higher priority is increased; thus, the overall quality of video transmission improves considerably. Because the size of the queuing buffer at each node is limited, if frames of varying priorities are provided the same level of protection, the node would be busy retransmitting low-priority frame packets, and its queue would have insufficient space for high-priority packets, thereby resulting in poor performance. Therefore, this study adjusted the priority level of various packets, and reduced the number of retransmission attempts for low-priority packets when the network is in congestive or highly competitive conditions. The packets are discarded once the maximum retransmission number is attained so that subsequent packets can enter the transmission queue. The probability of successfully transmitting packets is significantly improved, as is the transmission bandwidth and quality of video transmission.

According to the retransmission mechanism algorithm description, the number of packet retransmissions is adjusted based on the priority of various packets. Important I-frames are always assigned four retransmissions, regardless of the wireless network condition; for second-level P-frames, the number of retransmissions is set at two in busy wireless network conditions and at four in slower periods. For low-priority B-frames, the number of retransmissions is set at one in congested wireless network conditions and four in slower periods. Therefore, the proposed retransmission mechanism is activated only when the wireless network is busy to assign more retransmissions to the higher order packets, thereby increasing the possibilities of successful transmission for higher order packets and effectively improving the video transmission quality.

Results and Analysis of the Simulated Experiment

Settings for the Simulated Parameters

In this section, a simulated experiment is conducted using NS2 [11, 12] network simulation software. In the experiment, a multi-hop wireless network environment is simulated, which consists of four wireless network nodes. Suppose that the sequence of these four nodes is AP 1, AP 2, AP 3, and AP 4. Direct data transfer may occur between adjacent nodes, and non-adjacent nodes do not interfere with one another. In other words, if data are to be transferred from AP 1 to AP 4, they must first be transferred AP 2, then to AP 3, and finally to AP 4; if data are to be transferred from AP 1 to AP 3, they must be transferred to AP 2 first, and then transferred from AP 2 to AP 3. Data can be directly transferred from AP 1 to AP 2 because these two nodes are adjacent nodes and capable of direct data transfer. The simulative parameters of the experiment are set as shown in Table 3.

Table 3 Simulated parameter settings

Parameter	Description	Set value
threshold	Threshold of network congestion	25
default_TTL	Default TTL value	32
qmax	Total queue length at the node	50
normal	Normal number of retransmissions	4
less	Fewer number of retransmissions	2
min	Minimum number of retransmissions	1

Results and Analysis of the Simulated Experiment

In the experiment, a data stream was added to AP 1, which will be transferred to AP 4; data streams are also added to AP 2 and AP 3, which act as sources of interference during data transfer. After the data are transferred, the video frames transmitted in this experiment are decoded. By analyzing the peak-to-signal noise ratio (PSNR) of the decoding process, the proposed mechanism can be assessed for its improvement of the transmission quality of MPEG-4 compressed videos.

This simulated experiment generates three sets of data. The first set of data is obtained through the process of video transmission using the existing IEEE 802.11 mechanism; the second set is obtained from the experiment that incorporates the TTL mechanism into the 802.11 mechanism; and the third set of data is obtained from the simulated experiment that incorporates both the TTL and retransmission mechanisms.

Table 4 presents the Foreman video data. A total of 400 frames were extracted from the source video, including 51 I-frames, 83 P-frames, and 266 B-frames. The statistics of the video frames transmitted through the IEEE 802.11 mechanism (Table 5) show that 21 I-frames were received by the receiving end, 30 I-frames were lost; 50 P-frames were received and 33 were lost; 144 B-frames were received and 132 were lost.

The statistics of video frames transmitted by using the TTL mechanism (Table 5) show that after the TTL mechanism was incorporated, 31 I-frames were received by the receiving end, which is 47.6 % higher than the transmission with IEEE 802.11 mechanism. The number of P-frame transmissions increased to 55 frames from the previous 50 frames using IEEE 802.11, which is a 10 % increase. Finally, 168 frames of B-frame packets were received compared to the 134 frames transmitted by using IEEE 802.11, indicating a 25.4 % increase.

The statistics of video frames transmitted using TTL and the retransmission mechanism (Table 5) show that after the TTL and retransmission mechanism was

Table 4 Foreman video data

Video	Format	Number of frames			Total number of frames
Foreman	QCIF	I	P	B	400
		51	83	266	

Table 5 Statistics of video frames transmitted using (a) IEEE802.11, (b) TTL mechanism and (c) TTL and retransmission mechanism

Mechanisms	IEEE802.11			TTL			TTL and retransmission		
Categories	I-frames	P-frames	B-frames	I-frames	P-frames	B-frames	I-frames	P-frames	B-frames
Total transmissions	51	83	266	51	83	266	51	83	266
Total receptions	21	50	134	31	55	168	35	50	154
Number of losses	30	33	132	20	28	98	16	25	112

Table 6 Average PSNR of the foreman video

	IEEE 802.11	TTL mechanism	TTL and retransmission mechanism
Average PSNR	21.292	23.363	23.518

used, 35 I-frames were received by the receiving end, a 66.7 % increase over the transmission using the IEEE 802.11 mechanism; the number of P-frame transmissions increased to 58 from the 50 frames transmitted using IEEE 802.11, showing a 16 % increase; finally, 154 B-frame packets were received. Only 134 frames were received when using IEEE 802.11, indicating a 14.9 % increase.

The results of the average PSNR presented in Table 6 show that the average PSNR obtained by using IEEE 802.11 was 21.29 after the discarded frames were counted and compared to the source data. After incorporating the TTL mechanism, the transmission in the same simulation environment generated an average PSNR of 23.363; a 9.7 % improvement over IEEE 802.11. After the TTL and retransmission mechanism was added, the average PSNR was raised to 23.518 under the same experimental conditions, indicating a 10.5 % improvement over the IEEE 802.11 mechanism.

Overall, the TTL and retransmission mechanism proposed in this study can effectively improve the quality of video transmission.

Conclusion and Research Suggestions

This study proposed two mechanisms for improving the quality of video transmission in multi-hop wireless networks: the queuing mechanism implemented by adjusting packets based on TTL analyses, and adjustment of retransmissions based on the varying priorities of differing packets. The analytical results of a simulated experiment conducted on an NS2 network confirmed that, compared to the traditional IEEE 802.11 mechanism, the two mechanisms proposed in this study can effectively improve the quality of video transmission in multi-hop wireless networks.

In this study, fixed nodes were employed in a multi-hop wireless network. Future studies should simulate actual scenarios occurring in multi-hop wireless networks, such as environments in which multiple nodes randomly move within a certain range, or in which specific nodes remain static but others move.

References

1. Shugong X, Saadawi T (2001) Does the IEEE 802.11 MAC protocol work well in multi-hop wireless ad hoc networks? IEEE Commun Mag 39(6):130–137
2. G. Bianchi (2001) Performance analysis of the IEEE 802.11 distributed coordination function. pp 535–547

3. Hsieh H-Y, Sivakumar R (2002) IEEE 802.11 over multi-hop wireless networks: problems and new perspectives. In: IEEE 56th VTC 2002
4. Hsieh H-Y, Sivakumar R (2001) Improving throughput and fairness in multi-hop wireless networks. In: Proceedings of ICN, Colmar, France, July 2001
5. Nguyen LT, Beuran R, Shinoda Y (2007) Performance analysis of IEEE 802.11 in multi-hop wireless networks. In: Proceedings of the 3rd international conference on Mobile ad-hoc and sensor networks, ser. MSN'07. Springer, Berlin, Heidelberg, pp 326–337
6. Mustapha I, Jiya JD, Monguno K (2010) Throughput analysis of IEEE 802.11 MAC protocol in multi-hop wireless Ad-Hoc network. J Sci Technol Res 9(1):113–122
7. IEEE (1999) Wireless LAN media access control (MAC) and physical layer (PHY) specifications. IEEE Std. 802.11
8. Crow BP, Widjaja I, Kim LG, Sakai PT (1997) IEEE 802.11 wireless local area networks. IEEE Commun Mag 35(9):116–126
9. Lee SH, Yoo C (2010) Contention avoidance with hop based priority in 802.11e multi-hop network. Consumer electronics (ICCE), 2010 digest of technical papers international conference, January 2010, pp 151–152
10. IEEE Std 802.11e-2005 (2005) Wireless LAN medium access control (MAC) and physical layer (PHY) specifications amendment 8: medium access control (MAC) quality of service enhancements
11. The network simulator ns-2 [Online]. Available: http://www.isi.edu/nsnam/ns/
12. Evaluation of video stream quality over IEEE 802.11e EDCF [Online]. Available: http://140.116.72.80/smallko/ns2/ns2.htm

Mapping IDCT of MPEG2 on Coarse-Grained Reconfigurable Array for Matching 1080p Video Decoding

Guoyong Li, Leibo Liu, Shouyi Yin, Changkui Mao and Shaojun Wei

Abstract Coarse-grained reconfigurable array (CGRA) can achieve flexible and highly efficiencies for computing-intensive application such as multimedia, baseband processing and etc. MPEG2 is a popular multimedia algorithm which suits for CGRA. IDCT takes around 29 % of total time for MEPG2 Decoding, which is one of main parts of MPEG2. IDCT belongs to computation-intensive which fits for CGRA. The paper explores the parallelism of IDCT algorithm, mapping it on coarse-grained reconfigurable array. The simulation result shows 693 clock cycles are needed to complete 8 × 8 IDCT on REMUS, the cycles needed is just 36 % of XPP, just 24.7 % of ARM. The method improves performance for MPEG2 decoding. The performance fulfils MPEG2 decoding for 1080p @30 fps streams when employs 200 MHz clock frequency.

Keywords IDCT · MPEG2 · CGRA · Mapping · REMUS

Introduction

Many applications are computing-intensive algorithm. In the traditional way, two approaches are used for implementation: one is using a general-purpose processor that it is flexible but not efficient. The other is using ASIC which is efficient, however, has no flexibility. To achieve flexibility and high efficiencies, coarse-grained reconfigurable array (CGRA) has been raised, like XPP [1], Adres [2], REMUS [3]. This architecture can be used to process wide arrange of application in high efficient. In this architecture, different sets of configuration information

G. Li · L. Liu (✉) · S. Yin · S. Wei
Research Center for Mobile Computing, Tsinghua University, Beijing 100084, China
e-mail: liulb@mail.tsinghua.edu.cn
Institute of Microelectronics, Tsinghua University, Beijing 100084, China

C. MaoInstitute of Microelectronics, Xi'an Jiaotong University, Xi'an, China

Y.-M. Huang et al. (eds.), *Advanced Technologies, Embedded and Multimedia for Human-centric Computing*, Lecture Notes in Electrical Engineering 260,
DOI: 10.1007/978-94-007-7262-5_62,

called context are used to make CGRA to realize different functions. Decoding progress of MPEG2 [4] consists of several sub algorithms, VLD\Inverse Scan\Inverse Quantization\IDCT\MC\AddBlock. They can sort into computing-intensive algorithm and control-intensive algorithm. VLD\Inverse Scan\Inverse Quantization belong to control-intensive. And IDCT\MC\AddBlock belong to computation intensive, these algorithms can be pipelined and parallelized. This character fits for coarse-grained reconfigurable array. Doing these algorithms on coarse-grained reconfigurable processor can obtain high performance. Just as shown below, the whole figure is MPEG2 algorithm, data intensive algorithm (IDCT\MC\AddBlock) are processed on coarse-grained reconfigurable array.

The overall application behavior of MPEG2 is researched [5]. The Figs. 1, 2 shows the relative performance of decoder for a 9 MB/s case. IDCT takes around 29 % of total time for MPEG2, it is one of the main parts of MEPG2. By speeding up this part can improve decoding performance effectively. There are some ways to speed up IDCT. Swamy et al. [6] use ASIC to accelerate IDCT, which executes very fast, but needs a special module and has no flexibility. Fang et al. [7] maps IDCT on GPU. Winger [8] use CPU SIMD execute IDCT, which reduces 22 % execution time. XPP [1] maps IDCT on CGRA. This paper implements sub-algorithm IDCT of MPEG2 on REMUS which is one kind coarse-grained reconfigurable SOC. Section Architecture of REMUS and RPU introduces the architecture of REMUS and RPU. Section Implementation IDCT on REMUS do task partion of MPEG2 on REMUS and analyses IDCT algorithm of MPEG2. Section Result proposes techniques to mapping IDCT on REMUS and does the simulation to get results. Section Conclusion gets the conclusion.

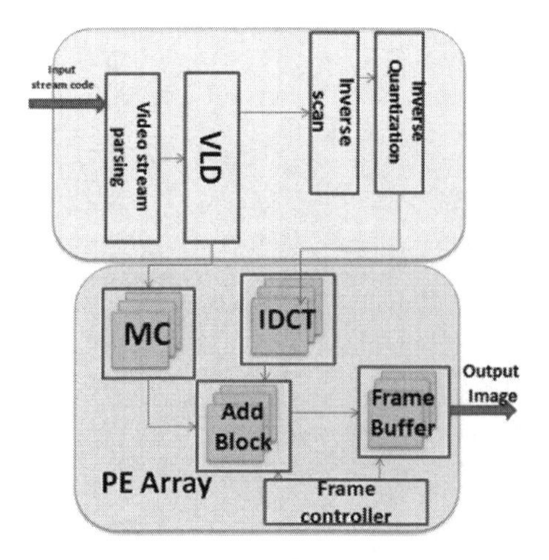

Fig. 1 MPEG2 decoding

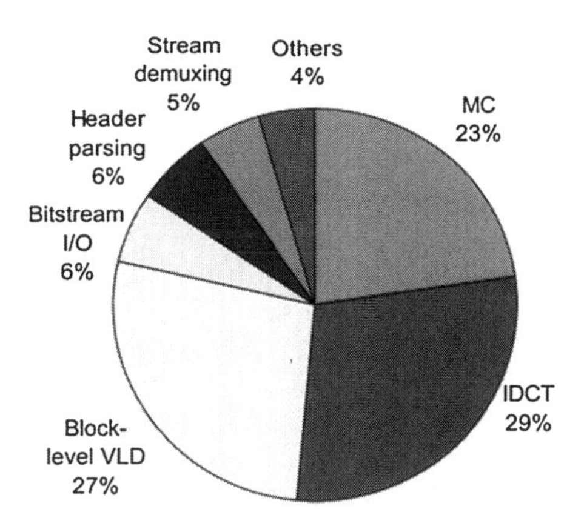

Fig. 2 MPEG2 decoding time on 3.06 GHz Pentium4

Architecture of REMUS and RPU

Architecture of REMUS

REMUS [3] is a reconfigurable Multi-media System, which is one kind of coarse-grained reconfigurable SOC. As shown below, REMUS contains RPU\u-PA\ARM7, Entropy Decoder (EnD), and some other modules. RPU contains four reconfigurable processing element array (PEA). The computing-intensive parts are processed on this part, which can achieve high efficiency and flexibility. UPA generates context for the RPU to realize different functions. The ARM7 is a RISC processor which is used for the application control. The EnD processes the entropy decoding such as Context Adaptive Variable-Length Decoding (CAVLD) and Context Adaptive binary Arithmetic Decoding (CABAD). The different parts are connected by AMBA bus, AHB bus is used to connect high speed modules, APB bus is used to connect low speed modules, dedicated high speed EMI bus is used to connect SSRAM with other modules (Fig. 3).

Architecture of RPU

As shown in Fig. 4, RPU consists of four processing element array (PEA), PEA controller, context interface, data interface, context memory, and data internal memory. The interface is used to connect RPU with outside bus. Memory is used to store context and data. The controller is used to control the progress of RPU. PEA is a powerful dynamic reconfigurable system; it consists of 8×8 processing elements (PE). Every PE can do kinds operation, like add, shift, multiply and so on. Context is used to control the function of PE. As shown below, PE consists ALU, MUX, and REG.

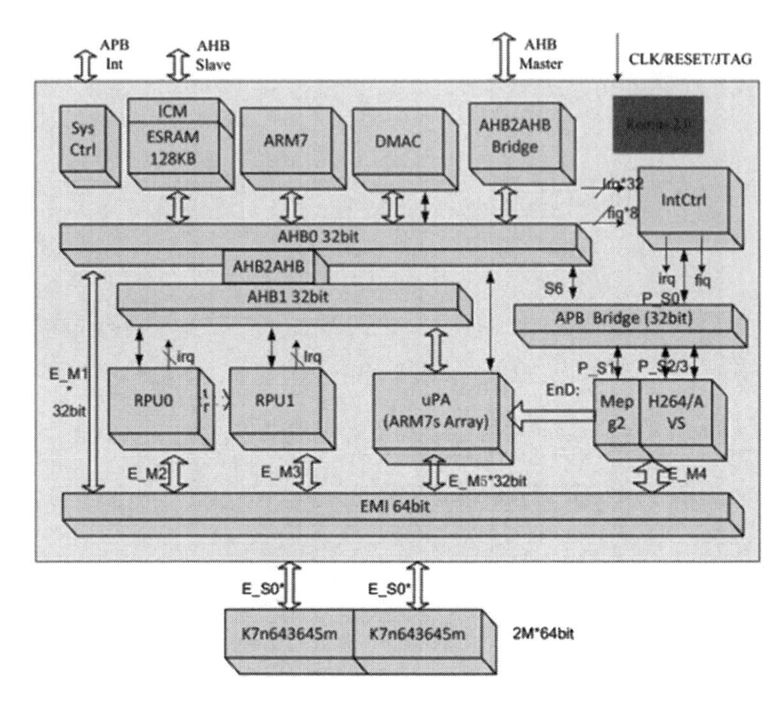

Fig. 3 Architecture of REMUS

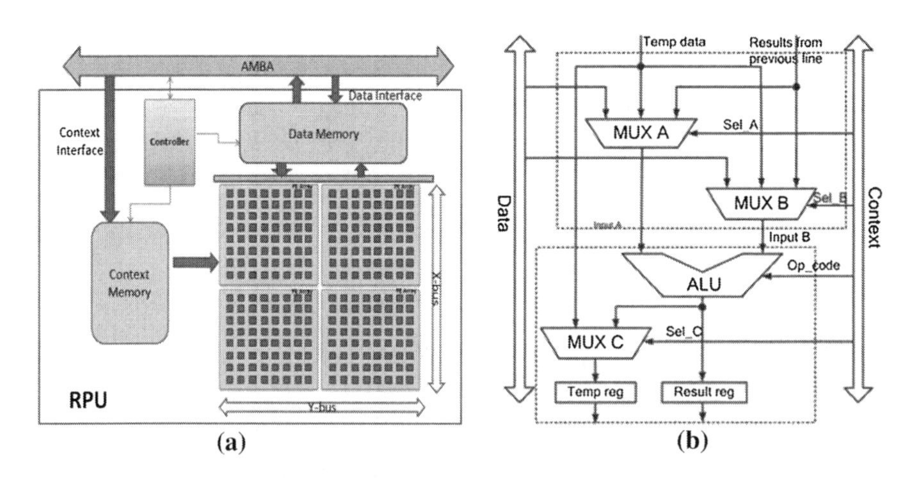

(a) (b)

Fig. 4 a Architecture of RPU. b Architecture of PE

Implementation IDCT on REMUS

Task Partion

Realizing MPEG2 algorithm on REMUS should depend on the characteristic of this algorithm. According to the previous architecture of RPU, RPU contains four PE array. The computation ability of RPU is so strong that it fits for the algorithm which are low data dependency and data can be parallelized and pipelined. So mapping computing-intensive sub algorithm on RPU. Mapping control-intensive sub algorithm on EnD. ARM control the whole progress, context used for RPU is generated by uPA. For details, main ARM responds for the decoder algorithm above slice level (include slice level), setting DMA, setting EnD, frame buffer management, and communication with outside systems. EnD responds for parsing the code stream, inverse scan\inverse quantization and so on, the characteristic of this part is that data dependency is high, belong to control-intensive, it is hard to be parallelism. The uPA responds for parsing command from ARM and MB from EnD, generates contexts for RPU. RPU responds for algorithm of computing-intensive, contain IDCT\MC\AddBlock, the characters of this part are parallel and pipelined.

IDCT Analysis

Discrete cosine transform (DCT) is a Fourier-related transform similar to the discrete Fourier transform (DFT) [9]. IDCT is the inverse algorithm of DCT. It belongs to the orthogonal transformation coded system. IDCT is used to wipe off the spatial redundancy of image data. The procedure is that transform coded image intensity matrix (Time domain signal) to coefficient space (Frequency domain signal), then process the frequency domain signal. IDCT is widely used in image processing, such as JPEG, mpeg2, H.264 and etc. For MPEG2, macro block is separated into four 8×8 sub-blocks, for every sub-block does 2-D IDCT. MPEG2 takes 8×8 block 2-D IDCT. So we can get the formula below, f(x, y) (x, y = 0, 1…7) is the pixel of the image, F(u, v) (u, v = 0,1..7) is the correspond coefficient of DCT.

$$f(x,y) = \frac{1}{4}\sum_{u=0}^{7}\sum_{v=0}^{7} c(u)c(v)F(u,v)\cos\left(\frac{2x+1}{16}u\pi\right)\cos\left(\frac{2y+1}{16}v\pi\right)$$
$$x,y,u,v = 0,1,\ldots,7 \quad c(u) = c(v) = \begin{cases} \frac{1}{\sqrt{2}} & u=0,\ v=0 \\ 1 & others \end{cases} \tag{1}$$

The computation of IDCT is very large. For 8×8 block, 2-D IDCT needs 8,192 multiply and 3,584 add. It is hard to directly map 2-D IDCT on PE array.

This paper takes the method of matrix decomposition. 2-D IDCT decomposes to two 1-D IDCT. IDCT can be written in matrix form: $f = C^T FC$, f is the matrix expression of f(x, y), F is the matrix expression of F(u, v).

$$f = C^T FC \Rightarrow Y = C^T F^T, f = C^T Y^T \tag{2}$$

$C^T C = I_N$, C is 8 × 8 matrix composed of the cosine function, C^T is the transposition matrix of C. From above formula, 2-D IDCT separates into two steps, first is $Y = C^T F^T$, second is $f = C^T Y^T$. In this way 2-D IDCT changes into two 1-D IDCT, one is IDCT-ROW ($Y = C^T F^T$), the other is IDCT-COL ($f = C^T Y^T$). The formula of IDCT-ROW is below, IDCT-COL is the same principle:

$$f(x) = \frac{1}{2}\left[\sum_{u=0}^{7} C(u)F(u)\cos\frac{(2x+1)u\pi}{16}\right] \quad C(u) = \begin{cases} \frac{1}{\sqrt{2}}, (u = 0) \\ 1, (u > 0) \end{cases}. \tag{3}$$

This paper adopts Chen-Wang [10] fast IDCT. Expressing above formula in matrix form. After some transformation, the formula presents in below, $ci = \cos(\frac{i \cdot \pi}{16}), \quad (1 \le i \le 7)$:

$$P = \begin{bmatrix} c4 & c2 & c4 & c6 \\ c4 & c6 & -c4 & -c2 \\ c4 & -c6 & -c4 & c2 \\ c4 & -c2 & c4 & -c6 \end{bmatrix} \cdot \begin{bmatrix} F0 \\ F2 \\ F4 \\ F6 \end{bmatrix} \tag{4}$$

$$Q = \begin{bmatrix} c1 & c3 & c5 & c7 \\ c3 & -c7 & -c1 & -c5 \\ c5 & -c1 & c7 & c3 \\ c7 & -c5 & c3 & -c1 \end{bmatrix} \cdot \begin{bmatrix} F1 \\ F3 \\ F5 \\ F7 \end{bmatrix} \tag{5}$$

$$\begin{bmatrix} f0 \\ f1 \\ f2 \\ f3 \end{bmatrix} = \frac{1}{2} \cdot P + \frac{1}{2} \cdot Q \tag{6}$$

$$\begin{bmatrix} f7 \\ f6 \\ f5 \\ f4 \end{bmatrix} = \frac{1}{2} \cdot P - \frac{1}{2} \cdot Q \tag{7}$$

Using Butterfly diagram to express calculation process, the Fig. 5 is shown:

Based on above analysis, IDCT algorithm is computing-intensive and the computations can be parallelized and pipelined.

IDCT Mapping

The Fig. 5 shows that IDCT algorithm is computing-intensive, and many computations can be parallelized and pipelined. These characteristics are very fit for processing on PE array. IDCT algorithm processes in 8 × 8 unit, this can implement on PEA. For 4:2:0 colour format, Y component exists 16 × 16, it can be decomposed to 8 × 8 base unit, then four PEA process 16 × 16. U\V component exists 8 × 8, these can be processed using 8 × 8 base unit. Therefore, IDCT of one macro block can be done by PEA. And the mapping is shown in (Fig. 6).

As previously introduced, 2-D IDCT can be decomposed to two 1-D IDCT, the algorithm of IDCT of MPEG2 works in three steps: IDCT-ROW, transposition, IDCT-COL. IDCT-ROW and IDCT-COL are the same principle, so the IDCT-ROW and IDCT-COL can be pipelined. Mapping 8 × 8 IDCT-ROW on the PE array. 8 × 8 IDCT is too complex to map the algorithm on one map, IDCT-ROW is decomposed into three maps. IDCT-COL uses the same three maps. The Fig. 7 maps the operation P (show in previous section, Eq. 4). In the Fig. 7 row[0]\-row[2]\row[4]\row[6] are the input F0\F2\F4\F6, w2\w4\w6 are correspond to c2\c4\c6, but done scale. The Figs. 8, 9 maps the operation Q (show in previous section, Eq. 5). The Fig. 10 maps the operation of (P + Q)/(P − Q) (Eq. 6) and

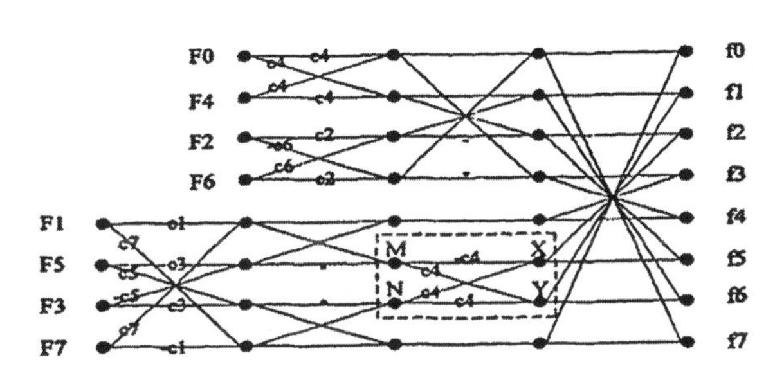

Fig. 5 Chen-Wang fast IDCT

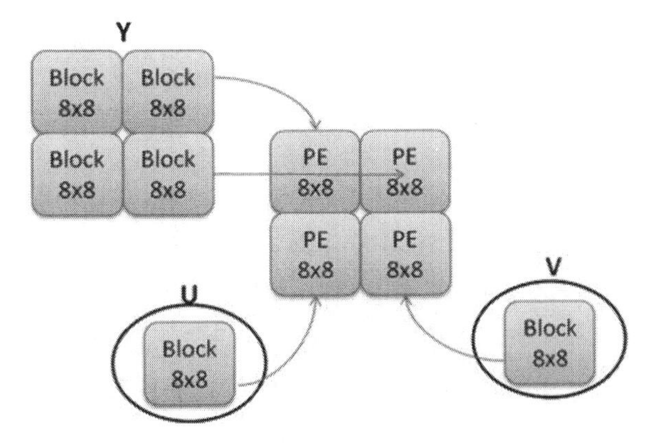

Fig. 6 Mapping 8 × 8 IDCT on PEA

transposition. The Figs. 7, 8, and 9 are mapping for one row. For 8 × 8 IDCT-ROW, 8 rows are needed, doing the operation in 8 loop, then doing maping 4, 8 × 8 IDCT-ROW is done, than continue to do 8 × 8 IDCT-COL in the same way.

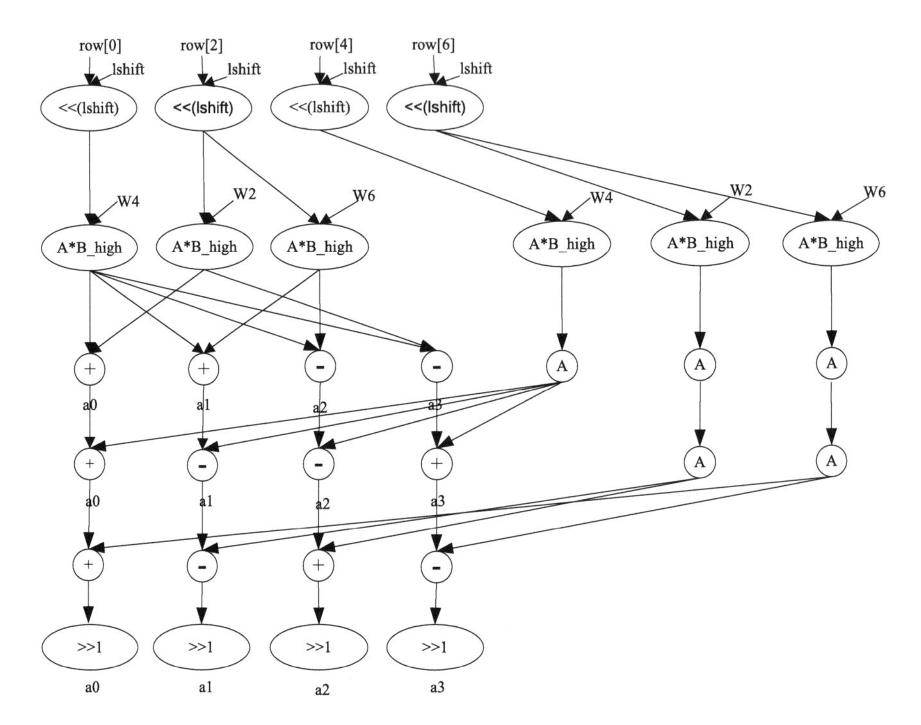

Fig. 7 IDCT mapping 1

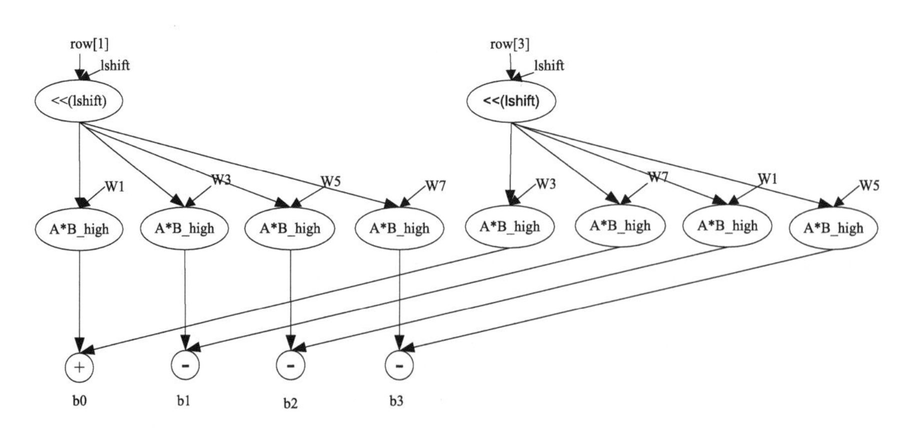

Fig. 8 IDCT mapping 2

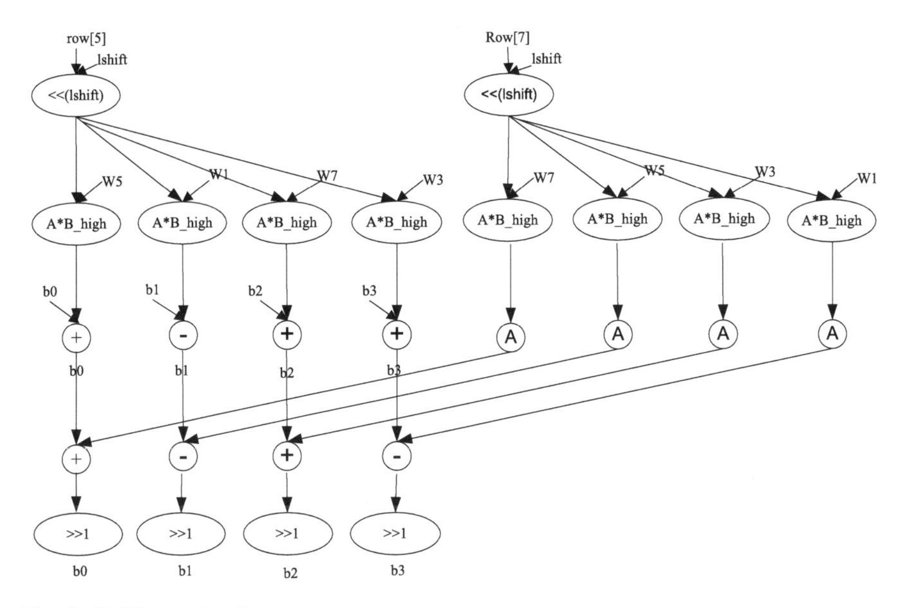

Fig. 9 IDCT mapping 3

Result

The simulation platform is based on REMUS. Implementation of the proposed mapping method for IDCT is verified with RTL simulation using Synopsys VCS. The frequency is set to 200 MHZ. Decades of 1080p MPEG2 code streams is run. Mapping computing-intensive sub algorithm on RPU, uPA generates context used for RPU. Mapping control-intensive sub algorithm on EnD, ARM control the whole progress. The result is shown below. The cycles needed for IDCT on REMUS is 693, just 36 % of XPP and just 24.7 % of ARM (Table 1).

To do more verification, test three 1080p code streams. MC and AddBlock belongs to computing-intensive are also mapping on CGRA. MC takes 589 cycles,

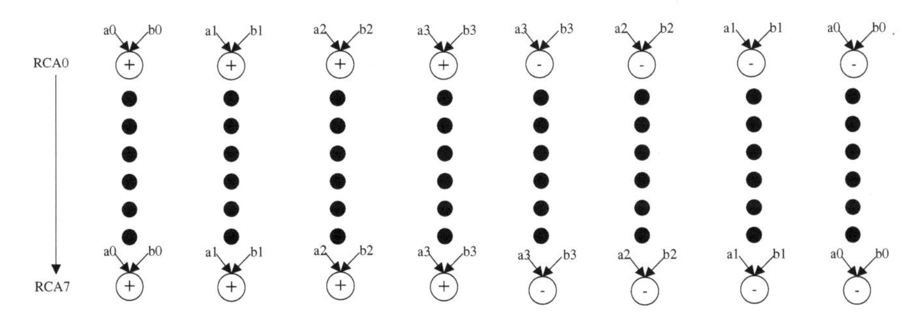

Fig. 10 IDCT mapping 4

Table 1 Cycles of IDCT

IDCT	REMUS	XPP [1]	ARM [11]
Cycles	693	1,700 + 192	2,796

Table 2 Performance of 1080p Decoding

Code streams	VanHelsing	benq2	Japan
Resolution	1,920 × 1,088	1,920 × 1,088	1,920 × 1,088
Performance (ms)	31.23479	32.16729	33.91178

and AddBlock takes 67 cycles. The overall cycles are reduced significantly compared to ARM. As shown in the figure below, the average of performance is around 32 fps. This performance meets the requirements of 1080p real-time decoding (Table 2).

Conclusion

For multimedia algorithm MPEG2, the sub-algorithms include computing-intensive and control-intensive. Mapping data-intensive on reconfigurable processing element array. This scheme can achieve both flexibility and high efficiency at the same time. IDCT is computing-intensive algorithm and it takes around 29 % of total time for MEPG2 decoding. Mapping it on PEA, parallelization of calculation and pipelined structure significantly reduces the execution cycle. The simulation result shows that 8 × 8 IDCT just needs 693 cycles on REMUS. The cycles needed is just 36 % of XPP, just 24.7 % of ARM. For the whole system, REMUS supports MPEG2 1080p @30 fps stream at 200 MHz.

Acknowledgments This work is supported in part by the China National High Technologies Research Program (No. 2012AA012701), the Tsinghua Information S&T National Lab Creative Team Project, the International S&T Cooperation Project of China grant (No. 2012DFA11170), the Tsinghua Indigenous Research Project (No. 20111080997), the Special Scientific Research Funds for Commonweal Section (No. 200903010), the Science and Technology Project of Jiangxi Province (No. 20112BBF60050) and the NNSF of China grant (No. 61274131).

References

1. XPP-III Processor Overview White Paper (2006)
2. Veredas F-J, Scheppler M et al (2005) Custom implementation of the coarse-grained reconfigurable ADRES architecture for multimedia purposes. In: International conference on field programmable logic and applications, 2005
3. Zhu M, Liu L, Yin S et al (2010) A reconfigurable multi-processor SoC for media applications. In: IEEE international symposium on circuits and systems, 2010

4. "MPEG-2 White Paper (2000)
5. Holliman M, YK Chen (2003) MPEG decoding workload characterization. In; Proceedings of workshop on computer architecture evaluation using commercial workloads 2003
6. Swamy R, Khorasani M, Liu Y, Elliott D, Bates S (2005) A fast pipelined implementation of a two-dimensional in verse discrete cosine transform. In: Conference on electrical and computer engineering 2005
7. Fang Bo et al (2005) Techniques for efficient DCT/IDCT implementation on generic GPU. In: IEEE international symposium on circuits and systems 2005
8. Winger LL Source adaptive software 2D iDCT with SIMD. In: IEEE international conference on acoustics, speech, and signal processing 2000
9. Wikipedia [Online]. Available: http://en.wikipedia.org/wiki/Discrete_cosine_transform
10. Rettberg A et al (2001) A fast asynchronous re-configurable architecture for multimedia applications. In: 14th symposium on integrated circuits and systems design 2001
11. Smit LT et al (2007) Implementation of a 2-D 8 × 8 IDCT On the Reconfigurable Montium Core", International Conference on Field Programmable Logic and Applications, 2007

Green Master Based on MapReduce Cluster

Ming-Zhi Wu, Yu-Chang Lin, Wei-Tsong Lee, Yu-Sun Lin and Fong-Hao Liu

Abstract MapReduce is a kind of distributed computing system, and also many people use it nowadays. In this paper, the Green Master based on MapReduce is proposed to solve the problem between load balance and power saving. There are three mechanism proposed by this paper to improve the MapReduce system efficiency. First, a brand new architecture called Green Master is designed in the system. Second, Benchmark Score is added to each services in the cluster. In the last, an algorithm about how to distinguish the high score service and the low score service, and how to use them effectively.

Keywords MapReduce · Benchmark · Cloud network

M.-Z. Wu (✉) · Y.-C. Lin · W.-T. Lee
Department of Electrical Engineering, Tamkang University, Taipei City, Taiwan, Republic of China
e-mail: dean64188@hotmail.com

Y.-C. Lin
e-mail: bearlaker34@gmail.com

W.-T. Lee
e-mail: wtlee@mail.tku.edu.tw

Y.-S. Lin
Chung-Shan Institute of Science and Technology, Taoyuan, Taiwan, Republic of China
e-mail: amani.amani@mail.tbcnet.net

F.-H. Liu
National Defense University, Taipei City, Taiwan, Republic of China
e-mail: lfh123@gmail.com

Y.-M. Huang et al. (eds.), *Advanced Technologies, Embedded and Multimedia for Human-centric Computing*, Lecture Notes in Electrical Engineering 260,
DOI: 10.1007/978-94-007-7262-5_63,

Introduction

The algorithm in this paper will be used to improve the system efficiency based on MapReduce [1] of Hadoop. Hadoop is a kind of open source software that develop from Google MapReduce, and it can will create a cluster that connects each services. The cluster is used to make more computing resources called computing pool, and it can be expanded more and more. In the end, we can decide what we want to get or how to execute the program through coding the Map Function and Reduce Function.

As usual, in order to make the maximum computing resources, the services must keep the high-speed state, but it also has a lot of unnecessary waste. For example, service performance usually are not the same to each other, some of them are very high, but some of them are very low. if we allocate the same amount of work to all service, it must cause a part of service will complete the work early, but it still have to wait other service that performance is poor, and the waiting time means resources wastes. We will talk about how to make the service off if the performance is too low that seriously affects the system performance.

Related Works

Master of MapReduce

Master of MapReduce Master Node is the most important node on MapReduce which cannot be replaced by other nodes. It includes map function, reduce function and mapreduce runtime system. Master node manages receiving command from user and assigning tasks to task trackers, and it stores status of task trackers in database. The status is verified in three different types: Idel, In-processing and completed. The memory address and size of processing data in HDFS (GFS in Google, HDFS in Hadoop) are notified to Master node, and assign map function and task tracker to complete the task (Fig. 1).

Benchmark

Benchmark [2], generally speaking, is a value about something's performance or ability and make comparison. However, a performance comparison of virtualization technology for the moment is not very common, VM Benchmark is a new type of test methods. It is discussed virtual environment build through virtualization and virtual machine management VM resources (hard discs, memory). We have adopted Virtual Machine system build, and we introduce the mechanism of the Benchmark to distinguish the VMs' performance.

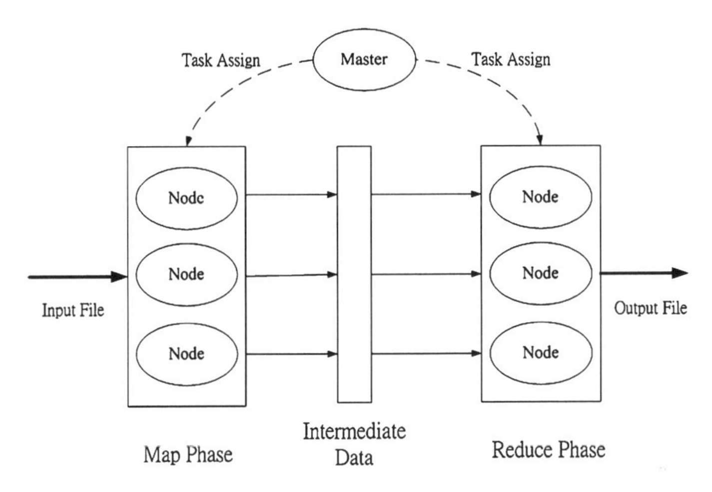

Fig. 1 MapReduce architecture

Implementation

In this section, the algorithm of Green Master will be explained how to implement. It includes Green Master System, Input File Index, Server Information, Queue, Record, Load Balance Optimization, Power Saving Algorithm, and Decision Algorithm. And we will discuss the detail at the following (Fig. 2).

Green Master System

The Green Master is a brand new architecture transformed from Hadoop's Master, and it can apply to each nodes that install the Hadoop. The brand new architecture called Green MapReduce System (GMS), and it can help users manage the node in the cluster to save the system consumption and service computing overhead. The Green Master does not change the Map Function and Reduce Function, it just changes the task allocation master according to server loading and server's Benchmark Score to achieve the goal about the energy saving.

It is not accepted that the system performance reduces caused by someone virtual machine low efficiency, especially in the Cloud Computing Network environment. It is not accepted that the system performance reduces caused by someone virtual machine low efficiency, especially in the Cloud Computing Network environment. In order to solve the above problems, Green Master is designed to delete the poor services and allocate the job distribution. Green Master is divides into eight blocks, and it includes Input File Index, Queue, Server Information, Record, Load Balance Optimization, Power Saving Algorithm and Decision Algorithm. Green Master has a strong adaptability to many systems, for

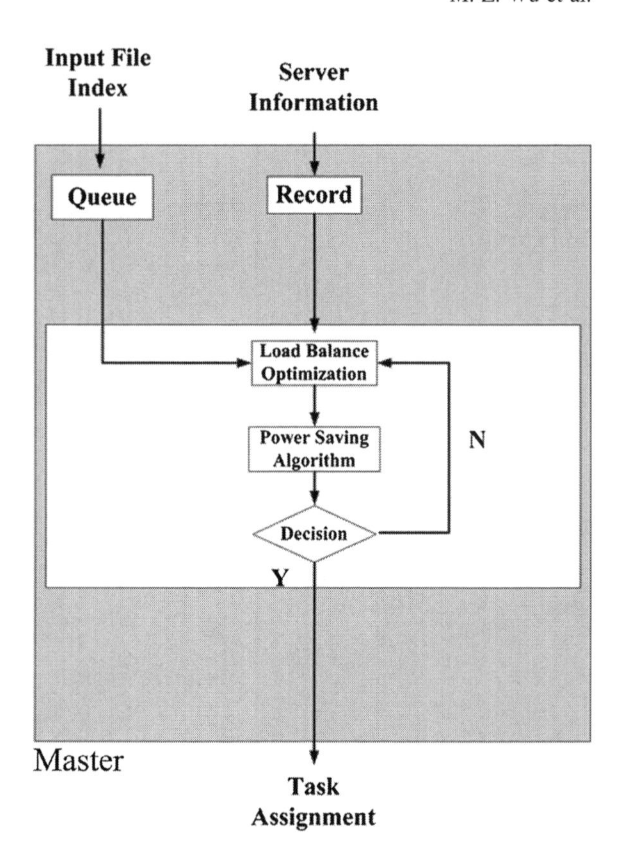

Fig. 2 Green master architecture

an instance, when we need a great amount of computing resources to calculate tasks, we can use Green Master to avoid energy wastes. For another instance, when the system equipment has a strong non-conformance, and the system can use the Benchmark Score in the Green Master to arrange the tasks allocation according to the services capability.

Server Information

The Server Information in the Green Master is to estimate the services' capability called Benchmark Score, and it will keep running and send the results to Green Master. In addition, whenever a new server join or quit the cluster, Benchmark Score will change. The range of the Benchmark Score is from zero to one hundred, and it is according to CPU computing performance, Memory read/write and Disk I/O rate to estimate the Benchmark Score. In other hand, the highest CPU response time, Memory read/write and Disk I/O is defined as 100 Benchmark Score. The definition of poorer virtual machines' Benchmark Score are based on the highest one.

$$BenchmarkScore = \frac{X_n}{Top_{VM}} \times 100\,\% \tag{3.1}$$

where the Top is the highest value of the virtual machine, and the x is the value of the virtual machine like CPU response time to be measured. Because of the CPU response time, Memory read/write and Disk I/O rate have to be considered in the formula, so we turn formulas evolution as follows:

$$Benchmark\,Score = \sum^{i} \frac{X_n}{Top_{VM}} \times W_i,\ W_i \in \{Measured\,Event\},\ i \in \{1, 2, \ldots, m\} \tag{3.2}$$

where the W_i is the event of Benchmark Score. In our case, the i of W_i is three, there are CPU response time, Memory read/write and Disk I/O rate respectively.

Recorder

Recorder is used for recording server information. Recorder refresh when it receives newer server information. A new recording table is established for information record when there is new node joins into the cluster. Servers update and refresh server information in recorder during the working time.

Load Balance Optimization

Load Balance [3] Optimization will allocate the work loading according to the information collecting from the above-mentioned blocks. The Benchmark Score is more higher, and the work loading is more; the Benchmark is lower, and the work loading is less. The job is allocated to VMs through Load Balance Optimization, and the formula is following:

$$Task\,Dsitribution\,Ratio = \frac{Local_{VM}}{\sum_{i=1}^{n} Score_i} \times 100\,\% \tag{3.3}$$

where Total Score is the sum of the VMs' Benchmark Score, and the Local Score is the VM's Benchmark what you want to estimate. In our experiment, we use six VM in the experiment environment and calculate the work loading ratio as following:

Power Saving Algorithm

In this paper, Power Saving Algorithm (PSA) [4, 5] will check the utilization of the server. In the first state, we allocate the work loading to VM according to the Benchmark Score, then the second state, we will determine the utilization of the VM. In Fig. 4, we can find that the huge difference of the work loading between Benchmark Score 100 and Benchmark Score 5, but they use almost same energy. This paper presents PSA to discuss how to get the balance between efficiency and energy management.

$$T_n = \sum_{i=1}^{n} a_i \times \frac{B_i}{\sum_{j=1}^{n} B_j} + \varepsilon \tag{3.4}$$

$$E_n = T_n \times N_{VM} \times P_{VM}, N_{VM} \in \{1, 2, \ldots, n\} \tag{3.5}$$

$$V_n = \frac{E_{n+1} - E_n}{T_{n+1} - T_n}, \ \mathrm{n} \geq 2 \tag{3.6}$$

where T_n is system computing time, and a_i is the system time which one virtual machine completed alone, and the B_i is the Benchmark Score of one virtual machine, and the ε is the error time. E_n is the energy (J) of virtual machine. P is the power (W) of virtual machine. V_n is the ratio of energy consumption.

Decision Algorithm

Decision Algorithm [6] in GMS is to judge the result which is from PSA reasonable or not. The formula is as following:

$$\alpha \leq \gamma \tag{3.7}$$

where α is the system consumption through PSA, and γ is without PSA. If γ is greater than α, then the system will back to Load Balance Optimization (Fig. 3).

Simulation Result

Figure 3 shows the highest performance virtual machine in the experiment environment of this paper.

Fig. 3 Experiment environment

test1

PHORONIX-TEST-SUITE.COM — Phoronix Test Suite 4.4.1

Value	Component
Intel Xeon E3-1230 V2 @ 3.30GHz (1 Core)	Processor
Intel 440BX	Motherboard
Intel 440BX/ZX/DX	Chipset
1024MB	Memory
34GB Virtual disk	Disk
VMware SVGA II	Graphics
Intel 82545EM Gigabit	Network
Ubuntu 10.04	OS
2.6.32-45-generic (i686)	Kernel
GNOME 2.30.2	Desktop
X Server 1.7.6	Display Server
vmware 10.16.9	Display Driver
2.1 Mesa 7.7.1	OpenGL
GCC 4.4.3	Compiler
ext4	File-System
800x600	Screen Resolution
VMware	System Layer

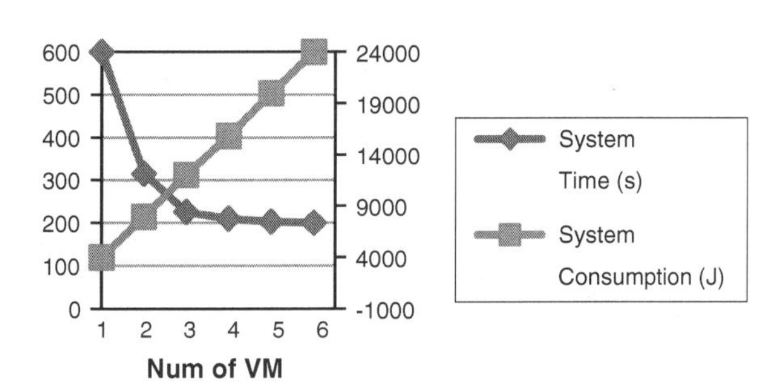

Fig. 4 System time and consumption

The Relationship Between System Computing Time and System Consumption

In Fig. 4, we can find that the cross point between the system time and system consumption is between two VMs and three VMs. In fact, the number of VM of the best performance in our experiment is three VMs.

Comparison Between Original and Green Master

In Figs. 5, 6 and 7, we take several different sizes of test file in our experiment environment, we can clearly find the original system time is less than Green Master, but system consumption is almost twice larger than Green Master.

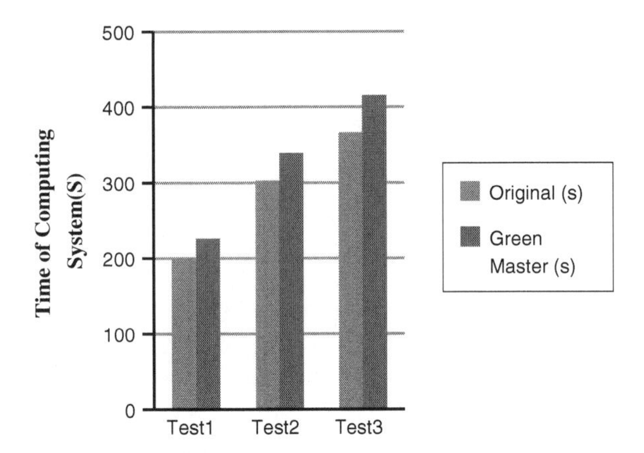

Fig. 5 System computing time

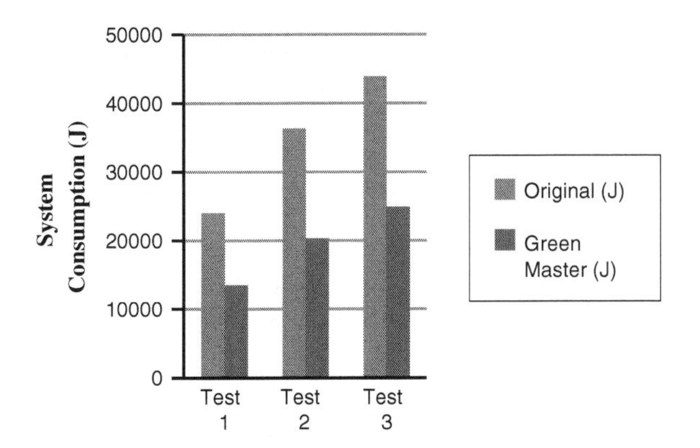

Fig. 6 Power consumption saving

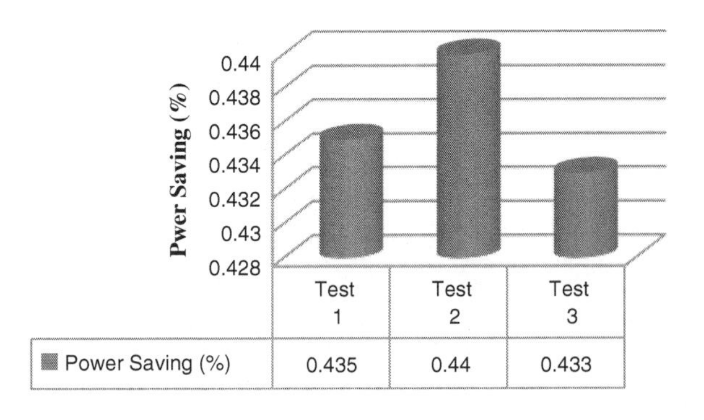

Fig. 7 Ratio of power saving

Conclusion

The idea of Green Master optimizes system power consumption by lower the performance slightly. In this paper, we provide a appropriate trade-off between power saving and performance loses, and improves energy conservation of the system.

References

1. Dean J, Ghemawat S (2004) MapReduce: simplified data processing on large clusters OSDO 2004
2. Xie X, Chen Q, Cao W, Yuan P, Jin H (2010) Benchmark object for virtual machines, 2010 second international workshop on education technology and computer science
3. Kolb L, Thor A, Rahm E (2012) Load balancing for mapreduce-based entity resolution, 2012 IEEE 28th international conference on data engineering
4. Zhu T, Shu C, Yu H (2011) Green scheduling: a scheduling policy for improving the energy efficiency of fair scheduler, 2011 12th international conference on parallel and distributed computing, applications and technologies
5. Sandhya SV, Sanjay HA, Netravathi SJ, Sowmyashree MV, Yogeshwari RN (2009) Fault–tolerant master-workers framework for mapreduce applications, 2009 international conference on advances in recent technologies in communication and computing
6. Liu C, Zhou S (2011) Local and global optimization of mapreduce program model, 2011 IEEE world congress on services
7. He Q, Li Z, Zhang X (2010) Study on cloud storage system based on distributed storage systems, computational and information sciences (ICCIS), 17–19 Dec 2010, pp 1332–1335
8. Bowers KD, Juels A, Oprea A (2009) HAIL: a high availability and integrity layer for cloud storage, computer and communications security (*CCS*), Nov 2009, pp 187–198
9. Luo M, Liu G (2010) Distributed log information processing with Map-Reduce: a case study from raw data to final models, information theory and information security (*ICITIS*), 17–19 Dec 2010, pp 1143–1146

Research on Graphic Digital Text Watermarking Research Framework

Jin Zhang, Xiaowei Liu, Xiaoli Gong, Rui Lu and Zhenlu Chen

Abstract This paper proposes a research framework for graphic digital text Watermarking algorithms and methods. We summarized graphic text watermarking after a brief review of digital text watermarking, and propose the research framework which consists of 8 levels, pixel, line, etymon, character, row, paragraph, page, and chapter. And we give the details of pixel level details and review most graphic text watermarking of this level as the example. At last we also figure out distribution of nowadays algorithms and discuss further research's possibility.

Keywords Text watermarking · Watermarking conceptual framework · Graphic digital text watermarking

Introduction

Digital Text watermarking technology is usually supposed to solve copyright protection and content authentication problems effectively [1]. For some reasons, it does not apply to practice systematically. However, protection measures of digital

J. Zhang (✉) · X. Liu · X. Gong
College of Information Technical Science, Nankai University, 300071 Tianjin, China
e-mail: Nkzhangjin@nankai.edu.cn

X. Liu
e-mail: Xiaoweiliu@nankai.edu.cn

X. Gong
e-mail: Gongxiaoli@nankai.edu.cn

R. Lu · Z. Chen
Tianjin Institue of Software Engineering, 300387 Tianjin, China
e-mail: ruilu@nankai.edu.cn

Z. Chen
e-mail: Chenzl@tjise.edu.cn

Y.-M. Huang et al. (eds.), *Advanced Technologies, Embedded and Multimedia for Human-centric Computing*, Lecture Notes in Electrical Engineering 260,
DOI: 10.1007/978-94-007-7262-5_64, © Springer Science+Business Media Dordrecht 2014

text content can be seen everywhere. Those are similar to digital text watermarking, such as printing a logo on the background of text content, or adding additional and invisible information after the end of segments. In principle, these both belong to the category of digital text watermarking technology. Reference to digital watermarking [2], digital text watermarking is defined as follow:

Digital text watermarking technique, which embeds some sign information of the text into the content or format of digital text production without affecting originality's value or use, is an available protection technique to authenticate the ownership and the integrity of text content. It can't be perceived by people unless extracted by a dedicated detector or reader. Generally, watermark information was encoded with "0" and "1" respectively.

Generally, digital content is composed of words, symbols, graphics, images and other elements according to some certain rules. And digital text or document is the digital content packaged by format. Content is mainly used for ideographic expression and format is used to organize the content to display text content and significance. Except for visible appearance, Format also include invisible edit control part of the background, while similar with format, in addition to visible text, images and other elements, content also include the significance which is non-intuitive visualization and needs to understand and learn. The difference is, even if the format is not visible in the control section, it is also severed to visible format appearance and therefore it is indirectly visible; while the meaning of the text is implicitly featured by elements such as text, image and format, it belongs to the comprehension visible. Like shaped and righteousness in the linguistics, text can also be understood as two parts of the text appearance (shape) and text meaning (sense).

Therefore there are four kinds of digital text watermarking technologies: Document Format Text Watermarking [3], Image Text Watermarking [4], Feature Coding Watermarking [5] and Natural Language Watermarking based on Semantics [6].

Document Format Text Watermarking embeds the watermark by changing the text format slightly, such as line and word shifting. Image Text Watermarking treats the digital document as an image. For example, embedding the watermark through the conversion of Spatial-Temporal Domain and Transform Domain, like image watermarking regardless of the feature of text, and changing the pixel on the unobtrusive part of text. The feature coding is more complex by its special relevance, such as Chinese and Hindi. The Natural Language Watermarking embeds the watermark based on the imperceptibility of semantics. This kind of watermark is vision visible, but it does not affect the understanding of the text, thus it is difficult to perceive. The former three kinds of watermarking belong to the text graphical watermarking based on the text appearance. They embed the watermarks by changing the text appearance slightly, which do not have a large visual deformation. However, the last one belongs to the text semantic watermarking, which changes the text content slightly in the case of retaining the original meaning.

Graphic Text Watermarking Framework

Framework Description

Text graphical watermarking is classified further considering with the position and method of watermark embedded. Because of the restrictions of language familiarity, the new classification introduced is based on simplified Chinese, which is adjusted to other languages look like Chinese. As for English, Arabic and others, the classification needs to be modified partially according to the characteristics.

As Fig. 1 shows graphic watermarking is divided into eight levels by the embedding position: pixel, line, etymon, character, row, paragraph, page, and chapter. Every level is composed of a certain number of the upper level units according to a certain rule. Wherein, the pixel is the basic unit of the text graphics. Any text consists of a myriad of pixels. The chapter is the highest level of the text, which is the first sight of a digital document and is associated with a variety of text formats with the content inside. There are different embedding methods for each level, such as, changing the gray value (changing the self-specificity); replacing the font with another similar one (similar unit replacement); adjusting the gap between etymon of one word (adjusting the combination rule). Three typical characteristics of each level are showed from the above methods: self-characteristics, similarity and structural feature.

Overall, the self-specificity includes inherited, including and unique attributes. As for the inherited attribute, the attribute of the next higher level is inherited; as for the including attribute, the structural feature of one level is the internal structural feature of the higher level because of the layered structure; the unique attribute is comprised of its self-specificity. In general, the similarity of graphical text

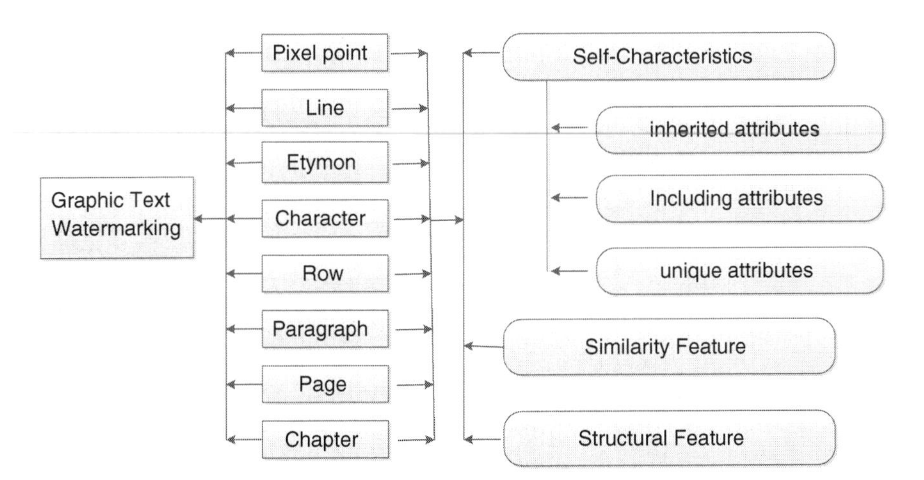

Fig. 1 This shows a graphic text watermarking framework figure consisting 8 embedding position levels

watermark is that an object is replaced by another similar one by all means on the premise of undetectable or imperceptible change. The common similarity replacement includes the similarity of fonts, languages, special symbols and error combinations. The structural feature is a structure which is constituted of different levels, and each level is constituted of a certain number of lower levels by certain rules. This feature is mainly reflected in two aspects: the number of rules. Apart from the pixel level, the amount and elements of other levels are fixed, as well as the theoretical rules (or the new unit will be created). We can only change the rules slightly, such as slight twist, tilt and stretching.

Framework Details

According to the definition of level, there are inclusion relations among the levels of graphical text watermark, that is to say, each chapter contains pages, and each page contains paragraphs, and so on. In other words, the pixel is the basic unit, which constitutes a line, then constitutes a etymon by some certain rules. The roots compose characters, which are put into a row. Many rows constitute a paragraph, and paragraphs build a page. Finally, all the pages are combined into a chapter with layout information, which is the common text we can see. All the levels are progressive and contained. As shown in Fig. 2.

Similarly, the inheritance between attributes and the transfer from characteristics to attributes embody the relationships among all the levels, that is, inheriting and inherited, transfer and inclusion. As the progressive inclusion relationship among levels, the characteristics and attributes is relevant closely, as well as that between the attributes. There are inclusion relationships among levels, which is the structural feature of each level. Therefore, all levels are linked closely together by the structural feature, and the structural feature is communicable, that is, the structural feature of one level can be the self-characteristic of the lower level. That is what called the transfer and inclusion relationship from characteristic to attribute, while some attributes are inheritable. To take the gray value as an example,

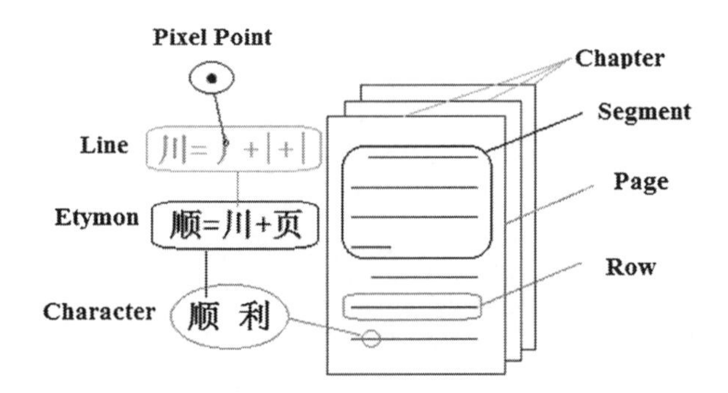

Fig. 2 This shows an example for graphic text watermarking framework embedding position

the gray value of all the layers are inherited from the pixel layer, which is inheriting and inherited relationship.

The inclusion and inheritance relations are different. The inclusion relation comes from the level definition. The higher level is formed by the structure of the lower level, which leads to a result that the internal attribute is the structural feature of the lower one. The point is combination method. However, the inheritance relation lays particular emphasis on the self-specificity itself. Some attributes can be inherited from the lower level, as the self-specificity of the higher one.

Pixel Point Level Study

Pixel Point Level Description

For the limitation of this paper, we only discuss the lowest and basic level-pixel point level, as shown in Fig. 3.

For the operation of the pixel gray value is divided into two aspects, one is to follow the image watermarking technology in the time-space domain and transform domain conversion algorithm; another more consideration text feature, combined with text characters, etymon and some of higher level characteristics or features, the use of certain rules relating to select the grayscale adjustment

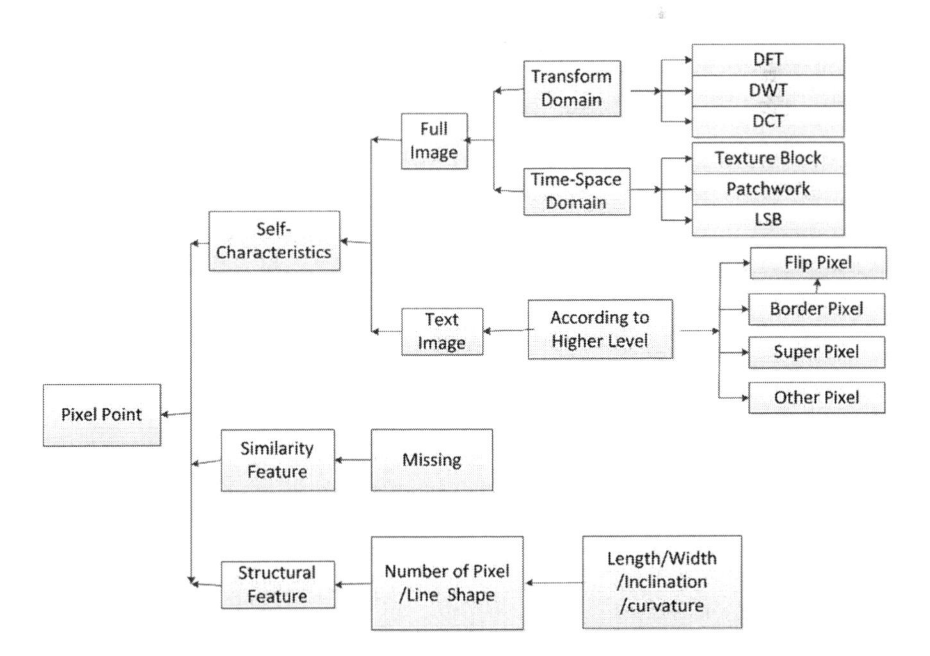

Fig. 3 This shows how pixel point level present under graphic text watermarking framework

imperceptible pixels flip. Time-space domain, transform domain watermark embedding technology from the image watermark, the current development is relatively mature, remarkable achievements in the study of image watermarking technique. Our work focuses on more text watermarking than image watermarking. The pixels constitute the basic unit of the text apart from the slightly itself level changes.

In addition, a different number of pixels according to a certain permutation or combination rules constitute the basic unit of the character—line, which forms the pixel level between the number (which determines the line length, width) and rules (which determines the inclination or curvature) structural characteristics.

Pixel Point Level Algorithms Review

Except inherited some of the traditional image watermarking algorithms, text graphical watermarking on pixel point level has also made a lot of results with considering text feature. Wu et al. [7, 8] introduced an algorithm that embedding the information by flipping the selected pixel called "flipped pixel". The selected pixels were chose with a 3×3 window on the text block. So the eight neighbor pixels around the selected ones will be checked to keep connectivity and gliding property after flipped. There are some similar algorithms, for example, Alex in [9] use 3×3 window to check 5×5 neighboring points and minimum flip distance to ensure the quality of the text appearance; Mei in [10] proposed to use the eight communicating boundary in a connection region of character to embed watermarking information. Gou and Wu [11] proposed a concept of "super pixel", which embedding the information through thickening or thinning a certain part of characters. It changes the structural feature of pixel point level.

As to modify the pixel values will cause distortion more or less. In order to enhance the imperceptibility of the embedded information, Anthony introduced an concept of Curvature Weighted Distance Difference, which picks flipped pixel by measuring the distortion after flip the pixel [12]. On this basis, Khen and Makur [13] take further study for improvement of coding efficiency.

Besides flipped pixel and super pixel, [14] proposed to add or delete a group of pixels in some characters, and the group of pixels relative to the character is central symmetry. In that paper, in order to reduce the loss of information by photocopying, it also introduced a packet-based synchronization method, thus greatly reducing the false detection rate.

The edge pixels are also used to hide information usually. Tirandaz et al. [15] embedded the secret watermark information into the edge pixels which will cause the minimum distortion, analyzes the edge pixels of character and defines edge to edge pixel distance. In addition, there are some algorithms with modifying the edge region of characters [16, 17]. Zhang et al. [18] modified the gray value of pixels with another rule. First, it selected words for information embedded with area of the words. Then, it checked the pixel in these words to get pixels whose

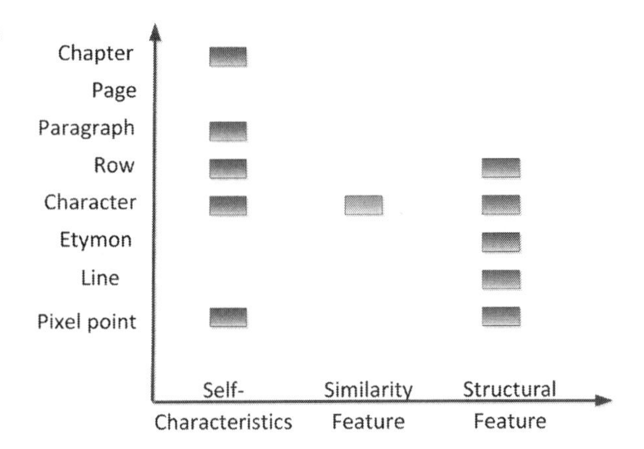

Fig. 4 This shows nowadays graphic digital text watermarking algorithms' distribution

gray value is closest to the average gray value of its eight neighboring pixels. Zhaoxing Yang proposed to adjust the stair edges of non-horizontal strokes and non-vertical strokes in characters in order to change the ratio of the sum of black pixels in the upper and lower halves of each character image for the purpose of embedding a watermark bit [19].

Finally, Qi in [20] proposed a method for binary text image digital watermarking algorithm. Under the assumption of multiplicative transformation model, this method flips boundary pixels of character image with a boundary point flip policy so that the watermarking embedded in the image is difficult to be modified traces. On this basis, [21] takes further study and makes the complexity of the characters as the standard to select embedded characters. Based on invariant after print or scan, it flips pixel with a quantization function to embed a number bits of information. It reduces visual distortion caused by the flip pixels effectively.

Conclusion

Under the framework in Graphic Text Watermarking Framework, we found nowadays graphic digital text watermarking algorithms' distribution map as Fig. 4 shows.

Most researchers is focus on self-characters and structural feature, while similarity is ignored. And the blank zone in Fig. 4 maybe the aspects for further research.

In this paper we propose a research framework for Graphic Digital Text Watermarking algorithms and methods. It consists of 8 levels, pixel, line, etymon, character, row, paragraph, page, and chapter; and 3 aspects, self-characteristics, similarity and structural feature. We hope it could help researchers find new research interests and ideas. In the future, we will try to build the relationship between this framework and application scene.

Acknowledgments This work is supported by a grant from the Research Fund for the Doctoral Program of Higher Education of China (Granted No.20110031120032), the Fundamental Research Funds for the Central Universities and the National Science Foundation of China (Grant No: 61103074).

References

1. Brassil JT, Low S, Maxemchuk NF (1999) Copyright protection for the electronic distribution of text documents. In: Proceedings of the IEEE. 87, IEEE Press, New York, pp 1181–1196
2. Cox I, Miller M, Bloom J et al (2007) Digital watermarking and steganography. Morgan Kaufmann Publishers In; 2nd revised edition
3. Maxemchuk NF (1994) Electronic document distribution. AT&T Tech J 6:73–80
4. Kim Y, Oh I (2004) Watermarking text document images using edge direction histograms. Pattern Recogn Lett 11:25
5. Shirali-Shahreza MH, Shirali-Shahreza M (2006) A new approach to persian/arabic text steganography, computer and information science. In: 5th IEEE/ACIS international conference, IEEE Press, New York, pp 310–315
6. Topkara U, Topkara M, Atallah MJ (2006) The hiding virtues of ambiguity: quantifiably resilient watermarking of natural language text through synonym substitutions. In: ACM multimedia and security conference, ACM Press, New York, Geneva
7. Wu M, Tang E, Liu B (2000) Data hiding in digital binary images. In: international conference on multimedia and expositions, IEEE Press, New York, pp 393–396
8. Wu M, Liu B (2004) Data hiding in binary image for authentication and annotation. Multimedia IEEE Trans 6(4):528–538
9. Yang H, Kot AC, Liu J (2005) Semi-fragile watermarking for text document images authentication. In: 2005 IEEE international symposium on circuits and systems, vol 4, IEEE Press, New York, pp 4002–4005
10. Mei Q, Wong EK, Memon ND (2001) Data hiding in binary text documents. In: 2001 SPIE 4314, IEEE Press, New York, pp 369–375
11. Gou H, Wu M (2007) Improving embedding payload in binary images with "Super-Pixels". In: 2007 IEEE international conference on image processing, vol 3, IEEE Press, New York, pp III-277–III-280
12. Ho ATS, Puhan NB, Makur A, Marziliano P, Guan YL (2004) Imperceptible data embedding in sharply-contrasted binary images. In: 2004 international conference on control, automation, robotics and vision, vol 2, IEEE Press, New York
13. Khen TV, Makur A (2006) A word based self-embedding scheme for document watermark. In: TENCON 2006. 2006 IEEE region 10 conference, IEEE Press, New York, pp 1–4
14. Varna AL, Rane S, Vetro A (2009) Data hiding in hard-copy text documents robust to print, scan and photocopy operations. In: 2009 IEEE international conference on acoustics, speech and signal processing, IEEE Press, New York, pp 1397–1400
15. Tirandaz H, Davarzani R, Monemizadeh M, Haddadnia J (2009) Invisible and high capacity data hiding in binary text images based on use of edge pixels. In: 2009 international conference on signal processing systems, IEEE Press, New York, pp 130–134
16. Yu X, Wang A (2009) Chain coding based data hiding in binary images. In: fifth international conference on intelligent information hiding and multimedia signal processing, IEEE Press, New York, pp 933–936
17. Zhang X, Liu F, Jiao L (2003) A new effective document water-marking technique based on outside edges. Syst Eng Electron 25(05):612–616
18. Zhang X, Liu F, Jiao L (2003) An effective document watermarking technique. J Commun 24(05):21–28

19. Zhao X, Sun J, Li L (2008) Watermarking of text images using character step edge adjustment. J Comput Appl 28(12):3175–3182
20. Qi W, Li X, Yang B, Cheng D (2008) Document watermark scheme for information tracking. J Commun 29(10):183–190
21. Guo C, Xu G, Niu X, Li Y (2011) High-capacity text watermarking resistive to print-scan process. J Appl Sci 29(02):140–146

A Predictive Method for Workload Forecasting in the Cloud Environment

Yao-Chung Chang, Ruay-Shiung Chang and Feng-Wei Chuang

Abstract Cloud computing provides powerful computing capabilities, and supplies users with a flexible pay mechanism, which makes the cloud more convenient. People are getting more and more usage of the cloud environment due to a steady increase of data. In order to improve the performance and energy saving of the cloud computing, the efficiency of resource allocation has become an important issue. In this study, a neural network model with learning algorithm is applied to predict the workload of the cloud server. The resource manager deployed on the cloud server provides the service of managing the jobs with a resource allocation algorithm. With this prediction mechanism, cloud service providers can forecast the following time workload of cloud servers in advance. The experimental results show that resources can be allocated efficiently and become load balanced by proposed mechanism. Therefore, the cloud server can avoid the problem of inadequate resources.

Keywords Cloud computing · Predictive workload · Neural network · Learning algorithm

Y.-C. Chang (✉)
Department of Computer Science and Information Engineering, National Taitung University, Taitung, Taiwan, Republic of China
e-mail: ycc@nttu.edu.tw

R.-S. Chang · F.-W. Chuang
Department of Computer Science and Information Engineering, National Dong Hwa University, Hualien, Taiwan, Republic of China
e-mail: rschang@mail.ndhu.edu.tw

F.-W. Chuang
e-mail: m9921043@ems.ndhu.edu.tw

Y.-M. Huang et al. (eds.), *Advanced Technologies, Embedded and Multimedia for Human-centric Computing*, Lecture Notes in Electrical Engineering 260,
DOI: 10.1007/978-94-007-7262-5_65,

Introduction

Numbers of physical and virtual resources provide a powerful computing ability in the cloud computing environment. When a user requests to execute the particular application, the resource manager of cloud computing environment will create a new virtual machine, and assign jobs to the host which operating under servers. Virtual resources can be added or removed at any time in the cloud computing environment. This characteristic makes the cloud platform more flexible and efficient. However, this feature also brings some issues, such as resource allocation imbalance and energy consumption in equable. On the other hand, the workload of the cloud environment may change frequently in a short period due to the large number of jobs or when only a small amount of new jobs is received from users. Furthermore, when the cloud server receives a large number of new jobs, it also results in the same situation in a short period of time. This varying workload may result a situation where there are not enough resources to be allocated or cause a resource imbalance. Therefore, a prediction method for cloud environments to allow the cloud provider to avoid this situation is needed. In this study, a prediction method to forecasting the workload of cloud environments based on a neural network is proposed. The workload information from the cloud server was collected and calculated with the prediction method. The cloud provider can provide more efficiency management of the resource allocation by this method.

The rest of this study is organized as follows: section Related Works introduces an overview and related works of prediction methods. Section Proposed Method presents the proposed method and learning algorithm to forecast the workload of cloud environments. Data collection and implementation results of proposed method are sent out and analyzed in section Implementation. Finally, section Conclusions concludes the paper.

Related Works

Workload characterization and prediction have been studied for many years. Many algorithms and proposed models have a somewhat different focus. Some algorithms are focused on virtual machine placement, which is related to the workload of the virtual machine and virtual machine allocation. While other algorithms are focused on energy consumption which seek to allocate a maximum number of resources and attempt to ensure that all the Service Level Agreements (SLA) are satisfied. Finally, some algorithms focus on the future workload of the cloud server which is also this study's focus. The following section will introduce the different algorithms for workload prediction.

Time Delay Neural Network (TDNN)

The time delay neural network [1] is a feed forward neural network which uses time delay information combined with the input layer. At any given time n, the input of the network consists of the vector:

$$y(n) = \sum_{j=1}^{m} w_j \phi\left(\sum_{i=0}^{p} w_j(i) \times (n-1) + b_j\right) + b_0 \tag{1}$$

The output of the TDNN is the predicted next value in the time series which is computed as the function of the past values in the time series. To achieve this predictive method, the inputs need a large number of continuous data. It is not appropriate to predict the workload on the new server.

Regression Model

Regression analysis [2] is used for estimating the relationships between a dependent variable and one or more independent variables. Regression analysis helps to understand how the typical value of the dependent variable changes when each of the independent variables is varied, while the other independent variables are held fixed. A general form of the non-linear model is provided as:

$$y_i = a_0 + a_i x_i + a_i x_i^2 \cdots + a_i x_i^m + \delta_i \quad (i = 1, 2, \ldots, n) \tag{2}$$

where ‘a’ is constant and represents successive numeric quantities. δ represents the random prediction error in the model. The value of m depends on the degree of freedom of the model to be deduced.

Predictive Data Grouping and Placement

The method [3] is focused on the load balance problem on the server. The strategy is to group the data which have a similar workload, and then determine how data is placed in different servers according to capacity and the amount of category thresholds. The grouping method is to place related and popular contents on the same server. The placement method consolidates any pair of under loaded servers, whose aggregate workload remains under the maximum desired load and reassigns the dispatching probabilities.

Statistic Based Load Balance

The SLB [4] is to solve load imbalance and guide the host selection while the virtual machine starts. SLB uses historical performance data of virtual machines to estimate the resource requirement of each virtual machine. The virtual machine load forecast method is defined as the formula:

$$L_i(VM_{cpu}) = E(l_i(VM_{cpu})) = \frac{\sum_{j=1}^{n} l_{ij}(VM_{cpu})}{n} \tag{3}$$

where L_i is the predicted load at time *i* in one day, l_i is the load value at time *i*, l_{ij} is the load value at time *i* in the *j*th day.

Proposed Method

The workload definition is the amount of processing slot that the computer has been given to do in the cloud environment. The workload consists of the amount of applications running in the computer and the number of users connected to or interacting with the applications. This study focuses on the amount of applications running in the computer. Besides, a recurrent neural network model is applied in the proposed system based on the prediction of total workload for the cloud severs. The workload data is collected from the NCHC Hadoop public experiment cluster.

The proposed system architecture consists of a client, resource master, predictor, servers and VMs which are shown in Fig. 1. The definition of resource manager is the service of managing the jobs with a resource allocation algorithm. The operation consists of the following steps:

- **Client submits a job**. The job in the cloud environment will split into many processes. The workload of the cloud environment is composed of these processes.

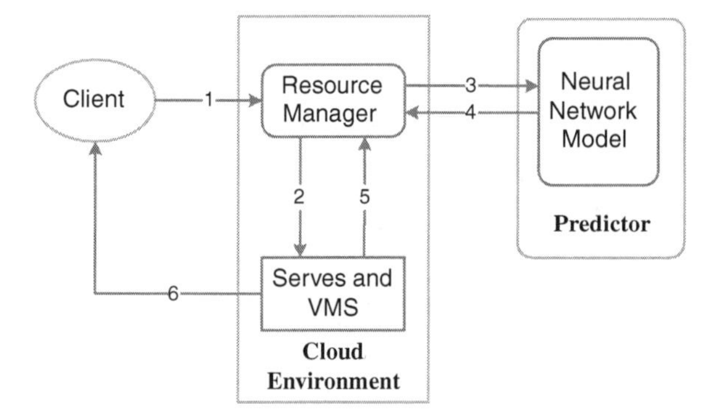

Fig. 1 System architecture

- **Choose available servers**. The resource manager will select the servers and create numbers of VMS based on the requirements from the job.
- **Send workload information to Predictor**. When the resource manager dispatches the processes to the servers and VMs, the predictor can get the workload information from the resource manager.
- **Send results to Resource Manager**. The predictor uses previous workload information to calculate the workload information for the following time. When the predictor has calculated the result of the next time's workload information, it will send the result to the resource manager. The result will help the resource manager to modify the resource allocation algorithm. The allocation algorithm will be suitable for the cloud environment.
- **Send resource information to resource manager**. When the servers or VMs have completed a job, it will send back the resource information and workload information.

The prediction of recurrent neural network model is illustrated in this work. The neural network consists of three layers with an input layer, one output layer and one hidden layer. In the input layer, the neural network receives the information of workloads from X_1 to X_a at various time steps of one complete sequence that constitutes the first epoch. At each time step the output is feedback to be employed as the input k for the next time step. At any given time t, the input of the neural network composed of the vector is shown as:

$$\mathrm{x}(\mathrm{t}) = [\mathrm{x}(\mathrm{t}), \mathrm{x}(\mathrm{t} - 1), \ldots, \mathrm{x}(\mathrm{t} - p)]^{\mathrm{T}} \tag{4}$$

where p is the number of selected delay line memory. The vector means the neural network can remember the numbers of workloads. For example, people usually can remember what they have already done recently. The vector is like the memory of people. For this vector, it works to select the number of p and decides how long the neural network can remember the workloads in the past.

For a neural network with one hidden layer, the output of a single neuron j in the hidden layer is given by the equation:

$$\mathrm{y_j}(\mathrm{t}) = f\left(\sum\nolimits_{\mathrm{i} \in \mathrm{A} \cup \mathrm{B}} \mathrm{w_{ji}}(\mathrm{t} - 1)\mathrm{k_i}(\mathrm{t} - 1) + \mathrm{b_j}\right. \tag{5}$$

$$\mathrm{k_i}(\mathrm{t}) = \begin{cases} \mathrm{x_i}(\mathrm{t}) & \text{if} \quad \mathrm{i} \in \mathrm{A} \\ \mathrm{y_i}(\mathrm{t}) & \text{if} \quad \mathrm{i} \in \mathrm{B} \end{cases} \tag{6}$$

where $f()$ is the activation associated with the neuron j, $\mathrm{w_{ji}}$ is the weight, $\mathrm{b_j}$ is the bias.

In the Eqs. (8), if $i \in A$, it represents that $\mathrm{k_i(t)}$ is real input $\mathrm{x_i(t)}$; If $\mathrm{i} \in \mathrm{B}$, it represents that $\mathrm{k_i(t)}$ is the feedback value from output $\mathrm{y_i(t)}$. It means a set of A is the workload information which is currently obtained from cloud servers; and a set of B is the past workload information. Therefore the output of a neural network with m neurons in the hidden layer is shown as

$$y(t) = \sum_{j=1}^{m} w_j y_j(t) \tag{7}$$

From Eq. (7) and (9), the output vector of the neural network maps to the input vector is shown as:

$$y(t) = \sum_{j=1}^{m} w_j f\left(\sum_{i \in A \cup B} w_{ji}(t-1) k_i(t-1) + b_j\right) + b_0 \tag{8}$$

After the calculated function was found for the neural network, the learning algorithm is needed to define the activation function. Besides, the error function and error update must be defined. The error function $E(t)$ is defined by selecting a output neuron $S_a(t)$ of the hidden state neurons which at time t is shown as:

$$\begin{cases} \text{state : on} & S_a(t) > 1 - \beta \quad \text{if} \quad \text{accepted} \\ \text{state : off} & S_a(t) < \beta \quad \text{if} \quad \text{accepted} \end{cases} \tag{9}$$

where β is the tolerance of the response neuron.

There are two error situations which result from the definition, the first is the network fails to reject a negative string, i.e., $S_a(t) > \beta$; the second is the network fails to accept a positive string, i.e., $S_a(t) < 1 - \beta$. The error function is defined as:

$$E(t) = \frac{1}{2} \sum_{a=1}^{a} (d_a(t) - S_a(t))^2 \tag{10}$$

where $d_a(t)$ is the target response value for the response neuron $S_a(t)$.

The steepest descent method [3] was used to calculate the weight update rule:

$$\Delta w_{kj} = -\alpha \frac{\partial E(t)}{\partial w_{kj}(t)} = \alpha (d_a(t) - S_a(t)) \frac{\partial S_a(t)}{\partial w_{kj}(t)} \tag{11}$$

where α is the learning rate.

Implementation

The experimental data is collected from the NCHC public environment. This environment is constructed from Hadoop and there are registered 3194 users with a total number of 160,447 jobs having already been run. Before the experiment, the number of neurons and delay times in the hidden layer for the proposed model is needed to find. From the results of experiment, the neural network had 10 neurons and 2 delay times in the hidden layer.

Figure 2 shows the simulation results. The learning effect of the algorithm is divided into three parts for analysis. The first part is the time before 600. It is noted that the experimental result does not seem better than the regression method and TDNN. This is because the proposed model does not have enough data to calculate the more accurate predicted results. The second part is the time between 600 and 1,600. This experimental result is better than previous data because the predicted

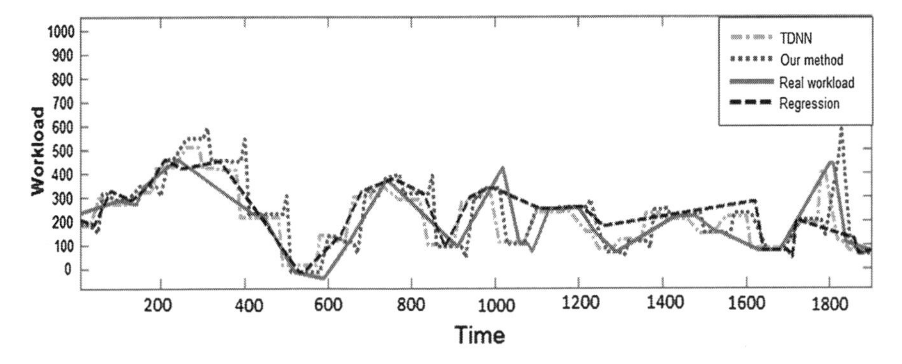

Fig. 2 The simulation result compare with regression and TDNN

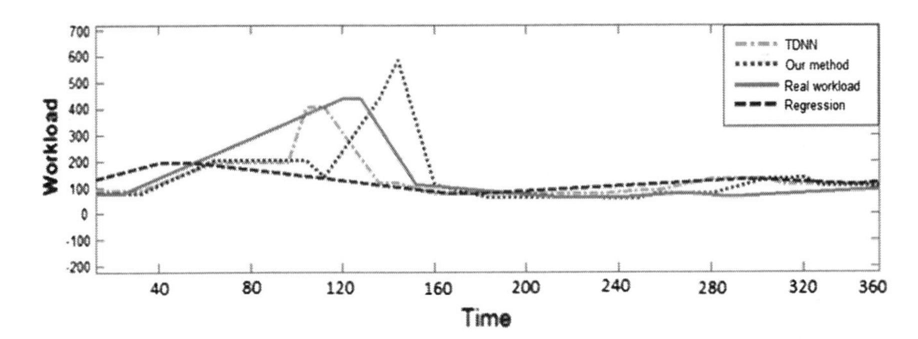

Fig. 3 The enlargement of simulation results compared with regression and TDNN

Table 1 Comparison between TDNN, regression and proposed method

Model	Mean squared error (MSE)
TDNN	557415.88979
Regression	7233256.47972
Proposed method	510293.67772

number is close to the real workload number. The last part is the time after 1,600. The proposed model is also better than the regression method. Even though there is some error, the model still represents a predicted line, which is close to the real line.

Figure 3 shows that the proposed algorithm and the regression method both have some errors in their predictions, but the learning effect of proposed method is clearly better than the regression method, which cannot show rapid changes in the data. The proposed method presents a trend similar to the real workload while the regression method does not. Therefore, the result of proposed method is more accurate than the regression one.

Mean Squared Error (MSE) [5] is used to calculate the predicted results. The calculation of the MSE is shown as below.

$$\text{MSE} = \frac{1}{N}\sum\nolimits_{i=1}^{N}(p-t)^2 \tag{12}$$

p = Predicted Workload Value, t = Total Workload Value, N = Number of data.

A comparison between TDNN, Regression and Proposed Method is given in Table 1. The result of proposed method is better than TDNN and regression method. Although the proposed method is only a little better than TDNN, TDNN requires a large amount of input data as the training set. Therefore, TDNN is not suitable to run in the new environment. The regression algorithm has high accuracy for prediction problems, but it is only suitable for long time analysis or stable system analysis. It is clearly to see this characteristic from the figures. If a time series over a month or a year are selected, the regression algorithm may not have more error than our algorithm. Consequently, the unit of time is an important factor for the prediction problem. The result shows that the proposed algorithm is suitable for predicting workloads over a short period of time, and the learning effect of proposed method clearly shows the trend of the real workload.

Conclusions

This study proposes the neural network model for predicting the workload of cloud environments. This work helps the cloud provider to avoid some unexpected situations such as a large number of resource requirements from users or applications. The experiment results show that the proposed model is able to predict the workload over a short period of time. Compared to the regression method, the proposed method is more accurate for workload prediction. Hence, the cloud provider can decide the resource allocation management more efficiently by applying the proposed method.

Acknowledgments The authors would like to thank the National Science Council of the Republic of China, Taiwan for financially/partially supporting this research under Contract No. NSC101-2221-E-143-005-, NSC101-2221-E-259-003- and NSC101-2221-E-259-005-MY2.

References

1. Zhang Q, Cherkasova L et al (2007) A regression-based analytic model for dynamic resource provisioning of multi-tier applications. IEEE Int Conf Auton Comput (*ICAC*)
2. Tirando JM, Higuero D, Isaila F, Carretero J (2011) Predictive data grouping and placement for cloud-based elastic server infrastructures. IEEE/ACM international conference on cluster, cloud and grid computing, pp 285–294
3. Mell P, Grance T (2009) The NIST definition of cloud computing. Nat Inst Stand Technol 53(6):50
4. Zhang Z, Wang H, Xiao L, Ruan L (2011) A statistical based resource allocation scheme in cloud. Cloud and service computing (*CSC*), 2011 international conference, pp 266–273

5. Gallant S (1993) Neural network learning and expert systems. MIT Press, Cambridge
6. Peterson C, Södeberg B (1993) Artificial neural networks. Modern heuristic techniques for combinatorial problems. In: Reeves CR (ed) Advanced topics in computer science, Oxford Scientific Publications, New York, pp 197–242
7. Abramson D, Buyya R, Giddy J (2002) A computational economy for grid computing and its implementation in the Nimrod-G resource broker. Future Gener Comput Syst 18(8):1061–1074
8. Picht SW (1994) Steepest descent algorithms for neural network controllers and filters. IEEE Trans Neural Networks 198–212
9. Haykin S (2008) Neural networks and learning machines: a comprehensive foundation, 3rd ed. Prentice Hall
10. Apache Hadoop. http://hadoop.apache.org/
11. Borthakur D (2009) The Hadoop distributed file system: architecture and design. http://hadoop.apache.org/common/docs/current/hdfs-design.html

Implementation of Face Detection Using OpenCV for Internet Dressing Room

Li-Der Chou, Chien-Cheng Chen, Chun-Kai Kui, Der-Ching Chang, Tai-Yu Hsu, Bing-Ling Li and Yi-Ching Lee

Abstract Human face detection is an important technology that is positively developed by industry. Therefore, this paper proposes a dressing application that adopts a face detection technology for better human life. Furthermore, this paper detects human face by using the combination of OpenCV and Internet camera. When the proposed application detect a human face, the proposed application dresses target on a screen, such as putting one selected cloth on the correspond position of human. Finally, the proposed application has a dressing ability for customers who are shopping clothes on Internet, and the dressing ability improves the sales performance of Internet store.

Keywords OpenCV · Internet camera · Face detection

Introduction

Stay-at-home-economy is a huge consumption market. In November 2012, the statistics of Taiwan Institute for Information Industry (III) shows that the gross value of e-commerce in 2012 reach to NT 660.5 billion dollars have 17.4 % growing a year. The III also predicts the gross value of e-commerce will break-through NT 1 trillion dollars in 2015. People save the time for better usage, and geek "Otaku" like to stay home, even those people who lives in remote districts have a problem to shop in urban will use Internet shopping to buy daily living equipment. Clothes are the most common things for customers purchasing via Internet. Internet is very convenient to browse some popular clothes. Customers can quickly search their favorite one from a pile of clothes. However, purchasing

L.-D. Chou (✉) · C.-C. Chen · C.-K. Kui · D.-C. Chang · T.-Y. Hsu · B.-L. Li · Y.-C. Lee
Department of Computer Science and Information Engineering, National Central University, Jhongli, Taiwan
e-mail: cld@csie.ncu.edu.tw

Y.-M. Huang et al. (eds.), *Advanced Technologies, Embedded and Multimedia for Human-centric Computing*, Lecture Notes in Electrical Engineering 260,
DOI: 10.1007/978-94-007-7262-5_66,

clothes via Internet has one problem: customers cannot try clothes before buy them. Therefore, customers always feels disappoint when they receive their packs. Customers may concern about whether the purchased cloth is fitness or not. Moreover, cloth manufactures have different size model. One cloth is fit but the other is not, even these two clothes are the same size. According to these reasons, customers usually doubt about shopping on Internet.

This paper proposes a concept about the dressing system module which could be a part of a smart store for online shops. In an Internet dressing room, customers can browse popular clothes conveniently. They can not only see the model demonstrate the cloth, but also experience when they put clothes on. The proposed application makes some lazy customers can try new clothes comfortably and conveniently when they shopping on Internet. Furthermore, Internet dressing room do attract more popularity and reduce the cost of physical shop. Since the Internet dressing room, clothes can cost down, and Internet dressing room promotes the aspiration of shopping on Internet.

The proposed application is implemented by using an internet camera and face detection function in OpenCV. The application locates the human face by detection technology. After calculate the location of the face, the application will get the position of the human body. Finally, the application puts the chosen cloth on the body. The rest of this paper is organized as follows. Section Related Work describes related works about OpenCV. In section Implementation, we propose the proposed application. Section Conclusion concludes the paper.

Related Work

OpenCV is developed by Intel cooperation. Open represents Open Source and CV is Computer Vision [1]. OpenCV is an Open Source Image Process Library and it can be used to make image, video, matrix operations, statistics, graphic, data storing programed by C language. OpenCV can be used for image processing, computer vision, pattern recognition, computer graphics, game design, and so on. The most famous application of OpenCV is face detection, it can also use for integrate different picture format's matrix operation. Apply OpenCV in static picture (i.e. BMP, JPG, TIF, PNG), and image processing of dynamic Webcam. OpenCV provides an easy GUI interface. OpenCV can be integrated into Visual C++ and C++ Builder. OpenCV can support Windows and Linux. OpenCV has equipped powerful graphic and matrix operations [1]. However, OpenCV is not supported by Intel cooperation, so it cannot offer some latest algorithm. But it has been revision recently. In new vision, some new function, like Scale-invariant feature transform (SIFT) for computer vision, has been added.

Through the combination of Kinect and OpenCV, color image, 3D depth image, and voice signal can be catch in one time. Kinect equipped with 3 shots [2]. The middle shot is use for recognizes the identification and detects facial expression of the user. Kinect will turn the 3D depth image into skeleton tracing system.

This system can detect at most 6 people in one time. Including the trunk, arms, legs, even fingers are in tracing area. Microsoft also implement machine learning and large image database for understand users motivation.

Implementation

The system flow chart diagram is shown on Fig. 1. When users start the program, the application shows a cloth catalog provided by cloth vendors. After user choose a dress he/she like, the application judges whether the computer is equipped with camera or not. The application executes cvGrabFrame() function after judgment. The cvGrabFrame() function mainly grabs a image from the camera. Then, the application detects whether there is a human face in this image. If yes, this function will return the position of human face. Therefore, the proposed application can predict where the human body is. Finally, the chosen cloth picture is shown on the predicted position of body.

System Architecture

The proposed application provides user a cloth menu that is offered by cloth vendors. Users may choose some cloths they have will to buy. System confirms whether a camera is ready. If yes, the next procedure is using the ready camera to take an image and decides where the human face is. According the position of human face on an image, system deduces the position of body by comparing the face position with depth image. System automatically puts one chosen cloth picture on body and shows the composite portrait on a screen. Finally, system provides shopping advices to customer. The detail of system architecture is shown on Fig. 2.

Select the Cloth

This function provides a menu that lists all clothes of one store for customers. In Fig. 3a, customers can browse clothes and choose one cloth they like. After customers make a selection from the three cloths image on a screen, system go to next step. The application can increase products database based on salesman's whosh in the future.

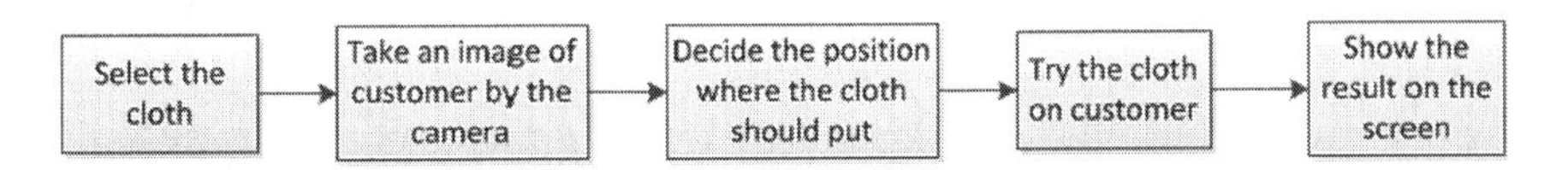

Fig. 1 System flow chart diagram

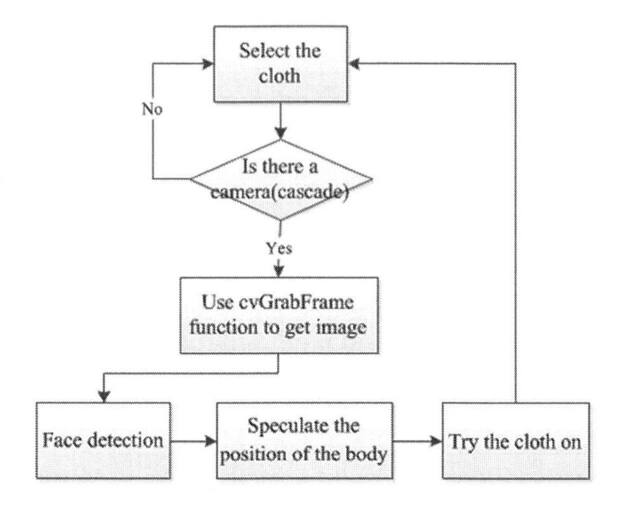

Fig. 2 System architecture

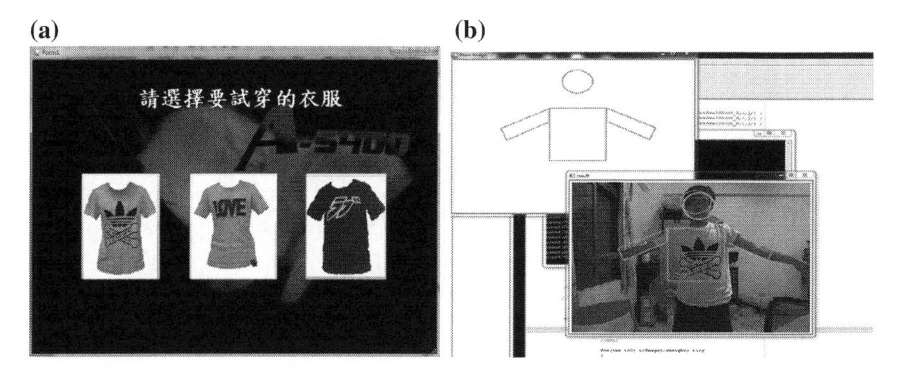

Fig. 3 **a** Select the cloth; **b** Face detection

Face Detection

This function detects human face, if users make a selection. A camera catches an image. The image is processed by AdaBoost Leaning with Haar-Like Features algorithm for face detection.

The first step of face detection is to define some Haar-Like features. Then, inputting some sample. For example, system detects the face features according to some face samples come from white men, black girl, and yellow boy. Then, this paper uses AdaBoost learning algorithm to find out some representative Haar-Like features. In this case, this function picks up some face features. Each feature represents a class. All features constitute a class array which called "strong classifier." Each class is used to analyze the input image whether contains a face or

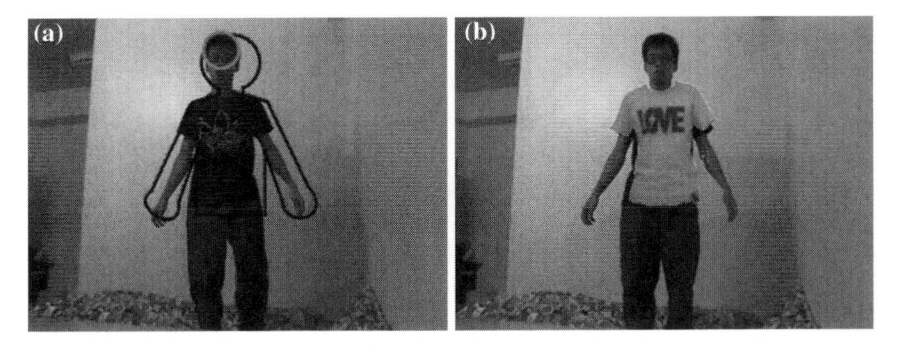

Fig. 4 **a** Speculate the position of the body; **b** Try the cloth on

not. If all classifiers return true to this image, then this image will be identified to be a face. In Fig. 3b user's face can be detected accurately.

Speculate the Body Position

This function begins to speculate the body position after the face detection function returns the coordination of human face. This function determines where the cloth should be put on, as shown in Fig. 4a. Through the image taken by the camera, this function recognizes the posture of the body. The posture will be compared with depth image and find out the most similar human body model. Finally, the application will get the real post of human and recognize every part of human body.

Try the Cloth On

This function decides the location of the cloth based on the function of speculate the body position. The application puts the chosen cloth on the position dynamically. Finally, accomplish the function of try the cloth on human.

Conclusion

This application detects face successfully by OpenCV and camera. After figure out the coordinate of a human face, the application will speculate where the body is and put the dress on the body position. Before this application, customers cannot try on the cloth unless they purchase on real store. According to this reason, a lot of customers don't like to buy the dress on the internet. By this application, customers' requirement could be satisfied easily. They can try on the cloth without span a lot of time to go to the physical store. Furthermore, this application could promote the developing of shopping on Internet. This application can not only increase the sales but also decrease the cost of real store.

In this thesis, the application merge camera, real-time face detection technology and body position speculate algorithm. This system makes a win-win situation to salesman and customers.

References

1. Intel Corporation, Open source computer vision library. Reference manual, Copyright © 1999–2001. Available. http://software.intel.com/en-us
2. http://www.techbang.com/posts/2936-get-to-know-how-it-works-kinect
3. Wren C, Azarbayejani A, Darrell T, Pentland A (1997) Pfinder: real-time tracking of the human body. IEEE Trans Pattern Anal Mach Intell 19(7):780–785
4. Lorsakul A, Suthakorn J Traffic sign recognition for intelligent vehicle/driver assistance system using neural network on OpenCV. International conference on ubiquitous robots and ambient intelligence
5. Haj MAI, Amato A, Roca X, Gonzàlez J (2007) Face detection in color images using primitive shape features. Computer Recognition Systems 2, vol 45
6. Haj MA et al (2009) Robust and efficient multipose face detection using skin color segmentation. Proceedings of the 4th Iberian conference on pattern recognition and image analysis
7. Chen L-F, Liao H-YM, Lin J-C, Han C–C (2001) Why recognition in a statistics-based face recognition system should be based on the pure face portion: a probabilistic decision-based proof. Pattern Recogn 34(5):1393–1403
8. Lapedriza A, Marin-Jimenez MJ, Vitria J (2006) Gender recognition in non-controlled environment. 18th IEEE international conference on pattern recognition, vol 3, pp 834–837
9. Hu M (1962) Visual pattern recognition by moment invariants. IRE Trans Inf Theory IT-8 179–187
10. Teague MR (1980) Image analysis via the general theory of moments. J Opt Soc Am 70:920–930
11. Khotanzad A, Hong YH (1990) Invariant image recognition by Zernike moments. IEEE Trans Pattern Anal Mach Intell 489–497
12. Ahonen T, Hadid A, Pietikäinen M (2004) Face recognition with local binary patterns. ECCV 2004

A Comparative Study on Routing Protocols in Underwater Sensor Networks

Jian Shen, Jin Wang, Jianwei Zhang and Shunfeng Wang

Abstract Underwater sensor networks (UWSNs) are a class of emerging networks that experience variable and high propagation delays and limited available bandwidth. Because of the different environment under the ocean, the whole protocol stack should be re-designed to fit for the surroundings. In this paper, we only focus on the routing protocols in the network layer. We survey the state-of-the-art routing protocols and give a comparison of them with respect to the important challenging issues in UWSNs. The pros and cons are discussed and compared for the routing protocols.

Keywords Underwater sensor networks (UWSNs) · Routing protocols · Range technology

Introduction

As the network communications technology developing, a new type of networks has appeared in the daily life which named underwater sensor networks (UWSNs). This research area has more attractiveness than ground-based networks because of its distinctive characteristics and the comprehensive applications. UWSNs are an

---

J. Shen (✉) · J. Wang
School of Computer and Software, Jiangsu Engineering Center of Network Monitoring, Nanjing University of Information Science and Technology, Nanjing 210044, China
e-mail: s_shenjian@126.com

J. Zhang
College of Mathematics and Statistics, Nanjing University of Information Science and Technology, Nanjing 210044, China

S. Wang
College of Bin Jiang, Nanjing University of Information Science and Technology, Nanjing 210044, China

Y.-M. Huang et al. (eds.), *Advanced Technologies, Embedded and Multimedia for Human-centric Computing*, Lecture Notes in Electrical Engineering 260,
DOI: 10.1007/978-94-007-7262-5_67,

occasionally connected network and experience frequent and long-duration partitions as well as long delay.

Moreover, UWSNs are very interesting in the ocean exploration applications and very important in military applications. In general, UWSNs are envisioned to enable applications for oceanographic data collection, pollution monitoring, offshore exploration, disaster prevention, assisted navigation and tactical surveillance applications. Multiple UUVs (unmanned underwater vehicles) and AUVs (autonomous underwater vehicles) equipped with underwater sensors, will also find application in exploration of natural undersea resources and gathering of scientific data in collaborative monitoring missions [1]. Sensors and vehicles under water manage and organize by themselves in an autonomous network which can adapt to the characteristics of the ocean environment in order to carry out a great variety of exploration and research missions.

The different environments under the ocean and such distinct features compared with the ground-based networks pose a number of technical challenges in designing the whole protocol stack [2]. In this paper, we only focus on the routing protocols in the network layer. We survey the state-of-the-art routing protocols and give a comparison of them with respect to the important challenging issues in UWSNs. The routing protocols are classified into three categories: proactive, reactive and geographical routing protocols. The pros and cons are discussed and compared for the routing protocols.

The rest of this paper is organized as follows. In the following section, the key properties of UWSNs are reviewed. Routing issues in UWSNs are presented in section Routing Issues in UWSNs. The existing routing protocols are summarized in terms of major features and characteristics in section Routing Protocols, and they are compared and discussed in section Comparison. Finally, the conclusions of this paper are covered in Conclusion.

Key Properties of UWSNs

There are some important key properties of UWSNs which have a great deal of discrepancy from the ground-based networks. They are briefly reviewed in this section.

Acoustic Wireless Communication

It is well-known that radio waves, optical waves and acoustic waves are three different kinds of information carriers. However, radio waves propagate at long distances through conductive sea water only at extra low frequencies. Meanwhile, optical waves do not suffer from the high attenuation but are affected by scattering. Therefore, acoustic wireless communication becomes the best choice in UWSNs.

Variable and High Propagation Delays

Variable and high propagation delays are the fundamental properties of UWSNs. The transmission under water experiences intermittent connectivity, and the transmit rate may be considerably low. Of course, the propagation delay is very high under water because of the underwater environment and asymmetric connection [3].

Limited Available Bandwidth

The available bandwidth is severely limited because of the underwater transmission. In general, the sensors in UWSNs are mobile and battery-operated with wireless connection and, thus, they have limited bandwidth.

Severely Impaired Channel

The multi-path transmission and the transmission fading severely affect the underwater channel. Acoustic waves do not pass through obstructions very well. In addition, even though the beam may be well focused at the transmitter, there is still some divergence during transmission [4].

High Bit Error Rates and Limited Battery Power

High bit error rates and temporary losses of connectivity can be experienced, due to the extreme characteristics of the underwater channel. Battery power is limited and usually batteries cannot be recharged, also because solar energy cannot be exploited.

Fouling and Corrosion

Underwater sensors are prone to failures because of fouling and corrosion. There are lots of corrosive chemical elements under the ocean such as sulfur, nitrogen, chlorine, bromine and so on. Hence, the sensors are easily destroyed or broken down by oxidative decomposition.

Routing Issues in UWSNs

The peculiar properties of UWSNs inevitably raise a number of interesting issues which are summarized in this section [5].

Routing Objective

The most important routing objective in UWSNs is to minimize the transmission delay. To minimize the communication signaling overhead and establish energy efficient paths are also important routing objectives. Because of the complex environment under water, as we known, the communication is asymmetric.

Reliability

For the reliable delivery of data in UWSNs, any routing protocol should have some acknowledge, which can ensure successful and stable delivery of data. For example, when a message correctly reaches to a destination, some acknowledge messages should be sent back from destination to source for later use.

Energy

Because of the severe environment underwater, it is impossible to change the sensor nodes frequently. It is also difficult to connect the underwater sensor nodes to the power station. Hence, the nodes in UWSNs are usually lack of energy. During the message routing lots of energy should be consumed for sending, receiving and storing messages as well as performing computation. Therefore, the energy-efficient design of routing protocols is of importance.

Security

Security is always an important issue not only in the UWSNs but also in all the traditional networks. The cryptographic techniques may be beneficial for secure end-to-end routing. The security in routing protocols is still an open issue to be studied.

Routing Protocols

In general, the existing routing protocols are usually divided into three categories, namely proactive, reactive and geographical routing protocols. Proactive routing protocols are traditional distributed shortest-path protocols, which maintain routes between every host pair at all time. Based on periodic updating, the routing overhead is very high by utilizing proactive routing protocols. Reactive routing protocols determine route if and when needed, in which source initiates route discovery. Geographical routing protocols establish source–destination path by the localization information, in which the sender uses a location service to determine the position of the destination. And the routing process at each node is based on the destination's location within the packet header and the location of the forwarding node's neighbors.

Proactive Routing Protocols

The proactive routing protocols attempt to minimize the message latency induced by route discovery, by maintaining up-to-date routing information at all times from each node to every other node. This is obtained by broadcasting control packets that contain routing table information. These protocols provoke a large signaling overhead to establish routes for the first time and each time the network topology is modified because of mobility or node failures, since updated topology information has to be propagated to all the nodes in the network. This way, each node is able to establish a path to any other node in the network, which may not be needed in UWSNs. For this reason, proactive protocols maybe not suitable for underwater networks.

Destination-Sequenced Distance-Vector Protocol

DSDV is a proactive hop-by-hop distance vector routing protocol. Every host maintains a routing table for all the possible destinations and the number of hops to each destination. Meanwhile, each host broadcasts routing updates periodically in order to achieve the newest and the most accurate routing table [6].

Wireless Routing Protocol

In wireless routing protocol, sensor nodes keep each other informed of all link changes through the use of update messages [7]. Obviously, WRP is a table-driven protocol with the goal of maintaining routing information among all sensor nodes.

Reactive Routing Protocols

In reactive routing protocols, a node initiates a route discovery process only when a route to a destination is required. Once a route has been established, it is maintained by a route maintenance procedure until it is no longer desired. These protocols are more suitable for dynamic environments but incur a higher latency and still require source-initiated flooding of control packets to establish paths. Reactive protocols are deemed to be unsuitable for UWSNs as they cause a high latency in the establishment of paths, which may be even amplified underwater by the slow propagation of acoustic signals. Furthermore, links are likely to be asymmetrical, due to bottom characteristics and variability in sound speed channel. Hence, protocols that rely on symmetrical links, such as most of the reactive protocols, are unsuited for the underwater environment.

Dynamic Source Routing Protocol

In brief, source routing means the source node is in charge of the whole transmitting and determines the path based on the topology of the network before the message gets into the node. The source must know all the intermediate nodes to be traversed from the source to a destination.

Ad Hoc On-demand Distance Vector Routing Protocol

AODV is an improvement on DSDV because it minimizes the number of the required broadcasts by creating routes on demand basis [8]. It carries out the route discovery by using on-demand mechanism and maintains from DSR.

Temporarily Ordered Routing Algorithm

TOAR is a highly adaptive loop-free distributed routing algorithm. It minimizes reaction due to topological changes while exhibits multipath routing capability [7].

Geographic Routing Protocols

These protocols establish source–destination path by the localization information. Each node selects its next hop based on the position of its neighbors and of the destination node. In fact, fine-grained localization usually requires strict synchronization among nodes, which is difficult to achieve underwater due to the variable propagation delay. Virtual circuit routing techniques can be considered in

UWSNs. In these techniques, paths are established a priori between each source and sink, and each packet follows the same path. Localization schemes are the most important issues in geographic routing protocols [9].

Routing Protocols with Respect to Infrastructure-Based Localization Schemes

Infrastructure-based (anchor-based) localization systems are similar to the GPS scheme. The distance to multiple anchor nodes is computed by using the propagation time of the sound signals between the sensor or the AUV and the anchors.

Routing Protocols Based on Distributed Positioning Schemes

In distributed positioning schemes, nodes are able to communicate only with their one-hop neighbors and compute the distances to their one-hop neighbors by making RSSI or ToA measurements.

Routing Protocols with Respect to Mobile Beacon-Based Schemes

In this scheme, a mobile beacon traverses the sensor network while broadcasting beacon packets, which contain the location coordinates of the beacon. RSSI measurements of the received beacon packets are used for ranging purposes.

Routing Protocols with Respect to Hop Count-Based Schemes

In brief, the anchor nodes are placed at the corners or along the boundaries of a square grid. In order to estimate the distance to the landmark by approximately knowing one hop distance, each node maintains and updates a table in term of hop number.

Routing Protocols Based on Centroid Scheme

In this scheme, anchor nodes are placed to form a rectangular mesh. The location of the node is then estimated to be the centroid of the anchor nodes that it can receive beacon packets.

Table 1 Comparison of the three different kinds of routing protocols

	Flexibility	Route acquisition	Resource usage	Multipath support	Flood for route discovery	Latency	Overhead	Routing table	Effectiveness
DSDV	Bad	Compute a priori	High	No	No	Short	High	Yes	Bad
WRP	Bad	Compute a priori	High	No	No	Short	High	Yes	Bad
DSR	Normal	On-demand	Normal	Not explicitly	Yes	Long	Normal	Yes	Bad
AODV	Normal	On-demand	Normal	Not directly	Yes	Long	Normal	Yes	Bad
TORA	Normal	On-demand	Normal	Yes	Yes	Long	Normal	Yes	Bad
Geographic routing protocols	Good	Compute a priori	Low	No	No	Normal	Low	No	Good

Routing Protocols with Respect to Area-Based Localization Scheme

In very large and dense wireless sensor networks, a coarse estimate of the sensors' locations may suffice for most applications. ALS [10] and APIT [11] approximate the area in which a node is located, rather than the exact location.

Comparison

We compare the state-of-the-art routing protocols proposed so far in the literature. First, we evaluate the three kinds of routing protocols such as proactive routing protocols, reactive routing protocols and geographic routing protocols, in terms of various characteristics including important performance metrics. Flexibility, route acquisition, resource usage, multipath support, flooding for route recovery, latency, overhead, routing vector/table, and effectiveness are studied in the comparative analysis. Table 1 summarizes the comparison results.

From the comparison table, some conclusive comments can be inferred: The geographic routing protocols are most suitable for underwater communication. Of the geographic routing protocols, ALS can be primarily chosen thanks to its many outstanding features although it has not accurate information.

Conclusion

Routing in UWSNs is a new area of research, with a limited but rapidly growing set of research results. The routing protocols have the common objective of trying to increase the delivery ratio while decreasing the resource consumption and latency. In this paper, we have presented a comparative survey of various routing techniques in UWSNs. Meanwhile, we have studied about some localization schemes on geographic routing protocols. The advantages and disadvantages of the routing protocols have been discussed with comparison results as well. Although many routing protocols have been studied so far, there are still many challenges to be solved. For example, we should make the routing protocol more scalable for a large network. Loop freedom, energy conservation, and efficient resource usage should be also addressed. Our future work is to design a robust routing protocol with good localization schemes for harsh operational environments.

Acknowledgments This work was supported by the research fund from Nanjing University of Information Science and Technology under Grant No. S8113003001, National Science Foundation of China under Grant No. 61272421.

References

1. Akyildiz F, Pompili D, Melodia T (2004) Underwater acoustic sensor networks: research challenges
2. Rahman RH, Benson C, Frater M (2012) Routing protocols for underwater ad hoc networks. OCEANS, 2012—Yeosu, pp 1–7
3. Proakis JG, Rice JA, Sozer EM, Stojanovic M (2003) Shallow water acoustic networks. Encyclopedia of Telecommunications
4. Tanenbaum AS (2003) Computer networks. Fourth edition, pp 104
5. Yang X, Ong KG, Dreschel WR, Zeng K, Mungle CS, Grimes CA (2002) Design of a wireless sensor network for long-term, in situ monitoring of an aqueous environment. Sensors, pp 455–472
6. Perkins CE, Bhagwat P (1994) Highly dynamic destination-sequenced distance-vector routing (DSDV) for mobile computers. ACM SIGCOMM Commun Rev
7. Cordeiro CM, Agrawal DP (2006) Ad Hoc and sensor networks, pp 23–24
8. Perkins C, Belding-Royer E, Das S (2003) Ad hoc on-demand distance vector (AODV) routing. Internet RFCs
9. Chandrasekhar V, Seah WKG (2006) Localization in underwater sensor networks—survey and challenges. In: International conference on mobile computing and networking, pp 33–40
10. Chandrasekhar V, Seah WKG (2006) Area localization scheme for underwater sensor networks. In: the IEEE OCEANS Asia Pacific Conference
11. He T (2003) Range-free localization schemes for large scale sensor networks. In: 9th ACM international conference on mobile computing and networking (Mobicom2003)

A Novel Verifiably Encrypted Signature from Weil Pairing

Jian Shen, Jin Wang, Yuhui Zheng, Jianwei Zhang and Shunfeng Wang

Abstract A novel efficient verifiably encrypted signature (VES) which makes use of the Weil pairing is presented in this paper. VES can be used in optimistic contract signing protocols to enable fair exchange. Compared with the previous schemes, our proposed VES scheme is more efficient.

Keywords Verifiably encrypted signature · Weil pairing · Fair exchange

Introduction

The verifiably encrypted signature (VES) was first proposed by Asokan et al. [1], which can keep well the fairness of the trading. The realization of VES relies on the trusted third party which needs not to join the exchange protocol in the online mode. Verifiably encrypted signatures are used in optimistic contract signing protocols [1, 2] to enable fair exchange. For example, by using VES, a user Alice can give Bob a signature on a message encrypted by a third party's public key. Meanwhile Bob can verify the encrypted signature after he receives it but cannot deduce any information of the ordinary signature. In 2003, Boneh et al. [3] first proposed a practical verifiably encrypted signature scheme based on bilinear maps.

J. Shen (✉) · J. Wang · Y. Zheng
School of Computer and Software, Jiangsu Engineering Center of Network Monitoring, Nanjing University of Information Science and Technology, Nanjing 210044, China
e-mail: s_shenjian@126.com

J. Zhang
School of Mathematics and Statistics, Nanjing University of Information Science and Technology, Nanjing 210044, China

S. Wang
College of Bin Jiang, Nanjing University of Information Science and Technology, Nanjing 210044, China

Y.-M. Huang et al. (eds.), *Advanced Technologies, Embedded and Multimedia for Human-centric Computing*, Lecture Notes in Electrical Engineering 260,
DOI: 10.1007/978-94-007-7262-5_68, © Springer Science+Business Media Dordrecht 2014

Recently, in [4], Ming proposed a security model of verifiably encrypted signature schemes. In his security model, a trusted third party (TTP) called adjudicator is used which generates a pair of public key and private key. The public key serves as a public parameter of the system and the corresponding private key kept secretly by the adjudicator is used to resolve the possible dispute between two trading parties.

In this paper, we develop a more efficient verifiably encrypted signature from the Weil pairing. Our scheme requires less computation complexity compared with the previous schemes.

Preliminaries

In this section, we briefly introduce the Weil pairing which is necessary for description of our signature scheme.

Weil pairing: Let p be a prime number such that $p = 12q - 1$ for some prime number q and E a super singular elliptic curve defined by the Weierstrass equation $y^2 = x^3 + 1$ over F_P. The set of rational points $E(F_p) = \{(x, y) \in F_p \times F_p : (x, y) \in E\}$ forms a cyclic group of order $p + 1$. Furthermore, because $p + 1 = 12q$ for some prime number q, the set of points of order q in $E(F_p)$ form a cyclic subgroup, denoted as G_1. Let $\mathcal{G}$ be the generator of G_1. Let G_2 be the subgroup of F_{p^2} containing all elements of order q. The modified Weil pairing [5] is a map:

$$\widehat{e} : G_1 \times G_1 \rightarrow G_2,$$

which has the following properties:

- Bilinear: For any $\mathcal{P}, \mathcal{Q} \in G_1$ and a, $b \in Z$, we have $\widehat{e}(aP, bQ) = \widehat{e}(P, Q)^{ab}$.
- Non-degenerate: if $\mathcal{G}$ is a generator of G_1, then $\widehat{e}(\mathcal{G}, \mathcal{G}) \in F^*_{p^2}$ is a generator of G_2.
- Computable: Given $\mathcal{P}, \mathcal{Q} \in G_1$, there is an efficient method to compute $\widehat{e}(P, Q) \in G_2$.

Efficient Verifiably Encrypted Signature from Weil Pairing

There are three entities in our signature scheme: user, verifier and adjudicator. Our signature scheme runs in seven algorithms: *KeyGen*, *Sign*, *Verify*, *AdjKeyGen*, *VESSign*, *VESVerify*, and *Adjudication*. The algorithms are described as follows:

- *KeyGen*: A user randomly picks $x \in Z_p^*$ as its private key. The user's public key is computed as: $x \cdot P = P_{pub} = (x_p, y_p)$.

- *Sign*: Given the private key x of the user, the message $m \in (0,1)^l$, the hash function H and the public key $P_{pub} = (x_p, y_p)$, the use signs a signature σ on m: $\sigma = \frac{H(m)}{x+x_p} \cdot P$
- *Verify*: Given the public key $P_{pub} = (x_p, y_p)$, the message m and the signature σ, the verifier verifies whether $\hat{e}\left(\sigma, \frac{P_{pub}+x_p \cdot P}{H(m)}\right) = \hat{e}(P,P)$.
- *AdjKeyGen*: The adjudicator randomly picks $y \in Z_p^*$ as its private key, and computes the corresponding public key as $P_{pub}{}' = y \cdot P$.
- *VESSign*: Given the user's private key x, the message m and the adjudicator's public key $P_{pub}{}'$, the user computes the verifiably encrypted signature σ_{VES} as $\sigma_{VES} = \frac{H(m)}{x+x_p} \cdot P_{pub}{}'$.
- *VESVerify*: Given the verifiably encrypted signature σ_{VES}, the message m, the public key of user $P_{pub} = (x_p, y_p)$, and the public key of adjudicator $P_{pub}{}'$, the verifier verifies whether $\hat{e}\left(\sigma_{VES}, \frac{P_{pub}+x_p \cdot P}{H(m)}\right) = \hat{e}\left(P_{pub}{}', P\right)$ can hold.
- *Adjudication*: Given the verifiably encrypted signature σ_{VES}, the message m, and the private key of adjudicator, the adjudicator extracts the original signature on the message m as: $\sigma = \frac{\sigma_{VES}}{y}$.

Validity Analysis

Validity requires that verifiably encrypted signatures are able to be successfully verified as ordinary signature and adjudicated verifiably encrypted signatures are also able to be successfully verified as ordinary signature. This means *VESVerify*(*m*, *VESSign*(*m*)) and *Verify*(*m*, *Adjudication*(*VESSign*(*m*))) hold for all m and for all properly generated key pairs and adjudicator key pairs.

For a verifiably encrypted signature σ_{VES} on a message m, the validity is easily proven as follows:

$$\begin{aligned}
\hat{e}\left(\sigma_{VES}, \frac{P_{pub} + x_p \cdot P}{H(m)}\right) &= \hat{e}\left(\frac{H(m)}{x + x_p} \cdot P_{pub}{}', \frac{P_{pub} + x_p \cdot P}{H(m)}\right) \\
&= \hat{e}\left(\frac{H(m)}{x + x_p} \cdot P_{pub}{}', \frac{x \cdot P + x_p \cdot P}{H(m)}\right) \\
&= \hat{e}\left(\frac{H(m)}{x + x_p} \cdot P_{pub}{}', \frac{x + x_p}{H(m)} \cdot P\right) \\
&= \hat{e}\left(P_{pub}{}', P\right)
\end{aligned}$$

Hence, *VESVerify*(*m*, *VESSign*(*m*)) = 1 holds. On the other hand,

Table 1 Performance comparison

	Boneh's scheme	Ming's scheme	Our scheme
Size	320 bits	320 bits	320 bits
VESSign	3 M	3 M	1 M
VESVerify	3e	3 M + 1e	1 M + 1e
Adjudication	1 M	1 M	1 M

$$\begin{aligned}\widehat{e}\left(\sigma,\frac{P_{pub}+x_p\cdot P}{H(m)}\right)&=\widehat{e}\left(\frac{H(m)}{x+x_p}\cdot P,\frac{P_{pub}+x_p\cdot P}{H(m)}\right)\\&=\widehat{e}\left(\frac{H(m)}{x+x_p}\cdot P,\frac{x\cdot P+x_p\cdot P}{H(m)}\right)\\&=\widehat{e}\left(\frac{H(m)}{x+x_p}\cdot P,\frac{x+x_p}{H(m)}\cdot P\right)\\&=\widehat{e}(P,P)\end{aligned}$$

Therefore, *Verify*(m, *Adjudication*(*VESSign*(m))) = 1 holds.

Performance Analysis

We compare our proposed signature scheme with previous schemes [3, 4] in Table 1. Note here that M means a scalar point multiplication and e indicates a pairing computation. In the proposed scheme, the size of the VES is 320 bits, which has the same security level compared with 1,024-bit RSA signature. As described above, there are seven algorithms in our scheme including *KeyGen*, *Sign*, *Verify*, *AdjKeyGen*, *VESSign*, *VESVerify*, and *Adjudication*. The costs of the first three algorithms in our scheme are almost same as that in the previous schemes. Hence, we focus only on comparing the cost of *VESSign*, *VESVerify*, and *Adjudication*. It is worth noting that, in our scheme, we can pre-compute $\widehat{e}(P_{pub}', P)$ in *VESVerify* phase. Therefore, there is only one pairing computation. From Table 1, we can conclude that our scheme is more efficient than Boneh's and Ming's schemes.

Conclusion

Verifiably encrypted signature is very important and useful cryptographic primitives. We proposed a new verifiably encrypted signature scheme based on bilinear pairings. We show that our scheme is more efficient than previous schemes.

Acknowledgments This work was supported by the research fund from Nanjing University of Information Science and Technology under Grant No. S8113003001 and Natural Science Foundation of Jiangsu Province under Grant No. BK2011825.

References

1. Asokan N, Shoup V, Waidner M (2000) Optimistic fair exchange of digital signature. IEEE J. Selected Areas in Comm. 18(4):593–610
2. Bao F, Deng R, Mao W (1998) Efficient and practical fair exchange protocols with offline TTP. In: IEEE symposium on security and privacy, pp 77–85
3. Boneh D, Gentry C, Lynn B, Shacham H (2003) Aggregate and verifiably encrypted signatures from bilinear maps. In: Eurocrypt'03, LNCS 2656, pp 416–432
4. Ming Y, Wang Y (2009) An efficient verifiably encrypted signature scheme without random oracle. Int J Network Security 8(2):125–130
5. Boneh D, Franklin M (2001) Identity-based encryption from the Weil pairing. In: Crypto'01, Santa Barbara, CA, USA, pp 213–229

Location-Aware Routing Protocol for Underwater Sensor Networks

Jian Shen, Jin Wang, Jingwei Wang, Jianwei Zhang and Shunfeng Wang

Abstract As the network communications technology developing, a new type of networks has appeared in the daily life which is named underwater sensor networks (UWSNs). Routing protocols in UWSNs should ensure the reliability of message transmission, not just decrease the delay. In this paper, we propose a novel routing protocol named Location-Aware Routing Protocol (LARP) for UWSNs, where the location information of nodes are used to help the transmission of the message. Simulation results show that the proposed LARP outperforms the existing routing protocols in terms of packet delivery ratio and normalized routing overhead.

Keywords Underwater sensor networks (UWSNs) · Location-aware · Anchor node

Introduction

Underwater sensor networks (UWSNs) are a class of emerging networks that experience variable and high propagation delays and limited available bandwidth. Compared with ground-based networks, UWSNs has more attractiveness due to its

J. Shen (✉) · J. Wang · J. Wang
School of Computer and Software, Jiangsu Engineering Center of Network Monitoring, Nanjing University of Information Science and Technology, Nanjing 210044, China
e-mail: s_shenjian@126.com

J. Zhang
School of Mathematics and Statistics, Nanjing University of Information Science and Technology, Nanjing 210044, China

S. Wang
College of Bin Jiang, Nanjing University of Information Science and Technology, Nanjing 210044, China

Y.-M. Huang et al. (eds.), *Advanced Technologies, Embedded and Multimedia for Human-centric Computing*, Lecture Notes in Electrical Engineering 260,
DOI: 10.1007/978-94-007-7262-5_69,

distinctive characteristics and the comprehensive applications. UWSNs are very interesting in the ocean exploration applications and very important in military applications, such as oceanographic data collection, pollution monitoring, offshore exploration, disaster prevention, assisted navigation and tactical surveillance applications [1]. In addition, Multiple unmanned underwater vehicles (UUVs) and autonomous underwater vehicles (AUVs) equipped with underwater sensors will also find application in exploration of natural undersea resources and gathering of scientific data in collaborative monitoring missions [2].

UWSNs have great potential and contain enormous values in economic and social field [3]. Sensors and vehicles under water manage and organize by themselves in an autonomous network which can adapt to the characteristics of the ocean environment in order to carry out a great variety of explore and research missions [4]. Because of the different environment under the ocean, the routing protocol should be re-designed to fit for the surroundings. However, the different environments under the ocean and such distinct features compared with the ground-based networks pose a number of technical challenges in designing the routing protocol [5]. In this paper, we propose a novel routing protocol named Location-Aware Routing Protocol (LARP) for UWSNs, where the location information of nodes are used to help the message transmission. Resort to a range-finding technique called received signal strength indicator (RSSI) [6], a node can easily obtain its location information [7]. Simulation results show that the presented LARP outperforms the existing routing protocols in terms of packet delivery ratio and normalized routing overhead.

The rest of this paper is organized as follows. In the following section, related works on routing protocols in UWSNs are briefly discussed. A Novel Location-Aware Routing Protocol for UWSNs is described in detail in section A Novel Location-Aware Routing Protocol for UWSNs. Simulations and results are presented in section Simulations and Results. Finally, the conclusions of this paper are covered in section Conclusions.

Related Works

Compared with ground-based networks, UWSNs has the following key properties: (1) acoustic wireless communication, (2) variable and high propagation Delays, (3) limited available bandwidth, (4) severely impaired channel, (5) high bit error rates and limited battery power, (6) fouling and corrosion. Some researchers have made a lot of effort in designing new protocols in this area. In general, the routing protocols in UWSNs are classified into three categories: proactive, reactive and geographical routing protocols.

Proactive Routing Protocols

The proactive routing protocols attempt to minimize the message latency induced by route discovery. This is obtained by broadcasting control packets that contain routing table information. These protocols provoke a large signaling overhead to establish routes for the first time. The representative of this category is destination-sequenced distance-vector (DSDV) [8] protocol.

Reactive Routing Protocols

In reactive routing protocols, a node initiates a route discovery process only when a route to a destination is required. Once a route has been established, it is maintained by a route maintenance procedure until it is no longer desired. These protocols are more suitable for dynamic environments. The representative of this category is ad hoc on-demand distance vector routing (AODV) [9] protocol.

Geographic Routing Protocols

These protocols establish source–destination path by the localization information. Each node selects its next hop based on the position of its neighbors and of the destination node. Localization schemes are the most important issues in geographic routing protocols [10]. The representative of this category is area localization scheme (ALS) [11] for UWSNs.

A Novel Location-Aware Routing Protocol for UWSNs

In this section, LARP is described in detail. We utilize anchor nodes to estimate the location information of general nodes. We determine the best next hop to relay the message by location information. In our protocol, anchor nodes equipped with GPS traverse the sensor network and broadcast beacon packets, which contain the location coordinates. RSSI measurements of the received beacon packets are used for ranging purposes. General nodes estimate the location information by cooperating with at least three anchor nodes. Every node stores its own location information. When an anchor node is situated in the transmission range of a certain general node, the information of this general node can be stored in this anchor node.

The proposed routing protocol has two steps. At the beginning of routing, the location information of destination node should be obtained by the source first. Suppose that there is a message transmitted from the source (S) to the destination

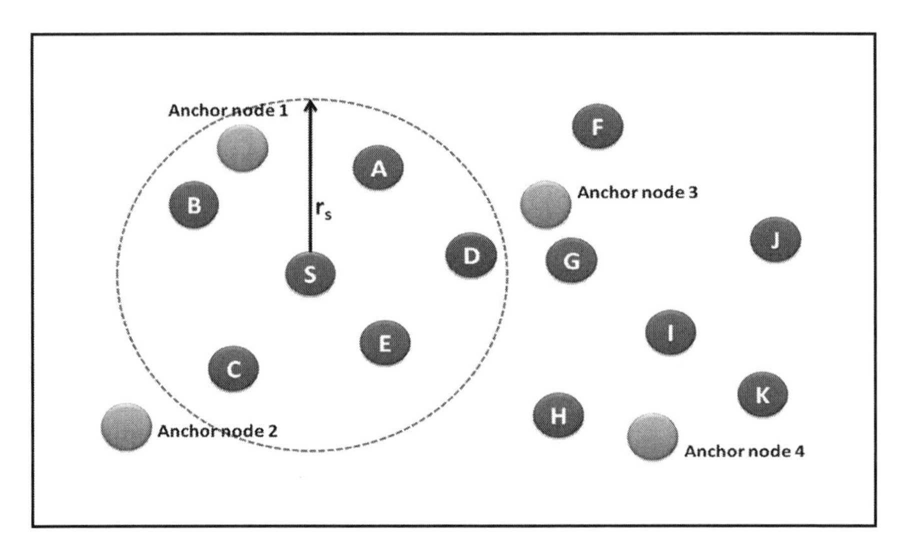

Fig. 1 Node S broadcasts "destination location" request

(D). If there is an anchor node in the transmission range of node S, S can request the anchor node to find the destination's location. Otherwise, node S will wait until an anchor node appears in its transmission range. After this anchor node broadcasts the ID of the destination node, all the other anchor nodes will check their lists to find the destination node. If one anchor node finds the destination node, the source can obtain the information about it.

After node S getting the information of the destination D, the second step is determining the next hop for this transmission. In the beginning, node S broadcasts "destination location" request. As shown in Fig. 1. If node D is in the transmission range of S, then D replies to S before S directly transmitting the message to D. Otherwise, no node replies to S, and node S broadcasts the "moving direction" request. All the information of nodes in the transmission range of S is collected by S through directly communicating with these nodes. As shown in Fig. 2, node A's moving direction is same as the message's transmission direction, so node A replies to S and the message is delivered to A immediately. That is to say, A becomes the best next hop. Note that the moving direction information can be easily calculated by the location information at different times. If two nodes have the same moving direction as the message's transmission direction, then the node with higher speed can only become the next hop. If all the moving directions of nodes in the transmission range of S are different from the message's transmission direction, then no node replies to S. Therefore, node S will wait.

Finally, the best next hop is decided. The message is delivered and stored in this intermediate node, which continues to determine the next hop until the message successfully arriving at the destination node.

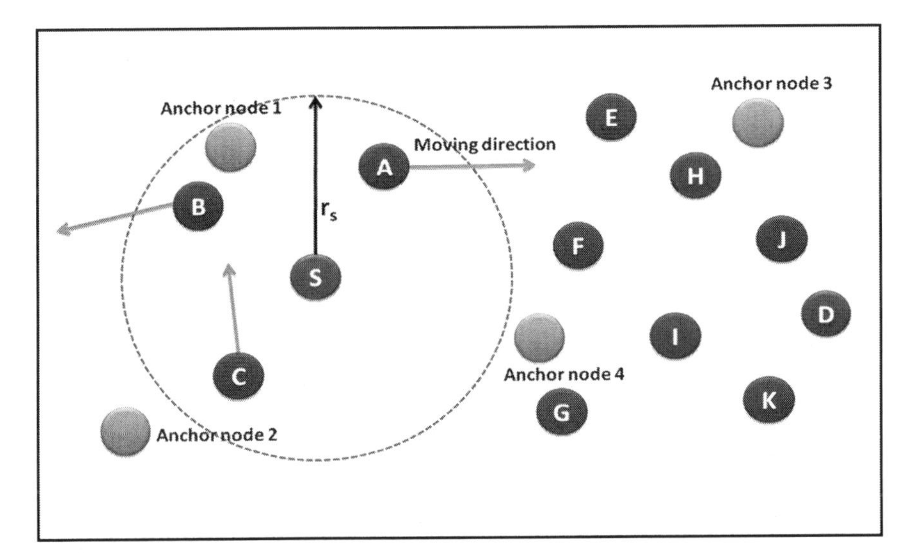

Fig. 2 Node A replies to node S

Simulations and Results

Simulation Environment

We implemented LARP by using the ns-2 simulator [12]. IEEE 802.11 [13] Medium Access Control (MAC) protocol is implemented. We model 50 nodes (including 10 % anchor nodes) in a square area 1,000 × 1,000 m during the simulation time 1,000 s. Each node moves with a speed uniformly distributed between 0 and 5 m/s. The radio transmission range is assumed to be 250 m and a two ray ground propagation channel is assumed. Most other parameters use ns-2 defaults. Three performance metrics of packet delivery ratio, delivery delay and normalized routing overhead are compared. For measuring the three metrics, two simulation factors of the pause time and the transmission rate are varied in a meaningful range.

Results and Discussion

We present a comparative simulation analysis of LARP with DSDV, AODV and ALS. We first analyze the packet delivery ratio. As shown in Fig. 3, the packet delivery ratio gradually increases as the pause time increasing. The curve of LARP shows that the packet delivery ratio persistently increases as the pause time raising. Figure 4 describes the change of packet delivery ratio as the transmission rate

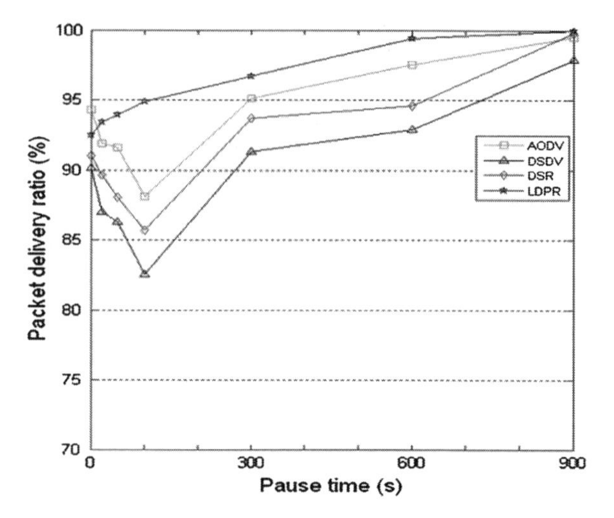

Fig. 3 Packet delivery ratio versus pause time

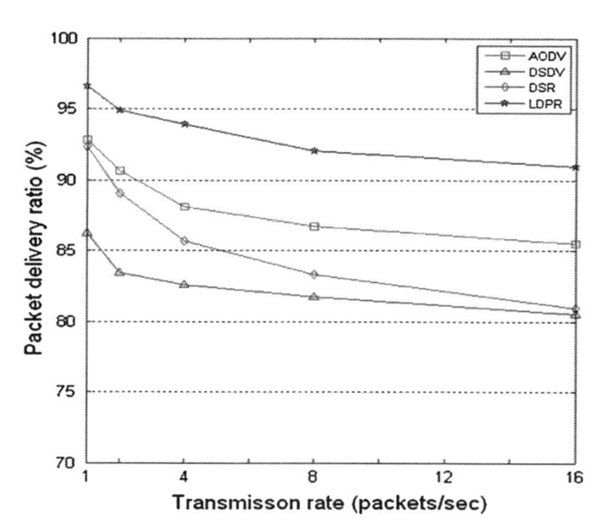

Fig. 4 Packet delivery ratio versus transmission rate

increasing. The packet delivery ratio reduces as the transmission rate raising. Here, LARP performs the highest packet delivery ratio under various transmission rates.

Another critical aspect we investigated is the normalized routing overhead. Fig. 5 shows that the routing overhead decreases with the pause time increasing. In particular, the change of LARP's curve is small. Note that when the pause time is more than 300 s, the routing overhead of LARP is a little higher than AODV and ALS. That's because the simulation environment trends to be static and the mobility of node declines. Figure 6 presents that the routing overhead increases as the transmission rate increasing. In addition, seen from the shape of the LARP's curve, the routing overhead of LARP always maintains a low level.

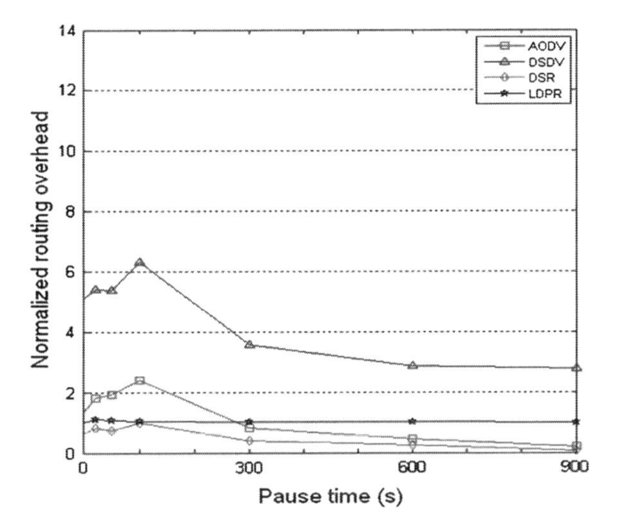

Fig. 5 Normalized routing overhead versus pause

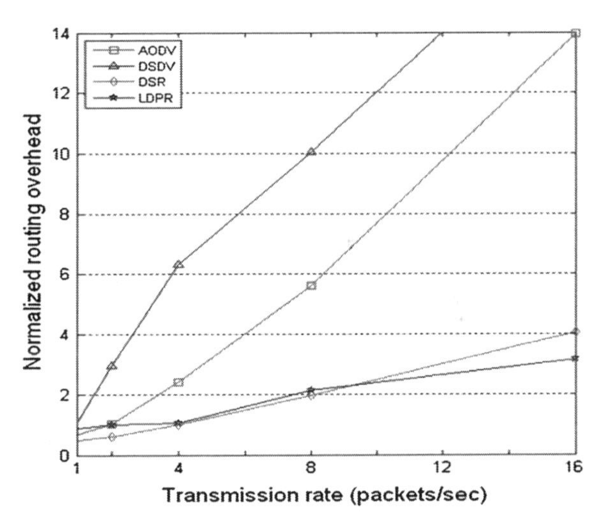

Fig. 6 Normalized rout-ing overhead versus trans-mission rate

It is still of interest to consider the end-to-end delay. Figures 7 and 8 show that the packet delivery delay of LARP is longer than other three routing protocols. In LARP, nodes are required enough time to obtain the information of location. The feature of packet delivery delay in LARP determines that LARP can be only implemented in the environment which focuses on the validity and integrity of the message rather than delivery delay.

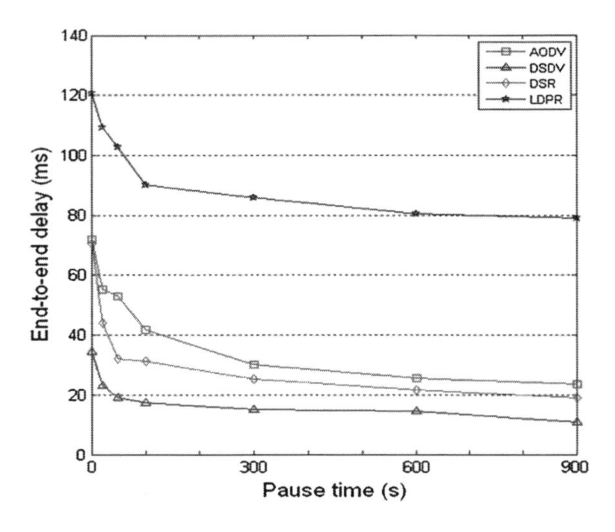

Fig. 7 End-to-End delay versus pause time

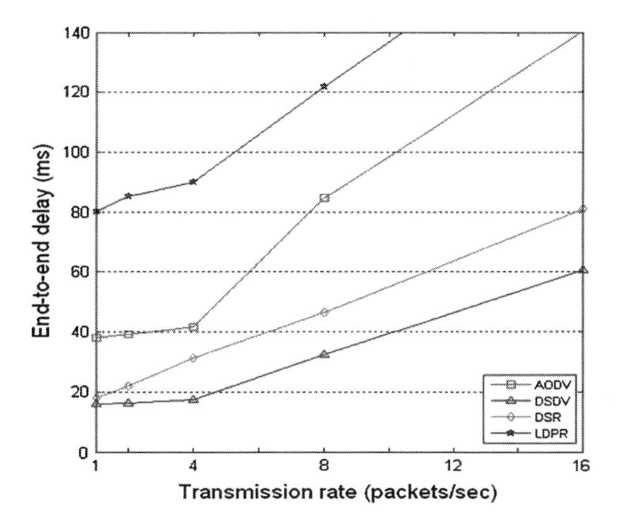

Fig. 8 End-to-End delay versus transmission rate

Conclusions

In this paper, we have proposed a routing protocol named LARP for UWSNs, which utilizes the location information of nodes to transmit a message. Resort to a range-finding technique RSSI, a node can easily obtain its location information. The simulation experiments have shown that LARP is able to ensure the reliability of message transmission. It is worth noting that LARP can be implemented in the environment which focuses on the reliability and validity of message transmission rather than delivery delay.

Acknowledgments This work was supported by the research fund from Nanjing University of Information Science and Technology under Grant No. S8113003001 and Natural Science Foundation of Jiangsu Province under Grant No. BK2011825.

References

1. Akyildiz F, Pompili D, Melodia T (2004) Underwater acoustic sensor networks: research challenges
2. Tanenbaum, A.S.: Computer networks. Fouth edition, pp.104 (2003)
3. Proakis JG, Rice JA, Sozer EM, Stojanovic M (2003) Shallow water acoustic networks. Encyclopedia of Telecommunications
4. Rahman, R.H., Benson, C., Frater, M.: Routing protocols for underwater ad hoc networks. OCEANS, 2012 - Yeosu, pp. 1-7 (2012)
5. Yang X, Ong KG, Dreschel WR, Zeng K, Mungle CS, Grimes CA (2002) Design of a wireless sensor network for long-term, in situ monitoring of an aqueous environment. Sensors, pp 455–472
6. He, T.: Range-free Localization Schemes for Large Scale Sensor Networks. Proceedings of the 9th ACM International Conference on Mobile Computing and Networking (Mobicom2003) (2003)
7. Cordeiro CM, Agrawal DP (2006) Ad Hoc and sensor networks, pp 23–24
8. Perkins CE, Bhagwat P (1994) Highly dynamic Destination-Sequenced Distance-Vector routing (DSDV) for mobile computers. ACM SIGCOMM Commun Rev
9. Perkins C, Belding-Royer E, Das S (2003) Ad hoc On-Demand Distance Vector (AODV) routing. Internet RFCs
10. Chandrasekhar V, Seah WKG (2006) Localization in underwater sensor networks: survey and challenges. In: International conference on mobile computing and networking, pp 33–40
11. Chandrasekhar V, Seah WKG (2006) Area Localization Scheme for Underwater Sensor Networks. Proceedings of the IEEE OCEANS Asia Pacific Conference (2006)
12. CMU Monarch Project. The CMU Monarch Project's wireless and mobility extensions to ns. ftp.monarch.cs.cmu.edu/pub/monarch/wireless-sim/ns-cmu.ps (1999)
13. IEEE Computer Society. Wireless LAN Medium Access Control (MAC) and Physical Layer (PHY) Specifications (1997)

Efficient Key Management Scheme for SCADA System

Jian Shen, Jin Wang, Yongjun Ren, Jianwei Zhang and Shunfeng Wang

Abstract Currently Supervisory Control And Data Acquisition (SCADA) system intends to be connected to the open operating environment. Thus, protecting SCADA systems from malicious attacks is getting more and more attention. A key management scheme is essential for secure SCADA communications. In this paper, we propose an efficient key management scheme for SCADA systems with good security properties and performance.

Keywords Supervisory control and data acquisition (SCADA) · Key management · Secure SCADA communications

Introduction

In order to deliver critical services, such as water, sewerage and electricity distribution, nations are increasingly depends on Supervisory Control And Data Acquisition (SCADA) systems. As the change of the operating environment in SCADA system from close to open, the risk of SCADA incidents occurring is increasing. Nowadays, SCADA system has been exposed to a wide range of network security problems. If the SCADA system is damaged from the attacks,

J. Shen (✉) · J. Wang · Y. Ren
School of Computer and Software, Jiangsu Engineering Center of Network Monitoring, Nanjing University of Information Science and Technology, Nanjing 210044, China
e-mail: s_shenjian@126.com

J. Zhang
School of Mathematics and Statistics, Nanjing University of Information Science and Technology, Nanjing 210044, China

S. Wang
College of Bin Jiang, Nanjing University of Information Science and Technology, Nanjing 210044, China

Y.-M. Huang et al. (eds.), *Advanced Technologies, Embedded and Multimedia for Human-centric Computing*, Lecture Notes in Electrical Engineering 260,
DOI: 10.1007/978-94-007-7262-5_70,

this system can have a widespread negative effect to society. One critical security requirement for SCADA systems is that communication channels need to be secured. Secure keys need to be established before cryptographic techniques can be used to secure communications.

Note that un-encrypted data communication via networks is vulnerable to several types of attacks. Therefore, secure data communication between each device is required to secure the SCADA system. Secure key management is essential for data encryption. In this paper, we focus on the key management scheme for SCADA systems and propose an efficient key management scheme (EKMS) with good security properties. Compared with the previous schemes, the presented key management scheme is more efficient in terms of the communication cost. Our scheme is based on a symmetric balanced incomplete block design (SBIBD), which can provide the authentication service and resist different key attacks. The structure of SBIBD makes the computation of a common conference key for each remote terminal unit (RTU) quite convenient.

The rest of this paper is organized as follows: In the following section, related work is briefly introduced. The proposed key management scheme is described in detail in section Efficient Key Management Scheme for SCADA System. Security analysis and performance analysis of our scheme are presented in section Security Analysis and Performance Analysis. Finally, the conclusions of this paper are covered in section Conclusions.

Related Work

A SCADA system consists of three types of equipment communicating with each other: (1) human–machine interface (HMI) that operators interact with; (2) master terminal unit (MTU) that provides supervisory control of an RTU; and (3) the remote terminal unit (RTU) that interacts with the physical environment. In this paper, the term node will be used to refer to any entity in the system. The structure of SCADA systems is based on master–slave structure, which is shown in Fig. 1. The structure of a SCADA system will normally include one central MTU, which communicates with a hierarchy of other nodes, including Sub-MTU and RTUs. Master stations and sub-master stations, are computers with resources at least as plentiful as a modern desktop computer.

Recently, SKE [1] was proposed by Sandia, where the MTU has to encrypt data with each key of the RTUs individually to broadcast a message. After that, SKMA [2] was proposed, where two types of keys must be managed by an MTU or RTUs. The long-term node-key distribution center (KDC) key is shared between the KDC and a node. The other key is the long-term node–node key shared between two nodes. Later, ASKMA [3] proposed a key-management scheme suitable for secure SCADA communication using a logical key hierarchy to support broadcast communication and multicast communication, but it may be less efficient.

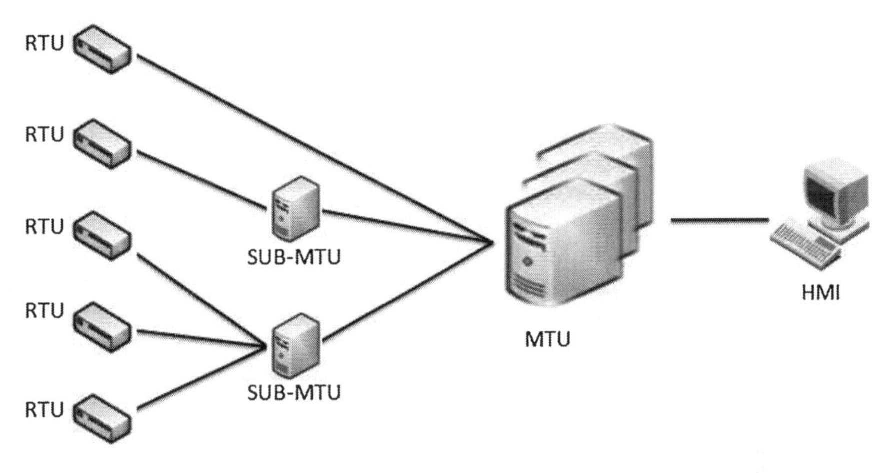

Fig. 1 SCADA system architecture

Due to the constrains of low-rate data transmission and real-time processing in different operational environment, satisfying the security requirements of confidentiality, integrity and availability in a SCADA system is really a challenging issue. In this paper, resort to a symmetric balanced incomplete block design (SBIBD), we design a novel key management scheme for SCADA systems with good security properties and performance.

Efficient Key Management Scheme for SCADA System

In this section, we propose an efficient key management scheme for SCADA system. By our scheme, the communication among Sub-MTUs can be secure and efficient, so can the communication among RTUs. The process of key management among RTUs is described as follows. Note that the process of key management among Sub-MTUs is similar to that of RTUs.

Each RTU registers to the Sub-MTU and gets their private key. After that, every RTU can process the key agreement to compute the common conference key. First of all, the Sub-MTU chooses two prime order group G_1 and G_2 and a modified Weil pairing map $\widehat{e}$ defined in [4]. Next, the Sub-MTU selects two one-way hash functions $H : \{0,1\}^* \to G_1$ and $h : \{0,1\}^* \to Z_q^*$ where H maps its arbitrary length to a nonzero point of G_1 while h maps its input with arbitrary length to a nonzero integer. At last, the Sub-MTU picks a random integer $s \in Z_q^*$ as its private key, computes its public key $P_{pub} = s\mathcal{G}$, and publishes $(p, q, G_1, G_2, \mathcal{G}, \widehat{e}, P_{pub}, H, h)$, but keeps s secret. Each RTU U_i's identity is $ID_i \in (0,1)^*$. The Sub-MTU computes U_i's public key $Q_i = H(ID_i)$ and then U_i's private key $S_i = sQ_i$ which is issued to U_i via a secure channel.

Fig. 2 (7 × 7) incidence matrix corresponding to the (7, 4, 2)-design

$$\mathcal{L} = (\ell_{ij}) = \begin{bmatrix} 1 & 1 & 0 & 1 & 0 & 0 & 1 \\ 1 & 1 & 1 & 0 & 1 & 0 & 0 \\ 0 & 1 & 1 & 1 & 0 & 1 & 0 \\ 0 & 0 & 1 & 1 & 1 & 0 & 1 \\ 1 & 0 & 0 & 1 & 1 & 1 & 0 \\ 0 & 1 & 0 & 0 & 1 & 1 & 1 \\ 1 & 0 & 1 & 0 & 0 & 1 & 1 \end{bmatrix}$$

The common conference key among RTUs is calculated by employing SBIBD, where the number of blocks is the same as that of participants. We choose a (7, 4, 2)-design. Let a finite set X $= \{1, 2, 3, 4, 5, 6, 7\}$, then $B_1 = \{1, 2, 4, 7\}$, $B_2 = \{1, 2, 3, 5\}$, $B_3 = \{2, 3, 4, 6\}$, $B_4 = \{3, 4, 5, 7\}$, $B_5 = \{1, 4, 5, 6\}$, $B_6 = \{2, 5, 6, 7\}$, $B_7 = \{1, 3, 6, 7\}$. Accordingly, a (7 × 7) incidence matrix L is depicted in Fig. 2. The rows and columns of the matrix correspond to the blocks and the elements, respectively. The entry l_{ij} in the ith row and the jth column of L is a 1 if the block i contains the element j and is a 0 otherwise.

For computing the common conference key among RTUs, two rounds are required in our scheme.

1. Each RTU U_i selects a random number r_i as secret key by itself for every session and then calculates $m_i = \hat{e}(\mathcal{G}, r_i S_i)$. Simultaneously, U_i calculates $T_i = r_i Q_i$. Let $D_i = \{m_i, T_i\}$. RTU i receives message D_j from RTU j in case $l_{ij} = 1$ and $j \neq i$, namely $j \in B_i - \{i\}$. m_i is used for generating conference key while T_i is used for authentication. We now describe the key agreement process from the viewpoint of RTU 1. U_1 receives D_2, D_4, D_7 from U_2, U_4, U_7 and then makes

$$\begin{aligned} c_{11} &= m_2 \cdot m_4 \cdot m_7 = \hat{e}(\mathcal{G}, r_2 S_2 + r_4 S_4 + r_7 S_7), \\ c_{12} &= m_1 \cdot m_4 \cdot m_7 = \hat{e}(\mathcal{G}, r_1 S_1 + r_4 S_4 + r_7 S_7), \\ c_{14} &= m_1 \cdot m_2 \cdot m_7 = \hat{e}(\mathcal{G}, r_1 S_1 + r_2 S_2 + r_7 S_7), \\ c_{17} &= m_1 \cdot m_2 \cdot m_4 = \hat{e}(\mathcal{G}, r_1 S_1 + r_2 S_2 + r_4 S_4), \\ W_{12} &= T_1 + T_4 + T_7, \\ W_{14} &= T_1 + T_2 + T_7, \\ W_{17} &= T_1 + T_2 + T_4, \end{aligned}$$

where $c_{ij} = \prod_{x \in B_i - \{j\}} m_x$ and $W_{ij} = \sum_{x \in B_i - \{j\} \text{ and } j \neq i} T_x$. In the viewpoint of RTU 1, we have that $c_{1j} = \prod_{x \in B_1 - \{j\}} m_x$ and $W_{1j} = \sum_{x \in B_1 - \{j\} \text{ and } j \neq 1} T_x$. Simultaneously, other RTUs do the same process.

2. Let $E_{ji} = \{c_{ji}, W_{ji}\}$. RTU i receives E_{ji} from RTU j in case $l_{ji} = 1, j \neq i$. Here, similar to that in round 1, c_{ji} is used for generating conference key while W_{ji} is used for authentication. Particularly, U_1 receives E_{j1} from RTU j, if $l_{j1} = 1$,

User	Round 1	Round 2
1	$c_{11} = m_2 \cdot m_4 \cdot m_7$ $c_{12} = m_1 \cdot m_4 \cdot m_7$ $c_{14} = m_1 \cdot m_2 \cdot m_7$ $c_{17} = m_1 \cdot m_2 \cdot m_4$	$\mathcal{K} = {m_1}^2 \cdot c_{11} \cdot c_{21} \cdot c_{51} \cdot c_{71} = \hat{e}(\mathcal{G}, 2\sum_{i=1}^{7} r_i S_i)$
2	$c_{22} = m_1 \cdot m_3 \cdot m_5$ $c_{21} = m_2 \cdot m_3 \cdot m_5$ $c_{23} = m_2 \cdot m_5 \cdot m_1$ $c_{25} = m_2 \cdot m_3 \cdot m_1$	$\mathcal{K} = {m_2}^2 \cdot c_{22} \cdot c_{12} \cdot c_{32} \cdot c_{62} = \hat{e}(\mathcal{G}, 2\sum_{i=1}^{7} r_i S_i)$
3	$c_{33} = m_2 \cdot m_4 \cdot m_6$ $c_{34} = m_2 \cdot m_3 \cdot m_6$ $c_{36} = m_2 \cdot m_3 \cdot m_4$ $c_{32} = m_3 \cdot m_4 \cdot m_6$	$\mathcal{K} = {m_3}^2 \cdot c_{33} \cdot c_{23} \cdot c_{43} \cdot c_{73} = \hat{e}(\mathcal{G}, 2\sum_{i=1}^{7} r_i S_i)$
4	$c_{44} = m_3 \cdot m_5 \cdot m_7$ $c_{45} = m_3 \cdot m_4 \cdot m_7$ $c_{47} = m_3 \cdot m_4 \cdot m_5$ $c_{43} = m_4 \cdot m_5 \cdot m_7$	$\mathcal{K} = {m_4}^2 \cdot c_{44} \cdot c_{14} \cdot c_{34} \cdot c_{54} = \hat{e}(\mathcal{G}, 2\sum_{i=1}^{7} r_i S_i)$
5	$c_{55} = m_4 \cdot m_6 \cdot m_1$ $c_{56} = m_4 \cdot m_5 \cdot m_1$ $c_{51} = m_4 \cdot m_5 \cdot m_6$ $c_{54} = m_5 \cdot m_6 \cdot m_1$	$\mathcal{K} = {m_5}^2 \cdot c_{55} \cdot c_{25} \cdot c_{45} \cdot c_{65} = \hat{e}(\mathcal{G}, 2\sum_{i=1}^{7} r_i S_i)$
6	$c_{66} = m_5 \cdot m_7 \cdot m_2$ $c_{67} = m_5 \cdot m_6 \cdot m_2$ $c_{62} = m_5 \cdot m_6 \cdot m_7$ $c_{65} = m_2 \cdot m_6 \cdot m_7$	$\mathcal{K} = {m_6}^2 \cdot c_{66} \cdot c_{36} \cdot c_{56} \cdot c_{76} = \hat{e}(\mathcal{G}, 2\sum_{i=1}^{7} r_i S_i)$
7	$c_{77} = m_1 \cdot m_3 \cdot m_6$ $c_{71} = m_3 \cdot m_7 \cdot m_6$ $c_{73} = m_7 \cdot m_1 \cdot m_6$ $c_{76} = m_7 \cdot m_1 \cdot m_3$	$\mathcal{K} = {\mathcal{M}_7}^2 \cdot c_{77} \cdot c_{17} \cdot c_{47} \cdot c_{67} = \hat{e}(\mathcal{G}, 2\sum_{i=1}^{7} r_i S_i)$

Fig. 3 Generating a common key

$j \neq 1$. Therefore, U_1 receives E_{21}, E_{51}, E_{71} from U_2, U_5, U_7 and derives c_{21}, c_{51}, c_{71}. Then the common conference key K is calculated as $K = m_1 \times c_{11} \times c_{21} \times c_{51} \times c_{71} = \widehat{e}\left(\mathcal{G}, 2\sum_{i=1}^{7} r_i S_i\right)$, where $c_{21} = \widehat{e}(\mathcal{G}, r_2 S_2 + r_3 S_3 + r_5 S_5)$, $c_{51} = \widehat{e}(\mathcal{G}, r_4 S_4 + r_5 S_5 + r_6 S_6)$, and $c_{71} = \widehat{e}(\mathcal{G}, r_3 S_3 + r_7 S_7 + r_6 S_6)$.

Then, following our scheme, the process for calculating the common conference key among all the RTUs is shown in Fig. 3.

In our scheme, we take advantage of RTUs' identity information for authentication.

1. Let $D_i = \{m_i, T_i\}$, RTU i receives D_j from RTU j in case $l_{ij} = 1$ and $j \neq i$. We now describe the authentication process from the viewpoint of RTU 1. U_1 receives D_2, D_4, D_7 from U_2, U_4, U_7 and makes

$$\widehat{e}(P_{pub}, T_2) = \widehat{e}(s\mathcal{G}, r_2Q_2) = \widehat{e}(\mathcal{G}, r_2sQ_2) = m_2,$$
$$\widehat{e}(P_{pub}, T_4) = \widehat{e}(s\mathcal{G}, r_4Q_4) = \widehat{e}(\mathcal{G}, r_4sQ_4) = m_4,$$
$$\widehat{e}(P_{pub}, T_7) = \widehat{e}(s\mathcal{G}, r_7Q_7) = \widehat{e}(\mathcal{G}, r_7sQ_7) = m_7,$$

Hence, U_1 can authenticate the entity of U_2, U_4, U_7 only if $\widehat{e}(P_{pub}, T_2) = m_2$, $\widehat{e}(P_{pub}, T_4) = m_4$, and $\widehat{e}(P_{pub}, T_7) = m_7$, respectively. Generally speaking, if $\widehat{e}(P_{pub}, T_i) = m_i$, then U_j can authenticate counterpart's entity.

2. Let $E_{ji} = \{c_{ji}, W_{ji}\}$ and $W_{ji} = \sum_{x \in B_j - \{i\} \text{ and } j \neq i} T_x$. RTU i receives E_{ji} from RTU j in case $l_{ji} = 1$, $j \neq i$. Particularly, in the viewpoint of RTU 1, $W_{j1} = \sum_{x \in B_j - \{1\} \text{ and } j \neq 1} T_x$ and $E_{j1} = \{c_{j1}, W_{j1}\}$. U_1 receives E_{21}, E_{51}, E_{71} from U_2, U_5, U_7, then derives W_{21}, W_{51}, W_{71} and calculates

$$\widehat{e}(P_{pub}, W_{21}) = \widehat{e}(s\mathcal{G}, T_2 + T_3 + T_5) = m_2 \cdot m_3 \cdot m_5 = c_{21},$$
$$\widehat{e}(P_{pub}, W_{51}) = \widehat{e}(s\mathcal{G}, T_4 + T_5 + T_6) = m_4 \cdot m_5 \cdot m_6 = c_{51},$$
$$\widehat{e}(P_{pub}, W_{71}) = \widehat{e}(s\mathcal{G}, T_3 + T_7 + T_6) = m_3 \cdot m_7 \cdot m_6 = c_{71},$$

Therefore, the RTU of U_2, U_5, U_7 can pass the authentication by U_1 only if $\widehat{e}(P_{pub}, W_{21}) = c_{21}$, $\widehat{e}(P_{pub}, W_{51}) = c_{51}$, $\widehat{e}(P_{pub}, W_{71}) = c_{71}$, respectively. Broadly speaking, if $\widehat{e}(P_{pub}, W_{ji}) = c_{ji}$, then U_i can authenticate counterpart's entity.

Security Analysis and Performance Analysis

A passive adversary tries to learn information about the conference key by eavesdropping on the broadcast channel. We show that an eavesdropper cannot get any information about the secret key r_i of U_i due to Weil Diffie-Hellman (WDH) problem [5] in $(G_1, G_2, \widehat{e})$ and discrete algorithm problem (DLP) in elliptic curves. In active attack, an adversary not only just records the data, but also can alter, inject, intercept and replay messages. Our protocol can be able to provide the authentication service by sending a special message T_i and W_{ji} in first round and second round, respectively. Our scheme has the security properties of known session key security, perfect forward secrecy, key-compromise impersonation resistance and no key control.

The communication cost of previous schemes are all $O(n^2)$, while the communication cost of our scheme is only $O(n\sqrt{n})$even though the communication round is 2.

Conclusions

SCADA system is a significantly important system that plays a very important role in national infrastructure, such as electric grids and water supplies. However, SCADA system is becoming increasingly vulnerable to adversarial manipulation due to the extreme operational environment. In this paper, we present a novel key management scheme for SCADA systems with good performance and security properties. We believe that our scheme must be promising in the secure communication in SCADA system in the future.

Acknowledgments This work was supported by the research fund from Nanjing University of Information Science and Technology under Grant No. S8113003001, National Science Foundation of China under Grant No. 61272421.

References

1. Beaver C, Gallup D, Neumann W, Torgerson M (2002) Key Management for SCADA. Available: http://www.sandia.org/ scada/documnets/013252.pdf
2. Colin RD, Boyd C, Manuel J, Nieto G (2006) SKMAA key management architecture for SCADA systems. In: 4th Australasian information security workshop, pp 138–192
3. Choi D, Kim H, Won D, Kim S (2009) Advanced key management architecture for secure SCADA communications. IEEE Trans Power Del 24(3):1154–1163
4. Boneh D, Franklin M (2001) Identity-based encryption from weil pairing. In: Advances in Cryptology-CRYPTO01, Lecture Notes in Computer Science, vol 2139, pp 213–229
5. Kim Y, Perrig A, Tsudik G (2004) Group key agreement efficient in communication. IEEE Trans Comput 53(7):905–921

Printed by Publishers' Graphics LLC